Estuarine Science

ESTUARINE SCIENCE

A Synthetic Approach to Research and Practice

Edited by John E. Hobbie

ISLAND PRESS

Washington, D.C. • Covelo, California

Library of Congress Cataloging-in-Publication Data

Estuarine science : a synthetic approach to research and practice / edited by John Hobbie.
    p. cm.
  Includes bibliographical references and index.
  ISBN 1-55963-699-8 (acid-free paper : cloth) — ISBN 1-55963-700-5 (acid-free paper : paper)
  1.Estuarine oceanography. 2. Estuarine ecology. I. Hobbie, John E.

GC97 .E785 2000
551.46'09—dc21

                                   99-053797

Printed on recycled, acid-free paper ∞ ✪

Manufactured in the United States of America
10 9 8 7 6 5 4 3 2 1

Contents

Contributors

Barry, Karen L., Fisheries and Oceans Canada, West Vancouver Laboratory, 4160 Marine Drive, West Vancouver, British Columbia, V7V 1N6 Canada

Billen, Gilles, Systems Hydriques Continentaux, Laboratoire de Geologie Appliqueé, Tour 26, 5eme étage, 4 Place Jussieu, 75005 Paris, France

Boesch, Donald F., University of Maryland, Center for Environmental Sciences, P.O. Box 775, Cambridge, MD 21613

Boynton, Walter R., University of Maryland, Chesapeake Biological Laboratory, P.O. Box 38, Solomons, MD 20688-0038

Brandt, Stephen B., Great Lakes Environmental Research Laboratory, 2205 Commonwealth Avenue, Ann Arbor, MI 48105

Burger, Joanna, Biological Science, Rutgers University, Piscataway, NJ 08855

Cerco, Carl F., U.S. Army Engineer Research and Development Center, 3909 Halls Ferry Road, Vicksburg, MS 39180

Chalmers, Alice, University of Georgia, 220 Marine Sciences Building, University of Georgia, Athens, GA 30602-3636

Correll, David, Smithsonian Environmental Research Center, Box 28, Edgewater, MD 21037

Costanza, Robert, University of Maryland, Center for Environmental Science, Institute for Ecological Economics, P.O. Box 38, Solomons, MD 20688-0038

D'Elia, Christopher F., Vice President for Research, Administration 227, University at Albany, State University of New York, Albany, NY 12222

Dame, Richard, Department of Marine Science, Coastal Carolina University, Conway, SC 29526

Deegan, Linda A., The Ecosystems Center, Marine Biological Laboratory, Woods Hole, MA 02543

Demers, Eric, Conor Pacific Environmental Technologies Inc., 300–1727 West Broadway, Vancouver, British Columbia, Canada V6J 4W6

Dortch, Quay, Louisiana Universities Marine Consortium, 8124 Highway 56, Chauvin, LA 70344

Fisher, Thomas R., University of Maryland, Center for Environmental Science, Horn Point Laboratory, P.O. Box 775, Cambridge, MD 21613

Geyer, W. Rockwell, Woods Hole Oceanographic Institution, Department of Applied Ocean Physics and Engineering, MS 12, Woods Hole, MA 02543

Giblin, Anne E., The Ecosystems Center, Marine Biological Laboratory, Woods Hole, MA 02543

Hobbie, John E., The Ecosystems Center, Marine Biological Laboratory, Woods Hole, MA 02543

Hodson, Robert, Department of Marine Science, University of Georgia, Athens, GA 30602-2206

Hofmann, Eileen E., Center for Coastal Physical Oceanography, Old Dominion University, Norfolk, VA 23529

Hollibaugh, James T., Department of Marine Science, University of Georgia, Athens, GA 30602-3636

Hopkinson, Charles S. Jr., The Ecosystems Center, Marine Biological Laboratory, Woods Hole, MA 02543

Houde, Edward D., University of Maryland, Chesapeake Biological Laboratory, P.O. Box 38, Solomons, MD 20688-0038

Howarth, Robert W., 311 Corson Hall, Cornell University, Ithaca, NY 14853

Jaworski, Norbert, NOAA/NMFS, 28 Tarzwell Drive, Narragansett, RI 02882-1199

Jay, David A., Center for Coastal and Land-Margin Research, P.O. Box 91000, Portland, OR 97291-1000

Jech, J. Michael, Northeast Fisheries Science Center, 166 Water Street, Woods Hole, MA 02543

Justić, Dubravko, Coastal Ecology Institute and Department of Oceanography and Coastal Sciences, Louisiana State University, Baton Rouge, LA 70803

Kemp, W. Michael, University of Maryland, Center for Environmental Science, Horn Point Laboratory, P.O. Box 775, Cambridge, MD 21613

Kremer, James N., Department of Marine Sciences, University of Connecticut, Avery Point, Groton, CT 06340-6097

Montgomery, David R., Quaternary Research Center, University of Washington, Seattle, WA 98195

Morris, James T., Department of Biological Sciences and Belle W. Baruch Institute, University of South Carolina, Columbia, SC 29208

Peterson, Bruce J., The Ecosystems Center, Marine Biological Laboratory, Woods Hole, MA 02543

Prahl, Frederick G., Oregon State University, College of Oceanic and Atmospheric Sciences, Corvallis, OR 97331

Rabalais, Nancy N., Louisiana Universities Marine Consortium, 8124 Highway 56, Chauvin, LA 70344

Reed, Denise J., Department of Geology and Geophysics, University of New Orleans, New Orleans, LA 70148

Richey, Jeffrey E., School of Oceanography, WB-10, University of Washington, Seattle, WA 98195

Scavia, Donald, NOAA Coastal Oceans Program, 1315 East-West Highway, Silver Spring, MD 20910

Seitzinger, Sybil P., Rutgers University, Institute of Marine and Coastal Studies, Rutgers NOAA CMER Program, 71 Dudley Road, New Brunswick, NJ 08903-0231

Sen Gupta, Barun K., Department of Geology, Louisiana State University, Baton Rouge, LA 70803

Simenstad, Charles A., School of Fisheries, 324A Fishery Sciences, 1122 Boat Street, Box 355020, University of Washington, Seattle, WA 98195-5020

Swaney, Dennis, Department of Systems Ecology, University of Stockholm, S-106 91 Stockholm, Sweden

Townsend, Alan, University of Colorado, INSTAAR, Campus Box 450, University of Colorado, Boulder, CO 80309

Turner, R. Eugene, Coastal Ecology Institute and Department of Oceanography and Coastal Sciences, Louisiana State University, Baton Rouge, LA 70803

Valiela, Ivan, Boston University Marine Program, Marine Biological Laboratory, Woods Hole, MA 02543

Voinov, Alexey, University of Maryland, Center for Environmental Science, Institute for Ecological Economics, P.O. Box 38, Solomons, MD 20688

Vörösmarty, Charles J., Institute for the Study of Earth, Oceans and Space, University of New Hampshire, Durham, NH 03824

Wiegert, Richard, Institute of Ecology, 711 Biological Sciences Building, University of Georgia, Athens, GA 30602-2602

Wiseman, William J. Jr., Coastal Studies Institute and Department of Oceanography and Coastal Sciences, Louisiana State University, Baton Rouge, LA 70803

CHAPTER 1

Estuarine Science: The Key to Progress in Coastal Ecological Research

John E. Hobbie

Executive Summary

Situated at the interface between land and ocean, estuaries of coastal rivers are semi-enclosed bodies of sea-water measurably diluted by the fresh water that flows into them. Estuaries are characterized by high biotic diversity and high primary production. They yield large harvests of fish and shellfish and provide transportation routes and recreational opportunities for human populations worldwide. The rise in population and changing land use in coastal regions is inevitably affecting the flow of water, sediments, organic matter, and inorganic nutrients into the estuaries of the world. Successful management of estuaries and their watersheds for sustainable use in the future requires us to bring all applicable knowledge to bear on the development of practical models that predict the results of various strategies.

Despite the extent of research on estuaries, our ability to generalize and predict the consequences of change is primitive. One reason is the inherent complexity and variability of estuaries, which exhibit tremendous temporal and spatial variation in physical and chemical characteristics. Another reason is the restricted nature of the questions asked; most estuarine research has been focused on documenting and solving local and state management problems. Although such studies do answer questions, there is no way to transfer information to other sites and to avoid repetitious data collection. A final reason is the lack of funding for basic research and synthesis of data from estuarine studies. Priorities for marine research have emphasized the open ocean. As a result, there have been few opportunities to develop long-term and intensive studies in coastal regions.

Such studies are necessary for synthesis, which we define as the bringing together of existing information in order to discover patterns, mechanisms,

FIGURE 1-1   The Childs River estuary and Waquoit Bay (upper right) are important recreational resources on Cape Cod, Massachusetts.

and interactions that lead to new concepts and models. Synthetic perspectives and methods are necessary to discover the general relationships and develop the models that can be applied to all estuaries. To date, the synthetic aspects of estuarine research have been neglected.

With support from the U.S. Scientific Committee for Problems of the Environment (SCOPE), the National Science Foundation (NSF), and the National Oceanic and Atmospheric Administration (NOAA), forty estuarine scientists met in 1995 to put together the case for synthesis of estuarine data and to show the capabilities of synthetic methods of research. The meeting featured twelve plenary talks that documented a variety of successful approaches to conducting synthetic studies in estuaries. Attendees also took part in five workshops that were charged with identifying important areas for synthesis in the next decade, along with specific ideas about the kinds of process studies or models that would be needed.

The plenary talks and workshop reports from this meeting comprise chapters 2–18 of this book. They are organized into five parts, each of which begins with a short introduction.

Part I takes up the measurement and prediction of changes in the amounts of water, sediment, and nutrients that flow from watersheds into estuaries. These amounts vary widely from estuary to estuary. Resource managers have developed correlations between types of landscape and the runoff of nutrients and sediment that are entirely adequate to describe changes that occur when, for example, land use changes from forests to farming or urban development.

The correlation approach, however, cannot predict changes in nutrient and sediment runoff when climate and vegetation type change drastically. Scientists need to take the next step and develop mechanistic models containing cause-and-effect relationships that combine vegetation and soil processes with hydrology models based on climate and the landscape topography.

Part II presents efforts to couple information on the physical environment of estuaries with biogeochemical and ecological data. Changes in currents, salinity, and sea level in estuaries are correlated with the productivity, survival, and distribution of plants and animals. The survival of postlarval shrimp can be predicted, for example, with the help of a physical model of the changes in salinity and temperature that the shrimp experience as they are transported throughout an estuary.

Although scientists are capable of developing detailed physical models of each estuary, it is more useful to develop general models that allow us to predict changes in physical conditions that affect the ecology of estuaries from information on bathymetry, river flow, climate, and so forth. The conclusion of the workshop reported in this section is that estuarine research over the next decade should emphasize quantitative comparisons among estuaries, combining observations with advanced modeling approaches to address the complex and variable processes within and among interacting estuarine systems.

Part III deals with biogeochemical processes in estuaries and their linkages to the food web. Processes such as denitrification can be studied in a single estuary over time or comparatively in a number of estuaries. The resulting information can be used to create models at the global scale. The effects of nutrients entering estuarine and coastal waters have been studied through measurements of a number of processes. Examples given include the response of biological systems in the Gulf of Mexico to nutrients and of effects on Chesapeake Bay systems of long-term changes in nutrient and water flow.

While links can be made between biogeochemical factors and the lower levels of the food web, links to fish at the top of the food web are extremely difficult to quantify at present. Recommendations for future research and synthesis are to gather more long-term data sets and to construct more mass balances of nutrients, to carry out manipulative experiments to identify causal links, and to bring together information from all sources to construct models reflecting the current understanding. The modeling of linked processes will answer many management questions.

Part IV discusses controls on distribution and abundance of organisms in the estuary; biologists at the workshop reported in this section concluded that good management necessitates an extension of the commonly held concepts of habitat to include the environment of an organism throughout its entire life. Estuaries are so changeable and dynamic that organisms encounter a wide range of environments throughout their life histories.

Although no single model exists of these changing environments, it is possible to simulate some crucial aspects. For example, the dynamics of sediment deposition can explain the distribution of rooted aquatic plants. Furthermore, the productivity and survival of algae and sea grasses in Chesapeake Bay can be simulated through use of coupled physical and biological models of the movement of nutrients throughout the entire bay. Remote sensing can be used to develop a picture of the distribution and abundance of small fish across transects of the estuary; this information allows a bioenergetics model to identify the habitats where predatory fish can survive and grow.

Part V addresses the need for synthesis that would improve the scientific management of estuaries directly. In one example, an integrated model of ecological and economic factors in the Patuxent watershed in Maryland incorporates economic forces as drivers for ecological changes that take place. Workshop participants concluded that the scientific management of estuarine systems would be greatly improved by the formal application of a series of steps. They include:

- designating a lead agency within each estuarine management program to coordinate the activities of other agencies;

- communicating among stakeholders in government, the public, and the scientific community;

- implementing a scientific advisory process that incorporates synthesis of data and comparisons across a variety of systems;

- providing sustained support at the national level for basic research on estuarine processes;

- encouraging comparison and synthesis within national networks of projects and programs to produce generalized models for managers.

Introduction

Why Are Estuaries Important?

Situated at the interface between land and ocean, the estuaries of coastal rivers provide vital habitat for fish, shellfish, and waterfowl as well as transportation routes and recreational opportunities for human populations.

Estuaries are, by definition, semi-enclosed bodies of seawater measurably diluted with fresh water from watersheds. This fresh water carries with it sediments, organic matter, and inorganic nutrients from terrestrial sources. The combination of high nutrients and the stratification and circulation patterns characteristic of estuaries result in high biotic diversity, high primary productivity, and the creation of harvest and nursery areas for fish and shellfish.

The human population has increased rapidly in coastal regions throughout the world in recent years. Worldwide, 61% of the population lives near a coast (Alongi 1998). In the United States, more than half of the population lives in coastal counties, including those bordering the Great Lakes, and the rate of population increase in these counties is higher than that of the country as a whole (Land-Margin Ecosystem Research Coordinating Committee 1992). A 1994 report from the U.S. National Research Council (NRC) titled "Priorities for Coastal Ecosystem Science" lists some of the major effects that the rise in human population has had on estuarine ecosystems:

- Nutrients, especially nitrogen, have increased manyfold in rivers entering estuaries, causing harmful algal blooms and the depletion of oxygen, especially in bottom waters.

- Intertidal and tidal habitats, such as salt marshes vital to the young of commercial fishery species, have been dredged and filled.

- Landscape alterations, water diversion, and damming of rivers have changed the amount and seasonal patterns of fresh water reaching estuaries as well as the amount of transported sediments.

- Overexploitation of natural resources has eliminated some shellfish and fish stocks.

- Industrial pollution has left toxic materials such as polycyclic aromatic hydrocarbons (PAHs) and polychlorinated biphenyls (PCBs) in some estuaries.

- Introduction of nonindigenous animals and plants into estuaries has resulted in loss of biodiversity, degradation of habitats, and reduction in fisheries production.

The NRC report also points out the susceptibility of coastal and estuarine ecosystems to changes in climate and weather patterns. While the effects of sea-level rise have been well publicized, changes in precipitation and estuarine salinity, circulation patterns, and riverine transport of nutrients may be expected as well. Shifts in weather patterns can alter the direction and strength of waves, causing drastic shoreline erosion.

The report concludes that these issues require strategies other than the regulatory and management approaches that are generally taken to problems such as point-source discharge, coastal land use, and spills of toxic chemicals. It notes that ". . . concern is shifting from problems amenable to single-factor risk assessment paradigms to multiple-factor risk assessment and regulatory strategies that take into account indirect, cascading, and scale-related effects that require an ecosystem perspective (for example, eutrophication, hydrologic and hydrodynamic modification, resource sustainability, loss of biodiversity)" (NRC 1994).

Why Is Synthesis Needed for Improved Scientific Understanding?

Synthesis may be defined as the bringing together of existing information in order to discover patterns, mechanisms, and interactions that lead to new concepts and models. An increased emphasis on synthesis is needed in estuarine research for a number of reasons.

Estuarine research suffers from a lack of integration of knowledge that can be applied across sites. Estuaries are dynamic, complex, and difficult to study. One result is that scientists generally concentrate on studying a single estuary. Little effort is made to integrate data into a coherent whole that makes optimal use of previous estuarine research. We study too many estuaries and treat each as if it were unique.

This approach is extremely costly as each estuary has to be studied in detail, starting with years of observations. The result is that research seldom advances beyond the observational level to answer questions about cause-and-effect relationships and ecosystem processes. Research on single systems needs to be extended to include a search for patterns common to all estuaries, and researchers need to adopt a comparative approach so that predictions can be made about unstudied estuaries.

The winter 1998 newsletter of the U.S. Environmental Protection Agency's (EPA) National Estuary Program (NEP) calls for better use and transfer of information. "There are approximately 130 estuaries in the United States. Do they all need to be part of the NEP? The EPA recognizes that it may not be appropriate, or even necessary, to designate all of these estuaries as NEPs. What may be more important and effective is to transfer the lessons learned within the NEP to other areas" (U.S. EPA 1998).

The results of most estuarine studies are not published in reviewed journals. Most estuarine research projects are aimed at solving local and state management problems. These studies, which are both necessary and important, account for most of the $227 million that federal agencies spent on research in the coastal zone in 1993, the last year for which we have complete numbers. But a focus on solving a particular problem in a particular place is generally associated with a failure to report results to a broader audience. Articles in peer-reviewed journals are the accepted method of reporting scientific data and management solutions. When results are not disseminated, costly research winds up being repeated again and again.

Understanding the complex interactions of physical, chemical, and biological factors is essential to answering many questions about estuarine systems. Concerns about the effect of changes in sea level and freshwater input on estuarine ecosystems require that we understand the interactions of physical properties and processes with the biota. Analyzing these interactions requires the use of models. Similarly, we need to look at the complex interactions of food webs rather than studying single species if we are to understand the impact of human activity on estuarine systems. Relatively simple questions, such as the relationship between nutrient loading and eutrophication, have been well studied. They are being replaced by more complex issues, such as the effects of habitat loss and species replacement on fish production, that require an understanding of whole food webs.

Estuarine research must be able to predict the consequences of future changes in climate or land use in watersheds. Predictions about the future are only possible with synthetic models that incorporate mechanistic understanding of ecological relationships. The key word here is "future." Empirical models are perfectly adequate and cost-effective ways to explain the effects of currently observed phenomena, for example, the present-day loss of sediment from a drainage basin. In these models, the analysis is based on field observations, not on experimental data or mechanistic understanding. These analyses make use of regressions, some of which involve integrated relationships, such as the one between nutrient flux into an estuary and fish production.

If predictions are required about the effects of future changes in climate or water and sediment input, however, the relationships derived by regression may no longer apply. A mechanistic or process-based model can predict outcomes if new inputs or new climate conditions are postulated. These modeling efforts will be successful only if a synthetic approach incorporates all the information necessary to construct and test the models.

The U.S. SCOPE Meeting on Synthesis in Estuaries

Despite the extent of research on estuaries, our ability to generalize and predict the consequences of change is primitive. One reason is the inherent complexity and variability of estuaries. In any estuary, there is tremendous temporal and spatial variability in physical and chemical factors such as salinity or oxygen levels.

Another reason is the restricted nature of the questions asked; most research has been focused on specific sites with the aim of documenting and solving local and state management problems. While this kind of research does answer questions, it is not efficient because there is no way to transfer information to other sites and to avoid repetitious data collection. As noted above, such studies seldom result in publications in reviewed journals of science or management.

A final problem is the lack of funding for basic research and synthesis of data from estuarine studies. Priorities for marine research have emphasized studies of the open ocean. As a result, there have been few opportunities to develop long-term, intensive, and comparative studies of coastal regions. Such studies are essential to discover the general relationships and develop the models that would apply to all estuaries. In short, the synthetic aspects of estuarine studies have been neglected.

In February 1995, forty estuarine scientists met at the National Academy of Science's Beckman Center in Irvine, California, to put together the case for synthesis of estuarine data and to show the capabilities of synthetic methods of research. The sponsors were the U.S. SCOPE, the NSF's divisions of Ocean Sciences and Environmental Biology, and NOAA's Coastal Ocean Program.

The meeting featured twelve plenary talks that documented a variety of successful approaches to conducting synthetic studies in estuaries. They ranged from studies of a single process, such as denitrification, across a wide range of estuaries to a study of eutrophication in the entire Chesapeake Bay that made use of a coupled physical-biological model. Attendees also took part in five workshops that were charged with identifying important areas for synthesis in the next decade along with specific ideas about the kinds of process studies or models that would be needed.

The plenary talks and workshop reports from this meeting comprise the subsequent chapters of this book. They are organized into five parts, each of which begins with a short introduction.

Types of Estuarine Synthesis

A number of synthetic approaches were represented in the presentations at the meeting. One way of classifying them is according to their level of complexity. Synthesis of descriptive data is the simplest approach (Type I). Correlations of ecological data with changes in environmental conditions represent a more complex approach (Type II), and studies that use mechanistic models and attempt to make large-scale predictions are still more complex (Types III and IV). Several of these approaches can be combined in one synthetic effort (Type V).

Examples of these five kinds of estuarine synthesis include:

Type I. Descriptions of changes in the characteristics of ecosystems (rates, biomass, and concentrations) and accompanying physical factors (river flows, climate). In chapter 10, Rabalais, Turner, Justić, Dortch, Wiseman, and Sen Gupta describe the response of oxygen levels and photosynthesis in the Gulf of Mexico to changes over time in the influx of nutrients from the Mississippi River.

Type II. Statistical and mathematical correlations between biological responses and environmental factors. An example is the report in chapter 11 by Boynton and Kemp on regression analyses of the response of ecosystem processes to river flow and nutrient loading in Chesapeake Bay.

Type III. Mathematical descriptions of controls of a process in a single system or by comparisons of environments across sites. In chapter 9, Seitzinger describes a relationship between nitrogen loading rates and denitrification rates that she found by comparing many sites.

Type IV. Mathematical simulation models of processes or of a number of interrelated processes in a single system or region. In chapter 14, Cerco reports simulations from the Chesapeake Bay model, in which biological processes such as algal growth and bacterial respiration over the entire bay are linked to results from a coupled physical-biological model.

Type V. Mathematical simulation models that combine process descriptions with correlation determinations. In chapter 3, Vörösmarty and Peterson describe how they combined results from physical models of the processes of water and energy flux with correlation data on nutrients in rivers to produce estimates of regional riverine transport of material to the coastal waters.

Scientific Synthesis and Management

The participants in the meeting shared a conviction that one of the most important reasons for estuarine research is to manage estuaries for sustainable use in the future. They also emphasized that management practices must be founded on a strong scientific understanding of estuarine biota and their physical environment. These convictions are not universally accepted, however. Alongi (1998, p. 347) presents two opposing views. "One school suggests that the best way to tackle the problem of providing research relevant for management is to perform intense, detailed research on the basic ecological conditions of the systems to provide a better level of understanding and better management. The other school states that traditional research does not provide the relevant information necessary for better management and that research should be targeted to specific problems and to better monitoring."

Both points of view have serious errors. It makes no sense to delay considering management questions until a high level of basic understanding is reached. It is equally without sense to reject the traditional approaches to understanding that must include scientific observations, hypotheses, and synthesis. The solution is to focus on the questions asked and on the minimum amount of scientific knowledge necessary to answer the question. In chapter 18, Boesch, Burger, D'Elia, Reed, and Scavia discuss these issues in detail. Some of the accomplishments of the use of scientific synthesis in management are evident in projects located in the Great Lakes, Chesapeake Bay, San Francisco Bay, and Florida Bay.

The synthetic perspective leads to a search for general patterns, relationships, and models. Our goals are to integrate data into a coherent whole that makes optimal use of previous estuarine research and to predict what will happen in the future in a particular estuary or in estuaries not yet studied. Although scientific management must be based on this synthetic view of knowledge about estuaries, it is important to remember that the way to answer a particular management question may be through an empirical and practical approach that deals only with a small part of the whole.

The Organization of the Book

This book is organized into five parts reflecting major topics in estuarine synthesis:

I. The prediction of inputs from drainage basins

II. The coupling of physical and ecological factors

III. The links between biogeochemistry and food webs

IV. The controls of estuarine habitats

V. The role of synthesis in addressing issues of management

Each part begins with an introduction that describes the issue being considered and introduces the reader to important points to keep in mind. Next, several chapters report plenary talks illustrating case histories of successful syntheses. Finally, each part closes with a workshop report presenting an overview of the topic, goals for further synthesis, and recommendations about the data, knowledge, and models necessary to attain these goals.

References

Alongi, D. M. 1998. *Coastal ecosystem processes.* Boca Raton, FL: CRC Press.

CENR. 1995. *Setting a new course for U.S. coastal ocean science.* National Science and Technology Council Committee on Environment and Natural Resources. Silver Spring, MD: NOAA Coastal Ocean Office.

LMER Coordinating Committee. 1992. Understanding changes in coastal environments: The LMER program. *EOS* 73 (45): 481–85.

National Research Council. 1994. *Priorities for coastal ecosystem science*. Washington,
DC: National Academy Press.

U.S. EPA. 1998. The National Estuary Program: A ten-year perspective. *Coastlines*
8(1) (http://www.epa.gov/owow/estuaries/coastlines/winter98/index.html)

PART ONE

Drainage Basin Synthetic Studies

RESEARCH ON ESTUARIES HAS TRADITIONALLY BEGUN at the interface between land and ocean, where the fresh water meets the saltwater. But the quality and quantity of water, nutrients, and sediment that flow into the estuary from the land are among the most important factors regulating the structure and functioning of estuarine ecosystems; events in the watershed are too important to be ignored. Resource managers need to be able to predict changes in the quality and amount of runoff into the estuary if changes occur in climate or in land use in the watershed. They must be prepared to predict the effects of alternative management strategies on nutrient flux into the estuary.

Managers frequently use empirical models that are based on observations of water flow and on correlations between type of land use and the runoff of nutrients and sediment. In empirical models, such as the one described by Howarth et al. in chapter 2, the concentrations of sediments and nutrients in streams are determined by measurement or extracted from the literature. The amount of water in a river is calculated with a physical model that uses data on meteorology, soil type, slope, and land use to estimate soil moisture, ground-water recharge, overland flow, and erosion. Fluxes of sediment, organic matter, and nutrients are simply estimated by assigning concentrations to levels of water flow.

These empirical models work well for questions where the environmental conditions are within the range of variation in the phenomena measured to construct the model. When new conditions exceed the range studied, however, as happens when questions are asked about the impact of changes in climate or an entirely new form of land use, this type of model is of little use. For example, the number of inhabitants in a watershed may be important for a model predicting nutrient concentrations in runoff from watersheds dominated by agriculture, but this variable may be of little importance in assessing watersheds dominated by dwellings with a citywide central sewer system. Correlation models of this type are concerned only with the accuracy and precision with which an observed system input models an observed system output; they are of limited use for developing ideas about cause and effect.

Despite years of research, we do not yet have enough information on processes in watersheds, including biological processes in plants and roots, microbial and chemical processes in soil, and the processes by which nutrients are transferred into streams, to construct predictive models. One reason has to do with the biological cycles of nutrients. In the case of nitrogen, for example, large quantities are taken up by plants in watersheds each year. Nearly equal amounts of nitrogen are released from plant material by decomposition. A small amount, the difference between these two large quantities, leaves the watershed in stream water. Modeling the processes responsible for these small differences is a difficult task.

Although we have not yet succeeded in creating a model based entirely on processes, we have useful models that combine processes and correlations. In chapter 3, Vörösmarty and Peterson present a model that predicts river runoff and the nutrients, sediment, and organic matter that it carries. Water flow estimates are derived from a physical model of the water and energy budget. The quantity of materials transported in streams and rivers is derived from correlations among flow, type of land use, and concentrations of each substance. A geographic information system (GIS) is used to extrapolate results from a local to regional and global scales. The combination of a model and a GIS can be used to predict what could happen to water flux and material transport under potential future climate conditions.

The next step is to put together all the relationships necessary to build process-based models that combine climate, vegetation, and soil processes with hydrology models based on climate and landscape.

CHAPTER 2

Some Approaches for Assessing Human Influences on Fluxes of Nitrogen and Organic Carbon to Estuaries

Robert W. Howarth, Norbert Jaworski, Dennis Swaney,
Alan Townsend, and Gilles Billen

Abstract

Inputs of organic carbon and nitrogen to estuaries and coastal seas from non-point sources in the landscape are critically important in regulating the metabolism of these coastal marine ecosystems. This chapter summarizes two approaches we have used to investigate the controls on such fluxes. The first is the use of a model, the Generalized Watershed Loading Function (GWLF) model, to evaluate sources of organic carbon to the tidal, freshwater Hudson River estuary. This ecosystem is highly heterotrophic, and heterotrophic respiration is driven largely by allochthonous inputs of organic matter from nonpoint sources. GWLF suggests that the major source of this allochthonous carbon is erosion from agricultural fields. The Hudson Valley was once largely forested, reached a peak of agricultural activity in the early 1900s, and has steadily become more forested again as agricultural land has been abandoned. GWLF indicates that these land-use changes are likely to have resulted in large changes in the metabolism of the freshwater river, with the greatest heterotrophy coinciding with the greatest agricultural activity. The model also suggests that climate change may affect allochthonous inputs to the estuary, since drier soils result in substantially less erosion of agricultural soils.

The second approach we describe is a large-scale comparative analysis of nitrogen fluxes in rivers to the estuaries and coastal seas of the North Atlantic Ocean. For this analysis, we divided the watersheds of the North Atlantic Basin into fourteen large regions. Nonpoint sources of nitrogen dominate the flux for all regions. Nitrogen inputs to a region from human activity (import of food, use of inorganic nitrogen fertilizer, nitrogen fixation by agricultural crops, and atmospheric deposition of oxidized nitrogen originating from fossil-fuel combustion) are linearly related to nitrogen export to the coast from the region. On average, only 20% of the human-controlled inputs to a region are exported from the region to coastal waters, and the rest are processed or retained in the landscape. Human activity has probably had substantial effects on increased nitrogen inputs to the coast for some regions, perhaps increasing inputs by some fifteen-fold for the North Sea and tenfold for the northeastern United States. Regression analysis suggests that the deposition of oxidized nitrogen from fossil fuel combustion may be particularly important as a source of nitrogen flowing from the landscape to estuaries; per unit nitrogen input, such deposition appears to contribute disproportionately to riverine flows in comparison to agricultural sources. Regression analysis also suggests that the deposition of ammonia and ammonium from the atmosphere may be a good surrogate measure of the leakage of nitrogen from agricultural systems to surface waters.

Introduction

The metabolism of estuaries is strongly influenced by the external inputs of organic carbon and nutrients (NRC 1993; Nixon 1995; Nixon et al. 1996; Kemp et al. 1997). In most estuaries, the majority of nutrient and organic matter inputs come from land, with upstream advection of nutrients from oceanic sources being relatively minor (Nixon et al. 1995, 1996; Howarth et al. 1995), although for some estuaries with limited watersheds the oceanic source can dominate metabolism (Smith et al. 1991). Despite the obvious importance of fluxes of nitrogen, phosphorus, and organic carbon from the landscape to estuaries, there has been relatively little study on the control of these fluxes and surprisingly few syntheses (but see Jaworski 1997; Meybeck 1982; Billen et al. 1991; Hopkinson and Vallino 1995). While sewage inputs dominate in some estuaries, nonpoint-source fluxes are often greater (Nixon and Pilson 1983; NRC 1993; Howarth et al. 1996a, 1996b). Much evidence points to increased inputs of nutrients and organic carbon to estuaries from human activity in the landscape (Meybeck 1982; Pacés 1982; Larsson et al. 1985; Turner and Rabalais 1991; Howarth et al. 1996a, 1996b). In this chapter, we summarize two approaches we have used to estimate the extent of this human influence and to predict future changes: (1) a modeling approach for examining sources of organic matter input to the Hudson River estuary; and (2) a comparative regional analysis of nitrogen fluxes from land to the estuaries and coastal seas of the North Atlantic Ocean.

Modeling Sources of Organic Matter Input to the Hudson River

For the past decade, we have been studying the metabolism of the Hudson River estuary. The tidal, freshwater Hudson River (a stretch of river approximately 150 km long) is highly heterotrophic, with ecosystem respiration exceeding gross primary production (GPP) by almost twofold (Howarth et al. 1992, 1996a). This heterotrophic metabolism is fueled largely by organic carbon inputs to the river from nonpoint sources in the landscape; these nonpoint inputs exceed inputs of organic carbon in sewage by tenfold (Howarth et al. 1996a). Many other large rivers and estuaries also receive significant inputs of allochthonous organic matter from their watersheds (Schlesinger and Melack 1981; Meybeck 1982; Richey et al. 1991), and many other river and estuarine ecosystems are also heterotrophic (Kempe et al. 1991; Smith et al. 1991; Smith and Hollibaugh 1993; Hopkinson and Vallino 1995). These observations lead to some interesting questions, such as: To what extent are these fluxes of organic carbon from the landscape natural, and to what extent are they influenced by human activity? Is there an important climatic control on carbon inputs to rivers and estuaries, and how may climate change affect organic carbon fluxes from the landscape? To address these questions, we have adapted the GWLF model to estimate fluxes of organic carbon to the Hudson River from its watershed (Howarth et al. 1991; Swaney et al. 1996). GWLF was originally developed by Haith and Shoemaker (1987) to estimate fluxes of nitrogen and phosphorus from watersheds of mixed land use in the northeastern United States.

The GWLF model as we have used it relies on a simple mass-balance approach to estimate surface hydrologic and groundwater fluxes of water (figure 2-1). Daily climatic data from NOAA weather stations drive the hydrologic part of the model; precipitation is divided into rainfall or snow based on temperature; melting of the snowpack, if any, is estimated from temperature; and evapotranspiration is estimated from temperature and the number of daylight hours. Groundwater is treated in two reservoirs: a surface zone which is usually not saturated and a deeper, permanently saturated zone; water is added to the surface zone from rainfall and snowmelt and is removed through evapotranspiration and percolation into the lower saturated zone. Percolation is a first-order function of the water content in the surface zone; when the surface zone becomes saturated, any further rainfall and snowmelt becomes surface runoff. Groundwater movement from the lower saturated zone to streams and rivers is a first-order function of the water content in the lower zone. In our first effort with the model, we used spatially averaged daily weather inputs for the entire Hudson River watershed for the meteorological inputs (Howarth et al. 1991); Swaney et al. (1996) improved on this approach by dividing the Hudson watershed into eight sub-basins and using averaged daily weather inputs for each sub-basin.

The GWLF model uses the universal soil loss equation (Stewart et al. 1976) to estimate erosion from forests and agricultural lands and the STORM model (Hydraulic Engineering Center 1977) to estimate erosion from urban and suburban areas (figures 2-2 and 2-3). For urban and suburban areas, erosion is a relatively simple function of the surface runoff of water estimated from

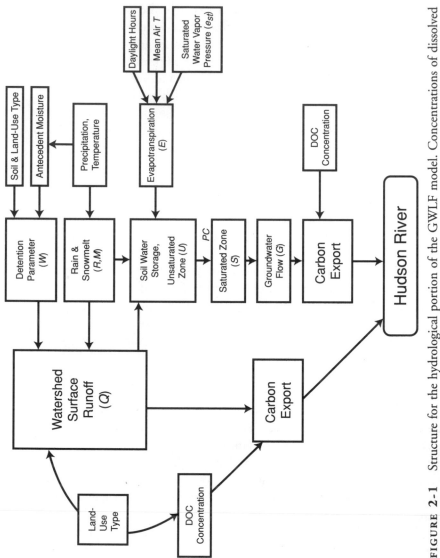

FIGURE 2-1 Structure for the hydrological portion of the GWLF model. Concentrations of dissolved organic carbon (DOC) are assigned to groundwater flows and to surface-water runoff as a function of land use. Reprinted from Howarth et al. (1991) by permission.

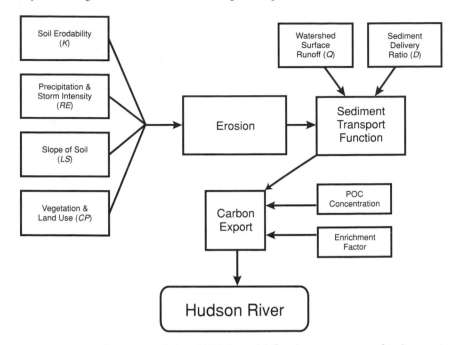

FIGURE 2-2 Structure of the GWLF model for the movement of sediment in forests and agricultural systems. Structure is based on the universal soil-loss equation. The concentration of particulate organic carbon is assigned to eroded soils as a function of land use. Reprinted from Howarth et al. (1991) by permission.

GWLF's hydrologic submodel; all soil eroded from urban and suburban lands is assumed to reach surface waters. For forests and agricultural lands, erosion is treated as a function of precipitation and intensity of storms, the type of land use, the erodability of soil, and the slope of the land. The model assumes that only a small amount of the soil eroded from forest and fields actually reaches surface waters, while most is retained in the landscape. The proportion that reaches surface waters is estimated from the "sediment delivery ratio," which in turn is estimated from the area of the watershed; the larger the watershed, the smaller the sediment delivery ratio. The transport of eroded soils to surface waters is estimated on a monthly basis using a power function of the surface runoff estimate derived from GWLF's hydrologic submodel (Haith and Shoemaker 1987; Howarth et al. 1991; Swaney et al. 1996).

Although GWLF was developed as a tool to estimate fluxes of nitrogen and phosphorus to surface waters (Haith and Shoemaker 1987), the model has no explicit inclusion of biogeochemical processes. Rather, concentrations of dissolved and particulate nutrients (Haith and Shoemaker 1987) or concentrations of dissolved and particulate organic carbon (Howarth et al. 1991) are

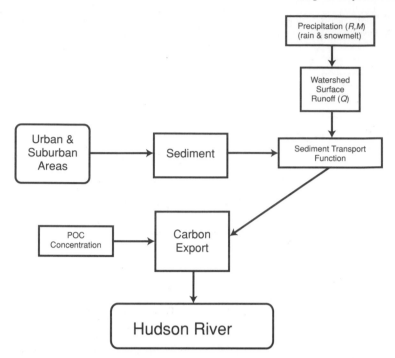

FIGURE 2-3 Structure of the GWLF model for movement of sediment in urban and suburban areas. Reprinted from Howarth et al. (1991) by permission.

assigned to flows of surface water, groundwater, and sediments as a function of land-use type. Also, although actual land-use data are used in the GWLF model, there is no spatial reality to the model and land areas are aggregated. That is, the model has no explicit recognition to whether water leaving agricultural fields flows through riparian wetlands or goes directly into surface waters.

The GWLF model does a very good job of estimating freshwater discharge and a good job of estimating fluxes of sediment to the Hudson River, particularly when using meteorological inputs at the finer scale of eight sub-basins (Swaney et al. 1996). Compared with freshwater discharge and sediment flux estimates based on U.S. Geological Survey (USGS) data from the Green Island gauging station, which includes the upper Hudson and Mohawk Rivers (two tributaries which together comprise two-thirds of the freshwater discharge for the entire Hudson River), the model underestimates average annual discharge by 9% and average annual sediment flux by 34% (Swaney et al. 1996). The model also does a good job of catching the seasonal variability in both freshwater discharge and sediment flux over a 3-year period (figures 2-4 and 2-5).

Evaluating the GWLF-derived estimate for organic carbon input to the Hudson River estuary is more difficult, as organic carbon fluxes have not been

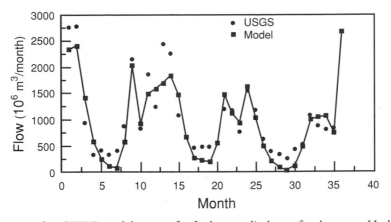

FIGURE 2-4   GWLF model output for freshwater discharge for the upper Hudson and Mohawk Rivers compared with data from the USGS gauging station at Green Island. The model run is for 36 months beginning in April 1983. USGS data are solid points. Model output is indicated by the line. Reprinted from Swaney et al. (1996) by permission.

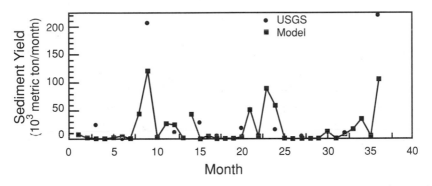

FIGURE 2-5   GWLF model output for sediment fluxes for the upper Hudson and Mohawk Rivers compared with data from the USGS gauging station at Green Island. The model run is for 36 months beginning in April 1983. USGS data are solid points. Model output is indicated by the line. Reprinted from Swaney et al. (1996) by permission.

routinely measured at the Green Island gauging station by the USGS. However, the USGS has occasionally measured organic carbon in many tributaries within the Hudson River watershed. Gladden et al. (1988) regressed the log of the flux of organic carbon per watershed area against the log of water discharge per watershed area for all times and places within the Hudson River watershed for which data were available for the early 1980s. Using this regression and estimates for total freshwater discharge, Gladden et al. (1988) estimated the flux of organic carbon from the watershed of the Hudson to be

3.1 g C m^{-2} yr^{-1} for an average year with typical freshwater discharge. A mass balance of organic carbon in the freshwater Hudson River for 1988 gives great credibility to this approach (Howarth et al. 1996a); 1988 was a drought year with low freshwater discharge; for that year, the approach of Gladden et al. (1988) yields an estimate of 2.4 g C m^{-2} yr^{-1} for allochthonous inputs of organic carbon to the Hudson; summing this estimate with other inputs gives a total estimate of carbon inputs to the Hudson that agrees to within 2% with the sum of carbon sinks for that same year (Howarth et al. 1996a). The years we modeled with GWLF had average precipitation, and so our best estimate of allochthonous inputs of organic carbon to the Hudson from the landscape for these years would be 3.1 g C m^{-2} yr^{-1} (Howarth et al. 1996a). Compared with this, the GWLF model estimate is low by 44% (Swaney et al. 1996; see also table 2-1). That GWLF underestimates organic carbon fluxes to a greater extent than total sediment fluxes may be due to the lack of consideration of in-stream and in-river processing of organic carbon by the model (Howarth et al. 1991).

The GWLF model estimates that in the Hudson watershed agricultural lands export tenfold more organic carbon per area than do forests with urban and sub-urban lands exporting an intermediate amount of carbon (table 2-1). The Hudson River watershed is largely forested (65% of total area), and only 28% of the land is in agricultural use, yet the GWLF model suggests that agricultural lands contribute 74% and forested lands only 18% of the total inputs of organic carbon to the Hudson from nonpoint sources (Swaney et al. 1996). Urban and suburban lands, which comprise 7% of the area, are estimated to contribute 8% of the flux of nonpoint-source organic carbon (Swaney et al. 1996).

TABLE **2-1**

Estimates of inputs of organic matter from nonpoint sources in the watershed to the tidal, freshwater Hudson River.

	g C m^{-2} yr^{-1}
Regression analysis—average for all of Hudson River	3.1
Model result—average for all land uses	1.7
Model result—forest lands only	0.47
Model result—agricultural lands only	4.5
Model result—urban and suburban lands only	1.9

The regression analysis is based on Gladden et al. (1988) and uses average freshwater discharge for the Hudson River and a regression of the log of water discharge per area vs. the log of organic carbon export per area for various tributaries of the Hudson (Howarth et al. 1996a). The model-derived estimates are from runs of GWLF reported by Swaney et al. (1996). All estimates are for years of average precipitation and discharge.

If these estimates of organic carbon input are accurate, then carbon loading to the Hudson River has changed dramatically as land use has changed. Before European settlement, the Hudson River watershed was largely forested; running GWLF with 1980s climatic data but assuming that 100% of the land is forested, the model estimates that organic carbon inputs would be only one-third of modern values (Swaney et al. 1996). This suggests that in its pristine state, the Hudson River would have been much less heterotrophic than at present (Howarth et al. 1996a). On the other hand, in the early 1900s most of the land in the Hudson River watershed was actively farmed, and the current situation of 65% forest use is a result of gradual agricultural abandonment and forest regrowth (Rod et al. 1989). Running GWLF with 1980s climatic data but assuming land use as it existed at peak agricultural use, the model estimates that nonpoint loading of organic carbon to the Hudson may once have been almost twice as large as at present (Swaney et al. 1996); the actual situation may have been even worse since we assumed modern agricultural practices for the model run but actual practices probably resulted in more erosion. Thus, the Hudson was probably more heterotrophic earlier in the 1900s than at present (Howarth et al. 1996a). This conclusion should perhaps be tempered because of changes in sewage inputs over time due to population growth and changes in sewage treatment technology; while sewage inputs are currently only 10% of the nonpoint source allochthonous inputs of carbon from the landscape (Howarth et al. 1996a) and are unlikely ever to have been the major input, the organic carbon in sewage is more labile and more easily respired than is the carbon in allochthonous inputs from eroded soils. Nonetheless, heterotrophic respiration in the freshwater Hudson currently is driven largely by the allochthonous inputs from nonpoint sources.

Gladden et al. (1988) noted that year-to-year variability in climate probably results in substantial interannual variability in organic carbon loading to the Hudson River. Hence, climate change could have a profound influence on the metabolism of the Hudson and other rivers and estuaries in the northeastern United States; models of global warming suggest a generally drier climate for the northeastern United States, which might be expected to decrease inputs of organic carbon from land to these ecosystems (Moore et al. 1997). Our runs with the GWLF model, however, indicate that the most important climatic factors regulating organic carbon fluxes from the landscape are not average annual discharge or precipitation, but rather day-to-day and seasonal patterns of precipitation and discharge (Howarth et al. 1991). Thus, even if climate change results in a generally drier climate, allochthonous fluxes of organic carbon might increase in years with periods of intense summer storms.

The GWLF model has contributed greatly to our understanding of carbon flows through the Hudson River watershed. GWLF has also been used to explore sources of nitrogen to tributaries of the Delaware River (Haith and

Shoemaker 1987). However, as noted above, the model relies entirely on hydrology, erosion, and concentrations of carbon and nutrients assigned as functions of land use, and does not contain any explicit representation of biogeochemical processes. We believe strongly that the inclusion of biogeochemical processes—both in-stream and in the terrestrial ecosystems— is necessary to make models more useful in exploring sources of nutrients and organic carbon for rivers and estuaries. We are aware of no model that has yet linked biogeochemical processes in terrestrial systems with hydrologic and erosional processes, and with in-stream processing of organic carbon or nutrients, but such linkages should be possible. The Century model has proven quite powerful and versatile for studying a variety of carbon and nitrogen biogeochemical processes in terrestrial ecosystems (Parton et al. 1987, 1996; Schimel et al. 1994) and shows promise as a tool for understanding export to aquatic ecosystems (Baron et al. 1994). Other terrestrial biogeochemical models have been developed which may be even more useful for modeling nitrogen fluxes from agricultural systems (Scholefield et al. 1991; Engel and Priesack 1993; Grant et al. 1993; Müller et al. 1997). We believe that linking a terrestrial biogeochemical model and a hydrological/erosion model such as GWLF could prove quite powerful in predicting fluxes of both carbon and nitrogen from the landscape. Further refinements could include processing of materials in riparian wetlands and in streams and rivers. For example, in-stream processing of nitrogen can perhaps be modeled simply by estimating retention as a function of water depth and residence time (Kelly et al. 1987; Howarth et al. 1996b). Similarly, in-stream processing of organic carbon can perhaps be modeled from information on "spiraling distances" or turnover times in streams and rivers of different sizes (Newbold 1992; Howarth et al. 1996a).

A Comparative Regional Analysis of Nitrogen Fluxes from Land to the North Atlantic Ocean

Nitrogen is frequently the element most limiting to primary productivity in estuaries and coastal seas in the temperate zone, and the management of coastal eutrophication demands that nitrogen inputs to estuaries be better controlled (Howarth 1988; NRC 1993; Nixon 1995; Nixon et al. 1996; Downing 1997). In many estuaries, nonpoint sources dominate inputs of nitrogen (Nixon and Pilson 1983; NRC 1993), and for the North Atlantic Ocean as a whole, nonpoint-source fluxes of nitrogen from the landscape exceed sewage inputs by ninefold (Howarth et al. 1996b). Thus, better management of coastal eutrophication requires an understanding of the nonpoint sources of nitrogen to surface waters from the surrounding landscape.

Anthropogenic activity has greatly accelerated nitrogen cycling and has caused the rate of nitrogen fixation in the terrestrial biosphere to at least double over natural biological rates (Galloway et al. 1995; Vitousek et al. 1997). This increase in nitrogen fixation, which is due largely to the manufacturing of inorganic nitrogen fertilizer but also includes substantial incidental nitrogen fixation from the combustion of fossil fuels, has been dramatic over the past few decades and continues to increase at an exponential rate (Vitousek et al. 1997). One should perhaps expect that the increased nitrogen fixation on land might result in larger fluxes of nitrogen to estuaries, and indeed several papers have reported increased fluxes of inorganic nitrogen as nitrate in rivers over time. For the Mississippi River, nitrate concentrations have more than doubled in the last 30 years (Turner and Rabalais 1991). Nitrate concentrations in many rivers of Central Europe (Pacés 1982) and the northeastern United States (Jaworski et al. 1997) have increased four- to tenfold during the 1900s.

While it is reasonable to deduce that both the increased use of inorganic fertilizer and increased atmospheric deposition of nitrogen from fossil-fuel combustion contribute to elevated riverine fluxes of nitrogen to the coast, estimating the relative contribution of each of these sources is a difficult challenge (Howarth et al. 1996b). Consider the relatively simple case of determining the fate of inorganic nitrogen fertilizer. If the fertilizer is applied to a grassland on a loam soil, only 3 to 10% of the application typically is leached as nitrate, but if the fertilizer is applied to tilled cropland on a sandy soil, from 25 to 80% of the fertilizer is generally leached (Howarth et al. 1996b). Even in the absence of fertilization, converting pastureland to tilled cropland appears to greatly increase nitrogen leaching, and the extent of this leaching varies with soil type (Howarth et al. 1996b). Despite this complexity, the general patterns of nitrogen leaching from agricultural systems are relatively well understood, and estimates of such leaching can be made for large regions (Howarth et al. 1996b). However, leaching of nitrate from agricultural fields is not the only pathway whereby fertilizer can reach surface waters; significant amounts can be volatilized as ammonia, with the ammonium later being deposited back onto the landscape where it may be nitrified and leached. Also, one must consider the fate of nitrogen in harvested crops; on average, about half of nitrogen fertilizer applied to tilled croplands is removed in the crop when it is harvested (Bock 1984; Nelson 1985). Much of the crop is then fed to animals, which can result in significant volatilization of nitrogen to the atmosphere or direct leaching of nitrogen to surface waters. Thus, to determine the fate of nitrogen from added fertilizer, the correct unit of study is not agricultural field plots but rather the entire food-agricultural system (Bleken and Bakken 1997a, 1997b; Isermann and Isermann 1998). At this scale, the linkage between fertilizer use and nitrogen inputs to surface waters remains poorly documented.

Estimating the fate of nitrogen in atmospheric deposition onto the land-scape is at least as challenging as determining the fate of nitrogen fertilizer. Consider deposition onto forests. Ecological theory would suggest that forest age is a major determinant of the fate of nitrogen deposition, with young aggrading forests retaining nitrogen and old-growth forests with zero rates of net ecosystem production exporting nitrogen (Vitousek and Reiners 1975; Hedin et al. 1995). However, even young, aggrading, nitrogen-limited forests can export large amounts of nitrogen from atmospheric deposition; factors such as highly permeable, sandy soils and nitrogen loss during a winter dor-mant season can contribute to nitrogen loss from young forests (Lajtha et al. 1995). Most forests of the temperate zone are nitrogen limited (Vitousek and Howarth 1991), but at high rates of atmospheric deposition, nitrogen inputs can become sufficient to "saturate" nutritional needs and other elements may become limiting. Under these conditions of nitrogen saturation, nitrogen leaching to downstream aquatic ecosystems can greatly increase (van Breemen et al. 1982; Aber et al. 1989; Schulze 1989; Gundersen and Bashkin 1994). This linkage between nitrogen deposition, nitrogen saturation of an ecosys-tem, and nitrogen export has received a tremendous amount of study, yet the relationship between nitrogen inputs and nitrogen export from forests remains unclear. In a review of data for forests in the United States, Johnson (1992) found no relationship between nitrogen inputs and nitrogen exports; however, few if any of the studies reviewed by Johnson included export of organic nitrogen; organic nitrogen export can be large and if included might result in clearer relationships between input and export (Hedin et al. 1995; Howarth et al. 1996b). For European forests, nitrogen leaching tends to be either quite small or quite large (Hauhs et al. 1989). Recently, Gundersen et al. (1997) have demonstrated that nitrogen export from European forests, while not correlated with nitrogen deposition, is well correlated with variables indicative of the nitrogen status in forests, such as the C:N ratio of the forest floor.

Much progress is being made in understanding the cycling of nitrogen at the scale of individual ecosystems, yet given the complexity of nitrogen cycling at this scale, the uncertainty in scaling up nitrogen export from individual ecosystems to larger scales is likely to remain great. To determine if more pre-dictability in nitrogen export from the landscape can be gained from analysis at larger spatial scales, we undertook an analysis of nitrogen flux from land to the North Atlantic Ocean at the scale of large regions (Howarth et al. 1996b). We divided up the watersheds of the North Atlantic Ocean into fourteen large geographical areas; for each region, we estimated nitrogen flux in rivers and in sewage to estuaries and coastal seas. We also estimated the inputs of nitrogen from human activity into each region. This allowed us to perform a mass bal-ance of human-derived nitrogen for each region. This effort was part of the first activity of the International SCOPE Nitrogen Project: an evaluation of the nitrogen cycle of the North Atlantic Ocean (Howarth 1996; Galloway et al. 1996). The International SCOPE Nitrogen Project was established to assess how humans have altered the nitrogen cycle at regional and global scales, and to determine the consequences of this alteration.

In our regional analysis, we used the best available data for each region and therefore used a variety of approaches for estimating the export of nitrogen in rivers to the coast. For many regions, good data on actual riverine nitrogen flows exist. This is particularly true for large regions dominated by a single river, as in the Amazon and Mississippi River Basins. For the northeastern United States, the southwestern coast of Europe, and the watersheds of the North Sea, nitrogen flows from the landscape to the coast occur in numerous rivers, but good data on nitrogen flows exist for most of these rivers (Howarth et al. 1996b). For some regions where data for individual rivers are scarce, we used hydrologic budgets and extrapolation of nitrogen concentrations from some smaller watersheds within the region. For some other regions, we used previously published estimates for regional nitrate export and estimated organic nitrogen export from estimates of organic carbon export and assumed C:N ratios (Howarth et al. 1996b).

The flux of nitrogen in rivers and sewage to the coastal areas of the North Atlantic Basin varies some twentyfold among the regions (Howarth et al. 1996b). The highest fluxes are in the regions with the highest population densities such as the North Sea area, northwestern Europe, and the northeastern United States. Peierls et al. (1991) reported that the log of nitrate fluxes from large rivers of the world are correlated with the log of population densities in their watersheds, and the same relationship holds for the North Atlantic regional analysis of total nitrogen fluxes, with the log of total nitrogen fluxes per area correlated with the log of population densities (Howarth et al. 1996b). When expressed on a linear-linear rather than a log-log scale, population density is even a better predictor of total nitrogen export from regions of the North Atlantic Basin ($r^2 = 0.77$, as opposed to 0.45 for the log-log relationship; Howarth et al. 1996b).

From the analysis of Peierls et al. (1991) for large rivers of the world, Cole et al. (1993) suggested that population density is correlated with nitrate fluxes because sewage is the major source of nitrogen to rivers in regions with higher population densities. However, sewage alone cannot explain the pattern between population density and total nitrogen fluxes for the regions of the North Atlantic Basin. For our analysis, we estimated the contribution of human sewage to nitrogen export from each region, using both population-based estimates and data on the extent of sewers in each region. Sewage is not the major contributor to riverine export of total nitrogen for any of the regions, and nonpoint sources of nitrogen dominated the riverine nitrogen flux in all regions surrounding the North Atlantic Ocean (Howarth et al.

1996b). Sewage is most important in the watersheds of the North Sea, where it contributes 34% of the riverine nitrogen flux, and is least important in the Amazon River Basin, where population density is very low and sewage contributes only 0.01% to the riverine nitrogen flux. On average, sewage contributes some 11% of the total nitrogen flux from land to the North Atlantic Ocean.

To analyze mass balances of human-derived nitrogen for each of the regions of the North Atlantic Basin, we considered the following as inputs of nitrogen to a region: the use of inorganic nitrogen fertilizer, nitrogen fixation by agricultural crops, the net movement of nitrogen into or out of a region in food and feedstocks, and the deposition of oxidized nitrogen compounds from the atmosphere (NO_y). To avoid double accounting, we did not consider sewage or the use of organic manure fertilizers as inputs of nitrogen to a region; both sewage and manure can contribute substantially to nitrogen inputs to surface waters, but the sewage and manure are a recycling of nitrogen within a region and not a new input. The nitrogen in the sewage and manure comes from regional inputs of nitrogen such as import of food/feedstocks or use of inorganic fertilizer to grow the food and feedstocks within the region which are then eaten by humans or livestock. Similarly, we did not consider the atmospheric deposition of ammonia/ammonium as an input to a region. Since most ammonia moves only fairly short distances through the atmosphere, ammonia deposition largely reflects nitrogen recycling within a region rather than a new input (Howarth et al. 1996b). The majority of the ammonia/ammonium from atmospheric deposition originated as volatilization of ammonia within the same region, and came from sources such as manure and feedlots; at the scale of large regions, the original source of the nitrogen in this ammonia is from the import of food, the use of inorganic fertilizer, nitrogen fixation by crops, or the deposition of oxidized nitrogen. In contrast to ammonia/ammonium deposition, the deposition of oxidized nitrogen (NO_y) to a region needs to be considered a net input of nitrogen to the region as most comes from a source not otherwise accounted for: the combustion of fossil fuels (Howarth et al. 1996b).

Sources of data for nitrogen inputs to a region are described in detail in Howarth et al. (1996b). Estimates for NO_y deposition due to human activity were derived from models and include a subtraction of background natural levels for each region from the estimate of current deposition (Prospero et al. 1996; Howarth et al. 1996b). For most regions, inorganic fertilizer is the largest single input of nitrogen, but NO_y deposition dominates in the northeastern United States and in the St. Lawrence Basin (Howarth et al. 1996b). For the regions as a whole, inorganic nitrogen fertilizer is the largest input and makes up almost two-thirds of all inputs (table 2-2).

Per area, the total input of nitrogen from human activity varies more than

TABLE 2-2

Anthropogenic nitrogen budgets per area of land for the northeastern United States and for all temperate watersheds of the North Atlantic Basin ($kg\ N\ km^{-2}\ yr^{-1}$). Data are from Howarth et al. (1996b).

	Northeastern United States	All Temperate Watersheds of North Atlantic Ocean
Inputs		
NO_y deposition	1,200	490
N fertilizer	600	1,400
N fixation in agriculture	750	315
Net import of N in food	1,000	0
Total inputs	3,550	2,205
Sinks and exports		
Net export of N in food	0	255
Export in rivers and sewage	1,070	490
Storage in the landscape and denitrification (by difference)	2,480	1,460
Total sinks and exports	3,550	2,205
Agricultural and wastewater fluxes within regions		
Wastewater N flux	280	80
Volatilization and deposition of NH_x	190	240

thirtyfold among regions (Howarth et al. 1996b). Total net inputs (where export of nitrogen in food/feedstocks from a region is taken as a negative input) of human-controlled nitrogen are almost twice as great for the North Sea region as for any other region when expressed per land area. The lowest inputs per area are to the northern Canada region and to the Amazon River Basin (Howarth et al. 1996b). For the temperate regions of the North Atlantic Basin, the riverine nitrogen export per area of landscape is well correlated with the sum of the human-controlled inputs of nitrogen (figure 2-6). Compared to the temperate regions, tropical areas such as the Amazon River Basin export somewhat more nitrogen per unit of human-controlled nitrogen input to a region (Howarth et al. 1996b), perhaps because the terrestrial biosphere of the tropics is less nitrogen limited than in the temperate zone (Vitousek and Howarth 1991). The tropical regions of the North Atlantic Basin are not included in figure 2-6.

For the regression of riverine nitrogen export on human-controlled nitrogen inputs to regions, the y-intercept suggests a riverine nitrogen export of approximately 100 $kg\ N\ km^{-2}\ yr^{-1}$ in the absence of human inputs of nitrogen to a region. The 95% confidence interval around the intercept is large (from a negative value to +400 $kg\ N\ km^{-2}\ yr^{-1}$), but an export of approximately 100 kg N

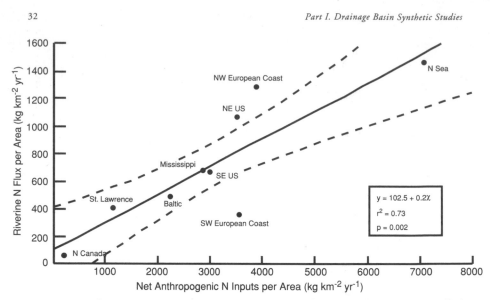

FIGURE 2-6   Riverine nitrogen export per area as a function of human-controlled nitrogen inputs for the temperate-zone regions of the North Atlantic Basin. Reprinted from Howarth et al. (1996b) by permission.

km^{-2} yr^{-1} from pristine temperate-zone systems is not far from other estimates for fluxes from pristine temperate systems compiled by Howarth et al. (1996b): 75 to 230 kg N km^{-2} yr^{-1}. The intercept-based estimate for pristine nitrogen exports suggests that current fluxes from the northeastern United States and from the North Sea region may be tenfold and from fifteenfold greater, respectively, than would be the case absent human activity. The slope of the regression illustrates that on average only 20% of the human-controlled inputs of nitrogen are exported to estuaries and the coast. The rest of this nitrogen must be processed or retained in the landscape; we suspect that most of this "missing" nitrogen is denitrified (Howarth et al. 1996b).

Atmospheric deposition of NO_y may be proportionately more important as a source of nitrogen to rivers and the coast than are the other nitrogen inputs to regions; multiple regression analysis suggests that per unit nitrogen input per area, NO_y deposition is some sevenfold better at predicting riverine nitrogen flows than are agricultural sources of nitrogen (Howarth et al. 1996b; Howarth 1998). Even though NO_y deposition is the largest nitrogen input to only two regions, it alone is a very good predictor of riverine nitrogen export from the temperate-zone regions of the North Atlantic Basin (figure 2-7). Inorganic nitrogen fertilizer is the largest nitrogen input to all of the other temperate-zone regions of the North Atlantic Basin and makes up two-thirds of the total inputs on average for all the regions (table 2-2); while nitrogen fertilizer is significantly correlated with riverine nitro-

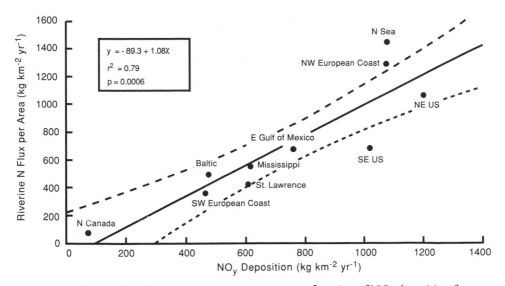

FIGURE 2-7   Riverine nitrogen export per area as a function of NO_y deposition for the temperate-zone regions of the North Atlantic Basin. Most of the nitrogen in NO_y deposition originates in the combustion of fossil fuels. Reprinted from Howarth et al. (1996b) by permission.

gen export, it is far less predictive than NO_y deposition for riverine flux (figure 2-8). This suggests the importance of controlling fossil-fuel sources of atmospheric nitrogen pollution, particularly in the northeastern United States, where these are the dominant nitrogen inputs to the region.

Even though inorganic fertilizer use is only a fair predictor of the export of nitrogen from a region, agricultural sources and practices clearly affect nitrogen export. Some recent analyses suggest that the major environmental impacts from nitrogen in agricultural systems come from "surplus" nitrogen that results in "leaky spots" (Bleken and Bakken 1997a, 1997b; Isermann and Isermann 1998). At the scale of the large regions of the North Atlantic Basin, the deposition of ammonia/ammonium may provide a good estimate of nitrogen flows into surface waters from these leaky spots in the agricultural system, as indicated by regressing ammonia/ammonium deposition plus NO_y deposition against riverine nitrogen fluxes from nonpoint sources (Howarth 1998). The nonpoint-source component of the riverine nitrogen flux for the regions of the North Atlantic Basin can be estimated by subtracting the estimated sewage input from the total riverine flux estimate for each region (Howarth 1998). We include NO_y deposition in the analysis because of its demonstrated relationship to riverine nitrogen export from regions. Estimates for ammonia/ammonium deposition are from Howarth et al. (1996b) and are

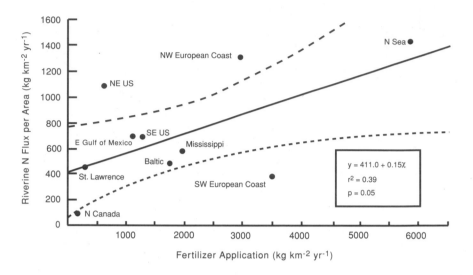

FIGURE 2-8   Riverine nitrogen export per area as a function of use of inorganic nitrogen fertilizer for the temperate-zone regions of the North Atlantic Basin. Reprinted from Howarth et al. (1996b) by permission.

based on models of emission strength and atmospheric transport. The regression of total inorganic nitrogen deposition, including wet and dry deposition of ammonia and ammonium, against the nonpoint-source flux of nitrogen from regions is an astonishingly good fit ($r^2 = 0.93$; figure 2-9).

Note that this analysis of total inorganic nitrogen deposition differs from the mass-balance approach we described above; ammonia/ammonium deposition is not an input of nitrogen to the region, but rather a recycling of nitrogen primarily from agricultural sources within the region (Schlesinger and Hartley 1991; Howarth et al. 1996b). However, ammonia/ammonium deposition appears to be a good measure of the agricultural component of nitrogen in a region that contributes to riverine fluxes to the coast. We stress that the regression shown in figure 2-9 need not indicate that ammonia/ammonium deposition is itself a major source of nitrogen to the riverine flux. An alternative explanation is that at the scale of these large regions, ammonia/ammonium deposition is a good surrogate measure for general leakiness of nitrogen from agriculture, and the factors that control export of nitrogen from agricultural sources to surface waters are strongly related in a linear fashion to the factors that regulate ammonia volatilization to the atmosphere; most ammonium deposition within a region comes from ammonia volatilization from agricultural sources within the same region since the transport of ammonia and ammonium through the atmosphere is limited (note that this assumption would not hold true for smaller watersheds). Thus, both ammonium deposition and the movement of nitrogen from agricultural

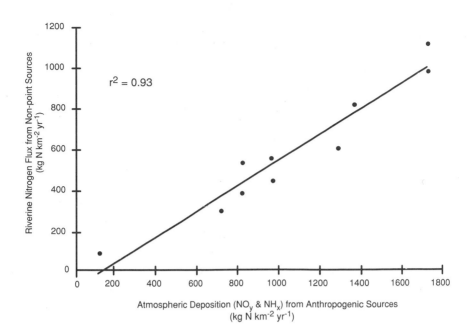

FIGURE 2-9   Riverine export of nitrogen from nonpoint sources as a function of the sum of wet and dry atmospheric deposition of oxidized and reduced forms of nitrogen from human-controlled sources. Reprinted from Howarth (1998) by permission.

sources into surface waters may be direct functions of the "surplus nitrogen" in agriculture (Howarth 1998). Ammonia/ammonium volatilization and deposition tends to be less than NO_y deposition in almost all regions of the North Atlantic Basin (Howarth et al. 1996b; see also table 2-2). Nonetheless, since ammonia/ammonium deposition may be a surrogate measure for agricultural nitrogen that moves directly into surface waters from sources such as leaching from feedlots and not just through volatilization to the atmosphere, the actual contribution of nitrogen from agriculture to the regional nitrogen exports could in theory be greater than that from NO_y deposition. Whether or not the pathway from agricultural sources to surface waters involves ammonia volatilization and redeposition, our result shows the importance of reducing the "surplus" nitrogen in agricultural and food systems (Bleken and Bakken 1997a, 1997b; van der Voet et al. 1996; Isermann and Isermann 1998).

Our large-scale regional analysis has proven a powerful tool for exploring controls on the movement of nitrogen from the landscape to estuaries and coastal seas. The mass-balance approach shows clearly that nonpoint sources of nitrogen dominate the movement of nitrogen to coastal waters and also indicate that only one-fifth of human-controlled inputs of nitrogen to regions reach estuaries and coastal seas. The fate of the majority of these human-controlled inputs of nitrogen to regions remains poorly known, although denitrification is likely to be the largest sink (Howarth et al. 1996b). Clearly, a critical research topic is to gain a better understanding of the fate of this "missing" nitrogen in the landscape. The correlational approach with the regional data needs to be interpreted with great care, as correlation cannot prove causation. Nonetheless, this approach suggests the great importance of fossil fuel combustion and subsequent deposition of NO_y onto the landscape as a major contributor to nitrogen inputs to coastal waters. The regression approach also suggests that the deposition of ammonia/ammonium may be a good surrogate, at the coarse scale of large regions, for nitrogen leakage from agricultural systems. Despite the power of working at large spatial scales for understanding nitrogen fluxes, much detail is lost. A better understanding of the controls on nitrogen flow through the landscape to surface waters probably requires inclusion of biogeochemical processes. We suggest that linking the regional approach to linked hydrological-biogeochemical models as we discussed for the Hudson River research may be a good approach to pursue next.

Acknowledgments

Our work described in this chapter is based largely on research funded by the Hudson River Foundation and by the Mellon Foundation. The views expressed here are those of the authors, not those of the foundations that funded the research. We thank Roxanne Marino for her review of the manuscript and for assistance in preparing the figures.

References

Aber, J. D., K. J. Nadelhoffer, P. Steudler, and J. M. Melillo. 1989. Nitrogen saturation in northern forest ecosystems. *BioScience* 39: 378–86.

Baron, J. S., D. S. Ojima, E. A. Holland, and W. J. Parton. 1994. Analysis of nitrogen saturation potential in Rocky Mountain tundra and forest: Implications for aquatic systems. *Biogeochemistry* 27: 61–82.

Billen, G., C. Lancelot, and M. Meybeck. 1991. N, P, and Si retention along the aquatic continuum from land to ocean. Pages 19–44 in R. F. C. Mantoura, J. M. Martin, and R. Wollast (editors), *Ocean margin processes in global change*. Chichester, England: John Wiley & Sons.

Bleken, M. A., and L. R. Bakken. 1997a. The anthropogenic nitrogen cycle in Norway. Pages 27–40 in E. Romstad, J. Simonsen, and A. Vatn (editors), *Controlling mineral emissions in European agriculture*. Wallingford, United Kingdom: CAB International.

————. 1997b. The nitrogen cost of food production: Norwegian society. *Ambio* 26: 134–42.

Bock, B. R. 1984. Efficient use of nitrogen in cropping systems. Pages 273–94 in R. D. Hauck (editor), *Nitrogen in crop production.* Madison, WI: American Society of Agronomy.

Cole, J. J., B. L. Peierls, N. F. Caraco, and M. L. Pace. 1993. Nitrogen loading of rivers as a human-driven process. Pages 141–57 in M. J. McDonnell and S. T. A. Pickett (editors), *Humans as components of ecosystems: The ecology of subtle human effects and populated areas.* New York: Springer-Verlag.

Downing, J. A. 1997. Marine nitrogen: Phosphorus stoichiometry and the global N:P cycle. *Biogeochemistry* 37: 237–52.

Engel, T., and E. Priesack. 1993. Expert-N, a building block system of nitrogen models as a resource for advice, research, water management, and policy. Pages 503–7 in H. J. P. Eijsackers and T. Hamers (editors), *Integrated soil and sediment research: A basis for proper protection.* Dordrecht, The Netherlands: Kluwer.

Galloway, J. N., W. H. Schlesinger, H. Levy, A. Michaels, and J. L. Schnorr. 1995. Nitrogen fixation: Atmospheric enhancement—environmental response. *Global Biogeochemical Cycles* 9: 235–52.

Galloway, J. N., R. W. Howarth, A. F. Michaels, S. W. Nixon, J. M. Prospero, and F. J. Dentener. 1996. Nitrogen and phosphorus budgets of the North Atlantic Ocean and its airshed. *Biogeochemistry* 35: 3–25.

Gladden, J. B., F. R. Cantelmo, J. M. Groom, and R. Shapot. 1988. Evaluation of the Hudson River ecosystem in relation to the dynamcis of fish populations. *American Fisheries Society Monograph* 4: 37–52.

Grant, R. F., M. Nyborg, and L. W. Laidlaw. 1993. Evolution of nitrous oxide from soil: 1. Model development. *Soil Science* 156: 259–65.

Gundersen, P., and V. Bashkin. 1994. Nitrogen cycling. Pages 255–83 in B. Moldan and J. Černý (editors), *Biogeochemistry of small catchments: A tool for environmental research.* Chichester, England: John Wiley & Sons.

Gundersen, P., B. A. Emmett, O. J. Kjønaas, A. Tietema, and K. R. Rasmussen. 1997. Nitrate leaching from European forests. *BIOGEOMON Journal of Conference Abstracts* 2: 189.

Haith, D. A., and L. L. Shoemaker. 1987. Generalized watershed loading functions for stream flow nutrients. *Water Resources Bulletin* 23: 471–78.

Hauhs, M., K. Rost-Siebert, G. Ragen, T. Pacés, and B. Vigeruse. 1989. Summary of European data. In J. L. Malnchuk and J. Nilsson (editors), *The role of nitrogen in the acidification of soils and surface waters.* Copenhagen, Denmark: Nordic Council of Ministers.

Hedin, L. O., J. J. Armesto, and A. H. Johnson. 1995. Patterns of nutrient loss from unpolluted, old-growth temperature forests: Evaluation of biogeochemical theory. *Ecology* 76: 493–509.

Hopkinson, C. S., and J. J. Vallino. 1995. The relationships among man's activities in the watersheds and estuaries: A model of runoff effects on patterns of estuarine community metabolism. *Estuaries* 18: 598–621.

Howarth, R. W. 1988. Nutrient limitation of net primary production in marine ecosystems. *Annual Review of Ecology & Systematics* 19: 89–110.

Howarth, R. W. (editor) 1996. *Nitrogen cycling in the North Atlantic Ocean and its watersheds: Report of the International SCOPE Nitrogen Project.* Dordrecht, The Netherlands: Kluwer. Reprinted from *Biogeochemistry* 35: 1–304.

Howarth, R. W. 1998. An assessment of human influences on inputs of nitrogen to the estuaries and continental shelves of the North Atlantic Ocean. *Nutrient Cycling in Agroecosystems* 52: 213–23.

Howarth, R. W., J. R. Fruci, and D. M. Sherman. 1991. Inputs of sediment and carbon to an estuarine ecosystem: Influence of land use. *Ecological Applications* 1: 27–39.

Howarth, R. W., R. Marino, R. Garritt, and D. Sherman. 1992. Ecosystem respiration in a large, tidally influenced river: The Hudson River. *Biogeochemistry* 16: 83–102.

Howarth, R. W., H. Jensen, R. Marino, and H. Postma. 1995. Transport to and processing of phosphorus in near-shore and oceanic waters. Pages 323–45 in H. Tiessen (editor), *Phosphorus in the global environment,* SCOPE No. 54. Chichester, England: John Wiley & Sons.

Howarth, R. W., R. Schneider, and D. Swaney. 1996a. Metabolism and organic carbon fluxes in the tidal, freshwater Hudson River. *Estuaries* 19: 848–65.

Howarth, R. W., G. Billen, D. Swaney, A. Townsend, N. Jarworski, K. Lajtha, J. A. Downing, R. Elmgren, N. Caraco, T. Jordan, F. Berendse, J. Freney, V. Kudeyarov, P. Murdoch, and Zhu Zhao-liang. 1996b. Regional nitrogen budgets and riverine inputs of N & P for the drainages to the North Atlantic Ocean: Natural and human influences. *Biogeochemistry* 35: 75–139.

Hydraulic Engineering Center. 1977. Storage, treatment, overflow, runoff model "STORM." Report No. 723-S8-L7520. Davis, CA: U.S. Army Corps of Engineers.

Isermann, K., and R. Isermann, 1998. Food production and consumption in Germany: N-flows and N-emissions. *Nutrient Cycling in Agroecosystems* 52: 289–301.

Jaworski, N. A., R. W. Howarth, and L. J. Hetling. 1997. Atmospheric deposition of nitrogen oxides onto the landscape contributes to coastal eutrophication in the northeast US. *Environmental Science & Technology* 31: 1995–2004.

Johnson, D. W. 1992. Nitrogen retention in forest soils. *Journal of Environmental Quality* 21: 1–12.

Kelly, C. A., J. W. M. Rudd, R. H. Hesslein, D. W. Schindler, P. J. Dillon, C. T. Driscoll, S. A. Gherini, and R. E. Hecky. 1987. Prediction of biological acid neutralization in acid-sensitive lakes. *Biogeochemistry* 3: 129–41.

Kemp, W. M., E. M. Smith, M. Marvin-DiPasquale, and W. R. Boynton. 1997. Organic carbon balance and net ecosystem metabolism in Chesapeake Bay. *Marine Ecology Progress Series* 229–48.

Kempe, S., M. Pettine, and G. Gauwet. 1991. Biogeochemistry of European rivers. Pages 169–212 in E. T. Degens, S. Kempe, and J. E. Richey (editors), *Biogeochemistry of major world rivers.* Chichester, England: John Wiley & Sons.

Lajtha, K., B. Seely, and I. Valiela. 1995. Retention and leaching losses of atmospherically-derived nitrogen in the aggrading coastal watershed of Waquoit Bay, MA. *Biogeochemistry* 28: 33–54.

Larsson, U., R. Elgren, and F. Wulff. 1985. Eutrophication and the Baltic Sea: Causes and consequences. *Ambio* 14: 9–14.

Meybeck, M. 1982. Carbon, nitrogen, and phosphorus transport by world rivers. *American Journal of Science* 282: 401–50.

Moore, M. H., M. Pace, J. Mather, P. S. Murdoch, R. W. Howarth, C. L. Folt, C. Y. Chen, H. F. Hemond, P. A. Flebbe, and C. T. Driscoll. 1997. Potential effects of climate change on the freshwater ecosystems of the New England/mid-Atlantic region. *Water Resources* 11: 925–47.

Müller, C., R. R. Sherloc, and P. H. Williams. 1997. Mechanistic model for nitrous oxide emission via nitrification and denitrification. *Biology and Fertility of Soils* 24: 231–38.

National Research Council. 1993. *Managing wastewater in coastal urban areas.* Washington, DC: National Academy Press.

Nelson, D. 1985. Minimizing nitrogen losses in non-irrigated eastern areas. Pages 173–209 in *Plant nutrient use and the environment.* Washington, DC: Fertilizer Institute.

Newbold, J. D. 1992. Cycles and spirals of nutrients. Pages 379–408 in P. Calow and G. E. Petts (editors), *The rivers handbook: Hydrological and ecological principles.* Oxford, England: Blackwell.

Nixon, S. W. 1995. Coastal marine eutrophication: A definition, causes, and future concerns. *Ophelia* 41: 119–219.

Nixon, S. W., and M. E. Q. Pilson. 1983. Nitrogen in estuarine and coastal marine ecosytems. In E. J. Carpenter, and D. G. Capone (editors), *Nitrogen in the marine environment.* New York: Academic Press.

Nixon, S. W., S. Granger, and B. Nowicki. 1995. An assessment of the annual mass balance of carbon, nitrogen, and phosphorus in Narragansett Bay. *Biogeochemistry* 31: 15–61.

Nixon, S. W., J. W. Ammerman, L. P. Atkinson, V. M. Berounsky, G. Billen, W. C. Boicourt, W. R. Boynton, T. M. Church, D. M. DiToro, R. Elmgren, J. H. Garber, A. E. Giblin, R. A. Jahnke, N. J. P. Owens, M. E. Q. Pilson, and S. P. Seitzinger. 1996. The fate of nitrogen and phosphorus at the land-sea margin of the North Atlantic Ocean. *Biogeochemistry* 35: 141–80.

Pacés, T. 1982. Natural and anthropogenic flux of major elements from central Europe. *Ambio* 11: 206–8.

Parton, W. J., D. W. Schimel, C. V. Cole, and D. S. Ojima. 1987. Analysis of factors controlling soil organic matter levels in the Great Plains grasslands. *Soil Science Society of America Journal* 51: 1173–79.

Parton, W. J., A. R. Mosier, D. S. Ojima, D. W. Valentine, D. W. Schimel, K. Weier, and A. E. Kulmala. 1996. Generalized model for N_2 and N_2O production from nitrification and denitrification. *Global Biogeochemical Cycles* 10: 401–12.

Peierls, B., N. Caraco, M. Pace, and J. J. Cole. 1991. Human influence on river nitrogen. *Nature* 350: 386–87.

Prospero, J. M., K. Barrett, T. Church, F. Detener, R. A. Duce, J. N. Galloway, H. Levy, J. Moody, and P. Quinn. 1996. Atmospheric deposition of nutrients to the North Atlantic Basin. *Biogeochemistry* 35: 27–73.

Richey, J. E., R. L. Victoria, E. Salati, and B. R. Forsberg. 1991. The biogeochemistry of a major river system: The Amazon case study. Pages 57–74 in E. T. Degens, S. Kempe, and J. E. Richey (editors), *Biogeochemistry of major world rivers*. Chichester, England: John Wiley & Sons.

Rod, S. R., R. U. Ayres, and M. Small. 1989. Reconstruction of historical loadings of heavy metals and chlorinated hydrocarbon pesticides in the Hudson-Raritan Basin, 1880–1980. Report to the Hudson River Foundation, New York.

Schimel, D. S., B. H. Braswell, E. A. Holland, R. McKeown, D. W. Ojima, T. H. Painter, W. J. Parton, and, A. R. Townsend. 1994. Climatic, edpahic, and biotic controls over storage and turnover of carbon in soils. *Global Biogeochemical Cycles* 8: 279–93.

Schlesinger, W. H., and J. M. Melack. 1981. Transport of organic carbon in the world's rivers. *Tellus* 33: 172–87.

Schlesinger, W. H., and A. E. Hartley. 1991. A global budget for atmospheric NH_3. *Biogeochemistry* 15: 191–21.

Scholefield, D., D. R. Lockyer, D. C. Whitehead, and K. C. Tyson. 1991. A model to predict transformations and losses of nitrogen in UK pastures grazed by beef cattle. *Plant and Soil* 132: 165–77.

Schulze, E-D. 1989. Air pollution and forest decline in a spruce (*Picea abies*) forest. *Science* 244: 776–83.

Smith, S. V., J. T. Hollibaugh, S. J. Dollar, and S. Vink. 1991. Tomales Bay metabolism, C-N-P stoichiometry, and ecosystem heterotrophy at the land-sea interface. *Estuarine and Coastal Shelf Science* 33: 223–57.

Smith, S. V., and J. T. Hollibaugh. 1993. Coastal metabolism and the oceanic organic carbon balance. *Reviews of Geophysics* 31: 75–89.

Stewart, B. A., D. A. Woolhiser, W. H. Wischmeier, J. H. Caro, and M. H. Frere. 1976. *Control of water pollution from cropland.* Vol. 1. EPA-600/2-75-026a. Washington, DC: U.S. EPA.

Swaney, D. P., D. Sherman, and R. W. Howarth. 1996. Modeling water, sediment, and organic carbon discharges in the Hudson-Mohawk Basin: Coupling to terrestrial sources. *Estuaries* 19(4): 833–47.

Turner, R. E., and N. N. Rabalais. 1991. Changes in Mississippi River water quality this century. *BioScience* 41: 140–47.

van Breeman, N., P. A. Burroughs, E. J. Velthorst, H. F. van Dobben, T. de Wit, T. B. Ridder, and H. F. R. Reijnders. 1982. Soil acidification from atmospheric ammonium sulphate in forest canopy throughfall. *Nature* 299: 548–50.

van der Voet, E., R. Kleijn, and H. A. Udo de Haes. 1996. Nitrogen pollution in the European Union—origins and proposed solutions. *Environmental Conservation* 23: 120–32.

Vitousek, P. M., and W. A. Reiners. 1975. Ecosystem succession and nutrient retention: A hypothesis. *BioScience* 25: 376–81.

Vitousek, P. M., and R. W. Howarth. 1991. Nitrogen limitation on land and in the sea: How can it occur? *Biogeochemistry* 13: 87–115.

Vitousek, P. M., J. Aber, S. E. Bayley, R. W. Howarth, G. E. Likens, P. A. Matson, D. W. Schindler, W. H. Schlesinger, and G. D. Tilman. 1997. Human alteration of the nitrogen cycle: Causes and consequences. *Ecological Issues* 1: 1–15.

CHAPTER 3

Macro-scale Models of Water and Nutrient Flux to the Coastal Zone

Charles J. Vörösmarty and Bruce J. Peterson

Abstract

This chapter reviews our current capacity to predict the transport of water and constituents through drainage basins as a boundary forcing to the coastal zone. Our emphasis is on modeling as a synthesis tool and we focus on continental- to global-scale fluxes. We briefly summarize the rationale for considering such fluxes at the macro scale, noting that they have received relatively little attention in the global change arena. We also review and critique some major modeling approaches at the macro scale. We find that there are major deficiencies in the community's ability to simulate the land-based hydrological cycle, including the accurate prediction of discharge. Further, the development of general aquatic ecosystem models to predict the mobilization, processing, and transport of materials is in its infancy. It will require several years to formulate and institute aquatic process models at continental and global scales. We offer a GIS-based framework for developing models in the near term that can be used to predict the flux of water and biogeochemical constituents over the broad domain. This framework is cast to permit drainage basin loadings to be integrated within the current generation of Earth Systems Models (ESMs) while simultaneously permitting the longer-term development of more process-based models in a geographically specific manner. The development of such models will improve our current capacity to quantify interconnections between the continental land mass and coastal ecosystems of the globe. This capability is important both for refining contemporary flux inventories as well as for uncovering potential feedbacks with the major climate and biogeochemical systems of the planet.

Introduction

The hydrologic cycle has received significant attention with respect to land-atmosphere exchanges at both continental and global scales. Indeed, general circulation models (GCMs) and, more recently, regional atmospheric models of increasing sophistication are widely employed to understand how complex landscapes, including those altered by humans, regulate surface water and energy fluxes (Sellers et al. 1996; Foley et al. 1996; Kalma and Calder 1995; Lean et al. 1995; Avissar and Pielke 1989). ESMs with interactive atmosphere-vegetation schemes are also under development (Thompson et al. 1993; Ciret and Henderson-Sellers 1995). Quantifying elements of the land-based hydrological cycle also has been the focus of major international observational and modeling programs (for example, GEWEX-GCIP 1993; IGBP-BAHC 1993; Shuttleworth 1991). With the advent of high-quality biophysical data sets, including those from remote sensing and operational weather forecasts, the scientific community is rapidly approaching a situation in which the hydrological cycle can be monitored over vast regions and in real time (Chahine 1992).

This research is of critical importance as we seek to understand the dimensions of greenhouse-induced climate change, now increasingly accepted as a matter of scientific fact (IPCC 1990, 1995). The impetus for such studies is derived from existing uncertainties concerning how, when, and where such global climatic changes, and the attendant alteration of weather and climatic extremes, will influence key natural systems upon which humans rely heavily—agriculture, forestry, and water resources.

An important component of the hydrologic cycle, namely fluvial transport of both water and waterborne constituents to the coastal zone, has received significantly less attention than land-atmosphere changes within a global change context. We find this paradoxical for several reasons. We know, for example, that there have been multifold increases in the global loadings of biotically active constituents within drainage basins. The production of inorganic phosphorus and nitrogenous fertilizers has increased by factors of at least 8 and 15, respectively, in only the last 40 years (Smil 1990). Point-source emissions are a significant additional source of loading to river systems (Nixon 1995). To put this in perspective, global CO_2 emissions in 1990 were 3.7 times their corresponding 1950 values (Marland et al. 1994) with a concomitant rise of 12% in ambient atmospheric concentration (Keeling and Whorf 1994). The magnitude of the CO_2 changes is markedly less than the changes in N and P loading to inland waters and the ocean.

There is no carefully maintained and clear biogeochemical record, akin to Mona Loa for atmospheric CO_2, by which to judge the global accumulation and impact of anthropogenically derived drainage basin loadings in freshwater and coastal receiving waters. In contrast to the well-mixed atmosphere, freshwater and coastal ecosystems have highly site- and region-specific physical and biogeochemical attributes that make such an assessment extremely difficult. And, given our incomplete monitoring of such systems, for all practical purposes an accurate assessment is currently impossible.

We do know from recent global inventories based on the synthesis of river-specific data sets (for example, Meybeck and Helmer 1989; Meybeck 1982; Wollast 1983) that there have been severalfold increases in the riverine transport of organic and inorganic N and P. As a consequence of this loading, it has been estimated that only a minority, about 20%, of the world's contemporary drainage basins maintain a nearly pristine water quality (personal communication, M. Meybeck, University of Paris). But uncertainty surrounds any such estimates for, as Meybeck (1993a) points out, at most only 40% of the global freshwater discharge has been sampled adequately to make such assessments.

The quantity and nature of these changes in riverine nutrient transport may have influenced the integrity of aquatic ecosystems at the global scale. From numerous river-basin and coastal-zone studies (for example, the Baltic region, Rosenberg et al. 1990; Mississippi River and Gulf of Mexico, Ortner and Dagg 1995, Turner and Rabelais 1991, 1994; Northern Adriatic, Justić et al. 1995; the Black Sea, Mee 1992) we know that excessive riverborne nutrients, shifts in nutrient limitation, and coastal eutrophication go hand in hand. An analysis by Justić et al. (1995) showed that the Si:N:P ratios of ten large river systems of the world have shifted in ways that lead to a reduction in N limitation, potentially increasing coastal productivity, toxic phytoplankton blooms, and bottom water hypoxia. Such changes in the transport of land-based constituents collectively may be capable of imparting a signal on biogeochemical cycles at the global scale. Tests using the C-N-P-X model of Mackenzie et al. (1993) showed that coastal ocean responses to such loadings occur well within the 100-year time frame.

These changes are more than of simple academic interest. Water quality is of tantamount importance to the sustainability of freshwater and coastal habitats, food webs, and commercial fisheries that serve as an important protein source for humans. It is believed that more than 90% of the world's fisheries catch depends in some way on the estuarine and coastal zone (IGBP-LOICZ 1993) and that recent landings have peaked because of overfishing, habitat destruction, and pollution (Hilborn 1990). Since more than 50% of the human population resides within 60 km of the global coastline (IGBP-LOICZ 1993), the coupling of the terrestrial land mass and coastal fringe through riverine transport constitutes a relevant global change and habitability issue.

The specific subject of this contribution is the large-scale terrestrial water cycle and associated delivery of water and constituents through river systems to the coastal zone. Our emphasis will be on modeling as a synthesis tool. We will briefly summarize some major modeling approaches, and then outline a strategy to be realized over the next several years for developing models that couple the transport of water and processing of biogeochemical constituents at the macro scale. We also offer a GIS-based framework suitable for progress in the nearer term. This framework permits drainage basin loadings to be integrated within the current generation of ESMs while simultaneously permitting the longer-term development of more process-based models in a geographically specific manner. The development of such models will improve our current capacity to quantify interconnections between the continental land mass and coastal ecosystems of the globe. This capability is important for refining contemporary flux inventories as well as for uncovering potential feedbacks with the major climate and biogeochemical systems of the planet.

The Drainage Basin Perspective at Continental to Global Scales

The drainage basin is a useful organizing concept through which to address the issue of water and constituent transport. Watershed analysis has served as the foundation for our understanding of process-level hydrology and land-water interactions at the local scale. In a literature that is vast but highly site-specific, the hydrological community has been able to identify as well as successfully model important mechanisms operating at hillslope and small catchment scales (see McDonnell and Kendall 1994; Likens et al. 1977; Correll et al. 1992; Swank and Crossley 1988; Bartell and Brenkert 1991; Federer 1993; IASH-UNESCO 1972). Paired watershed studies, in particular, have yielded valuable insights into how the water cycle responds to disturbances such as deforestation, how landscapes translocate constituents to aquatic ecosystems, and how biogeochemical cycles in terrestrial ecosystems are restructured during sucession.

The inherent advantages of such an approach should be realized as well over larger spatial domains. First and foremost, models of continental-scale drainage basins facilitate closure of water cycle budgets through an appropriate inventory of inputs and outputs across system boundaries. As summarized below, macro-scale hydrology models driven with land-based meteorological station data or remotely sensed precipitation fields can be validated against observed discharges. Similarly, airshed water balances can be established by computing the horizontal flux of water vapor and, over sufficiently long time periods, the computed convergence of water over large drainage basins can be validated against runoff determined from observed hydrographs.

The ability to adequately validate water budgets in this manner will greatly improve the quality of ongoing terrestrial ecosystem process modeling. Close linkage between terrestrial ecosystem productivity, nutrient cycling, and water availability (see McGuire et al. 1992; Raich et al. 1991; Running and Coughlan 1988) suggests that correct water balances must be computed to construct accurate models of terrestrial ecosystem processes, including the exchange of CO_2 with the atmosphere. Wetland and floodplain inundation, often controlled by the dynamics of runoff transported huge distances by rivers, further necessitates a drainage basin perspective. Accurate depiction of both the spatial and temporal distribution of trace gas fluxes is dependent on substrate wetness and the hydrologic status of open water systems (Bartlett and Harriss 1993; Keller 1990; Devol et al. 1990; Bartlett et al. 1989).

Analysis of the mobilization, transport, and deposition of particulate and dissolved constituents can take advantage of the natural integrating effect of drainage basins. Distributed runoff, particulate material, and dissolved substances are focused onto river corridors and delivered to the coastal zone in what amounts to a time-varying, point-source input. In continental- to global-scale studies, these inputs can be mapped with an increasingly high degree of spatial specificity and used to quantify ocean-basin and coastal sea loadings. The nature and quantities of loadings derived from the continental land mass are also important to assess from the standpoint of terrestrial donor systems. Estimating the anthropogenically induced transport of nutrients and organic materials in rivers provides an indirect measure of biotic impoverishment of terrestrial ecosystems as well as fertilization of the continental land mass through the application of soil amendments, deposition of atmospheric chemicals, and point-source pollution.

Modeling Approaches

The remainder of this review is devoted to the modeling of continental-scale hydrology and constituent transport. We begin with a summary of some commonly used approaches for computing water budgets that can ultimately be used to generate river discharge. We then summarize some key conceptual models of the mobilization and transport of constituents through drainage basins and their associated river systems. A strategy for near-term progress, coupling existing tools and databases, is then offered.

Models of River Discharge

Spatially distributed water balance models are the first step in calculating local runoff that can later be routed as discharge through large drainage basins. Continuing improvements in the current generation of climate models have spawned a plethora of land surface hydrology algorithms of increasing sophistication (Kalma and Calder 1995; IGBP/BAHC 1993). These models, so-called Soil-Vegetation-Atmosphere Transfer Schemes (SVATs; figure 3-1), either reside within an interactive atmospheric model or are driven "off-line" by atmospheric model outputs. A series of parameters is applied to each element of the simulation's domain, which characterizes inherent land

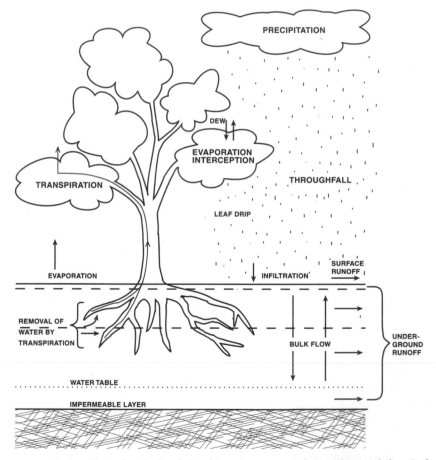

FIGURE 3-1 Typical, idealized land-based water cycle as depicted by Soil-Vegetation-Atmosphere Transfer schemes (SVATs) embedded within the current generation of general circulation and meso-scale atmospheric models. The accurate simulation of runoff has proven to be problematical. From Abramopoulos et al. 1988.

surface properties such as soil texture and vegetative cover and derived parameters such as available water-holding capacity, surface roughness, and albedo. These inputs are used with meteorological forcings by the SVATs and predictions are made of fluctuations of soil moisture and shallow groundwater, evapotranspiration, and runoff. Large disparities in computed fluxes have been noted in SVAT intercomparison studies and the depiction of runoff has been identified as an area requiring further, significant improvement (see Henderson-Sellers 1996; Shao et al. 1994).

In coupled atmosphere-land surface models, runoff has not routinely been used as a diagnostic tool for checking the accuracy of simulated water budgets. When used, it typically has been summed or routed using simple advection schemes to construct hydrographs at the mouths of large rivers (for example, Miller and Russell 1992; Dümenil and Todini 1992). Figure 3-2 shows the results of recent attempts to compare General Circulation Model output within a large drainage basin context. There are obvious opportunities for improvement in calculating both the overall water balance and the subannual dynamics of runoff. The accurate depiction of precipitation has been particularly troublesome in such studies, both for mean conditions and extreme events (IPCC 1990, 1995; Kite et al. 1994; Bass 1993). The same shortcomings are apparent in studies employing meso-scale atmospheric models (for example, Hostetler and Giorgi 1993), although steady improvements in meso-scale simulations offer encouragement that coupled models at the meso scale will eventually be able to adequately provide inputs to fluvial transport models (Browning 1989; Lyons and Bonell 1992). Attempts have also been made to characterize the statistical distributions (both spatial and temporal) of precipitation over drainage basins (for example, Rodriguez-Iturbe and Eagleson 1987; Gupta and Waymire 1990; Islam et al. 1990; Bass 1993) that can be used in improved rainfall-runoff simulations. Satellite-derived precipitation fields (see Barrett and Beaumont 1994; Wyss et al. 1990), in tandem with land surface hydrology models, can be used to generate estimates of runoff in real time.

A major and continuing intellectual challenge centers around the development of methods to adequately aggregate local runoff dynamics to sequentially larger scales associated within full drainage systems. Interested readers are directed to several publications outlining the issue (University of Western Australia et al. 1993; Famiglietti and Wood 1994; Beven 1989, 1992; Bathurst and O'Connell 1992). Much work remains to be done and, in our opinion, a coherent and generalizable set of scaling rules probably will take several years to develop.

A countervailing approach has been the adoption of conceptually simpler models with parameters defined or measured directly at the scale of interest, that is, the meso or macro scale (Dooge 1986). There have been some successful attempts using this approach. For example, basin-specific parameters for

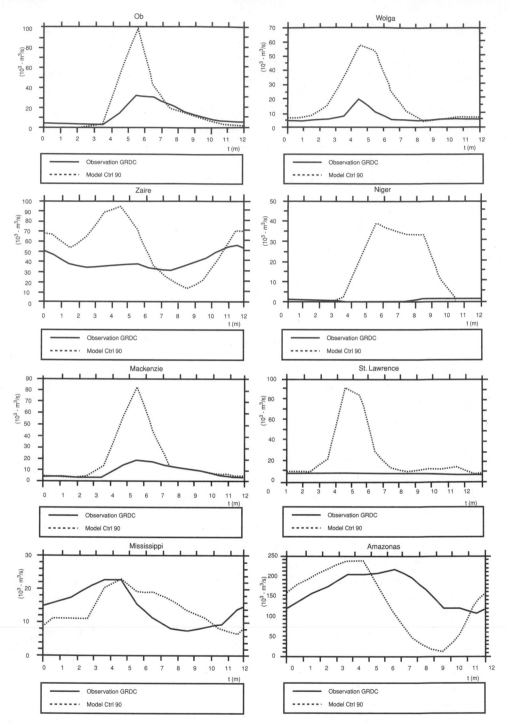

FIGURE 3-2 Hydrographs predicted by the coupled land-atmosphere GCM of Dümenil and Todini (1992) (dashed line) in relation to actual discharge measurements (solid line) for large rivers. Model performance is typical of the existing set of such coupled simulations and demonstrates the need for additional improvements to these algorithms. GRDC indicates Global Runoff Data Center (Koblenz, Germany).

infiltration and hydraulic conductivity have been applied effectively to models at approximately 4,500 km^2 (Georgakakos and Baumer 1994). Simple water balance models have been employed in watershed studies in France (315 to 5,560 km^2) with promising results (Makhlouf and Michel 1994). Over a much larger domain, Sugawara (1988) offered a lumped conceptual model of the Yangtze River, which, after optimization, produced predictions in excellent agreement with observed hydrographs. Strikingly, the entire system was functionally represented by a "tank" model operating on two sub-basins, each approximately 0.5×10^6 km^2.

Vörösmarty et al. (1989, 1991) used a simple parameterization of a distributed water balance and linked transport model (WBM/WTM) to analyze the closure of water budgets in the Amazon/Tocantins (6.9×10^6 km^2) and Zambezi (1.2×10^6 km^2) River systems (figure 3-3). The modeling was based on a data-rich approach employing ground-based station data to generate interpolated climate fields which were then processed by the water balance model to yield runoff on a grid-cell basis (0.5-degree latitude $\times$ longitude). The WTM component converted the calculated runoff into horizontal fluxes, which were then integrated through a drainage network of rivers that Vörösmarty et al. (2000) represented by a system of differential equations. The river network was organized as a cascade of cells at the same 0.5-degree resolution as that used in the WBM. A successful implementation of this approach is shown for Amazonia (figure 3-4) in which interannual variations in precipitation were used to predict flow rates that can be compared directly to observed hydrographs (Vörösmarty et al. 1996). The model successfully simulated the significant depression in peak flows associated with the 1982–1983 ENSO event. Similar approaches were taken by Costa and Foley (1997) and Marengo et al. (1994).

The success of this approach is dependent on recognizing and adequately simulating actual physical phenomena that operate in large river systems. For example, in the Amazon River discharge is strongly regulated by seasonal storage of water along the extensive varzea floodplains, which flank the river's mainstem (Meade et al. 1991). Predicted hydrographs are sensitive to variations in flooding parameters and do not bear resemblance to observed time series unless floodplain inundation is explicitly treated in the model (Vörösmarty et al. 1989; Richey et al. 1989). As another example, simulations of large boreal and arctic river systems may similarly need to capture the distortion of discharge hydrographs due to ice damming and melt-out in rivers (Ferrick et al. 1993; Ashton 1979). Furthermore, the pandemic construction and operation of water engineering works (Vorosmarty et al. 1997a; Dynesius and Nilsson 1994; Walling 1987) necessitates a consideration of the impact that flow regulation and water diversion have on the discharge of otherwise naturally flowing rivers.

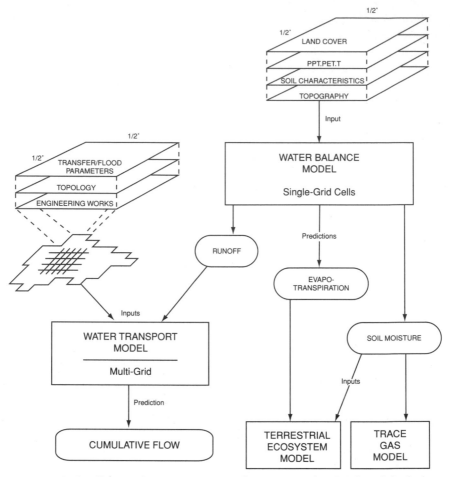

FIGURE 3-3 Schematic representation of a macro-scale, GIS-based hydrology model (Water Balance/Water Transport Model [WBM/WTM] of Vörösmarty et al. 1989, 1991, 1996). The WBM uses a series of biophysical data sets to compute evapotranspiration, variations in soil moisture, and runoff. Runoff is routed using the WTM to simulate discharge hydrographs at any point within a simulated basin. The fluvial component accommodates the accretion of tributary flows along the mainstem, floodplain inundation, and reservoir effects. From Vörösmarty et al. (1989). Copyright © American Geophysical Union.

Another strategy for predicting discharge is to use the aerological approach (Roads et al. 1994; Oki et al. 1995; Brubacker et al. 1993; Serreze et al. 1995), which estimates atmospheric moisture convergence (precipitation minus evapotranspiration or net available water) through horizontal water vapor fluxes and changes in atmospheric storage. These water balances are quantified through analysis of operational weather forecast data and derived through

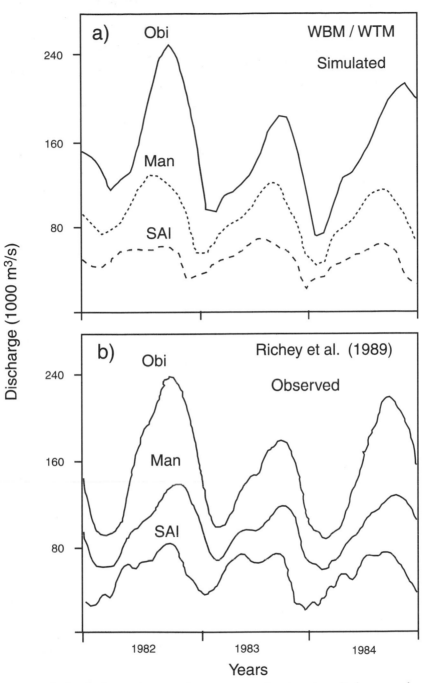

FIGURE 3-4 A time series over 3 water years showing river discharge at three points along the mainstem Amazon River as simulated by the WBM/WTM (top panel) and observed (bottom). Note the large reduction in peak discharge associated with the 1982–83 ENSO event. SAI = San Antonio do Ica, Man = Manacapuru, Obi = Óbidos. Water years begin in September. Predictions from Vörösmarty et al. (1996); observations from Richey et al. (1989). From Vörösmarty et al. (1996). Copyright © American Geophysical Union.

rawinsonde networks. Despite the relative data richness of the aerological approach, significant errors have arisen and the closure of large-scale water balances has sometimes proven to be problematic (figure 3–5). In principle, it is possible to estimate time-varying atmospheric vapor contents and convergence fields that, together with contemporary precipitation fields (from ground-based stations or satellite remote sensing), yield estimates of time-varying evapotranspiration. This evapotranspiration estimate can then be carried forward through an additional set of water balance and fluvial transport calculations (similar to those used by the SVATs) to estimate changes in soil water, groundwater, and runoff, thereby constituting an additional approach to the calculation of time-varying river flow.

River Chemistry Models

Central to the development of large-scale biogeochemical flux models will be an appreciation of how landscapes are linked to aquatic ecosystems and in turn how water fluxes regulate the transport and transformation of nutrient and organic materials. In the context of global change, we must also consider how humans have altered the natural rates of material fluxes through entire drainage basins. A long-term goal is to produce a process-based model of similar simplicity to current terrestrial ecosystem models (Rasool and Moore 1995; Melillo et al. 1993). An appropriate model structure would necessarily include three basic components: landscape mobilization and loss; physical transport; and in-stream, riparian, and lacustrine processing. Billen et al. (1991) present a heuristic model that considers the coupling of landscape and aquatic process simulations (figure 3-6). It offers a specific set of mobilization and retention mechanisms that sequentially alter nutrient biogeochemistry along the full cascade of pathways from land to ocean. We use this as a key organizing principle in the remainder of this discussion.

Terrestrial Mobilization

Natural ecosystems process and release biogeochemical constituents based on the interactions among soils, resident vegetation, underlying lithology, and history of weathering. In broad terms, undisturbed terrestrial ecosystems can be considered relatively "closed" and cycle water, carbon, and nutrients through conservative pathways that are self-organizing and self-sustaining (Moore et al. 1989; Vitousek 1983). Upon disturbance these cycles become relatively "open," throughputs become quantitatively more important, and materials are translocated in large quantities from the land into aquatic ecosystems. There is a well-established body of empirical evidence to support this contention at the local scale (Swank 1988; Likens et al. 1977; Dillon and

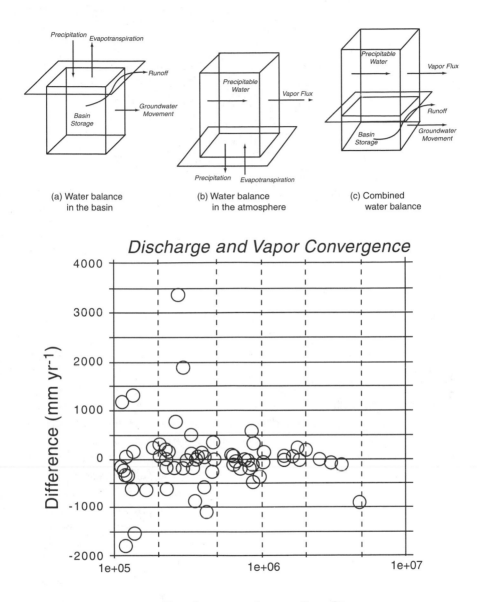

FIGURE 3-5 Summary of the aerological water budget approach as given by Oki et al. (1995). Differences relate to the disparity between atmospheric water vapor convergence (computed using European Center for Medium-range Weather Forecast data sets) and observed runoff for several river systems across the globe. Convergence is assumed to be roughly equivalent to runoff over long time periods.

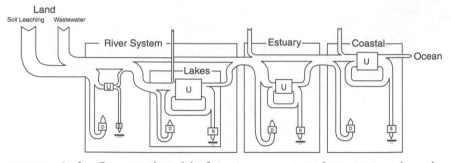

FIGURE 3-6   Conceptual model of riverine processing of constituents along the full cascade of landscape and aquatic ecosystems with eventual delivery to the coastal zone and ocean. Numerous compartments depicting physical movement, sources, transformations, and sinks must be considered in any such process-oriented perspective of whole drainage basins. (U = uptake, S = sink, D = denitrification) From Billen et al. 1991. Copyright John Wiley & Sons Limited. Reproduced with permission.

Kirchner 1975). The collective effects of such disturbances as deforestation, together with the conversion of natural systems to agriculture or grazing and point-source pollution, have created an important anthropogenic signal relative to natural rates of constituent discharge.

Constituent mobilization can be estimated from literature-based surveys employing empirical loss terms or derived from terrestrial ecosystem models (see Cole et al. 1993; Jordan and Weller 1996). Gildea et al. (1986) used a budgeting approach based on compilations of nutrient stocks and fluxes to estimate a nearly threefold increase, over the predisturbance condition, in total N loading from terrestrial systems within the full Mississippi River drainage system. Esser and Kohlmaier (1991) took a similar approach in computing the globally significant contribution of anthropogenic activities (namely tropical deforestation, fossil fuel emissions, fertilizer application, industrialization) to terrestrially derived loadings of organic C, inorganic N and P, and sulfate/sulfite. Nutrient-to-nutrient ratios are important indicators of the nature of terrestrial sources. Billen et al. (1991) showed the sequential reduction in N:P loadings (by atoms) from 100–200:1 to approximately 10:1 in passing from forest/agricultural lands to industrialized regions. Atmospheric deposition is also an important loading that must be quantified in contemporary and future broad-scale nutrient budgets (Galloway et al. 1994).

Inventory of the areal distribution of upland and wetland ecosystems, vegetation cover, and the nature and intensity of land use is also necessary because of the differential mobilization of constituents inherent within contrasting subsystems of whole drainage basins. Drainage basin models have been characterized spatially using regular grids (Cluis et al. 1979), irregularly shaped sub-basin polygons, or both (White et al. 1992). Such subgrid-scale variability can be inventoried through both optical (e.g., LANDSAT and AVHRR) and microwave remote sensing systems. Passive microwave (Sippel et al. 1994), as well as active radars (Imhoff et al. 1987) have been used to map the onset, duration, and extent of wetland inundation. Land surface classification schemes (Olson 1994a, 1994b; IGBP-GAIM/DIS 1994; Running et al. 1994) can give the distribution of important vegetation classes with characteristic rates of water and nutrient cycling. Nonpoint sources determined using this spatially explicit approach need to be augmented with point loadings, such as those from population centers and feedlots, that can be mapped using a geographic information system (GIS).

In-stream Processing and Transport

In-stream processing at the macro scale should rely on two well-established conceptual models of stream processing. The river continuum concept (Vannote et al. 1980) holds that the biotic components of lotic ecosystems are adapted to gradients of physical characteristics (for example, flow rate, depth, temperature, light penetration) that change with passage through sequentially larger stream systems (figure 3-7). Furthermore, systematic loadings of organic material, either derived from external sources or produced within the aquatic domain, result in a characteristic community structure organized to effectively exploit these available carbon and energy resources. The products of upstream processing are connected to downstream reaches through river discharge, thus defining a continuum of aquatic ecosystem dynamics.

Naiman (1983) tested this concept within a drainage basin context and found that indeed the metabolism of such ecosystems, across several stream orders, was highly dependent on channel geomorphology and hydrology. He also documented the dominant contribution of higher-order streams to the gross productivity of the overall drainage system. Howarth et al. (1996a) tested elements of the continuum concept in the lower reaches of the Hudson River and found that some of its general predictions were valid (for example, low production-to-respiration ratio; increased importance of allochthonous inputs). However, the continuum concept's prediction of metabolism in higher-order river ecosystems, supported by allochthonous inputs from upstream, will need modification for channels with large flanking floodplains that can contribute significant quantities of labile carbon more locally, thus increasing the relative importance of respiration in the river corridor (Sedell et al. 1989; Quay et al. 1992). This supports the idea that aquatic communities process nutrients and organic materials in response to a flood pulse in large rivers with associated floodplains (Junk et al. 1989; Bayley 1995), and a more accurate conceptual

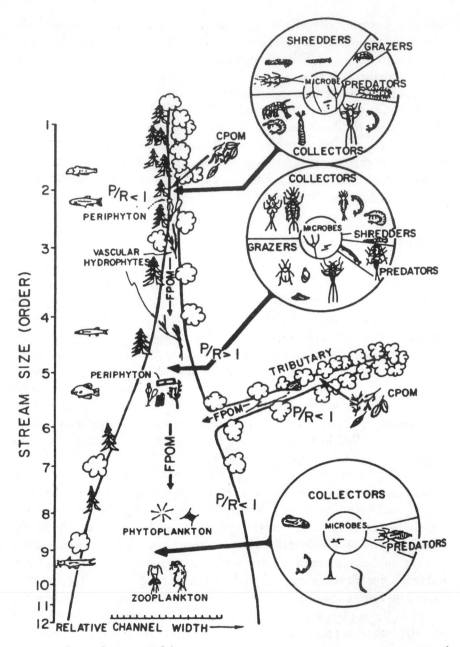

FIGURE 3-7 Summary of the river continuum concept, an important conceptual tool for organizing the development of drainage basin and aquatic processing models at the macro scale. From Vannote et al. 1980. Used with permission of NRC Research Press.

model of large river systems would include both the continuum approach for channel-dominated waters and the flood-pulse response for floodplain-dominated reaches (Johnson et al. 1995; Power et al. 1995). Despite the need for further refinements, we see the continuum concept as a useful set of first principles by which the major river systems of the world can be classified, mapped, and modeled at the global scale.

The second concept is that of nutrient and organic matter spiraling (Elwood et al. 1983; Newbold 1992). It encompasses the idea that biotic and chemically active constituents are introduced into aquatic ecosystems and are alternatively sequestered from and released into free water with eventual transport downstream, loss to the atmosphere, or permanent storage (figure 3-8). Sequestration and release encompass many physical, chemical, and biological processes. The speed of such transport is naturally dependent on characteristic residence times defined by the geometry of the open-water system, flow velocity, the mobility of different trophic levels, and the presence or absence of adjacent wetlands. Some measures of nutrient spiraling may be scaleable over several stream orders. Figure 3-8 shows this for mean spiraling length, with results obtained from an extrapolation to sequentially larger channels of kinetics derived from ^{15}N stable isotope tracer experiments in small-order streams in the Kuparuk River system on the North Slope of Alaska (B. Peterson and W. Wollheim, unpublished data). The extrapolation was based on changes in the hydraulics and geomorphological attributes of open-water channels across different stream orders. The spiraling concept, developed originally in the temperate zone, should be tested in additional climate regimes and biomes. One such test, a cross-site comparison of ammonium spiraling (Lotic Intersite Nitrogen Experiment—LINX), showed that differences in water depth and velocity explained much of the variation in spiraling length in a wide variety of streams including desert, tropical, temperate, and arctic streams (B. Peterson, unpublished data from the LINX project).

A related concept involves direct anthropogenic alteration of river discharge regimes through the construction of dams and associated impoundments. The increase in hydraulic residence time associated with impoundment influences a wide array of physical and biological variables as rivers, to varying degrees, become converted into reservoirs. The classic and dramatic example of this phenomenon is the near total elimination of sediment flux associated with the operation of the Aswan High Dam on the Nile River (Schamp 1983 in Kempe 1988). Serial discontinuities in many other ecosystem attributes (Ward and Stanford 1983; Gurnell et al. 1994; Petts 1984) are apparent at points of impoundment and these disrupt the natural continuum of characteristics associated with free-flowing streams. The severity of this disruption depends on both the nature of the impoundment and the stream order considered. Furthermore, impacts can be expected in association with the pandemic alteration

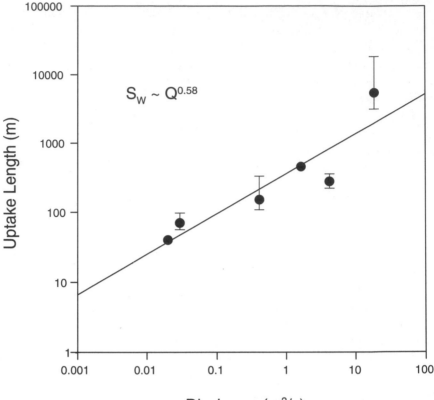

$$S_W \sim Q^{0.58}$$

FIGURE 3-8   The relationship between ^{15}N-ammonium uptake distance (indicated as S_w) and discharge for streams of the Alaska tundra. Ammonium is transported downstream for kilometers before being taken up in large streams but travels only tens of meters in smaller streams. (B. J. Peterson, unpublished data).

of natural rivers through stream channelization, interbasin transfers, and the consumptive use of water for irrigation (Dynesius and Nilsson 1994; Walling 1987). There are currently more than 35,000 impoundments registered globally (IWPDCA 1994; ICOLD 1988). Recent work by Vörösmarty et al. (1997a, 1997b) has shown a 700% increase in the standing stock of water, a tripling of the age of continental runoff in otherwise free-flowing rivers, and a likely 25–30% global sediment retention behind large dams.

Most of these concepts have been developed for single rivers, or reaches of rivers. They have never been tested collectively or coherently across multiple river systems at the broad scale. This constitutes a formidable undertaking that will require a significant infusion of funding and intellectual energy over the next several years. We suggest the integration of these ideas into a generalizable model, like that presented by Billen et al. (1991) and Billen (1993) (figure 3-6), that can be used to generate predictions of constituent delivery to the coastal seas at the regional scale and that can be validated with a stratified sampling program of representative river export fluxes (Meybeck and Ragu 1996). The model should be structured hierarchically and in a geographically specific manner to provide for the sequential coupling of terrestrial loadings through various aquatic domains with an eventual delivery to the coastal seas. Each component would require a set of inputs, some appropriate characterization of aquatic processing, and linkage through discharge to the adjacent downstream component. The use of emerging digitized or elevation model-derived stream networks offers an opportunity to integrate this scheme into a topologically organized set of processing compartments that together will constitute a biogeochemical model of whole drainage basins and river networks.

A Near-term Modeling Strategy

Again, a modeling strategy encompassing the numerous components described above represents a highly ambitious undertaking. Unfortunately, the models and scientific knowledge available today are simply not adequate to meet this longer-term goal of developing a fully articulated drainage basin biogeochemical model, although some attempts have been made (Billen et al. 1994). If the development of terrestrial ecosystem models is any indication, it realistically may require several years to formulate, parameterize, and validate such aquatic ecosystem simulations. Even the most advanced continental- and global-scale terrestrial ecosystem models (see Melillo et al. 1993; McGuire et al. 1997) consider only the dynamics of natural ecosystems and are only beginning to consider the critical issue of land-use and land-cover change. An interim modeling strategy is clearly needed. We suggest this strategy focus on improving our ability to perform contemporary global inventories of riverborne C, N, P, Si, and particulate material fluxes from land to ocean. By contemporary, we mean the last 30–40 years, a period during which substantial acceleration in drainage basin loadings and changes in water chemistry have been documented.

What we advocate is a GIS-based approach that exploits the large and growing number of regression-based models in tandem with geographically specific data sets that are actively being used in global change research (tables 3-1 and 3-2). The literature on lake eutrophication provides a rich legacy of relatively simple yet robust, empirical relationships that predict nutrient loads from observable characteristics within drainage basins (Dillon and Kirchner 1975; Vollenweider 1968). A series of more recent studies have extended this

TABLE 3-1

Key examples of the statistical approach used in estimating watershed constituent transport.

TSS	NO$_3^-$
Mozzherin (1991) (runoff, area, relief, soils) Jansen and Painter (1974) (climate, relief, land cover, lithology) Milliman and Syvitski (1992) (area runoff location) Meybeck (1993b) (runoff/temp. typology)	Cole et al. (1993) Jordan et al. (1994) (land use, fertilizer, point sources, and/or deposition) Meybeck (1982) (demophoric index)
TSS/TDS	**PO$_4^{-3}$**
Meybeck (1976) (runoff/river size)	Caraco (1994) Dillon and Kirchner (1975) (geology and land use) Meybeck (1982) (demophoric index)
POC/DOC	**SiO$_2$**
Meybeck (1993b) (runoff/temp. typology)	Bluth and Kump (1994) (runoff/lithology)

TABLE 3-2

Sample of databases available for use in global riverine flux inventories.

Watershed boundaries and river networks	Sewage collection rates and level of treatment
Digital topography	Water engineering works
Vegetation/land cover	Lake density
Potential, contemporary	Nitrogen deposition
Soils	Climate
Surficial geology	Temperature
Lithology, age	Precipitation
Wetlands	Radiation
Type, density, connectivity to rivers	Winds
Population	Vapor pressure
Human	Runoff
Livestock	Hydrogeology
Industrial/demophoric indices	
Fertilizer inputs	

These data represent basic biophysical information as well as georeferenced fields derived from modeling experiments (for example, computed runoff). The majority of these data sets are available in digitized format as either point, polygon, or gridded GIS coverages. Data are from a variety of sources including many available through the Internet. Distributed gridded data sets vary from 1 km to 2.5-degree resolution; scales on available map data vary from 1:10M to 1:30M. (References available in Vörösmarty et al. [eds.] 1997c.)

approach to nutrient fluxes in river systems across a wide variety of scales, from regions (for example, Smith et al. 1993) to the scale of the United States (Jordan et al. 1994) to the globe (Meybeck 1982). Such relationships have even been used to explore changes in N:P ratios due to differential loading of N and P in industrial fertilizer use (Caraco 1994). The regression approach also has been used in large-scale sediment flux studies (Sidorchuk 1994; Milliman and Syvitski 1992; Jansen and Painter 1974) based on one or more drainage basin characteristics such as discharge, relief, land use/cover, area, and the presence of lakes and impoundments. Table 3-1 gives a sampling of publications of empirically based estimates of constituent flux in rivers. An application and intercomparison of these and other relationships should be a first step toward refining contemporary flux estimates at the global scale. The blending of regression and basin processing models for nutrient flux shows great promise (see Smith et al. 1997; Seitzinger and Kroeze 1998; Howarth et al. 1996b).

Reliable data on riverine chemistry are essential to realizing this strategy. Two important efforts are currently under way (personal communication, M. Meybeck, University of Paris). The Global Environmental Monitoring System (GEMS) data sets from the WHO Collaboration Center at Environment Canada/CCIW (Burlington, Ontario) contain information on approximately 300 river stations worldwide. The database covers more than fifty countries in the developed, underdeveloped, and rapidly changing regions of the globe. A wide spectrum of drainage basins is represented, down to 10,000 km^2. Major ions, nutrients, and trace metals are represented with a sampling frequency of from six to twenty-four per year. Development of the GEMS/Global Register of River Inputs (GLORI) database will yield information on 400–600 river mouths for several important attributes including discharge, area, elevation, TSS, TDS, and DIC. For 100–200 rivers information will be available on NO_3^-, NH_4^+, PO_4^{-3}, total P, and SiO_2. DOC, POC, and TOC will likely be documented for about fifty rivers (see Meybeck and Ragu 1996).

The riverine chemistry data must be linked to key biophysical driving variables (table 3–2) within a common, geographically coregistered framework. A critical step is the delineation of a global river network topology, from which these data sets can be organized into discrete drainage basins (Vörösmarty et al. 2000). This has been achieved at 30-minute resolution, and approximately 4,800 rivers actively discharge fresh water to the world's oceans (figure 3-9). This data set was derived from a preliminary routing based on a global digital elevation model (ETOPO5; Edwards 1989) and a maximum-gradient path algorithm. Corrections were made manually using a computer program that superimposed the resulting 30-minute routing with digital line graphs depicting river systems at 1:3M scale (ArcWorld/ESRI, Redlands, Calif.) such that discrepancies between the two data sets were minimized and drainage basin

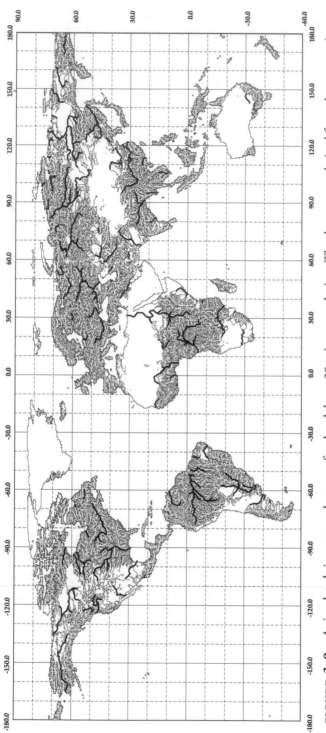
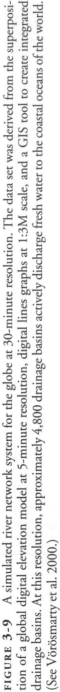

FIGURE 3-9 A simulated river network system for the globe at 30-minute resolution. The data set was derived from the superposition of a global digital elevation model at 5-minute resolution, digital lines graphs at 1:3M scale, and a GIS tool to create integrated drainage basins. At this resolution, approximately 4,800 drainage basins actively discharge fresh water to the coastal oceans of the world. (See Vörösmarty et al. 2000.)

areas corresponded to published accounts (UNESCO, various years; Meybeck and Ragu 1996).

Since water-quality monitoring of global river systems is incomplete, it will be necessary to develop a means to extrapolate across both space and time the knowledge that we do have on the factors controlling fluvial transport of carbon, nutrients, and suspended sediments. Because of differences in biophysical attributes such as topography, geology, vegetation, and climate, as well as various levels of exposure to anthropogenic disturbance, it is unlikely that a single set of predictors can be developed for all river systems. Further, the interactions of rivers and their coastal zones depend on factors such as the presence or absence of deltas, a near-shore shelf system, or proximity to oceanic trenches. It is therefore necessary to assign a "typology" by which drainage basins of the earth can be categorized. Several such typologies could be tested in this context (Mackenzie et al. 1991; Meybeck 1993b; Bluth and Kump 1994; Gurnell et al. 1994; Milliman and Syvitski 1992).

Class-specific statistical relationships are then constructed using a portion of the water chemistry data sets for calibration and a remaining subset for validation. The drainage basin categories are then mapped onto their respective river systems and associated coastal zones to make predictions in unmonitored watersheds. A template of rivers for which observed data exist, together with the extrapolation, will allow a geographically referenced inventory to be constructed. For any drainage basin class, spatial variations in driving factors (table 3-2) yield basin-to-basin differences in material flux. This approach has been used successfully in the early development of global-scale terrestrial ecosystem models for extrapolating the dynamics of biome-specific terrestrial ecosystem models (Raich et al. 1991). The method is applicable to regional, continental, and global-scale inventories.

An example of the approach is given in figure 3-10 and color plate 1 for sediment flux at the continental scale. The approach takes a series of empirical relations (Milliman and Syvitski 1992) and maps these according to simple geomorphological parameters that can be derived from a series of GIS base maps. The typology used is based on geographic location and the maximum elevation of landscape drained by a particular river system. This results in a detailed pattern of material flux that forms a first step in an eventual linkage of continental processes to coastal and open-ocean dynamics. Such analyses may also be of use in guiding future monitoring campaigns. The analysis shown indicated that for a 95% inventory of sediment flux across South America, fewer than seventy rivers would need to be monitored. A stratified sampling of representative rivers could further reduce the required number of monitored systems.

Remote sensing of the spectral signatures of river plumes along the coastal fringe may also offer opportunities for inventorying land-derived constituent

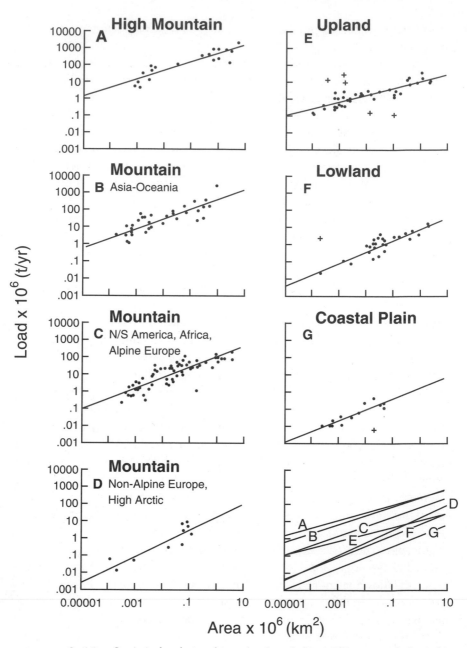

FIGURE 3-10 Statistical relationships developed by Milliman and Syvitski (1992) showing the dependence of suspended sediment load on drainage basin area. A typology based on geographic location and maximum elevation in the basin was employed. (See also color plate 1.)

fluxes for suspended sediments as well as for chlorophyll and dissolved organics (Stumpf and Goldschmidt 1992; Fischer et al. 1991; Muller-Karger et al. 1988). The five-band Sea-viewing Wide Field-of-view (SeaWIFS) sensor launched in 1997 is expected to significantly improve our current capabilities in this area, although there still remains to be developed a usable methodology to convert remotely sensed concentrations to fluxes per se (Campbell 1994).

Finally, human population growth and development have changed dramatically in recent decades and will likely continue to do so well into the future. There is thus an important time dimension to the constituent transport question. Time-varying forcing functions for population growth, industrialization, land use and land cover change, and greenhouse warming when applied to this modeling scheme form the basis of retrospective, contemporary, and future river flux analyses. What are required are data on the geographic distribution of such changes. Recent activities under the auspices of IGBP/HDP (1993) seek to predict the distribution of human population and land-cover change over time. Such initiatives lend hope that progressive changes in riverborne chemistry can be predicted in advance of significant deterioration and that sensitive freshwater and coastal ecosystems can be identified and protected.

Recent Progress and Emerging Institutional Support

Despite our overall assessment that river systems have not received their deserved place in the global change research agenda, there is a small but growing interest in the issue of large-scale riverine transport, both at the national and international levels. In the United States, the NSF programs, Land Margin Ecosystem Research (LMER) and Long-Term Ecological Research (LTER), have provided an integrative and long-term quantification of land-water interactions and subsequent processes in the aquatic domain extending to the coastal zone. The recent NSF/EPA joint program in water and watersheds research provides a much-needed commitment by the federal government to this issue. NOAA's Human Dimensions of Global Change and Nutrient Enhanced Coastal Ocean Productivity programs are additional examples.

Numerous international groups are also focusing on the issue. Three IGBP core projects ("Biological Aspects of the Hydrological Cycle" [BAHC], "Land-Ocean Interactions in the Coastal Zone" [LOICZ], and "Past Global Changes" [PAGES]) consider the river transport question as part of their foci activities. In addition, a joint workshop titled "Modeling the Delivery of Terrestrial Materials to Freshwater and Coastal Ecosystems"

(Vörösmarty et al. 1997c) was carried out under the auspices of the three Core Projects and many of the themes articulated here were discussed and adopted as part of an overall inter-core project strategy to catalyze international, multidisciplinary research in this realm. PAGES hosted in February 1994 a workshop titled "Land Use and Climate Impacts on Fluvial Systems during the Period of Agriculture." A newly formed IGBP Water Group will coordinate these efforts. The Joint Group of Experts on the Scientific Aspects of Marine Environmental Protection, representing several UN agencies, the World Health Organization, World Meteorological Organization, International Maritime Organization, and International Atomic Energy Agency, also has addressed issues relating to land-sea boundary fluxes (GESAMP 1987). The World Climate Research Program's Global Energy and Water Exchange Experiment (GEWEX) provides an important resource for developing methods to link observational and modeling capabilities in the analysis of large river systems. Such a perspective, applied to constituents, could provide a mechanism to eventually monitor and simulate biogeochemical fluxes in near real time.

A series of synthesis volumes have begun to appear in the literature. This U.S. SCOPE workshop report itself presents several themes of importance toward understanding the linkages between river basin processes and coastal zone health and management. Another recent SCOPE volume (Degens et al. 1991) provides a state-of-the-art synthesis on the geochemistry of world river systems. The contribution of Mantoura et al. (1991) considers numerous aspects of land-coastal zone linkages. Naiman (1992) provides a synthesis on watershed management that helps to further articulate the drainage basin perspective. Total nitrogen fluxes to the ocean were recently summarized for the North Atlantic as part of the International SCOPE N Project (Howarth et al. 1996b).

Conclusions

Establishing the specific linkages between water and constituent dynamics in river systems at the continental scale will be complex and challenging. It is hoped that this brief summary has provided a small sense of the scope of the problem. If we seek applications at the regional, continental, and larger, global scale, it is no exaggeration to state that existing models of terrestrial mobilization, riverine transport, and processing are in their infancy and currently at the conceptual stage. The first step in constructing such coupled process models will be to quantify the translocation of C, N, P, Si, and particulate matter from land to water using enhancements to a growing number of terrestrial ecosystem models and inventories of point source emissions. Next, discharge dynamics will need to be simulated in a manner that accurately reflects the true nature of river systems, for example, with an explicit treatment of floodplain inundation. Also required will be a formal integration of ecological principles into aquatic process models applied at successively larger spatial domains. Successful modeling will require both algorithm development and an associated field program to support both calibration and validation. Such models will take years to develop. In the interim, significant progress can be made by applying a more empirical, GIS-based approach that links drainage basin characteristics to water-quality data. Application of the resulting statistical relationships, together with a growing number of global biophysical data sets, will provide a geographically specific loading of nutrients, organic matter, and sediments into broad-scale models of estuaries and coastal seas. With the appropriate input data sets, such river transport models can be used to reconstruct retrospective time series of coastal zone loading, evaluate contemporary fluxes, and predict the dimensions of anthropogenic change into the next century. This prognostic capability is particularly important for managers as we seek methodologies by which to predict the progressive deterioration of global water quality, to identify sensitive freshwater and coastal ecosystems, and to protect and restore them.

Acknowledgments

This work was supported by NASA's Biological Oceanography Program (Grant #NAG5-6452), NASA-EOS (Grant #NJ5-6137), NSF's Division of Polar Programs (OPP-952470), and the U.S. Environmental Protection Agency (CR-816278). The authors wish to thank Balazs Fekete for valuable help in assembling computer codes and several of the biophysical data sets used in this analysis.

References

Abramopoulos, F., C. Rosensweig, and B. Choudhury. 1988. Improved ground hydrology calculations for global climate models (GCMs): Soil water movement and evapotranspiration. *Journal of Climate* 1: 921–41.

Ashton, G. D. 1979. Modeling of ice in rivers. Pages 14/1–14/26 in H. W. Shen (editor), *Modeling of rivers*. New York: John Wiley & Sons.

Avissar, R., and R. A. Pielke. 1989. A parameterization of heterogeneous land surfaces for atmospheric numerical models and its impact on regional meteorology. *Monthly Weather Review* 117: 2113–36.

Barrett, E., and M. Beaumont. 1994. Satellite remote rainfall monitoring: An overview. *Remote Sensing Reviews* 11: 23–48.

Bartell, S. M., and A. L. Brenkert. 1991. *A spatial-temporal model of nitrogen dynamics in a deciduous forest watershed.* In M. Turner and R. Gardner (editors), *Quantitative methods in landscape ecology.* New York: Springer-Verlag.

Bartlett, D. S., K. B. Bartlett, J. M. Hartman, R. C. Harriss, D. I. Sebacher, R. Pelletier-Travis, D. D. Dow, and D. P. Brannon. 1989. Methane emissions from the Florida Everglades: Patterns of variability in a regional wetland ecosystem. *Global Biogeochemical Cycles* 3: 363–74.

Bartlett, K. B., and R. C. Harriss. 1993. Review and assessment of methane emissions from wetlands. *Chemosphere* 26: 261–320.

Bass, B. (editor). 1993. Biospheric Aspects of the Hydrological Cycle (BAHC). Focus 4: The Weather Generator Project. Berlin, Germany: IGBP-BAHC.

Bathurst, J. C., and P. E. O'Connell. 1992. Future of distributed modelling: The Systeme Hydrologique Europeen. *Hydrological Processes* 6: 265–77.

Bayley, P. B. 1995. Understanding large river-floodplain ecosystems. *BioScience* 45: 153–58.

Beven, K. J. 1989. Changing ideas in hydrology—the case of physically-based models. *J. Hydrology* 105: 157–72.

Beven, K. J. 1992. The future of distributed modeling. *Hydrol. Processes* 6: 253–54.

Billen, G. 1993. A conceptual model of C, N, and P transformations in the aquatic continuum from land to sea. Pages 141–61 in R. Wollast, F. T. Mackenzie, and L. Chou (editors), *Interactions of C, N, P, and S biogeochemical cycles and global change*. NATO ASI Series, vol. 14. Berlin, Germany: Springer-Verlag.

Billen, G., J. Garnier, and P. Hanset. 1994. Modeling phytoplankton development in whole drainage networks: The RIVERSTRAHLER model applied to the Seine River system. *Hydrobiologia* 289: 119–37.

Billen, G., C. Lancelot, and M. Meybeck. 1991. N, P, and Si retention along the aquatic continuum from land to ocean. Pages 19–44 in R. F. C., Mantoura, J.M. Martin, and R. Wollast (editors), *Ocean margin processes in global change*. New York: John Wiley & Sons.

Bingham, G. E., D. S. Bowles, E. Kluzek, A. S. Limaye, and J. P. Riley. 1994. A nested model chain between GCM scale and river flow: A testbed for vegetation, erosion, and water yield scaling studies. Pages 26–30 in *Proceedings of the American Meteorological Society, Conference on Hydrology*.

Bluth, G. J., and L. R. Kump. 1994. Lithologic and climatologic controls of river chemistry. *Geochimica et Cosmoshimica Acta* 58: 2341–59.

Browning, K. A. 1989. The mesoscale database and its use in mesoscale forecasting. *Quarterly Journal of the Royal Meterological Society* 115: 717–62.

Brubaker, K. L., D. Entekabi, and P. S. Eagleson. 1993. Estimation of continental precipitation recycling. *Journal of Climate* 6: 1077–89.

Campbell, J. 1994. Remote sensing of coastal water masses. In *Preliminary proceedings of the IGBP Inter-Core Project workshop on modeling the delivery of terrestrial material to freshwater and coastal ecosystems*. IGBP-BAHC/PAGES/LOICZ. Durham: University of New Hampshire.

Caraco, N. 1994. Influence of humans on P transfers to aquatic ecosystems: A regional study using large rivers. In H. Tiessen (editor), *Phosphorus cycles in terrestrial and aquatic ecosystems*. SCOPE. Chichester, England: John Wiley and Sons.

Chahine, M. T. 1992. The hydrological cycle and its influence on climate. *Nature* 359: 373–80.

Ciret, C., and A. Henderson-Sellers. 1995. "Static" vegetation and dynamic global climate: Preliminary analysis of the issues of time steps and time scales. *Journal of Biogeography* 22 (4–5): 843–56.

Cluis, D. A., D. Couillard, and L. Potvin. 1979. A square grid model relating land use exports to nutrient loads in rivers. *Water Resources Research* 15: 630–36.

Cole, J. J., B. L. Peierls, N. F. Caraco, and M. L. Pace. 1993. *Human influence on river nitrogen*. Pages 141–57 in M. McDonnell and S. Pickett (editors), *Humans as components of ecosystems: The ecology of subtle human effects and populated areas*. New York: Springer-Verlag.

Correll, D., T. E. Jordan, and D. E. Weller. 1992. Nutrient flux in a landscape: Effects of coastal land use and terrestrial community mosaic on nutrient transport to coastal waters. *Estuaries* 15: 431–42.

Costa, M. H., and J. A. Foley. 1997. Water balance of the Amazon Basin: Dependence on vegetation cover and canopy conductance. *Journal of Geophysical Research* 102 (D20): 23,973–23,989.

Degens, E. T., S. Kempe, and J. E. Richey (editors). 1991. *Biogeochemistry of major world rivers*. SCOPE 42. New York: John Wiley & Sons.

Devol, A. H., J. E. Richey, B. R. Forsberg, and L. A. Martinelli. 1990. Seasonal dynamics in methane emissions from the Amazon River floodplain to the troposphere. *Journal of Geophysical Research* 95: 16, 417–16, 426.

Dillon, P. J., and W. B. Kirchner. 1975. The effects of geology and land use on the export of phosphorus from watersheds. *Water Research* 9: 135–48.

Dooge, J. 1986. Looking for hydrological laws. *Water Resources Research* 22: 46S–58S.

Dümenil, L., and E. Todini. 1992. A rainfall-runoff scheme for use in the Hamburg climate model. Pages 129–56 in J. P. O'Kane (editor), *Advances in theoretical hydrology: A tribute to James Dooge*. European Geophysical Society Series on Hydrological Sciences. 1. Amsterdam, The Netherlands: Elsevier.

Dynesius, M., and C. Nilsson. 1994. Fragmentation and flow regulation of river systems in the northern third of the world. *Science* 266: 753–62.

Edwards, M. O. 1989. *Global gridded elevation and bathymetry on 5-minute geographic grid (ETOPO5)*. Boulder, CO: NOAA.

Elwood, J. W., J. D. Newbold, R. V. O'Neill, and W. Van Winkle. 1983. Resource spiralling: An operational paradigm for analyzing lotic ecosystems. Pages 3–28 in T. D. Fontaine III and S. M. Bartell (editors), *Dynamics of lotic ecosystems*. Ann Arbor, MI: Ann Arbor Science.

Esser, G., and G. H. Kohlmaier. 1991. Modelling terrestrial sources of nitrogen, phosphorus, sulfur, and organic carbon to rivers. Pages 297–322 in E. T. Degens, S. Kempe, and J. E. Richey (editors), *Biogeochemistry of major world rivers*. SCOPE 42. New York: John Wiley & Sons.

Famiglietti, J. S., and E. F. Wood. 1994. Multiscale modeling of spatially variable water and energy balance processes. *Water Resources Research* 30: 3061–78.

Federer, C. A. 1993. *BROOK-90: A simulation model for evapotranspiration, soil water, and streamflow*. Durham, NH: USDA Forest Service.

Ferrick, M. G., J. E. Kutzbach, M. T. Coe, and S. Levis. 1993. Kinematic model of river ice motion during dynamic breakup. *Nordic Hydrology* 24: 111–34.

Fischer, J., R. Doerffer, and H. Grassl. 1991. Remote sensing of water substances in rivers, estuarine and coastal waters. Pages 25–55 in E. T. Degens, S. Kempe, and J. E. Richey (editors), *Biogeochemistry of major world rivers.* SCOPE 42. New York: John Wiley & Sons.

Foley, J. A., I. C. Prentice, N. Ramankutty, S. Levis, D. Pollard, S. Sitch, and A. Haxeltine. 1996. An integrated biosphere model of land surface processes, terrestrial carbon balance, and vegetation dynamics. *Global Biogeochemical Cycles* 10: 603–28.

Galloway, J. N., H. Levy II, and P. S. Kasibhatia. 1994. Year 2020: Consequences of population growth and development on deposition of oxidized nitrogen. *Ambio* 23(2): 120–23.

Georgakakos, K. P., and O. W. Baumer. 1994. Measurement and analysis of on-site soil moisture data. Tiburon, CA: NASA Soil Moisture Workshop, 25–28 January 1994.

GESAMP. 1987. *Land-sea boundary flux of contaminants: Contributions from rivers.* Joint Group of Experts on the Scientific Aspects of Marine Environmental Protection. Report No. 32. IMO, FAO, UNESCO, WMO, WHO, IAEA, UN, UNEP.

GEWEX-GCIP. 1993. *Implementation plan for the GEWEX continental-scale International Project (GCIP).* Vol. 1: *Data collection and operational model upgrade.* Washington, DC: IGPO. Publ. Series No. 6.

Gildea M. P., B. Moore, C. J. Vörösmarty, B. Berquist, J. M. Melillo, K. Nadelhoffer, and B. J. Peterson. 1986. A global model of nutrient cycling: 1. Introduction, model structure and terrestrial mobilization of nutrients. In D. Correll (editor), *Watershed research perspectives.* Washington, DC: Smithsonian Institution Press.

Gupta, V. K., and E. Waymire. 1990. Multiscaling properties of spatial rainfall and river flow distributions. *Journal of Geophysical Research* 95: 1999–2009.

Gurnell, A. M., P. Angold, and K. J. Gregory. 1994. The classification of river corridors: Issues to be addressed in developing operational methodology. *Aquatic Conservation* 4(3): 219–31.

Henderson-Sellers, A. 1996. Soil moisture simulation: Achievements of the RICE and PILPS intercomparison workshop and future directions. *Global Planet. Change* 13: 99–116.

Hilborn, R. 1990. Marine biota. Pages 371–85 in B. L. Turner et al. (editors), *The earth as transformed by human action.* Cambridge, England: Cambridge University Press.

Hostetler, S. W., and F. Giorgi. 1993. Use of output from high-resolution atmospheric models in landscape-scale hydrologic models: An assessment. *Water Resources Research* 29: 1685–95.

Howarth, R. W., R. Schneider, and D. Swaney. 1996a. Metabolism and organic carbon fluxes in the tidal, freshwater Hudson River. *Estuaries* 19: 848–65.

Howarth, R. W. et al. (14 coauthors). 1996b. Riverine inputs of nitrogen to the North Atlantic Ocean: Fluxes and human influence. *Biogeochemistry* 35: 75–139.

IASH-UNESCO. 1972. *Symposium on the results of research on representative and experimental basins.* IASH Publ. No. 97. Wellington, New Zealand: Deslanes.

ICOLD. 1988. *World register of dams: 1988 updating.* Paris: International Commission on Large Dams.

IGBP-BAHC. 1993. *Biospheric Aspects of the Hydrological Cycle (BAHC): The operational plan.* Stockholm, Sweden: IGBP.

IGBP-GAIM/DIS. 1994. *IGBP global modelling and data activities: 1994–98.* IGBP Report No. 30. Stockholm, Sweden: IGBP.

IGBP-HDP. 1993. *Relating land use and global land cover change.* Edited by B. L. Turner, D. L. Skole, and R. Moss. IGBP Report No. 24 and HDP Report No. 5. Stockholm, Sweden: IGBP/HDP.

IGBP-LOICZ. 1993. *Land-Ocean Interactions in the Coastal Zone (LOICZ): Science plan.* Stockholm, Sweden: IGBP.

Imhoff, M. L., Vermillion, C., Story, M. H., Choudhury, A. M., Gafoor, A., and F. Polcyn. 1987. Monsoon flood boundary delineation and damage assessment using space-borne imaging radar and Landsat data. *Photogramm. Engng. Remote Sens.* 53: 405–13.

IPCC. 1995, 1990. *Climate change: The IPCC scientific assessment.* WMO, UNEP. New York: Cambridge University Press.

Islam, S., D. Entekabi, R. L. Bras, and I. Rodriguez-Iturbe. 1990. Parameter estimation and sensitivity analysis for the modified Bartlett-Lewis rectangular pulses model of rainfall. *Journal of Geophysical Research* 95 (D3): 2093–2100.

IWPDCA. 1994. *International water power and dam constructions handbook.* Surrey, UK: Reed Business Publ.

Jansen, J. M. I., and R. B. Painter. 1974. Predicting sediment yield from climate and topography. *Journal of Hydrolology* 21: 371–80.

Johnson, B. L., W. B. Richardson, and T. J. Naimo. 1995. Past, present, and future concepts in large river ecology. *BioScience* 45: 134–41.

Jordan, T. E. and D. E. Weller. 1996. Human contributions to the terrestrial nitrogen flux. *BioScience* 46: 655–64.

Jordan, T. E., D. L. Correll, and D. E. Weller. 1994. The effects of agriculture on discharge of nitrogen from watersheds. In *Preliminary proceedings of IGBP Inter-Core Project workshop on modeling the delivery of terrestrial material to freshwater and coastal ecosystems.* IGBP-BAHC/PAGES/LOICZ. Durham: University of New Hampshire.

Junk, W. J., P. B. Bayley, and R. E. Sparks. 1989. The flood pulse concept in river-floodplain systems. *Canadian Special Publication of Fisheries and Aquatic Sciences* 106: 110–27.

Justic, D., N. N. Rabelais, and R. E. Turner. 1995. Stoichiometric nutrient balance and origin of coastal eutrophication. *Marine Pollution Bulletin* 30: 41–46.

Kalma, J. D., and I. R. Calder. 1995. *Land surface processes in large-scale hydrology.* Report to the Commission for Hydrology of the World. Meteorological Organization, Operational Hydrology Report No. 40. Geneva, Switzerland: WMO (with contributions by T. J. Lyons, M. Nunez, A. J. Pitman, and C. J. Vörösmarty).

Keeling, C. D., and T. P. Whorf. 1994. Atmospheric CO_2 records from sites in the

SIO air sampling network. Pages 16–26 in T. A. Boden, D. P. Kaiser, R. J. Sepanski, and F. W. Stoss (editors), *Trends '93: A compendium of data on global change.* ORNL/CDIAC-65. Oak Ridge, TN: Carbon Dioxide Information Analysis Center.

Keller, M. 1990. "Biological Sources and Sinks of Methane in Tropical Habitats and Tropical Atmospheric Chemistry." PhD diss., Princeton University,

Kempe, S. 1988. Estuaries—Their natural and anthropogenic changes. In T. Rosswall, R. G. Woodmansee, and P. G. Risser (editors), *Scales and global change.* SCOPE. New York: John Wiley & Sons.

Kite, G. W., A. Dalton, and K. Dion. 1994. Simulation of streamflow in a macroscale watershed using general circulation model data. *Water Resources Research* 30: 1547–59.

Lean, J., C. B. Bunton, C. A. Nobre, and P. R. Rowntree. 1995. *The simulated impact of Amazonian deforestation on climate using measured ABRACOS vegetation characteristics.* In J. H. C. Gash, C. A. Nobre, J. M. Roberts, and R. L. Victoria (editors), *Amazonian deforestation and climate.* Chichester, England: John Wiley & Sons.

Likens, G. E., F. H. Bormann, R. S. Pierce, J. S. Eaton, and N. M. Johnson. 1977. *Biogeochemistry of a forested ecosystem.* New York: Springer-Verlag.

Lofgren, B. M., and T. E. Croley. 1994. Validation of the Coupled Hydrosphere-Atmosphere Research Model (CHARM). Pages 26–30 in *Proceedings of the American Meteorological Society conference on hydrology.*

Lyons, W. F., and M. Bonell. 1992. Daily meso-scale rainfall variability in the tropical wet/dry climate of the Townsville areas, north-east Queensland during the 1988-89 wet season: Synoptic-scale airflow considerations. *International Journal of Climatology* 12: 655–84.

Mackenzie, F., L. M. Ver, C. Sabine, M. Lane, and A. Lerman. 1993. C, N, P, S biogeochemical cycles and modeling of global change. In R. Wollast, F. T. Mackenzie, and L. Chou (editors), *Interactions of C, N, P, and S biogeochemical cycles and global change.* NATO ASI Series, vol. 14. Berlin, Germany: Springer-Verlag.

Mackenzie, F. T., J. M. Bewers, R. J. Charlson, E. E. Hofmann, G. A. Knaver, J. C. Kraft, E. M. Nothig, B. Quack, J. J. Walsh, M. Whitfield, and R. Wollast. 1991. Group report: What is the importance of ocean margin processes in global change? Pages 433–54 in R. F. C. Mantoura, J. -M. Martin, and R. Wollast (editors), *Ocean margin processes in global change.* New York: John Wiley & Sons.

Makhlouf, Z., and C. Michel. 1994. A two-parameter monthly water balance model for French watersheds. *Journal of Hydrology* 162: 299–318.

Mantoura, R. F. C., J.-M. Martin and R. Wollast (editors). 1991. *Ocean margin processes in global change,* 19–44. New York: John Wiley & Sons.

Marengo, J. A., J. R. Miller, G. L. Russell, C. E. Rosenzweig, and F. Abramopoulus. 1994. Calculations of river-runoff in the GISS GCM: Impact of a new land surface parameterization and runoff routing model on the hydrology of the Amazon River. *Climate Dynamics* 10: 349–61.

Marland, G., R. J. Andres, and T. A. Boden. 1994. Global, regional, and national CO_2

emissions. Pages 505–84 in T. A. Boden, D. P. Kaiser, R. J. Sepanski, and F. W. Stoss (editors), *Trends '93: A compendium of data on global change*. ORNL/CDIAC-65. Oak Ridge, TN: Carbon Dioxide Information Analysis Center.

McDonnell, J., and C. Kendall (editors). 1994. *Isotope tracers in catchment hydrology*. Amsterdam, The Netherlands: Elsevier.

McGuire, A. D., J. M. Melillo, L. A. Joyce, D. W. Kicklighter, A. L. Grace, B. Moore III, and C. J. Vörösmarty. 1992. Interactions between carbon and nitrogen dynamics in estimating net primary productivity for potential vegetation in North America. *Global Biogeochemical Cycles* 6: 101–24.

McGuire, A. D., J. M. Melillo, D. W. Kicklighter, Y. Pan, X. Xiao, J. Helfirch, B. Moore III, C. J. Vörösmarty, and A. L. Schloss. 1997. Equilibrium responses of global net primary production and carbon storage to doubled atmospheric carbon dioxide: Sensitivity to changes in vegetation nitrogen concentration. *Global Biogeochemical Cycles* 11: 173–89.

Meade, R. H., J. M. Rayol, S. C. DaConceicao, and J. R. G. Natividade. 1991. Backwater effects in the Amazon River basin of Brazil. *Environmental Geology and Water Science* 18: 105–14.

Mee, L. D. 1992. The Black Sea in crisis: A need for concerted international action. *Ambio* 21: 278–86.

Melillo, J. M., A. D. McGuire, D. W. Kicklighter, B. Moore, C. J. Vörösmarty, and A. L. Schloss. 1993. Global climate change and terrestrial net primary production. *Nature* 363: 234–40.

Meybeck, M. 1976. Total mineral transport by major world rivers. *Hydrological Sciences Bulletin* 21: 265–84.

———. 1982. Carbon, nitrogen, and phosphorus transport by world rivers. *American Journal of Science* 282: 401–50.

———. 1993a. Natural sources of C, N, P, and S. Pages 163–93 in R. Wollast, F. T. Mackenzie, and L. Chou (editors), *Interactions of C, N, P, and S biogeochemical cycles and global change*. NATO ASI Series, vol. 14. Berlin, Germany: Springer-Verlag.

———. 1993b. Riverine transport of atmospheric carbon: Sources, global typology and budget. *Water, Air, and Soil Pollution* 70: 443–63.

Meybeck, M., and R. Helmer. 1989. The quality of rivers: From pristine to global pollution. *Global and Planetary Change* 75: 283–309.

Meybeck, M., and A. Ragu. 1996. *River discharges to the oceans: An assessment of suspended solids, major ions, and nutrients*. Geneva, Switzerland: UNEP/WMO.

Miller, J. R., and G. L. Russell. 1992. The impact of global warming on river runoff. *Journal of Geophysical Research* 97(D3): 2757–64.

Milliman, J. D., and J. P. M. Syvitski. 1992. Geomorphic/tectonic control of sediment discharge to the oceans. *Journal of Geology* 100: 525–44.

Moore, B., M. P. Gildea, C. J. Vörösmarty, D. L. Skole, J. Melillo, B. J. Peterson, E. Rastetter, and P. Steudler. 1989. Strategy for understanding changes in the metabolic system of the planet. In M. B. Rambler, L. Margulis, and D. Sagan (editors), *Global ecology*. New York: Academic Press.

Mozzherin, V. I. 1991. The recent global suspended sediment yield and prognosis of its change. Pages 63–85 in R. S. Chalov (editor), *Problems of erosion, fluvial and river mouth processes* (in Russian) Moscow: Izhevsk.

Muller-Karger, F. E., C. E. McClain, and P. L. Richardson. 1988. The dispersal of the Amazon's water. *Nature* 333: 56–59.

Naiman, R. 1983. The annual pattern and spatial distribution of aquatic oxygen metabolism in boreal forest watersheds. *Ecological Monographs* 53: 73–94.

Naiman, R. (editor). 1992. *Watershed management: Balancing sustainability and environmental change.* New York: Springer-Verlag.

Newbold, J. D. 1992. Cycles and spirals of nutrients. Pages 379–408 in P. Calow and G. E. Petts (editors), *The rivers handbook: Hydrological and ecological principles.* Oxford, England: Blackwell.

Nixon, S. 1995. Coastal marine eutrophication: A definition, social causes, and future concerns. *Ophelia* 41: 199–219.

Oki, T., K. Musiake, H. Matsuyama, and K. Masuda. 1995. Global atmospheric water balance and runoff from large river basins. Scale issues in hydrological modeling. *Hydrological Processes* 9 (5–6): 655–78.

Olson, J. S. 1994a. *Global ecosystem framework—definitions.* Sioux Falls, SD: USGS EROS Data Center.

———. 1994b. *Global ecosystem framework—translation strategy.* Sioux Falls, SD: USGS EROS Data Center.

Ortner, P. B., and M. J. Dagg. 1995. Nutrient-enhanced coastal ocean productivity explored in the Gulf of Mexico. *EOS* 76: 97, 109.

Petts, G. E. 1984. *Impounded rivers: Perspectives for ecological management.* New York: Wiley-Interscience.

Power, M. E., A. Sun, G. Parker, W. E. Dietrich, and J. J. Wootton. 1995. Hydraulic food-chain models. *BioScience* 45: 159–67.

Quay, P. D., D. O. Wilbur, J. E. Richey, J. I. Hedges, and A. H. Devol. 1992. Carbon cycling in the Amazon River: Implications from the ^{13}C composition of particles and solutes. *Limnology and Oceanography* 37: 857–71.

Raich, J. W., E. B. Rastetter, J. M. Melillo, D. W. Kicklighter, P. A. Steudler, B. J. Peterson, A. L. Grace, B. Moore, and C. J. Vörösmarty. 1991. Potential net primary productivity in South America: Application of a global model. *Ecological Applications* 4: 399–429.

Rasool, I., and B. Moore. 1995. Potsdam '95 NPP Modeling Workshop. Potsdam, Germany: Potsdam Institute for Climate Impact Studies, June 1995.

Richey, J. E., L. A. K. Mertes, T. Dunne, R. L. Victoria, B. L. Forsberg, A. C. N. S. Tancredi, and E. Oliveira. 1989. Sources and routing of the Amazon River flood wave. *Global Biogeochem. Cycles* 3: 191–204.

Roads, J. O., S. C. Chen, A. K. Guetter, and K. P. Georgakakos. 1994. Large-scale aspects of the United States hydrologic cycle. *Bulletin of the American Meteorological Society* 75(9): 1589–1611.

Rodriguez-Iturbe, I., and P. S. Eagleson. 1987. Mathematical models of rainstorm events in space and time. *Water Resources Research* 23: 181–90.

Rosenberg, R., R. Elmgren, S. Fleischer, P. Jonsson, G. Persson, and H. Dahlin. 1990. Marine eutrophication case studies in Sweden. *Ambio* 19: 102–8.

Running, S. W., and J. C. Coughlan. 1988. A general model of forest ecosystem processes for regional applications: 1. Hydrologic balance, canopy gas exchange and primary production processes. *Ecological Modeling* 42: 125–54.

Running, S. W., T. R. Loveland, and L. L. Pierce. 1994. A vegetation classification logic based on remote sensing for use in global biogeochemical models. *Ambio* 23: 77–81.

Schamp, H. 1983. Sadd el Ali, the High Dam of Assuan. 1, 2. *Geowissenschaft. in unserer Zeit* 1: 51–59 and 73–85.

Sedell, J. R., J. E. Richey, and F. J. Swanson. 1989. The river continuum concept: A basis for the expected ecosystems behavior of very large rivers? Pages 49–55 in D. P. Dodge (editor), *Proceedings of the international large river symposium.* Canadian Special Publication of Fisheries and Aquatic Sciences 106. Ottawa: Fisheries and Oceans, Communications Directorate.

Seitzinger, S. P., and C. Kroeze. 1998. Global distribution of nitrous oxide production and N inputs in freshwater and coastal marine ecosystems. *Global Biogeochemical Cycles* 12: 93–113.

Sellers, P. J., B. W. Meeson, J. Closs, J. Collatz, F. Corprew, D. Dazlich, F. G. Hall, Y. Kerr, R. Koster, S. Los, K. Mitchell, J. McManus, D. Myers, K.-J. Sun, and P. Try. 1996. The ISLSCP initiative: Global datasets—surface boundary conditions and atmospheric forcings for land-atmosphere studies. *Bulletin of the American Meteorological Society* 77: 1987–2005.

Serreze, M. C., M. C. Rehder, R. G. Barry, J. D. Kahl, and N. A. Zaitseva. 1995. The distribution and transport of water vapor over the Arctic Basin. *International Journal of Climatology* 15 (7): 709–28.

Shao, Y., R. D. Anne, A. Henderson-Sellers, P. Irannejad, P. Thornton, X. Liang, T. H. Chen, C. Ciret, D. Desborough, and O. Balachova. 1994. *Soil moisture simulation: A report of the RICE and PILPS Workshop.* WCRP/GEWEX, IGBP/GAIM. North Ryde, NSW, Australia: Macquarie University.

Shuttleworth, J. 1991. Insight from large-scale observational studies of land/atmosphere interactions. In E. Wood (editor), *Land surface—atmosphere interactions for climate modeling.* Dordrecht, The Netherlands: Kluwer.

Sidorchuk, A. A. 1994. Modeling of sediment budget through the fluvial system. In *Preliminary Proceedings of IGBP Inter-Core Project workshop on modeling the delivery of terrestrial material to freshwater and coastal ecosystems.* IGBP-BAHC/PAGES/LOICZ. Durham: University of New Hampshire.

Sippel, S. J., S. K. Hamilton, J. M. Melack, and B. J. Choudhury. 1994. Determination of inundation area in the Amazon River floodplain using SMMR 37 GHz polarization difference. *Remote Sensing of Environment* 48(1):70–76.

Smil, V. 1990. Nitrogen and phosphorus. Pages 423–36 in B. L. Turner, W. C. Clark,

R. W. Kates, J. F. Richards, J. T. Matthews, and W. B. Meyer (editors), *The earth as transformed by human action.* Cambridge, England: Cambridge University Press.

Smith, R. A., R. B. Alexander, G. D. Tasker, C. V. Price, K. W. Robinson, and W. White. 1993. Statistical modeling of water quality in regional watersheds. Pages 751–54 in *Watershed T93: A national conference on watershed management.*

Smith, R. A., G. E. Schwartz, and R. B. Alexander. 1997. Regional interpretation of water-quality monitoring data. *Water Resources Research* 33: 2781–98.

Stumpf, R. P., and P. M. Goldschmidt. 1992. Remote sensing of suspended sediment discharge in the western Gulf of Maine during the April 1987 100–year flood. *J. Coast. Res.* 8: 218–25.

Sugawara, M. 1988. *Runoff analysis of the Yangtze River.* First, second, and third reports. Tokyo, Japan: Institute for Disaster Prevention.

Swank, W. T. 1988. Stream chemistry responses to disturbance. Pages 339–57 in W. T. Swank and D. A. Crossley (editors), *Forest hydrology and ecology at Coweeta.* Ecological Studies 66. New York: Springer-Verlag.

Swank, W. T., and D. A. Crossley (editors). 1988. *Forest hydrology and ecology at Coweeta,* 339–57. Ecological Studies 66. New York: Springer-Verlag.

Thompson, S., R. Lassiter, J. Bergengren, and D. Pollard. 1993. GENESIS: The NCAR Earth System Model project. Athens, GA: EPA Global Change All-Investigators Meeting, March 16–18.

Turner, R. E., and N. N. Rabelais. 1994. Coastal eutrophication near the Mississippi River delta. *Nature* 368: 619–21.

Turner, R. E., and N. N. Rabelais. 1991. Changes in Mississippi River water quality this century. *BioScience* 41: 140–47.

UNESCO. Various years. *Discharge of selected rivers of the world.* Vols. 1, 2. Paris: UNESCO/IHP.

University of Western Australia. 1993. *Scale issues in hydrological/environmental modelling.* Workshop abstracts for the Roberston, New South Wales meetings. Perth: CSIRO Division of Water Resources, Water Research Foundation of Australia, and Australian National University.

Vannote, R. L., G. W. Minshall, K. W. Cummins, J. R. Sedell, and C. E. Cushing. 1980. The river continuum concept. *Canadian Journal of Fisheries and Aquatic Sciences* 37: 130–37.

Vitousek, P. M. 1983. The effects of deforestation on air, soil, and water. Pages 223–46 in B. Bolin and R. B. Cook (editors), *The major biogeochemical cycles and their interactions.* SCOPE 21. New York: John Wiley & Sons.

Vollenwieder, R. A. 1968. *The scientific basis of lake and stream eutrophication with particular reference to phosphorus and nitrogen as eutrophication factors.* Publ. DAS/CSI/68. 27. Paris: OECD.

Vörösmarty, C. J., B. Moore, A. L. Grace, M. P. Gildea, J. M. Melillo, B. J. Peterson,

E. B. Rastetter, and P. A. Steudler. 1989. Continental scale models of water balance and fluvial transport: An application to South America. *Global Biogeochemical Cycles* 3: 241–65.

Vörösmarty, C. J., B. Moore III, A. Grace, B. J. Peterson, E. B. Rastetter, and J. Melillo. 1991. Distributed parameter models to analyze the impact of human disturbance of the surface hydrology of a large tropical drainage basin in southern Africa. Pages 233–44 in van de Ven, F. H. M., D. Gutknecht, D. P. Loucks, and K. A. Salewicz (editors), *Hydrology for the water management of large river basins.* Publication No. 201. Wallingford, England: IAHS Press.

Vörösmarty, C. J., C. J. Willmott, B. J. Choudhury, A. L. Schloss, T. K. Stearns, S. M. Robeson, and T. J. Dorman. 1996. Analyzing the discharge regime of a large tropical river through remote sensing, ground-based climatic data, and modeling. *Water Resources Research* 32: 3137–50.

Vörösmarty, C. J., K. Sharma, B. Fekete, A. H. Copeland, J. Holden, J. Marble, and J. A. Lough. 1997a. The storage and aging of continental runoff in large reservoir systems of the world. *Ambio* 26: 210–19.

Vörösmarty, C. J., M. Meybeck, B. Fekete, and K. Sharma. 1997b. The potential impact of neo-Castorization on sediment transport by the global network of rivers. Pages 261–72 in D. Walling and J. -L. Probst (editors), *Human impact on erosion and sedimentation.* Wallingford, England: IAHS Press.

Vörösmarty, C. J., R. Wasson, and J. E. Richey (editors). 1997c. *Modeling the transport and transformation of terrestrial materials to freshwater and coastal ecosystems.* Workshop report and recommendations for IGBP Inter-Core Project collaboration. IGBP Report No. 39. Stockholm, Sweden: IGBP Secretariat.

Vörösmarty, C. J., B. M. Fekete, M. Meybeck, and R. Lammers. 2000. A simulated toplogical network representing the global system of rivers at 30-minute spatial resolution (STN-30). *Global Biogeochemical Cycles* (In press).

Walling, D. E. 1987. Hydrological processes. Pages 53–85 in K. J. Gregory, and D. E. Walling, (editors), *Human activity and environmental processes.* Chichester, England: John Wiley & Sons Ltd.

Ward, J. V., and J. A. Stanford. 1983. The serial discontinuity concept of lotic ecosystems. Pages 29–42 in T. D. Fontaine III and S. M. Bartell (editors), *Dynamics of lotic ecosystems.* Ann Arbor, MI: Ann Arbor Science.

White, D. A., R. A. Smith, C. V. Price, R. B. Alexander, and K. W. Robinson. 1992. A spatial model to aggregate point-source and nonpoint-source water-quality data for large areas. *Computers & Geosciences* 18: 1055–73.

Wollast, R. 1983. Interactions in estuaries and coastal waters. Pages 385–407 in B. Bolin and R. B. Cook (editors), *The major biogeochemical cycles and their interactions.* SCOPE 21. New York: John Wiley & Sons.

Wyss, J., E. R. Williams, and R. L. Bras. 1990. Hydrological modeling of New England river basins using radar rainfall data. *Journal of Geophysical Research* 95

CHAPTER 4

Synthesizing Drainage Basin Inputs to Coastal Systems

Thomas R. Fisher, David Correll, Robert Costanza, James T. Hollibaugh, Charles S. Hopkinson Jr., Robert W. Howarth, Nancy N. Rabalais, Jeffrey E. Richey, Charles J. Vörösmarty, and Richard Wiegert

Abstract

To address material transport from terrestrial basins to the coastal zone, it is necessary to synthesize existing data on watershed export and improve our ability to make predictions over both space and time. Synthesis of existing data on watershed export must account for the processes of transformation, storage, and removal which become important at larger scales. In this chapter we propose a set of general principles, major approaches, and priorities in order to understand material transport to coastal areas, and we review useful modeling approaches. We also suggest ways to remove obstacles to synthesis and propose selected projects that could provide significant advances toward understanding material transport to the coastal zone.

Why Synthesize?

Currently there is no coherent synthesis of material fluxes from river systems across spatial scales. Although experimental and modeling data are available on material fluxes at several scales, linkages between local, regional, and continental scales are lacking (figure 4-1); in other words, we do not know how material fluxes at the smallest watershed scales integrate to continental scale discharges, nor can we derive the basis of large river material fluxes at small scales. This is due to our lack of understanding of key process dynamics operating at different drainage-basin scales. This understanding results from independent site-specific experimentation and modeling.

A vast, but site-specific literature is available on the terrestrial mobilization of water, particulates, and biotically active constituents (for example,

MODELING SYSTEM FOR MULTILEVEL
HYDROLOGICAL ANALYSIS

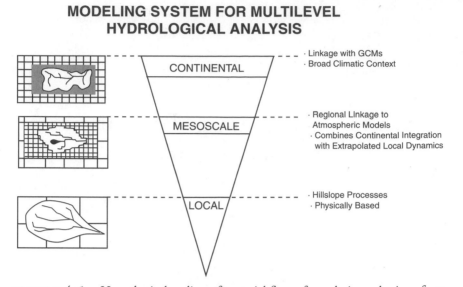

FIGURE 4-1   Hypothetical scaling of material fluxes from drainage basins of various areas. Many data are available at local (small) scales from experimental measurements of small or paired catchments, and this approach has led to a mechanistic understanding of transport at small scales. However, fewer data have been obtained from larger rivers at the regional scale, which includes linkages to meso-scale atmospheric inputs as well as transport and processing of diffuse and point sources from a variety of smaller sub-basins. At the largest (continental) scales, there are linkages with climate and large-scale atmospheric inputs, and material fluxes are coupled with ocean basins. At this largest scale, the fewest data are available. Modeling efforts have concentrated on the smallest and largest scales, and linkages between local, regional, and continental scales are poorly known. After Vörösmarty et al. (1993).

Beaulac and Reckhow 1982). This literature has clearly documented the impacts of geology, changing land cover, land management, and fertilization on constituent delivery (see figure 4-2). For example, it is well known that conservation tillage practices decrease field-scale erosion but increase the use and subsequent leaching of herbicides as well as leaching of P from undisturbed soils (Staver and Brinsfield 1994). Impacts of point-source pollution and water engineering works (for example, impoundment, irrigation, interbasin transport) have been detected when considering larger stream orders (Malone 1984). Attempts to extrapolate our understanding from monitoring of individual lower-order catchments or from plot-scale studies are confounded by the following processes, which become important within larger drainage systems:

* water withdrawals and consumption for agriculture, domestic, and industrial uses;

* the position and extent of intervening wetland and riparian areas;

- differential in-stream processing across stream orders;

- hyporheic effects;

- particulate deposition/resuspension and streambed changes;

- altered discharge dynamics due to reservoir operation; and

- interbasin transfers.

Synthesis of existing data on watershed export must account for these processes of storage, transformation, and removal which may become important at larger scales. It is clearly necessary to improve our understanding of the complex mechanisms of transport from terrestrial to coastal ecosystems as well as our ability to make predictions over both space and time. Such synthesis must take place not only at each relevant spatial scale (figure 4-1) but across contrasting spatial scales as well.

Each of the scales depicted in figure 4-1 serves a distinct yet integrative role. Site-specific studies of small watersheds serve as the raw material from which process-level understanding can be obtained (figure 4-2B). Regional-scale studies give broader, though less mechanistic, features of material transport and are useful for generating spatial patterns and correlates by which national and continental-scale assessments can be made (figure 4-2A). Knowledge of continental-scale material inputs is required for global models and may be required for future international policy development. Applying the constraints of mass-balance at both the regional and continental scales can lend insight into broad-scale sources and sinks (see Howarth et al., Seitzinger, this volume). Knowledge of these mechanisms is critical for rational management decisions within catchments to address water-quality problems. Data without synthesis (see figure 4-3) are inappropriate and inadequate for effective management.

Central Question

What is the effect of changes in land use, atmospheric deposition, climate, and water engineering works on delivery of water, particulate materials, and dissolved substances to estuaries and coastal areas?

General Principles

The above question can be addressed by synthesis and integrative studies. Synthesis is possible because the following general principles appear to be true in most catchments draining into the coastal zone.

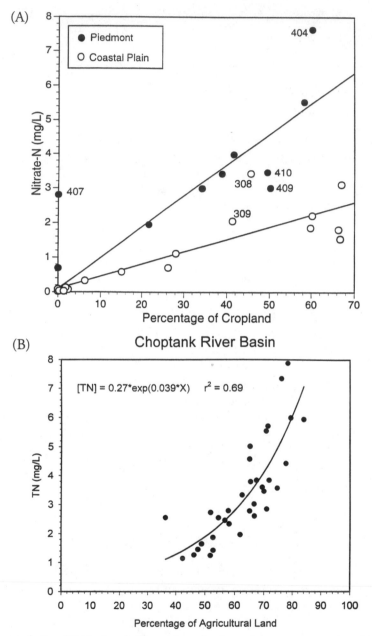

FIGURE 4-2 (A) Regional effects of cropland on nitrate concentrations in streams of piedmont and coastal plain watersheds of the mid-Atlantic region of North America (after Jordan et al. 1997a). (B) Effects of agricultural land on total N (TN) concentrations in streams of the Choptank River Basin on the Delmarva Peninsula (Norton and Fisher, in press).

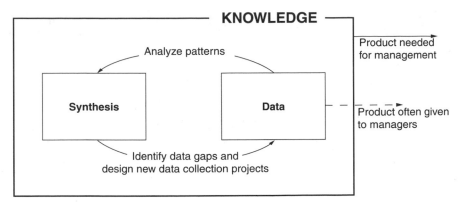

FIGURE 4-3 Relationships between data, synthesis, knowledge, and management in scientific investigations related to material fluxes from terrestrial basins to aquatic systems.

Riverine inputs to estuaries can be predicted from knowledge of diffuse and point sources, and in-stream processing. Water, dissolved substances, and particulate materials are exported from soils via overland flows and groundwater discharges from aquifers of varying depths (diffuse sources). In low-order streams, the chemistry and quantity of water from these variable sources are strongly influenced by soil type, land use, land cover, topography, temperature, antecedent moisture, and current precipitation. In higher-order streams, these diffuse sources are augmented by anthropogenic point and diffuse sources (for example, sewage, urban storm drains). Furthermore, in-stream processing by plankton and benthos, and within hyporheic zones and floodplains, may mobilize or immobilize dissolved and particulate materials on varying time scales. Transport of materials into the estuary or coastal zone typically occurs on time scales varying from days to months following precipitation, depending on the catchment size, and typically has a seasonal peak during late winter or spring following a period of low evapotranspiration in temperate zones or following the rainy season in the tropics.

What Are the Major Approaches and Priorities to Address the Central Question?

We have combined topics from the twenty-one scientific priorities identified on pages 3–6 of the NRC (1994) document with some of our own considerations:

- emphasize measurements of diffuse inputs from the land and atmosphere as the dominant sources in most nonurban catchments;

- study the coupling of watershed hydrology and material fluxes in low-order streams to understand the mechanisms of material transfer to stream channels;

- examine the effects of water management and the spatial patterns of land-use types on material fluxes from catchments;

- evaluate the role of wetlands, riparian zones, and floodplains on transport and transformation of particulate and dissolved materials;

- apply improved in-situ and remote sensing systems to the measurement of turbidity, pigments, DOM, and the like (for example, image analysis, improved algorithms);

- create or improve access to GIS databases (for example, a home page on the Internet) by creating metadata bases (databases of where to find data), particularly for meteorological, hydrologic, and terrestrial data (for example, relative humidity, temperature, precipitation, discharge, chemistry, land use, soil maps, topography); an example is the MERCURY software system at ORNL;

- evaluate the impacts of spatial and temporal resolution in databases (for example, number of land-use types, pixel size, and the like);

- link regional and national monitoring and add linkages between ecological and economic databases;

- relate riverine nutrient flux to estuarine and coastal ecosystem dynamics;

- evaluate the impacts of time lags in river basins on estuarine responses;

- develop coupled atmosphere-watershed-coastal models as management tools; and

- employ adaptive ecosystem management by learning from experimental management approaches and developing alternatives.

Models for Synthesizing Existing Data

Synthesis may be achieved by several methods. Two extremes are mechanistic models (process-based, using the interior structure of a catchment) and statistical models (correlation with easily observable features, with no direct mechanistic understanding). These are often combined to some degree. Experimental manipulation of watersheds is a useful tool to extract the effects of the manipulated variables, but it is also possible to learn from other approaches:

- time series analysis of one catchment with natural variations in input parameters;

- comparative analysis of multiple catchments at different spatial scales, across basins of similar spatial scales but with different characteristics (for example, disturbance, soils, topography), and by creating a typology (categorization) of basin types; and

- process-based spatial modeling using a GIS calibrated to a basin with monitored inputs and outputs.

Process-based Versus Correlation Models

As described above, models fall into two general classes—those which are process-based (mechanistic) and those which are based on correlations (statistical). These two extremes of model organization are represented by purely mechanistic (causal) models versus empirical (correlative) models, although many models have elements of both approaches. Process-based models are concerned with the explanations of why a given input to any compartment is mapped into the simulated output. They incorporate causal statements (mechanisms) in their equations or algorithms. In contrast, correlation models are concerned only with the accuracy and precision with which an observed system input is mapped into an observed system output. Such a correlative model can then be used to predict any output, given an observed input within the range of inputs used to construct the model. The parameters of the model may have no explanatory significance and may have been chosen solely for their utility in improving the fit of the model to the observed data sets. For instance, a correlation between mean annual values of rainfall and stream discharge reveals little about the routing of water through a drainage basin. However, the variables used in correlation models (for example, vegetation cover and soil drainage characteristics) may suggest plausible mechanistic connections between inputs (rainfall) and outputs (discharge).

Clearly, the choice of model will depend on the use to which the model will be put. Is the objective prediction? Do prior data exist on a suitable range of perturbations to systems similar to the one to be modeled? If prior data do not exist or if explanation of system behavior is desired, then mechanisms must be hypothesized (for instance, by using information from small-scale catchments) and used in the model.

The two classes of models described above may be applied to three types of problems in watershed dynamics: (1) those that address the sources of materials in terrestrial ecosystems; (2) those that deal with the routing, storage, and processing of materials in streams and rivers; and (3) those that include socioeconomic interactions with watershed function. All of these types of problems may include models that are process-based and models that are correlative.

1. Terrestrial source models. Many process-based models have been developed to estimate water and sediment fluxes from watersheds. Examples include the BROOK model for first-, second-, and third-order streams (Federer and Lash 1983) and the GWLF model for larger watersheds (Haith and Shoemaker 1987). It is important that the model includes both surface and groundwater trajectories and adequate treatment of evapotranspiration. GWLF has been applied successfully to watersheds as large as three million hectares by working with smaller sub-basins of approximately 200,000 to 1,000,000 hectares each (Howarth et al. 1991; Swaney et al. 1996; Lee et al. in press). Models such as BROOK and GWLF use data on meteorology, soil type, slope, and land use as input to estimate soil moisture, groundwater recharge, overland flow, and erosion. These models are based purely on physical mechanisms of water flow and erosion; fluxes of organic matter and nutrients are simply estimated by assigning concentrations to flows of water and particles (Haith and Shoemaker 1987; Howarth et al. 1991). There is a clear need to build more biogeochemical sophistication into these efforts, particularly if we are to address how changes in land use, atmospheric inputs, or climate influence fluxes to estuaries. One approach being explored is to link physically based watershed loading models such as GWLF and BROOK with biogeochemical models of element interactions, such as CENTURY (see Howarth et al., this volume). Such coupling of physical and biological processes within watershed models was identified as a research priority in the NRC (1994) report titled "Priorities for Coastal Ecosystem Science." Similar approaches based on different watershed loading and general ecosystem models are also being explored (Costanza and Voinov, this volume).

Correlative models are also used to predict fluxes of materials from terrestrial sources to estuaries and coastal waters. For example, nationally the NOAA Status and Trends program estimates nutrient loading to the coast from export coefficients which are a function of land use and population density. Jordan et al. (1997a) have shown that percentage of cropland and an index of baseflow are good predictors of nitrate concentrations in mid-Atlantic watersheds ($r^2 = 0.79$). Vörösmarty and Peterson (this volume) have used the regression analyses of Milliman and Syvitski (1992) based on location, topography, and drainage basin area in conjunction with a GIS to predict suspended sediment fluxes from 275 drainage basins at 30-minute resolution in South America. They documented the importance of small and medium-sized watersheds (<30,000 km^2), particularly those draining mountainous regions that have traditionally been under-represented in such inventories (color plate 1). Such tools to estimate material inputs to estuaries where primary data are lacking are extremely useful for synthesis. However, they are inherently unable to predict how changes in the watershed will affect fluxes to estuaries.

2. In-stream and in-river routing, storage, and processing. Streams and rivers are not merely conduits that carry materials from the land to the sea. The timing and volume of water transported varies significantly with watershed area because of scale-dependent processes such as groundwater recharge, interbasin transfers, and gauging errors (figure 4-4). For sediments, only a small percentage of material eroded from land surfaces actually reaches major rivers and estuaries (Renfro 1975; Sheridan et al. 1982; Howarth et al. 1991). For organic carbon and nutrients, significant alteration can occur not only through sedimentation but also through metabolism. A well-developed body of theory relates the processing of nutrients and organic matter in streams and rivers to downstream advection through the concepts of spiraling (Elwood et al. 1983) and the river continuum (Vannote et al. 1980). However, quantitative models that relate these concepts to transport processes in streams and rivers are lacking (Howarth et al. 1996). A high-priority research topic is to build process-based models that can use these principles to relate terrestrial source estimates to fluxes that actually reach the estuary.

Correlative models are also quite useful for estimating the in-river processing of nutrients. Examples of such models are those of Kelly et al. (1987) and Seitzinger (1988) for rates of denitrification in lakes and estuaries. These models estimate rates of denitrification as functions of such properties as depth and residence time. Howarth et al. (1996) have extrapolated the approach of Kelly et al. (1987) to estimate denitrification in rivers, and Billen et al. (1994) have modeled phytoplankton development using the statistical characteristics of river networks. Such correlative models could prove highly useful to estimate the in-river processing of other elements.

3. Socioeconomic aspects. It is also possible to integrate the socioeconomic aspects of the watershed with the ecological aspects in what has been called integrated assessment (Costanza and Voinov, this volume). This includes a broad range of human influences on the system from drainage structures such as dams, canals, and levees to agricultural practices and urban development. It is particularly necessary to engage in synthesis activities in order to integrate these factors, and they have a major influence on the behavior of the system.

Mid-Atlantic Coastal Plain and Piedmont Watersheds

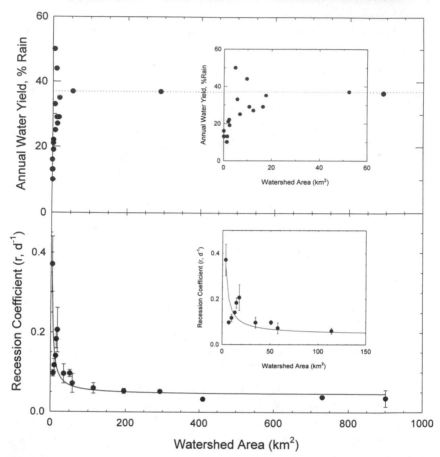

FIGURE 4-4   Annual water yields and recession coefficients for coastal plain and piedmont watersheds in the mid-Atlantic region of North America. Water yields are expressed as percentage of rainfall; data are from Jordan et al. (1997b) and Fisher et al. (1998). Variability of water yields in small basins is due to measurement errors and the increasing importance of groundwater flows beneath gauges of surface discharge. Recession coefficients (r, d^{-1}) are the exponential rate of decline of discharge following a rain event and are inversely related to watershed area; data from Lee et al. (in press). From Jordan et al. (1997a). Copyright © American Geophysical Union.

Overcoming Obstacles to Synthesis

Currently, limited funding opportunities and mechanisms exist for synthetic approaches in estuarine research. Institutional structures and feedback mechanisms primarily support research projects rather than synthetic studies. To overcome these obstacles, we must modify the traditional views of the importance of synthetic approaches within the academic community, facilitate the transfer of synthetic studies to resource managers, and convince providers of both basic and applied research funds of the importance of synthetic approaches.

Changes within the academic community may help overcome obstacles to synthesis. Multidisciplinary academic centers, such as those emphasizing environmental studies, accumulate expertise from several disciplines and often support and facilitate synthetic studies. Tenure and promotion decisions, within either science departments or multidisciplinary centers, should include consideration of synthetic research activities.

Research funds from federal agencies should also be targeted for synthetic analyses. A primary example of movement in this direction is the NSF Center for Ecological Analysis and Synthesis. Similar activities should become embedded in other research programs. Synthesis activities at the end of major research programs should be adequately funded and not just supported in theory. In some cases, the intention of program managers to support synthesis at the end of a study is thwarted by negative peer reviews. Greater recognition of the value of synthesis within our own ranks may facilitate addition of synthesis to research plans.

The research product often given to environmental managers may be a data compilation rather than knowledge derived from the synthesis of data (figure 4-3). While the data are needed to support the synthesis, the most useful tool to managers is the synthetic product, which often embeds resource management recommendations. A positive feedback mechanism that reinforces synthetic studies with further funding should become increasingly influential as the usefulness of synthetic information increases in resource management decisions.

Hypothetical Projects

How would these approaches and methods be employed specifically to address the central question given above? Below we list six possible projects that require synthesis and involvement from multiple groups of scientists. Several are interlinked and could potentially be addressed simultaneously at one meeting.

1. Metadata base of watershed drivers and characteristics. Assembling data from many disparate sources has been one of the major obstacles to performing broad synthesis studies in ecology. Broadly based, intelligent data access, visualization, and review are thus essential for a range of synthesis activities. While it is not appropriate or necessary to collect and store vast amounts of data in a single location, it is desirable to create a metadata base that can intelligently guide users to the locations of stored data. This will focus on providing practical access to the huge amounts of data already stored elsewhere and accessible over the Internet or in other ways. It will also provide the means to visualize and review these data easily by the entire range of potential users. The metadata base should focus on the issues of providing:

- comprehensive access through intelligent Internet links;

- better quality assessment and communication through new data grading systems; and

- advanced visualization using the latest computer technology.

This approach will allow interdisciplinary teams of researchers to access and view data they would never have seen before, to see the data in new ways, and to use the data effectively in synthesis activities. In addition, there are many data sources that have not yet been digitized or stored on the Internet, and providing access to these data would encourage their use in a wide range of studies.

As an example, significant insights into the influence of certain activities in watersheds can be learned through time-series analysis of changes (see below). We have seen examples of data that have been obtained from historical records of nutrient concentrations in selected rivers of the world (see Rabalais et al., this volume, and figures 4-5 and 4-6). Historical records of nitrate concentrations in the Mississippi River provided insight into the influence of fertilizer application to agricultural systems on nutrient export to the coastal zone, and historical records of nitrate concentrations in the Seine and Rhine Rivers have been attributed to both agriculturalization and increasing fertilizer use (Turner and Rabalais 1991; Bennekom and Wetsteijn 1990). Similar types of data exist for other watersheds and rivers of North America (see Smith et al. 1987; Fisher et al. 1998; figure 4-6). It is essential to locate, assemble, catalogue, and add these data records to the overall metadata base along with data on socioeconomic characteristics, land use, and the like so that researchers can begin to do correlative, time-series, and process modeling studies as described below. A current example of this approach is the TYGRIS-BAHC metadata base being created within the IGBP program (IGBP-BAHC 1998).

2. Community (generic) watershed process-based modeling. One approach to comparing different systems is to apply a common model to all. We propose this approach for estimating fluxes of water, sediment, organic carbon, and nutrients from land to estuaries. The purpose of this modeling comparison is

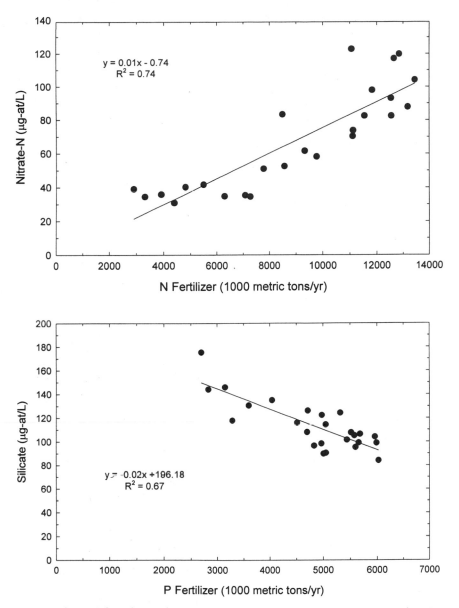

FIGURE 4-5 The relationship between fertilizer use and water quality in the Mississippi River Basin. (A) Nitrogen fertilizer applications (as N) from 1960–1985. (B) Phosphorus fertilizer applications (as P_2O_5) from 1950–1985. Modified from Turner and Rabalais (1991).

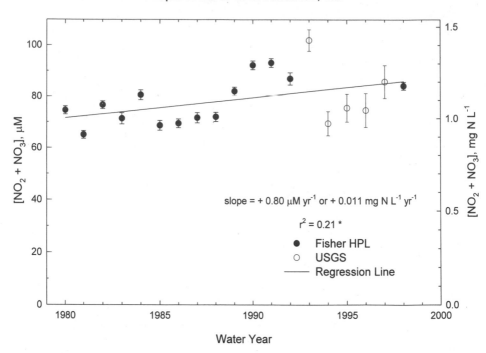

Choptank River at Greensboro, MD

FIGURE 4-6   Nitrite + nitrate concentrations (arithmetic mean ± SE) in the Choptank River from 1980 through 1998. Data sources: Fisher et al. (1998), USGS, and
Fisher et al. (unpublished).

threefold: (i) to compare modeled output with measured output, when available; (ii) to explore commonalities and differences in the controls on material
deliveries to different estuaries; and (iii) to improve the performance and predictability of the model. There are several candidates for the building blocks
of this generic model, including the GWLF, a model designed to estimate
nonpoint source loadings from moderate- to large-sized watersheds (Haith
and Shoemaker 1987; see discussion above) and the General Ecosystem
Model (GEM, Fitz et al. 1996), which was designed as a medium-complexity,
ecosystem process model including water, nutrients, sediments, and vegetation. GWLF has been applied to the Hudson River watershed, where it does
an excellent job of predicting freshwater discharge at time scales from weeks
to years, a very good job of predicting sediment delivery at time scales from
months to years, and a fair job of predicting annual fluxes of particulate and
dissolved organic carbon (Howarth et al. 1991; Swaney et al. 1996). A similar

application of GWLF to a gauged area of the Choptank River watershed resulted in predictions with RMS errors of <10% at decadal time scales, 10–50% at annual time scales, and 50–100% (within a factor of 2) at monthly time scales (Lee et al., in press). A generic model could be applied to a set of watersheds draining into well-characterized estuaries; the watersheds should be of a suitable size, in other words, with areas of 10^3–10^6 hectares. Suitable estuaries and watersheds might include the Parker River, the Rhode River, the Choptank River, the Patuxent River, the Everglades, and Tomales Bay. Linked to biogeochemical models within the landscape (such as GWLF, GEM, or CENTURY) or to in-stream and in-river processing models, these linked models could be applied and tested across this set of watersheds and estuaries. Identification of consistent deficiencies in the predictions of the models in well-characterized watersheds would be useful for future improvements in generic process-based modeling.

3. Scaling Case Study. A major objective of synthesis is to test our present ability to understand quantitatively the sources, transfers, and transformations of water and waterborne materials from a large, complex watershed to its estuary. Such an understanding is needed in order to improve management of the watershed and to predict the effects of various management options. One large watershed should be selected for a case study of our ability to identify key information gaps. This watershed should be one for which a relatively rich database already exists. It is likely that we may not be able to explain and/or predict the effects of spatial scaling from small watershed studies to larger, more complex systems because of important processes that may become dominant at larger scales (for instance, the effects of floodplains, in-stream processing, and the like). The synthesis case study would be aimed at identification of scaling problems and approaches to solving these obstacles.

One large watershed should be selected for a case study of our ability to identify key information gaps by synthesizing available data at the local and regional scale. The synthesis efforts would include:

- present atmospheric and terrestrial inputs to low-order stream channels;

- data relevant to in-stream transport and transformations (such as water routing kinetics, sediment transport, microbial transformations) and to other stream network interactions (for example, effects of reservoirs, floodplain and hyporheic interactions, water withdrawals, point-source inputs); and

• predictions of changes in inputs that would result from changes in natural and anthropogenic forcing functions.

4. Comparative tracers of the sources and fate of material transported from watersheds. Organic and inorganic materials measured in the main channel of a large river are a mixture originating from many sources. Materials from thousands of kilometers away in upland regions mix and react continuously with inputs from the adjacent floodplain and local low-order streams. Dissolved and particulate matter from all sources is subjected to within-channel transport and reactive processes before discharge into coastal waters. In order to understand the physical and chemical controls operating in large drainage basins, it is necessary to determine the terrestrial origin of the riverborne material and the dynamics of downstream routing. In the case of dissolved and particulate organic matter, the degree to which these materials have been diagenetically altered may allow us to understand the processes of carbon cycling, and to identify the biological and geographic sources and the marine sinks of these materials. This can be summarized in the following questions:

(i) How is the biogeochemical signature of organic materials that persists through the river system imparted by the (aggregated) land surfaces?

(ii) How is the land-derived signature of organic matter modified through transit within the river system?

(iii) What is the fate and impact of the riverborne organic materials in the coastal zone?

To address these questions, data are required not only on the bulk concentration of organic materials but also on specific chemical features that relate to the sources and condition of the organic matter in transport. The identification of a set of tracers would help characterize the origin and processing of organic matter along river systems from terrestrial sources to marine fate, and these tracers could be tracked through selected river/coastal systems. For dissolved organic matter, molecular weight distribution, metal-binding capacity, spectroscopic behavior, and elemental or isotopic composition can provide useful information. For particulate organics, elemental and isotopic composition can be diagnostic. Molecular-level tracers include total amino acids, carbohydrates, and lignin oxidation products (see Richey et al. 1991). Combined with bulk chemistry and hydrologically based routing models, this approach would provide a basis for comparison and synthesis between river/estuary systems.

5. A comparative study of the relationship between changes in watershed characteristics and response of estuaries. A comparative study of estuarine responses to changes in watershed characteristics would require that predictive watershed models are available and that the data needed for reliable hindcasting have been incorporated. The first step would be to characterize a number of estuaries as a community effort, as in hypothetical project 2 above. The general classes of variables that are likely to be important in the characterization include:

- magnitude and timing of material fluxes from the drainage basin;

- estuarine geomorphology;

- physics—circulation, mixing, and ocean exchange;

- characteristics of the coastal ocean;

- the biology of the system—both species and functional groups;

- factors that affect the chemistry of the estuary such as anoxia/hypoxia, [DOM], particle load, and characteristics of particles;

- society/economics of the community living around the estuary, which influences whether mitigation measures have been applied; and

- kinds and magnitudes of sources of materials of concern in the estuary or local watershed.

We would then compare the responses of a variety of estuaries to changes in their watersheds. Variables, parameters, and processes of interest include eutrophication, sedimentation rate, or the presence and impact of toxic substances. Of particular interest is the response of fisheries or species of fish. The question of fisheries' impacts is often of great interest and importance to the human societies living around estuaries. Furthermore, alterations to a watershed may cause fish (and fisheries) to respond in a manner that is different from that of geochemical variables.

The actual comparison between estuaries can be performed in two ways. One way is to perform a correlative analysis or cross-system comparison where steady state is assumed and the average conditions in a number of estuaries over some short time scale are compared (for example, Nixon et al. 1996). This approach will work best with a set of similar estuaries with different watershed characteristics and loading functions. The analysis might also be accomplished by an ANOVA approach, since it is unlikely that replication will be adequate. This approach will probably work best for physical properties because of time lags in the responses for other, especially biological, properties, if estuaries are still changing.

This proposed study would test the accuracy and robustness of our conceptual model of the coupling between an estuary and its watershed. For example, the response of an estuary to alterations of the watershed depends on the characteristics of the estuary (for example, water residence time) and sensitivity to changes in inputs. The study would help us to evaluate the relative importance of changes within the estuary in response to changes in its watershed (urbanization and shoreline development, discharges, dredging, introduced species, and the like).

6. Time-series analysis of the relationship between changes in watershed characteristics and response of estuaries as preserved in sediments. A geochronological sequence may provide a second way to test our concepts of watershed/estuary coupling. Changes through time in several estuaries could be compared with a hindcast prediction of the changes in the watershed inputs generated by the model described in hypothetical projects 2 or 3 above. Care must be taken in the selection of a site to be cored, or the analysis will lose resolution into the past.

The relatively recent record (10–20, to possibly 50–100 years) can be obtained from data records of the scientific community. Data for variables including nutrients, dissolved oxygen, chlorophyll, standing crops of plants or animals, fishery yield, and sedimentation rate could be obtained in this way. A record of the longer-term response of the estuary (>50 years) may be available from dated sediment cores, but temporal resolution will be limited because of such factors as bioturbation, dredging, and slumping. Furthermore, we will only be able to resolve temporal changes of a limited number of properties because often surrogates must be used. Time-series data on changes in sedimentation rate can be obtained fairly directly. Indices of the historical trends in salinity, metal delivery, and the like can also be obtained fairly directly. Biological silica, microfossils, or other biological remnants can be interpreted to indicate changes in the trophic status of an estuary and the composition of the estuarine biotic community. Pollen, geochemical tracers such as stable isotope ratios, lignin oxidation products, or specific compounds can be used to construct a record of changes in watershed vegetation and material delivery to the estuary (see Cooper and Brush 1991; Turner and Rabalais 1994).

References

Beaulac, M. N., and K. H. Reckhow. 1982. An examination of land use–nutrient export relationships. *Water Resources Bulletin* 18: 1013–24.

Billen, G., J. Garnier, and P. Hanset. 1994. Modeling phytoplankton development in whole drainage networks: The RIVERSTRAHLER model applied to the Seine River system. *Hydrobiol.* 289: 119–37.

Cooper, S. R., and G. S. Brush. 1991. Long-term history of Chesapeake Bay anoxia. *Science* 254: 992–96.

Elwood, J. W., J. D. Newbold, R. V. O'Neill, and W. VanWinkle. 1983. *Resource spiraling: An operational paradigm for analyzing lotic ecosystems.* Pages 3–67 in T. D. Fontaine and S. M. Bartell (editors), *The dynamics of lotic ecosystems.* Ann Arbor, MI: Ann Arbor Science.

Federer, C. A., and D. Lash. 1983. BROOK: A hydrologic simulation model for eastern forests. Water Resources Research Center Report no. 19. Durham: University of New Hampshire.

Fisher, T. R., K.-Y. Lee, H. Berndt, J. A. Benitez, and M. M. Norton. 1998. Hydrology and chemistry of the Choptank River basin. *Water Air and Soil Pollution* 105: 387–97.

Fitz, H. C., E. B. DeBellevue, R. Costanza, R. Boumans, T. Maxwell, and L. Wainger. 1996. Development of a general ecosystem model (GEM) for a range of scales and ecosystems. *Ecological Modeling* 88: 263–95.

Haith, D. A., and L. L. Shoemaker. 1987. Generalized watershed loading functions for stream flow nutrients. *Water Resources Bulletin* 23: 471–78.

Howarth, R. W., J. R. Fruci, and D. Sherman. 1991. Inputs of sediment and carbon to an estuarine ecosystem: Influence of land use. *Ecological Applications* 1: 27–39.

Howarth, R. W., R. Schneider, and D. Swaney. 1996. Metabolism and organic carbon fluxes in the tidal, freshwater Hudson River. *Estuaries* 19: 848–65.

Howarth, R. W., G. Billen, D. Swaney, A. Townsend, N. Jaworski, K. Lajtha, J. A. Downing, R. Elgren, N. Caraco, T. Jordan, F. Berendse, J. Freney, V. Kudeyarov, P. Murdoch, and Z.-L. Zhu. 1996. Regional nitrogen budgets and riverine N & P fluxes for the drainages to the North Atlantic Ocean: Natural and human influences. *Biogeochemistry* 35: 75–139.

IGBP-BAHC. 1998. Revised implementation plan. Project 6 (influence of climate change and human activities on mobilization and transport through riverine systems). IGBP Core Project BAHC (Biospheric Aspects of the Hydrologic Cycle). Stockholm, Sweden: IGBP.

Jordan, T. E., D. L. Correll, and D. E. Weller. 1997a. Relating nutrient discharges from watersheds to land use and streamflow variability. *Water Resources Research* 33: 2579–90.

Jordan, T. E., D. L. Correll, and D. E. Weller. 1997b. Effects of agriculture on discharges of nutrients from coastal plain watersheds of Chesapeake Bay. *Journal of Environmental Quality* 26: 836–48.

Kelly, C. A., J. W. M. Rudd, R. H. Hesslein, D. W. Schindler, P. J. Dillon, C. T. Driscoll, S. A. Gherini, and R. E. Hecky. 1987. Prediction of biological acid-neutralization in acid sensitive lakes. *Biogeochemistry* 3: 129–40.

Lee, K.-Y., T. R. Fisher, T. E. Jordan, D. L. Correll, and D. E. Weller. Modeling the hydrochemistry of the Choptank River basin using GWLF and GIS. *Biogeochemistry* (in press).

Malone, T. C. 1984. Anthropogenic nitrogen loading and assimilation capacity of the Hudson River estuarine system, USA. Pages 291–311 in V. S. Kennedy (editor), *The estuary as a filter*. New York: Academic Press.

Milliman, J. D., and J. P. Syvitski. 1992. Geomorphic/tectonic control of sediment discharge to the ocean: The importance of small mountainous rivers. *Journal of Geology* 100: 525–44.

National Research Council. 1994. *Priorities for coastal ecosystem science*. Washington, DC: National Academy Press.

Nixon, S. W., J. W. Ammerman, L. P. Atkinson, V. M. Berounsky, G. Billen, W. C. Boicourt, W. R. Boynton, T. M. Church, D. M. Ditoro, R. Elmgren, J. H. Garber, A. E. Giblin, R. A. Jahnke, N. J. P. Owens, M. E. Q. Pilson, and S. P. Seitzinger. 1996. The fate of nitrogen and phosphorus at the land-sea margin of the North Atlantic Ocean. *Biogeochemistry* 35: 141–80.

Norton, M. G., and T. R. Fisher. The effects of forest on stream water quality in two coastal plain watersheds of the Chesapeake Bay. *Ecol. Engin.* (In press)

Renfro, G. W. 1975. Use of erosion and sediment delivery ratios for predicting sediment yield. Pages 33–45 in *Proceedings of the sediment-yield workshops*. Oxford, MS: USDA.

Richey, J. E., R. L. Victoria, E. Salati, and B. R. Forsberg. 1991. The biogeochemistry of a major river system: The Amazon case study. Pages 57–74 in E. T. Degens, S. Kempe, and J. E. Richey (editors), *Biogeochemistry of major world rivers*. Chichester, England: John Wiley & Sons.

Seitzinger, S. P. 1988. Denitrification in freshwater and coastal marine ecosystems: Ecological and geochemical significance. *Limnology and Oceanography* 33: 702–24.

Sheridan, J. M., C. V. Booram, and L. E. Asmussen. 1982. Sediment delivery ratios for a small coastal plain agricutural watershed. *Trans. Am. Soc. Agricul. Eng.* 25: 610–15.

Smith, R. A., R. B. Alexander, and M. G. Wolman. 1987. Water-quality trends in the nation's rivers. *Science* 235: 1607–15.

Staver, K. W., and R. B. Brinsfield. 1994. The effect of erosion control practices on phosphorus transport from coastal plain agricultural watersheds. Pages 215–222 in P. Hill and S. Nelson (editors), *Proceedings of the 1994 Chesapeake Research Consortium Conference, Toward a Sustainable Costal Watershed: The Chesapeake Experiment*. Edgewater, MD: Chesapeake Research Consortium.

Swaney, D. P., D. Sherman, and R. W. Howarth. 1996. Modeling water, sediment, and organic carbon discharges in the Hudson/Mohawk basin: Coupling to terrestrial sources. *Estuaries* 19: 833–47.

Turner, R. E., and N. N. Rabalais. 1991. Changes in Mississippi River water quality this century. Implications for coastal food webs. *BioScience* 41: 140–48.

Turner, R. E., and N. N. Rabalais. 1994. Evidence for coastal eutrophication near the Mississippi River delta. *Nature* 368: 619–21.

van Bennekom, A. J., and F. J. Wetsteijn. 1990. The winter distribution of nutrients in the southern bight of the North Sea (1961–1978) and in the estuaries of the Scheldt and the Rhine/Meuse. *Netherlands Journal of Sea Research* 25: 75–87.

Vannote, R. L., G. W. Minshall, K. W. Cummins, J. R. Sedell, and C. E. Cushing. 1980. The river continuum concept. *Canadian Journal of Fisheries and Aquatic Sciences* 37: 130–37.

Vörösmarty, C. J., W. J. Gutowski, M. Person, T.-C. Chen, and D. Case. 1993. Linked atmosphere-hydrology models at the macro-scale. Pages 3–27 in W. B. Wilkinson (editor), *Macro-scale modeling of the hydrosphere*. Publication no. 214. Wallingford, England: IAHS Press.

PART TWO

Coupling of Physics and Ecology

THE DISTRIBUTION AND FATE of chemicals and organisms in estuaries is strongly influenced by physical processes acting at virtually every spatial and temporal scale. These processes, including the circulation of estuarine waters, the formation of salt and temperature stratifications, and turbulence at the sediment-water interface play a critical role in the biogeochemistry and biology of estuaries.

Physical processes are driven by a variety of factors. For example, the distribution and eventual disposal of nutrients in an estuary is related to the amount that enters the estuary from both land and ocean, the currents within the estuary, the oxygen levels in the deep waters, the chemical exchanges between water and sediments, and the length of time water parcels stay within the estuary. Tides and weather cause dramatic variations in physical processes every time the tide turns and storms move across the estuary. Droughts, seasonal changes in temperature, and fortnightly cycles of neap and spring tides alter physical processes.

It is certainly within our capability to develop detailed physical models of each estuary. Such models have been created for a number of estuaries as a means of investigating the transport of pollutants or the causes of low oxygen conditions. The challenge is to develop an understanding of processes and their controls that is comprehensive enough to develop general models in which the physics that affect the ecology of estuaries can be predicted from a knowledge of bathymetry, river flow, weather, and climate. These models could then be applied to new systems with a tremendous gain in efficiency. The workshop report in this part notes that research on physical features has proceeded on the basis of individual estuaries. A comparative approach that takes advantage of differences among estuaries is needed to understand and model the controls of physical processes.

In chapter 5, Morris shows how new relationships between sea-level anomalies and biological productivity emerge when a long-term database is examined. He links changes in primary productivity in *Spartina* marshes over 11 years to year-to-year mean annual sea-level variations of 2.9 cm and month-to-month changes of 24 cm. The mechanisms at work include changes in hydraulic gradients and drainage of nutrient-rich pore waters, changes in the area of marsh flooded, and resuspension of particles.

In chapter 6, Hofmann reviews several models that couple estuarine circulation with biological dynamics such as larval transport and changes in individual populations. She explains differences between Eulerian and Lagrangian models and shows how biological and physical models can be combined to study the recruitment success of penaeid shrimp larvae.

As these examples illustrate, the interactions of physics and biology in estuaries are complex. How are we to synthesize the data we have collected? One step is to classify estuaries into a limited number of systems based on physical properties. Information from one system can then be applied to others. One of the first classifications to be developed divided estuaries into well-mixed, partially mixed, fjord, and salt-wedge types. In chapter 7, Jay, Geyer, and Montgomery explore estuarine classifications aimed at including ecosystem processes. They first discuss the retention time of dissolved and particulate material in estuaries as an important physical feature that is linked to biological characteristics. Second, they take up a river-estuarine classification, similar to that developed by river geomorphologists, in order to deal with particle suspension and sedimentation.

Participants in the workshop discussed in chapter 8 concluded that research on the coupling of physical and biological processes in estuaries can best be advanced by emphasizing a comparative approach rather than studies of single estuaries. Among the tools needed are a classification with a finite number of types and numerical model simulations of coupled physical-biological processes. Workshop participants also concluded that a regional approach is necessary because coastal currents and plumes cause the output of one estuary to be a part of the input of another estuary.

CHAPTER 5

Effects of Sea-level Anomalies on Estuarine Processes

James T. Morris

Abstract

Several of the effects of sea-level anomalies on estuarine processes including primary production and nutrient cycling are illustrated using data collected at North Inlet, South Carolina. Interannual and seasonal variations in mean sea level affect material exchanges between the intertidal zone and open waters. Mean annual sea level varies from year to year along the southeast coast by an average of ±2.9 cm, which is nearly an order of magnitude greater than the annual change in long-term sea-level rise. Mean monthly sea level varies periodically over the solar annual cycle with an average range of 24 cm, though both the range and months of maximum water level vary from year to year. These anomalous tidal components change the frequency of flooding of intertidal salt marshes, the area of marsh flooded at high tide, and hydraulic gradients. One effect of these anomalies is a change in the salt balance of intertidal sediments. When sea level is anomalously low, salt-marsh primary production decreases because of an increase in pore-water salinity. The net aboveground production of the salt-marsh grass *Spartina alterniflora* varies yearly by a factor of 2 and is negatively correlated with summer pore-water salinity. Sea-level anomalies may also modify exchanges of nutrients and sediments between the vegetated intertidal zone and open water. The vegetated intertidal areas of North Inlet appear to be net sources of nutrients to the tidal creeks and can account for a large fraction of the net export of nutrients from North Inlet to adjacent coastal waters. Changes in hydraulic gradients and extent of area flooded brought on by variation in mean sea level should be ecologically significant and affect the nutrient budgets of estuaries.

Introduction

Ecosystems interact by exchanging materials that are transported in the air and water. Bioreactive materials are transformed during their passage among ecosystem components and between ecosystems. There are important interactions between ecosystems and external forcing functions that determine the flux of these materials as well as ecosystem structure and function. For instance, tidal exchanges of water and solutes are fundamental to the function of estuaries. These exchanges are modified by complex interactions among the hydrology and geomorphology of the watershed and estuary, sea level, and tides.

Tidal variation is a terminology that often connotes the high frequency variations in sea level, though there are regular periodic frequencies that range up to 18.6 years due to changes in the moon's orbit (Marmer 1954). There are aperiodic components of sea-level variation as well that result in anomalous behavior and long-term trends, for instance, eustatic sea-level rise. This chapter focuses largely on the effects of sea-level anomalies on important estuarine processes including primary production and exchanges of materials between intertidal sediments and open water. Exchanges of materials between the intertidal zone and open estuarine water are largely determined by the frequency and duration of flooding, and the size of the intertidal area. These are determined by relative mean sea level, tidal harmonics, and local geomorphology. The consequences of anomalous changes in mean sea level are important to the function of estuaries. The examples discussed in this chapter derive from research that has been done at North Inlet, South Carolina.

Materials and Methods

Nondestructive measurements of aboveground production of *Spartina alterniflora* Loisel. were begun in 1984 at a high marsh site on Goat Island (GIHM) at North Inlet, South Carolina (figure 5-1). This site is situated approximately 60 cm above mean sea level (MSL). A second site 2.8 km distant, but at the same elevation, was added during 1986 at Oyster Landing (OLHM). Both of these sites are in the middle of the marsh platform (the broad expanse of marsh situated near the mean high-tide elevation) and are dominated by a monoculture of the short form of *Spartina* grass.

Net above ground production was measured using a census technique, which was described in detail by Morris and Haskin (1990). At each site we selected six permanent plots. The dimensions of the plots were fixed so that each plot contained about fifteen stems. The sizes of plots were 1 and 2 dm² at GIHM and OLHM, respectively. All stems in these plots were tagged with

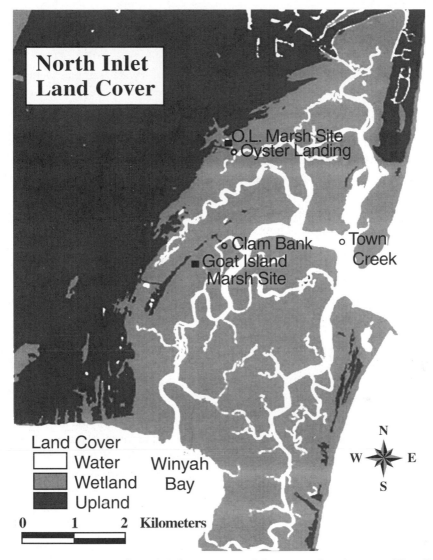

FIGURE 5-1   Map of North Inlet estuary with locations of study sites ■ (Goat Island and Oyster Landing) and LTER water sampling sites ○; Oyster Landing is the site referred to in subsequent figures and text as the landward site, Clam Bank as the mid-estuarine site, and Town Creek as North Inlet mouth.

labeled plastic bands. Each month the lengths of all tagged stems were measured from the ground surface to the tip of the longest leaf. Newly emerged stems within each plot also were tagged monthly. Individual stem weights were estimated from stem heights using regression equations derived earlier from destructive harvests. This technique gives an estimate of monthly stem growth that is not biased by spatial variability, as harvest techniques are prone to do, and it gives a wealth of information about the demographics of the stem population.

Pore-water salinity was measured usually biweekly from 1987 to 1993 by a temperature-compensated refractometer (Reichert). Water samples were collected by syringe from suction samplers that were permanently placed in the sediment at depths of 1, 3, 5, 6, 13, and 19 cm at both sites. Since 1993, pore-water chlorinity has been monitored monthly using diffusion samplers that were placed at several depths down to 1 m. Chlorinity was measured by coulometric titration (Buchler) and calibrated to salinity.

A variety of water quality parameters has been analyzed from data collected for >10 years during the Long-Term Ecological Research (LTER) program at North Inlet. Water samples were collected daily at 10 A.M. EST from three stations within North Inlet. These water samples were analyzed for organic and inorganic nutrients, chlorophyll, and suspended sediments. The documentation for these data can be accessed through the Internet via the address http://www.baruch.sc.edu/baruch/datadoc.html.

Summary statistics were computed using the SAS statistical package. Geometric means were computed for chemical data that typically have lognormal frequency distributions. Arithmetic means were computed for biomass production.

Physical Setting

North Inlet is a tidally dominated salt-marsh estuary with a watershed area of about 75 km² and minimal surface-water input. The terrestrial watershed is largely forested. North Inlet drains a 25 km² estuary of which approximately 82% is intertidal salt marsh, dominated by the grass *Spartina alterniflora* and mud flat, and 18% is open water. About 11.3×10^6 m³ of water floods the estuary (Nummedal and Humphries 1978) on an average tide with a range of 1.37 m (Finley 1975). Tidal water enters the estuary with a mean salinity of 33.97 g/l (the geometric mean salinity at slack high tide near the mouth at Town Creek) and leaves with a salinity of 32.95 g/l. The latter value is an average of the geometric mean salinities of water taken at slack high tide and low tide near the mouth of North Inlet (derived from 10 years of daily water samples at the Town Creek station). Since salt behaves conservatively, the freshwater input can be estimated by solving the mass-balance equation: 33.97 g/liter $\times$ $11.3 \cdot 10^9$ liter/tide = 32.95 g/liter $\times$ $(11.3 \cdot 10^9 + Y)$ liter/tide. The value of Y is $3.5 \cdot 10^8$ liter/tide or about $2.5 \cdot 10^8$ m³/yr. This is an approximate solution, because it ignores the complex spatial patterns of discharge across the mouth of North Inlet (Kjerfve and Proehl 1979).

Although the freshwater input is only about 3% of the tidal volume, it exceeds the freshwater supply to the watershed, which indicates that a source of water from outside the watershed is required to account for the decrease in salinity of water leaving the estuary. Total net precipitation is about 44 cm (precipitation less evapotranspiration) or 33×10^6 m^3 annually within the entire watershed (0.4% of the tidal exchange). This estimate is based in part on a calculation of potential evapotranspiration using the Thornthwaite method (Sellers 1965). Thus, the major freshwater supply must be groundwater and surface water that originates outside the watershed boundaries, such as a flow through the tidal creeks that connect North Inlet to Winyah Bay (see figure 5-1). However, because surface water and groundwater inputs of fresh water are small in comparison to tidal exchanges, the chemistry of the tidal water within the estuary is dominated by the magnitudes of material fluxes through the mouth, by exchanges of materials between the water column and intertidal sediments, and by the trophic dynamics inside the estuary.

The high marsh sites at Oyster Landing and Goat Island that have been the setting for the long-term productivity study differ in sediment physical characteristics. The Oyster Landing salt marsh is situated landward of Goat Island (figure 5-1) and its sediments contain significantly less organic matter (Morris 1988). The difference in organic matter profiles between these two marshes probably reflects differences in their ages. North Inlet salt marshes have slowly migrated inland with the rise in sea level (Gardner and Bohn 1980), and this has resulted in a chronosequence of marshes. The oldest marshes exist in the interior of North Inlet where sedimentation and organic matter accumulation have kept pace with sea-level rise. The Oyster Landing site, which is representative of a very young marsh, exists at the land margin on recently colonized, sandy forest soils. Goat Island marsh is older as indicated by the forest soil horizon, which occurs at a depth of about 50 cm below the present sediment surface, depending on location.

Results

Sea level. Relative mean annual sea level shows a long-term increasing trend as well as considerable interannual variation about the trend (figure 5-2a). The residuals of the time series of mean annual sea level recorded at Charleston Harbor, South Carolina and other sites, such as Galveston, Texas (figure 5-2b) or Boston, Massachusetts ($r^2 = 0.25$, $p = 0.0001$), are highly correlated, which indicates that this variation is at least regional in scale and not simply a result of local weather patterns or runoff effects. A portion of the interannual variation is accounted for by an 18.6-year lunar nodal cycle that is due to changes in the moon's orbit (Marmer 1954). The magnitude of these variations differs regionally as demonstrated by the slope (1) of the regression between Charleston and Galveston residuals (figure 5-2b). The greater

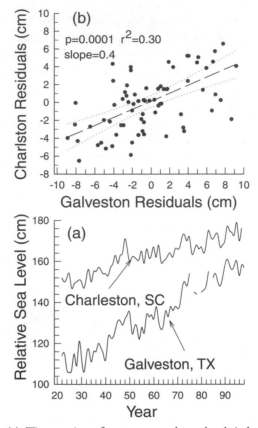

FIGURE 5-2   (a) Time series of mean annual sea level (relative to the station datum) in Charleston Harbor, South Carolina, and Galveston, Texas (NOAA data); (b) correlation between the residuals from linear regressions, shown in (a), of the time series of mean annual sea level in Charleston and Galveston.

variance in the Galveston data is consistent with the fact that disturbances have the greatest effects in shallow water bodies (Marmer 1954). The significance is that the effects of sea-level anomalies will differ regionally, because the variance differs regionally.

Mean monthly sea level from Charleston Harbor (figure 5-3) shows a distinctly bimodal pattern with an annual range of 24.3 cm (69-year mean range). In an average year, maximum sea level occurs in October, while the minimum occurs during January. This solar annual cycle of mean sea level is apparently accounted for largely by steric (specific-volume) changes in the ocean associated with temperature fluctuations in the upper 100 m (Pattullo et al. 1955). However, there is considerable variation in the solar annual cycle in both the timing of the cycle and its amplitude. Mean sea

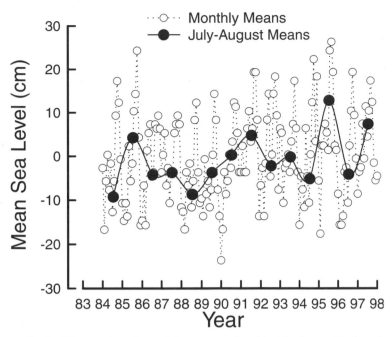

FIGURE 5-3 Time series of monthly mean sea level in Charleston Harbor, South Carolina (······○······) and the mean sea level from July through August (—•—).

level during a critical part of the growing season from July through August has differed by as much as 18 cm from one year to the next and by 22 cm during the 14 years since 1984 (figure 5-3).

Salt marsh responses. The annual rates of aboveground production at Oyster Landing (OLHM) and Goat Island (GIHM) were similar (figure 5-4) and highly correlated ($p = 0.03$), despite differences in the physical characteristics of sediments, marsh age, and the density of *S. alterniflora* stems. Stem density at GIHM is about twice as great as at OLHM (Morris and Haskin 1990). At GIHM annual production (g m^{-2} yr^{-1}, dry weight) was as low as 466±39 and 377±103 (±1 SE) in 1984 and 1988, and as great as 1173±144 and 1318±246 in 1993 and 1997, respectively. Aboveground production at OLHM was also lowest in 1988, 418±45, and greatest in 1995 and 1997 (1132±176 and 1052±181, respectively). Thus, although there are a number of important differences in the vegetation, like stem density and turnover, the temporal growth patterns are quite similar.

Earlier it was reported that there was a high correlation between annual aboveground production on these high marsh sites and mean sea level during summer months (Morris and Haskin 1990). This correlation remains

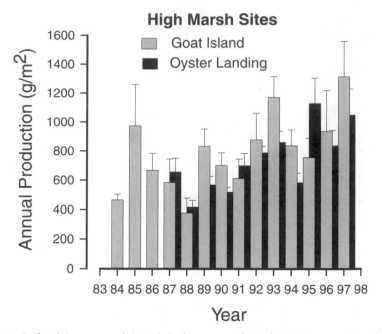

FIGURE 5-4 Mean annual (±1 SE) aboveground production at two sites (Goat Island and Oyster Landing) in salt marshes at North Inlet (see figure 5-1 for locations).

highly significant (figure 5-5). Mean sea level during the months of July and August, which are months of maximum growth, high light, and high temperature, explains about 35% of the temporal variation in annual aboveground production in these high marsh sites. Although the sea level effect as measured by the July–August mean is highly significant ($p < 0.002$), it is likely that a better metric exists, such as some weighted measure of flood duration and rainfall, analogous to the degree-day.

It was also hypothesized earlier (Morris and Haskin 1990) that the proximate cause of the sea-level effect on marsh production was related to changes in pore-water salinity. In fact, the correlation between annual production and mean pore-water salinity during the months of July and August is highly significant ($p < 0.007$) (figure 5-6). The summer months of maximum production are also months of maximum evapotranspiration and pore-water salinity. As flood frequency declines with negative anomalies in sea level, mean pore-water salinity and variation in salinity increase (Morris 1995). The increased variation in salinity that accompanies low sea level arises due to the susceptibility of exposed sediments to dilution of their pore water by precipitation.

The OLHM and GIHM marsh sites are situated near the mean high-tide elevation where small changes in mean sea level have a large effect on flood

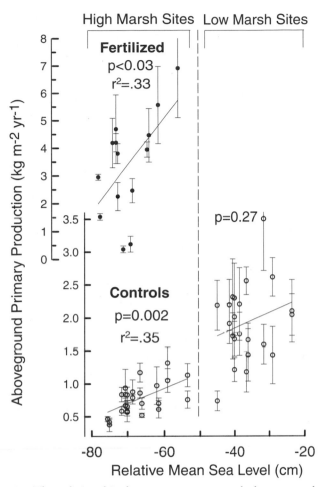

FIGURE 5-5 The relationship between mean annual aboveground production (±1 SE) of *Spartina alterniflora* (1984–1997) and the corresponding mean sea level during the months of July and August. Control high-marsh and low-marsh data include Oyster Landing and Goat Island sites. The high-marsh fertilized site is located at the Goat Island site. Mean sea level is relative to the elevation of the sites. High marsh sites average approximately 60 cm above MSL while low marsh sites are about 30 cm above MSL.

frequency. Using tidal constants from North Inlet, including the solar annual and semiannual cycles, it was estimated that a marsh situated at 60 cm above MSL should be flooded 7% of the time at an average frequency of 1.9 inundations per day during June through August. Dropping MSL by 5 cm reduced the calculated flood duration to 4%. Conversely, raising sea level 5 cm increased flood duration to 9%. More significantly, the simulation predicted that the maximum number

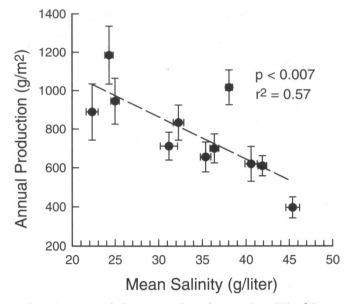

FIGURE 5-6   Mean annual aboveground production (± 1 SE) of *Spartina alterniflora* (1987–1997) from high marsh sites as a function of mean pore-water salinity (± 1 SE, n = 18 per datum) measured during the months of July and August. Pore-water salinity was averaged over samples collected from the upper 50 cm of sediment at Goat Island and Oyster Landing high marsh sites.

of consecutive days without a flood decreased from 12 to 8 days during the June through August period as relative mean sea level was raised from −5 to +5 cm. It is during the neap tidal cycle that pore-water salinity can escalate.

Water-quality parameters.   The spatial and temporal patterns of water-quality data at three stations within North Inlet (figure 5-1) suggest that there are important exchanges of water and nutrients occurring between the intertidal zone and water column in the interior of the estuary, and that the intertidal zone at North Inlet acts as a source of inorganic nutrients to the water column. For three stations along a transect inside North Inlet, there is a clear increase in the geometric monthly mean concentrations of phosphorus and nitrogen compounds in the water column from the mouth of the estuary at Town Creek to the land margin at Oyster Landing (figure 5-7). Even chlorophyll *a* concentrations increase landward, which suggests that there is a relationship between plankton production and enrichment of the water column by nutrients that presumably originate within the intertidal sediments or groundwater. Also note that reduced nitrogen, NH_4^+, dominates NO_3^- as the inorganic nitrogen species, which is consistent with a groundwater or pore-water source of nitrogen

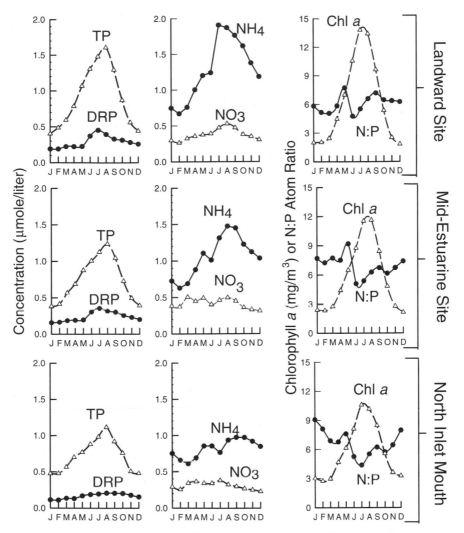

FIGURE 5-7 Geometric monthly mean concentrations of water-quality parameters that were computed from 10 years of samples collected daily from creek water at three sites within North Inlet: Oyster Landing at the land margin (top panel), in the interior of the estuary at Clam Bank (middle panel), and at Town Creek near the mouth of North Inlet (bottom panel). See figure 5-1 for location. Water-quality data include total phosphorus (TP), dissolved reactive phosphorus (DRP), ammonium (NH_4^+), nitrate-nitrite (NO_3^-), chlorophyll a (Chl-a), and the N:P atom ratio computed as $(NH_4^+ + NO_3^-)/DRP$.

to the estuary. By contrast, river-dominated estuaries are typically dominated by NO_3^- (see Magnien et al. 1992).

Evidence for the importance of pore water as a source of nutrients to the tidal creeks comes from a comparison of pore-water and surface-water nutrient concentrations (figure 5-8). Pore-water concentrations of DRP are about 40 times greater than surface-water DRP concentrations within North Inlet, while pore-water concentrations of NH_4^+ are more than 40 times greater than surface-water values. The seasonality of nutrients in pore water and surface water is also similar (figure 5-8), suggesting that the cycles of pore-water and surface-water nutrients could be related. The cycles are out of phase, with pore-water values reaching their maximum amplitude during August, one month later than surface-water values.

An analysis of surface-water chlorophyll and nutrient concentrations as a function of tide level or stage (figure 5-9) also supports a connection between surface-water chemistry and a nutrient supply at the land margin. When the 10-year daily water sample data were stratified by stage of tide and station, it was clear that nutrient-rich estuarine water is diluted by incoming floodwater from the coastal boundary layer (figure 5-9), which indicates that there is a nutrient source near the land margin. Taking NH_4^+ as an example,

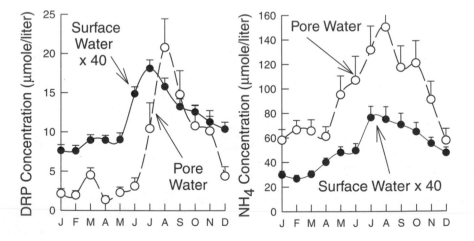

FIGURE 5-8 Geometric monthly mean concentrations of dissolved reactive phosphorus (left) and NH_4^+ (right) in marsh pore water (—○—) and tidal creek water (—●—). Values derived for tidal creeks are geometric means (and upper 95% CI) of data collected at three sampling stations (see figure 5-1) daily for 10 years. Creek-water values have been multiplied by 40. Pore-water values are geometric means ($n \approx 225$ samples per datum; + upper 95% CI) of samples collected monthly at four depths (25–100 cm in triplicate at four sites (low marsh and high marsh at Goat Island and Oyster Landing) from December 1993 through March 1998. Pore water was collected by diffusion samplers equilibrated for 1 month.

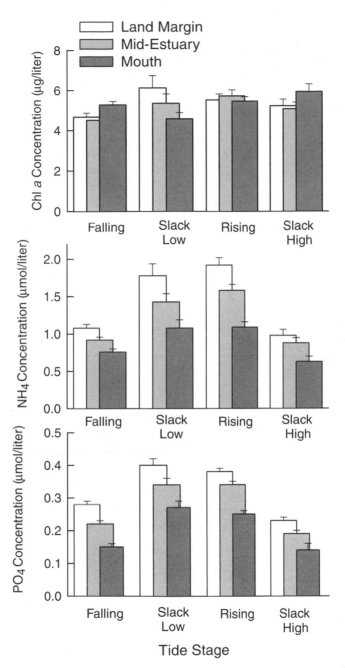

FIGURE 5-9 Geometric mean (and upper 95% CI) concentrations of chlorophyll *a* (top), NH_4^+ (middle), and dissolved reactive phosphorus (bottom) in tidal creek water taken from three stations (same as in figure 5-7) and stratified by tidal stage. Slack tide was defined as those portions of the tidal cycle having derivative values of less than 20 cm/hr. Data were averaged over all time (10 years).

floodwater enters the estuary with an average concentration of 0.63 µmol/l, which is the concentration at slack high tide at Town Creek near the mouth. This water mixes with water at Oyster Landing, having an average NH_4^+ concentration at slack low tide of 1.78 µmol/l, to produce water (at Oyster Landing) with a concentration of about 0.98 µmol/l during the high tide. Water leaves the estuary having an NH_4^+ concentration of about 0.9 µmol/l, which is the average of the slack high- and low-tide NH_4^+ concentrations at Town Creek near the mouth (figure 5-9). The intermediate nutrient concentrations observed at the mid-estuarine station at Clam Bank are suggestive of conservative mixing, though obviously neither NH_4^+ nor DRP is conservative. At every tidal stage, concentrations of both NH_4^+ and DRP at Clam Bank are intermediate as though from a mixture of water from the end members: nutrient-poor water at the mouth (Town Creek station) and nutrient-rich water at the land margin (Oyster Landing).

The net exports of NH_4^+ and DRP from the North Inlet estuary were calculated, making the simplifying assumptions that water enters the estuary with a composition like that at the mouth at high tide, and that water leaves the estuary with nutrient concentrations intermediate between those at the mouth at slack high and low tides. The geometric annual mean concentrations of DRP in incoming (slack high-tide) and outgoing (slack low-tide) water at the mouth were 0.14 and 0.27 µmol/l, respectively, while the corresponding concentrations of NH_4^+ were 0.63 and 1.08 µmol/l. The approximate exports can be calculated by setting up the following mass-balance equation (internal source + import = export) and solving for the internal sources of NH_4^+ or DRP, denoted by Y:

DRP: $Y_P + (0.14 \text{ µmol/l} \times 11.3 \cdot 10^9 \text{ liter/tide}) = 0.21 \text{ µmol/l} \times$
                   $(11.3 \cdot 10^9 + 3.5 \cdot 10^8 \text{ liter/tide})$

NH_4^+: $Y_N + (0.63 \text{ µmol/l} \times 11.3 \cdot 10^9 \text{ liter/tide}) = 0.86 \text{ µmol/l} \times$
                   $(11.3 \cdot 10^9 + 3.5 \cdot 10^8 \text{ liter/tide})$

The internal source term Y is the net export, and solutions for Y indicate that there are net exports of $2.9 \cdot 10^3$ mol/tide of NH_4^+ and $8.6 \cdot 10^2$ mol/tide of DRP. Earlier estimates from the outwelling study at North Inlet placed the annual export of NH_4^+ and DRP at $15 \cdot 10^3$ mol/tide and $1.8 \cdot 10^3$ mol/tide, respectively (Dame et al. 1986).

The pattern of chlorophyll data, stratified by tide stage, was different from that of nutrients (figure 5-9). Chlorophyll concentrations from the mouth of North Inlet were highest at slack high tide and lowest at slack low tide. Chlorophyll concentrations at the mouth of North Inlet averaged 5.9 µg/l at slack high tide and 4.6 µg/l at slack low, which gives a calculated net chlorophyll *a* consumption by the estuary of 5.5 kg/tide. The chlorophyll pattern at the land margin (Oyster Landing) is opposite that at the mouth. The slack low-tide concentration of chlorophyll at Oyster Landing was 6.1 µg/l (figure 5-9). This is diluted (and grazed) to 5.2 µg/l at slack high tide. These data suggest that there is significant phytoplankton production inside the estuary, but apparently this production is entirely grazed or sedimented *in situ* such that net export is zero. This is at odds with an earlier outwelling study at North Inlet in which a net annual export of phytoplankton was found, though there was a net export of phytoplankton during the summer and fall (Dame et al. 1986).

Discussion

Sea-level anomalies have a significant effect on salt-marsh productivity. For salt-marsh macrophytes, the effect is positive; that is, production increases when sea level is unusually high. The ±50% interannual variation in production is ecologically significant because of the importance of primary production to trophic dynamics, nutrient cycling, and sediment accretion. The correlation between sea-level anomalies and production is positive because of the effect of sea level on flood frequency and consequent changes in sediment chemical parameters (for example, pore-water salinity). At North Inlet, the marsh platform is situated near the mean high-tide elevation where, during neap tides, the marsh may not flood for several days. For a marsh situated near the mean high-tide elevation, the maximum number of consecutive days without a flood during the summer period of maximum growth varies from 8 to 12 corresponding to ±5 cm variations in mean monthly sea level. Flooding has an optimum frequency and duration with respect to salt-marsh production, and at elevations near mean high tide the flood frequency is suboptimal.

The magnitudes of interannual and monthly changes in mean sea level are far greater than the long-term trend (figures 5-2a and 5-3). The slope of relative mean annual sea level in Charleston Harbor is 0.36 cm/yr (1922–1997), but mean annual sea level differs by several centimeters from year to year. Variation in mean monthly sea level is even greater (figure 5-3). This raises an interesting issue about the time scales of responses of ecological processes in relation to the long-term trend and short-term variations in sea level. It appears from the response of primary production that the vegetation at North Inlet is responding positively to monthly and interannual variations in sea level superimposed on a long-term trend. If sediment accretion keeps pace with long-term sea-level rise, then it is likely that the long-term trend will proceed without affecting the vegetation. If accretion fails to keep pace, then it is plausible that production at the higher elevations of the marsh will actually increase until a threshold is reached; beyond that the duration or depth of flooding would exceed the optimum for the vegetation.

The sensitivity of marsh production to sea-level anomalies should be more pronounced at lower latitudes, because of the importance of evapotranspiration as a determinant of pore-water salinity (Morris 1995). Growth experiments with salt-marsh grasses (*Spartina* spp.; Phleger 1971; Haines and Dunn 1976; Bradley and Morris 1991) demonstrate a negative relationship between production and salinity. These experiments are also consistent with gas exchange measurements (Pearcy and Ustin 1984; Pezeshki 1991; Hwang and Morris 1994) that demonstrate a negative relationship between net photosynthesis and salinity. The importance of episodic changes in salinity to the success of salt-marsh vegetation, including *Spartina foliosa*, has also been described (Zedler and Beare 1986; Zedler et al. 1986). The North Inlet data confirm the importance of salinity to marsh production (figure 5-6) and show that the sea-level effect is most likely mediated by resultant changes in soil salinity.

Estuarine productivity and nutrient dynamics are linked to short- and long-term changes in sea level, because of the effects of changes in the hydraulic gradient on movements of groundwater (Nuttle and Portnoy 1992). For example, the vertical components of pore-water movement change with variations in the hydraulic gradient associated with the spring-neap tidal cycle (Carr and Blackley 1986). The hydrological cycle in estuaries includes a circulation of water between the intertidal sediments and surface waters. The sediments are recharged at high tide, and water drains back into the creeks at low tide. When mean sea level drops, the increased hydraulic gradient in the marshes should increase the drainage of nutrient-rich pore water into the creeks, provided that the flood tide continues to recharge the marsh sediments at high tide.

Variations in sea level may affect the nutrient dynamics of estuaries in other ways as well, because of the relationship between sea level and pore-water salinity. For instance, the NH_4^+ exchange capacity of sediments decreases as salinity increases (Simon and Kennedy 1987; Gardner et al. 1991; Seitzinger et al. 1991). At typical salt-marsh salinities the NH_4^+ exchange capacity of sediments is quite low, but the effect may be significant at the lower end of estuarine salinity gradients. In addition, spring-neap tidal cycles result in alternating periods of net reduction and oxidation in surface sediments of the high intertidal zone (Armstrong et al. 1985) that affect microbial processes, solubility, and very likely nitrification-denitrification. One additional interaction that is definitely important in salt- and brackish-water marshes concerns the efficiency of NH_4^+ uptake by vegetation and its relation to salinity. In *Spartina alterniflora* it was shown that NH_4^+ uptake decreases with increasing salinity due to competitive uptake inhibition from seawater cations (Bradley and Morris 1991). For these reasons marsh sediments should release NH_4^+ as salinity increases.

The potential significance of pore-water drainage to the nutrient budgets of North Inlet can be assessed by making a rough calculation of the rate of nutrient delivery from direct measurements of water advected out of the bottoms of tidal creeks and its nutrient content. High-salinity water, consistent with the salinity of pore water, is advected from the bottoms of tidal creeks in North Inlet at rates of $4-14$ l m^{-2} (tidal cycle)$^{-1}$ (Whiting and Childers 1989). A supply of pore water of this magnitude, given that the geometric mean concentrations of NH$_4^+$ and DRP in pore water are about 40 times greater than those found in tidal-creek water (figure 5-8), can be significant to the nutrient budget of the estuary. For example, the N:P atom ratio of creek water in North Inlet drops considerably during the plankton bloom (figure 5-7), indicating that nitrogen regeneration may sustain the bloom. Pore water from the intertidal marshes is enriched in nitrogen relative to phosphorus (figure 5-8), which probably helps to sustain the bloom. Based only on these few measurements (Whiting and Childers 1989), the volume of pore water advected into tidal creeks is equivalent to $18-63 \times 10^3$ m^3/tide, which is equivalent to a drainage rate from the salt marshes of about $0.9-3.1$ l m^{-2} day^{-1}. With mean pore-water concentrations of NH$_4^+$ and DRP of 88 and 5 µmol/l, the calculated drainage rate of these nutrients across the whole estuary should be $1.6 \cdot 10^3$ to $5.5 \cdot 10^3$ mol/tide of NH$_4^+$ and 90 to 315 mol/tide of DRP. This can account for the calculated $2.9 \cdot 10^3$ mol/tide export of NH$_4^+$ from the mouth of North Inlet, though drainage of DRP is less than the calculated DRP export of 860 mol/tide. Nevertheless, it is clear that the benthic flux of nutrients represents an important component of the overall nutrient balance of North Inlet. Any change in benthic flux, such as that due to variation in sea-level and hydraulic gradient, is likely to be significant to the ecology of the estuary. Presently, no direct measurements have been made that address the seasonality of this advective benthic flux or its relationship to productivity, sea level, or tidal dynamics.

Calculations of average annual export rates hide the fact that there is considerable seasonal variation. Seasonal cycles arise because of imbalances in the processes of mineralization or decomposition and consumption. The fact that concentrations of the NH$_4^+$ and DRP peak during the summer (figure 5-7) indicates that the regeneration of nutrients, which is sensitive to temperature, dominates consumptive processes (meaning assimilation of nutrients by organisms and adsorption by sediments) during the first half of the year, while consumptive processes dominate during the second half of the year. The cycles of surface-water and pore-water nutrients are not in phase, which indicates only that the sink terms acting on pore-water nutrients become dominant over the source terms about 1 month in advance of the sink terms acting on surface-water nutrients.

Other possible effects of sea-level anomalies on estuaries involve feedbacks between the function of benthic fauna and processes such as filtration, resuspension, and nutrient cycling. For instance, benthic filter feeders are important transformers of nutrients in North Inlet. Benthic filter feeders such as oysters and mussels are capable of filtering a volume of water at a rate that is equivalent in magnitude to the tidal exchange rate (Dame et al. 1980; Jordan and Valiela 1982). Not surprisingly, calculations of the internal nutrient cycles are particularly sensitive to assumptions about the density of these organisms (Childers et al. 1993). In North Inlet, oysters are important net consumers of N and P. However, they release large quantities of NH_4^+ annually, primarily during summer (Dame et al. 1989). The net effect of seasonal or interannual changes in flood duration on the function of these organisms is not known, but variations in sea level could affect recruitment, mortality, integrated filtration rates, and nutrient dynamics.

Finally, the nature of the interactions discussed here may differ qualitatively among estuaries. For example, nutrient concentrations decline in the Parker River, Massachusetts, during spring tides and rise during neap tides (Vörösmarty and Loder 1994). The decline that is coincident with the spring tides was attributed to increased flooding and nutrient uptake by the marsh surface. This contrasts with the function of wetlands at North Inlet that appear to serve as net sources of inorganic nutrients to the water column. Reported concentrations of inorganic N and P in the Parker River were an order of magnitude greater than in the creeks of North Inlet. It is possible that the net effect on water quality of an intertidal wetland can switch from that of source to sink, depending on the nutrient load in surface water and tidal dynamics. Insights could be gained from comparative studies that take advantage of the natural variation that exists among estuaries in tidal forcing and other determinants of estuarine structure and function.

Conclusions

Exchanges of materials within and among ecosystems are critical to ecosystem structure and function. These exchanges are often mediated by physical processes. For example, in estuaries, the exchanges of materials between intertidal sediments and open water (for example, salts and nutrients) are influenced by flood frequency and duration. Estuarine responses to sea-level changes must be frequency dependent. There are significant variations in relative mean sea level on monthly and annual time scales that have important consequences for estuarine processes. Responses to long-term sea-level rise will likely differ from responses to sea-level fluctuations that occur on seasonal or interannual time scales. An analysis of 70 years of sea-level data from Charleston Harbor shows that mean annual sea level has varied from year to year by an average of ±2.9 cm. This is nearly an order of magnitude greater than long-term sea-level rise of 0.36 cm/yr (figure 5-2a). Seasonal variations in mean sea level are even greater (figure 5-3). Mean monthly sea level varies periodically over the solar annual cycle with an average range of 24 cm, and there are significant interannual variations in both the range and months of maximum water level.

Fourteen years of measurements from two salt marshes at North Inlet, South Carolina, demonstrated that there is a positive correlation between net aboveground production of the salt-marsh grass *Spartina alterniflora* and mean summer sea level (figure 5-5). At a single high-marsh site in North Inlet, net aboveground production has varied from as little as 380 g m^{-2} yr^{-1} to as much as 1,320 g m^{-2} yr^{-1}. The proximal cause of the sea-level effect is a change in pore-water salinity. Salt-marsh primary production responds negatively to pore-water salinity (figure 5-6), which is regulated by flood frequency and evapotranspiration. At intertidal elevations near mean high tide, a decrease in mean sea level of 10 cm can approximately double the number of consecutive days without a flood event during summer neap tides.

North Inlet is a tidally dominated estuary with minimal inputs of fresh water. Thus, inputs of terrigenous materials in surface runoff are minimal. For this reason, the behavior of nutrients and chlorophyll in the open water of the estuary appear to be influenced by exchanges of materials between the vegetated intertidal zone and tidal creeks. Concentrations of nutrients in the open water increase landward inside the estuary (figures 5-7 and 5-9). This strongly suggests that the intertidal marshes at North Inlet are sources of nutrients to the tidal creeks, and that nutrient-rich estuarine water is diluted by exchanges with coastal waters through the mouth of North Inlet. Inorganic nitrogen in open waters of the estuary is dominated by ammonium, which is a further indication of the importance of a benthic nutrient supply. This exchange of nutrients between the salt marshes and open water should be sensitive to changes in hydraulic gradient and flood frequency, which change with relative mean sea level.

Acknowledgments

This research was supported by the NSF (LTER and LTREB programs). The Baruch Institute GIS lab provided estimates of landscape areas. I would like to thank Joy Zedler and P. V. Sundareshwar for their helpful reviews.

References

Armstrong, W., E. J. Wright, S. Lythe, and T. J. Gaynard. 1985. Plant zonation and the effects of the spring-neap tidal cycle on soil aeration in a Humber salt marsh. *Journal of Ecology* 73: 323–39.

Bradley, P. M., and J. T. Morris. 1991. The influence of salinity on the kinetics of NH$_4^+$ uptake in *Spartina alterniflora*. *Oecologia* 85: 375–80.

Carr, A. P., and W. L. Blackley. 1986. The effects and implication of tides and rainfall on the circulation of water within salt marsh sediments. *Limnology and Oceanography* 31: 266–76.

Childers, D. L., H. N. McKellar, R. F. Dame, F. H. Sklar, and E. R. Blood. 1993. A dynamic nutrient budget of subsystem interactions in a salt marsh estuary. *Estuarine Coastal and Shelf Science* 36: 105–31.

Dame, R., T. Chrzanowski, K. Bildstein, B. Kjerfve, H. McKellar, D. Nelson, J. Spurrier, S. Stancyk, H. Stevenson, and D. R. Zingmark. 1986. The outwelling hypothesis and North Inlet, South Carolina. *Marine Ecology Progress Series* 33: 217–29.

Dame, R. F., J. D. Spurrier, and T. G. Wolaver. 1989. Carbon, nitrogen and phosphorus processing by an oyster reef. *Marine Ecology Progress Series* 54: 249–56.

Dame, R., D. R. Zingmark, H. Stevenson, and D. Nelson. 1980. Filter feeding coupling between the estuarine water column and benthic subsystems. Pages 521–26 in V. Kennedy (editor), *Estuarine perspectives.* New York: Academic Press.

Finley, R. J. 1975. Hydrodynamics and tidal deltas of North Inlet, South Carolina. Pages 277–91 in E. Cronin (editor), *Estuarine research. Vol. 2.* New York: Academic Press.

Gardner, L. R., and D. M. Bohn. 1980. Geomorphic and hydraulic evolution of tidal creeks on a subsiding beach ridge plain, North Inlet, S.C. *Marine Geology* 34: M91–M97.

Gardner, W. S., S. P. Seitzinger, and J. M. Malczyk. 1991. The effects of sea salts on the forms of nitrogen released from estuarine and freshwater sediments: Does ion pairing affect ammonium flux? *Estuaries* 14: 157–66.

Haines, B. L., and E. L. Dunn. 1976. Growth and resource allocation responses of *Spartina alterniflora* Loisel to three levels of NH_4-N, Fe and NaCl in solution culture. *Botanical Gazette* 137: 224–30.

Hwang, Y. H., and J. T. Morris. 1994. Whole plant gas exchange responses of *Spartina alterniflora* (Poaceae) to a range of constant and transient salinities. *American Journal of Botany* 81: 659–665.

Jordan, T. E., and I. Valiela. 1982. A nitrogen budget of the ribbed mussel, *Geukensia demissa*, and its significance in nitrogen flow in a New England salt marsh. *Limnology and Oceanography* 27: 75–90.

Kjerfve, B., and J. A. Proehl. 1979. Velocity variability in a cross-section of a well-mixed estuary. *Journal of Marine Research* 37: 409–18.

Magnien, R. E., M. Summers, and K. G. Sellner. 1992. External nutrient sources, internal nutrient pools, and phytoplankton production in Chesapeake Bay. *Estuaries* 15: 497–516.

Marmer, H. A. 1954. Tides and sea level in the Gulf of Mexico. *Fishery Bulletin* 55: 101–18.

Morris, J. T. 1988. Pathways and controls of the carbon cycle in salt marshes. Pages 497–510 in D. D. Hook (editor), *The ecology and management of wetlands*. Vol. 1: *Ecology of wetlands*. London: Croom Helm.

———. 1995. The mass balance of salt and water in intertidal sediments: Results from North Inlet, South Carolina. *Estuaries* 18: 556–67.

Morris, J. T., and B. Haskin. 1990. A 5-yr record of aerial primary production and stand characteristics of *Spartina alterniflora*. *Ecology* 71: 2209–17.

Nummedal, D., and Humphries, S. M. 1978. Hydraulics and dynamics of North Inlet, South Carolina, 1975–76. GITI Report 16. Fort Belvoir, VA: Department of the Army Corps of Engineers, Coastal Engineering Research Center; and Vicksburg, MS: U.S. Army Engineer Waterways Experiment Station.

Nuttle, W. K., and J. W. Portnoy. 1992. Effect of rising sea level on runoff and groundwater discharge to coastal ecosystems. *Estuarine Coastal and Shelf Science* 34: 203–12.

Pattullo, J. W. Munk, R. Revelle, and E. Strong. 1955. The seasonal oscillation in sea level. *Journal of Marine Research* 14: 88–156.

Pearcy, R. W., and S. L. Ustin. 1984. Effects of salinity on growth and photosynthesis of three California tidal marsh species. *Oecologia* 62: 68–73.

Pezeshki, S. R. 1991. Population differentiation in *Spartina patens*: Gas-exchange responses to salinity. *Marine Ecology Progress Series* 72: 125–30.

Phleger, C. F. 1971. Effect of salinity on growth of a salt marsh grass. *Ecology* 52: 908–11.

Sellers, W. D. 1965. *Physical climatology*. Chicago, IL: University of Chicago Press.

Seitzinger, S. P., W. S. Gardner, and A. K. Spratt. 1991. The effect of salinity on ammonium sorption in aquatic sediments: Implications for benthic nutrient recycling. *Estuaries* 14: 167–74.

Simon, N. S., and M. M. Kennedy. 1987. The distribution of nitrogen species and adsorption of ammonium in sediments from the tidal Potomac River and estuary. *Estuarine Coastal and Shelf Science* 25: 11–26.

Vörösmarty, C. J., and T. C. Loder III. 1994. Spring-neap tidal contrasts and nutrient dynamics in a marsh-dominated estuary. *Estuaries* 17: 537–51.

Whiting, G. J., and D. L. Childers. 1989. Subtidal advective water flux as a potentially important nutrient input to southeastern U.S.A. saltmarsh estuaries. *Estuarine Coastal and Shelf Science* 28: 417–31.

Zedler, J. B., and P. A. Beare. 1986. Temporal variability of salt marsh vegetation: the role of low-salinity gaps and environmental stress. Pages 295–306 in D. Wolfe (editor), *Estuarine variability*. New York: Academic Press.

Zedler, J. B., J. Covin, C. Norby, P. Williams, and J. Boland. 1986. Catastrophic events reveal the dynamic nature of salt marsh vegetation. *Estuaries* 9: 75–80.

CHAPTER 6

Modeling for Estuarine Synthesis

Eileen E. Hofmann

Abstract

Coupled circulation-biological models provide a framework for integrating and synthesizing information about estuarine systems and for investigating hypotheses about controlling processes in these systems. Therefore, it is appropriate to briefly review some existing circulation-biological models that have been developed for estuarine systems. A selection of models that consider larval transport and dispersion, whole-ecosystem dynamics, and population dynamics of specific species are discussed in terms of their basic structure. Emphasis is placed on potential model shortcomings as a basis for recommending improvements to future models. These discussions highlight the need for circulation-biological models that are integrated with a range of observation systems and visualization techniques. It is through such systems that estuarine synthesis will be accomplished.

Introduction

The development of mathematical models for synthesis and integration of physical and biological measurements from estuarine systems is a challenge. Scale analysis of the equations governing circulation dynamics in estuarine systems usually results in nondimensional ratios that are of order one, which implies that all processes are important. Thus, the ability to eliminate certain ranges of motions or dynamics in estuarine circulation models is limited. Biological processes in estuarine systems are complex and include all of the dynamics that are known to be important in other marine systems. Moreover, the shallow nature of most estuarine systems results in tight linkages between benthic and pelagic systems. Hence, approaches for a priori elimination of

129

biological dynamics are not well developed. As a result, physical-biological models for estuarine systems can be as complex as or more complex than those developed for other marine systems.

Estuarine systems are strongly coupled to offshore coastal systems. Consequently, processes that occur in the coastal ocean can potentially affect circulation and biological production of these systems. Numerous examples exist of remote forcing of estuarine circulation and for some systems this may be the major exchange process (Weisberg 1976; Elliott 1978; Garvine 1985). Also, estuarine systems can play a role in regulating the circulation of the adjacent coastal waters. It is usually not feasible to include a coastal ocean circulation model as part of an estuarine model. Thus, offshore processes tend to be included via boundary conditions that prescribe the way in which the estuarine and coastal ocean environment interact. Consequently, specification of these boundary conditions and the databases on which they are based are critical in determining the realism of the estuarine model. Furthermore, many of the biological assemblages found in estuarine systems have characteristics of offshore coastal waters and vice versa. Hence, biological production of estuarine and coastal waters is often linked and this linkage occurs through the circulation patterns (Malone et al. 1980).

At the other boundary, estuarine systems are strongly coupled to the land system. Freshwater input via rivers and land runoff and nutrient inputs from land sources are important in regulating circulation and biological production of estuarine systems. As with the offshore environment, it is not usually feasible or even desirable to include a full watershed model to obtain these boundary effects. Thus, land-derived inputs are included through boundary conditions, and how these are specified is quite important.

For these reasons and many more, models that treat coupling between the physical and biological systems in estuaries are few, in spite of considerable interest in understanding how these systems operate. The few models that do exist are designed to investigate exchanges between the outer estuary and the offshore coastal system. The equally important exchanges that occur in the riverine-dominated portion of the estuary remain to be explored with coupled physical-biological models and represent an area worthy of future study. Advances in physical and biological models and expanding computer resources make the development of interdisciplinary models for estuarine systems a more attainable goal. Also, data sets that are adequate for developing interdisciplinary models for estuarine systems are now becoming available through programs such as the LMER initiative.

This chapter provides examples of the types of models that have been used to understand physical and biological interactions in estuarine systems. These examples illustrate the approaches that have been taken to study estuarine systems using mathematical models and as such provide guidance on how future models can be improved and in what areas resources should be allocated to further model development. However, before presenting these examples, some general comments on biological and physical models are first given. These issues are relevant to modeling physical and biological interactions in any marine environment. The final section of this chapter provides some suggestion for future work that may help to advance the state of modeling for estuarine synthesis.

General Comments on Biological/Ecological Models

Most biological models use a bulk approach to represent trophic levels, which removes the effects of size, stage, and age known to be important in structuring trophic levels and in the transfer of energy between trophic levels. Some models use functional groups (such as diatom and dinoflagellate-type organisms), which does preserve some of the characteristics of the different organisms but still represents a bulk approach. The development of models that include all of the dominant or important species may be desirable, but this is not feasible without a firm database that describes the processes and interactions that occur at the species level. Models that focus on a single species are limited and may not be representative of actual conditions. Hence, most biological models represent compromises between what is known and what can be done. However, the key issue is that the models be formulated with a level of complexity that is suitable to address the scientific question being asked.

Parameterizations of biological processes are usually based on empirical relationships that express correlations between variables that can be measured. The real ecological or physiological processes underlying the observed correlation are not explicit in these relationships. A persistent problem with biological models is that the available measurements are often not made in a way that is readily usable in models. Measurements of standing stock or relative abundance estimates are useful for model verification, but they are not sufficient for formulation of model processes. Measurements of the rate at which properties change and how biological processes vary with environmental conditions are needed.

Once developed, implementation of biological models is limited by the data sets that are available for model calibration and verification. Even multidisciplinary field programs do not necessarily provide data sets that are adequate for comparison with biological models. These programs tend to be focused on specific phenomena, places, or times and as a result do not usually include the breadth of measurements that are needed for biological models.

Comments on Circulation Models

Most of the circulation models that are used with biological models are numerical in nature. Analytical circulation models do not usually include sufficient representation of processes that underlie observed biological distributions. However, even the more complex numerical circulation models have limited dynamics because they are solved on a prespecified grid, which sets a limit on the scales of motion that can be resolved. Moreover, most circulation models are not designed from the start to be used with biological models and therefore usually lack the space and time resolution that is needed for modeling biological processes.

Types of Models

Eulerian Models

An Eulerian model provides a three-dimensional estimate of the distribution of properties at specific locations over time. For nonconservative substances, such as biological properties, the Eulerian equations are of the form:

$$\frac{\partial B}{\partial t} + \nabla \cdot \left(\vec{V}B\right) - \nabla \cdot \left(K \cdot \nabla B\right) = biological\ sources/sinks \qquad (1)$$

The first term represents the time (t) rate of change of the biological property, B. The next two terms represent the three-dimensional spatial distribution of B that results from the effects of advection and diffusion, respectively. The biological processes that result in the production or loss of B are given by the *source/sink* terms. The vector operator denotes changes in the x, y, and z directions. Advective velocities and diffusive rates in the three dimensions are denoted by $\vec{V}$ and K, respectively.

Implementation of equation (1) to model estuarine ecosystems typically has not been done. Rather, a truncated form of the equation is used, which usually consists of the elimination of one or more spatial dimensions, for instance, a vertical plane approach, or the assumption of steady state. It is only recently that three-dimensional, time-dependent models have been constructed for marine ecosystems. These models are solved on a predetermined grid using a variety of numerical methods.

A finite difference or finite element approximation to equation (1) assumes that processes are homogeneous over smaller space and time scales than those used for the discrete space intervals and the integration time interval. These smaller-scale processes are included through parameterizations, such as that used for diffusion. Also, Eulerian models do not retain information about the characteristics or history of individuals. Rather, biological properties are usually expressed in terms of concentration, for example, mg N l⁻¹.

The simulated distributions obtained with an Eulerian model can be compared directly with point measurements obtained from moored instruments, which provides verification of the time evolution of the biological distributions at a specific location. The horizontal and vertical distributions obtained from Eulerian models can be compared with distributions constructed from ship surveys or satellite measurements.

Lagrangian Models

Lagrangian models provide an approach that allows trajectories of particles to be tracked over time. This allows determination of likely transport pathways and of residence times. Lagrangian models are of the form

$$\frac{d\vec{X}}{dt} = \vec{U}(x,y,z,t) + \vec{U}_b(x,y,z,t) \tag{2}$$

where the time rate of change in the location $(\vec{X})$ of a particle is determined by the physical advective field $[\vec{U}(x,y,z,t)]$ and biological advection $[\vec{U}_b (x,y,z,t)]$. Biological advection consists of directed horizontal swimming, vertical migration, and sinking.

A Lagrangian modeling approach preserves information on the time history of individuals as they move along a trajectory. If equation (2) is modified such that an additional dimension such as animal age is included, then it is possible to track the development of an individual in space and time. Furthermore, this individual-based approach may be formulated to allow for interactions among individual organisms (DeAngelis and Gross 1992). This class of models is only now being applied to marine systems (Batchelder and Miller 1989; Madenjian and Carpenter 1991; Batchelder and Williams 1995) and to date has not been used in estuarine systems.

Combined Models

A combined model is one that uses an Eulerian approach to establish the distribution of some property through which a particle then moves. In most cases, an Eulerian model is used to provide circulation fields through which biological particles move. Eulerian models can also be used to provide distributions of passive (no behavior) biological quantities such as phytoplankton. Then nonpassive biological quantities, such as zooplankton or fish predators that have behaviors, can be tracked through the simulated phytoplankton (prey) distribution. This latter approach is used when modeling nonplanktonic populations. In either case, techniques that allow for interpolation of information from the Eulerian grid to the location of the Lagrangian particle are needed.

Examples of Models for Estuarine Systems

Eulerian Models

Narragansett Bay Model. The first comprehensive physical-biological model developed for an estuarine system was that for Narragansett Bay (Kremer and Nixon 1978). The biological portion of this model consisted of equations that described time-dependent biological interactions among nitrate, phosphate, silicate, phytoplankton (summer and winter species), herbivorous zooplankton, carnivorous zooplankton, and benthic infauna. The benthic-pelagic coupling that occurs in most estuarine systems was explicitly included in the biological model. The benthos affected the plankton portions of the model through reduction of phytoplankton standing stocks by filtering from benthic hard clam communities and through nutrient regeneration.

The physical environment included in the Narragansett Bay model consisted of seasonal temperature increases and decreases, seasonal changes in solar radiation, freshwater input, and tidal circulation. Wind and gravitational circulation forcing were not included. The first three effects were included through time-dependent empirical formulations that were constructed from observations. The tidal circulation input to the biological model consisted of spatial averages obtained from simulations done with a barotropic tidal model developed for Narragansett Bay. The velocity averages were assumed to be representative of the tidal flow in eight subregions of the bay and were used to specify exchanges between the regions (figure 6-1).

The averaged tidal velocities were used instead of explicitly solving for the circulation distribution on a finer grid because of limitations in available computer resources. However, the coarse scale over which the tidal velocities were averaged resulted in removal of all but the lowest frequency of motions and also gave exchanges between the bay subregions that were too fast as compared to observations.

The simulated biological distributions obtained from the model were verified by comparison with time-series observations made at several locations in Narragansett Bay (figure 6-2). For some state variables in certain areas, these comparisons were good. However, at other locations there was little correspondence between observed and simulated distributions. The mismatch in the observed and simulated time series was attributed in part to the representation of the circulation. This led to the conclusion that better representation was needed for the circulation. Other factors leading

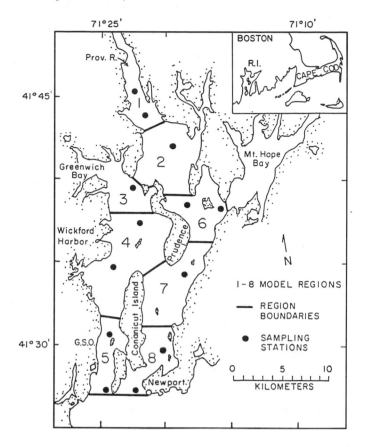

FIGURE 6-1   The eight spatial subsystems that were used to calculate averages of the simulated tidal flow fields for Narragansett Bay, Rhode Island. The black dots indicate stations where an annual time series of nutrients, phytoplankton, and zooplankton were collected for comparison with the simulated distributions. Figure from Kremer and Nixon (1978), used with permission.

to the mismatches were uncertainty in biological parameter estimation, the use of constant rather than seasonally varying formulations for some model parameters, and potentially missing dynamics.

Galveston Bay, Texas, Model.   A series of oyster models was developed to estimate the impact of deepening and widening the Houston Ship Channel on the oyster populations in Galveston Bay, Texas. A primary effect of the Houston Ship Channel is to provide a conduit for movement of higher-salinity Gulf of Mexico water into Galveston Bay. Since some factors that affect oyster growth and reproduction, notably the prevalence and intensity of the disease *Perkinsus marinus*, become more pronounced at high salinity, the

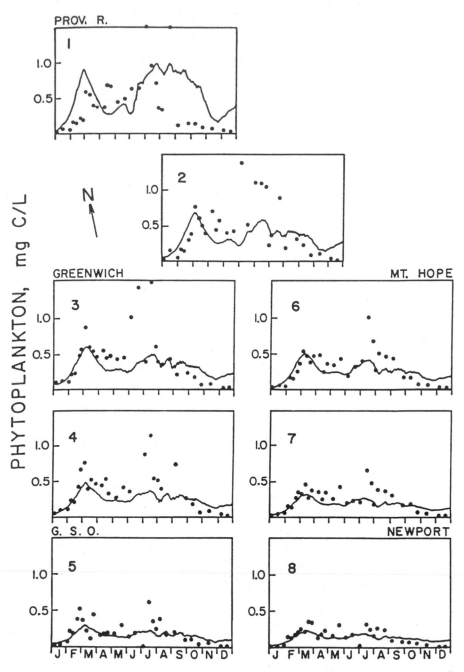

FIGURE 6-2 Comparison of simulated (solid line) and observed (•) time series of phytoplankton in the eight subregions of Narragansett Bay. Figure from Kremer and Nixon (1978), used with permission.

concern was that an enlarged ship channel would have deleterious effects on the Galveston Bay oyster population. Moreover, the effect of the ship channel could be altered, favorably or unfavorable for the oyster populations, by the approach taken by Texas in freshwater management over the next 50 years (1999–2049). Therefore, a series of simulations was undertaken that used the present channel configuration with three freshwater inflow regimes: mean, low, and high river-inflow conditions. The mean-flow conditions represent an average of the mean of sixty flows over 20 years of inflow data, determined from stream gauge records. The high- and low-flow conditions represent the average of the highest and lowest flows, respectively, over 20 years of inflow records. As a comparison, a second series of simulations was undertaken that used a Galveston Bay geometry that included the proposed enlarged ship channel as well as several islands that are to be constructed in the bay from dredge spoil.

The oyster-circulation model consists of several components, the first of which is a submodel for the growth of the post-settlement oyster population. This submodel also included a model that simulates the effects of the parasite *P. marinus* on the adult oyster. Thus, the host-parasite model consists of separate components for the dynamics of the post-settlement oyster population, the growth of *P. marinus*, and the influence of predators and competitors. The model components are coupled by relationships that describe the removal of oyster energy by the parasite to support its metabolic requirements, relationships that relate the rates of parasite cell division and mortality to host mortality, relationships that remove food from the oyster by mussel competitors, and relationships that produce mortality in the oyster population by *P. marinus*, crabs, and drills (*Thais*). In addition, the host-parasite model contains a larval-oyster component. The oyster population model is described in detail by Powell et al. (1992, 1994, 1995) and Hofmann et al. (1992, 1994). The oyster larval model is described by Dekshenieks et al. (1993) and the *P. marinus* model is described in Hofmann et al. (1995).

The Galveston Bay circulation model is a three-dimensional, finite element model that was developed by the U. S. Army Corps of Engineers, Waterways Experiment Station (WES) in Vicksburg, Mississippi. The advantage of this approach is that fine resolution can be placed where it is needed, such as in the portion of Galveston Bay that includes the ship channel. The domain of the finite-element circulation model extends throughout Galveston Bay and the near-coastal region (figure 6-3). In all, the bay was subdivided into 1996 elements. Of these model elements, only 408 cover regions of Galveston Bay that include oyster reef. The percentage of the total bay reef included in each of these elements was calculated. This was used to determine the regions of the bay to which the post-settlement oyster model was applied.

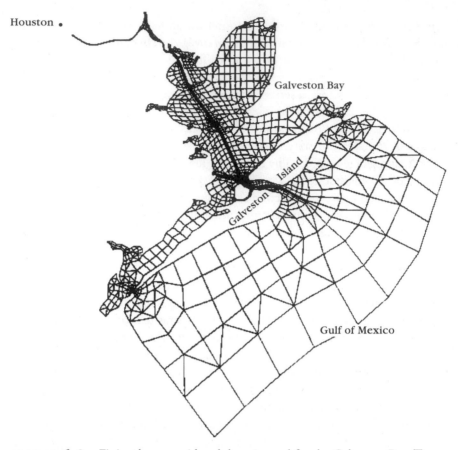

FIGURE 6-3   Finite element grid and domain used for the Galveston Bay, Texas, oyster-circulation model. The dark area through the center of Galveston Bay is the Houston Ship Channel. The ship channel is represented by many small finite elements in order to provide maximum spatial resolution in this region.

The governing equations for the circulation in the bay include the momentum equations, the volume-continuity equation, the advection-diffusion equation for salinity transport, and the equation of state relating water density to salinity (King 1985; King 1988; Berger et al. 1992; McAdory et al. 1995). Energy is input at the surface through surface wind stress and buoyancy forcing is provided by river inflows. Outputs of the circulation model are the velocity and salinity at the nodal points of each element. Hourly element-averaged values were calculated from these nodal values. For the post-settlement component, the time series of bottom velocities and salinities obtained from the WES circulation model were used. For the larval component, water-column averages of these time series were used.

Other environmental forcing consisted of temperature obtained at Houston, Texas, for the year 1984 and was input as hourly measurements. Food (from chlorophyll) and turbidity (total seston) were obtained from monthly measurements made throughout 1992 at thirty-two stations around Galveston Bay. These values were interpolated to the same resolution as the finite-element grid. Each simulation was run for 3 years with a time step of 1 hour. The simulated distributions from year 2 were used for analysis since these had come into equilibrium with the model forcing.

The simulated distributions show clearly the effect of changes in freshwater discharge on oyster populations in Galveston Bay. The simulated distribution of the post-settlement oyster population for mean-flow conditions shows that the structure of the populations is determined by food, turbidity, and salinity levels in the bay. Regions of low food, high turbidity, or high salinity, which favor increased prevalence and intensity of *P. marinus*, are characterized by low population abundance. The most productive region of the bay is along the axis of the Houston Ship Channel. Periods of high freshwater discharge result in simulated post-settlement oyster populations that are decreased by up to 200 relative to the mean freshwater conditions. Moreover, increased rates of low-salinity water shift the regions of productive oyster growth to the lower bay. Any mitigation of the deleterious effects of increased freshwater discharge results from the high-salinity water that enters the bay via the ship channel. Decreased freshwater inflow results in increases in the oyster population by up to 200 relative to the reference simulation. The isohalines move further into the bay, which allows former moderately productive regions to be more productive. *P. marinus* is more successful at higher salinities, but the mortality due to this organism is offset by the decrease in mortality due to low salinity, higher filtration rates, and increased larval survivorship.

Conception Bay, Newfoundland Model. One issue common to all estuarine systems is determination of transport pathways and residence times of larvae, especially those of commercially important species. Circulation models that track the distribution of a scalar quantity such as larval concentration provide one means for doing this. This approach was used to determine the residence time of capelin larvae in Conception Bay, Newfoundland (deYoung et al. 1994).

The larval capelin model consisted of the three-dimensional advection-diffusion equation into which were input velocities derived from a diagnostic circulation model. The circulation model (deYoung et al. 1993) provided a three-dimensional steady-state field that was calculated from a specified density field and surface wind stress. The model included the effects of vertical mixing and bottom friction. Horizontal resolution was about 1 km and vertical resolution was variable because of the use of sigma coordinates. Vertical resolution in shallow areas was about 1 to 2 m and increased to about 15 in the deep portions of the bay.

The density field used in the circulation model was obtained from hydro-graphic surveys of Conception Bay, which provided simulated circulation patterns for five periods: April, June, September, and October 1989 and June 1990. The flow fields for the four times in 1989 are shown in figure 6-4. Capelin larvae were introduced into different regions of the bay and advected and diffused by these simulated circulation fields. No animal behavior such as vertical migration was included in the model.

The vertically integrated capelin larvae distributions obtained for June 1989 are shown in figure 6-5. The larvae concentrations spread rapidly from the initial distribution toward the mouth of the bay and within a few days covered most of the bay. Larvae released at the head of Conception Bay had a residence time of 20 to 40 days, which places them within the bay waters during the critical period for growth. However, the rate of loss of larvae is quite dependent on the values chosen for the diffusion coefficients in the advection-diffusion equation.

Combined Eulerian-Lagrangian Models

Gulf of Carpentaria Model. The interaction of circulation and behavior in determining recruitment success of penaeid shrimp larvae in the Gulf of Carpentaria on the north-central coast of Australia was investigated with a combined Eulerian and Lagrangian modeling approach (Rothlisberg et al. 1983). Adults of this species spawn offshore and the larvae must then move across the continental shelf to nearshore nursery grounds. Hence, the primary objective of this study was to explain observed seasonal patterns in larval recruitment to estuarine nursery grounds around the gulf. Also, this model attempted to predict the time when spawning by offshore adult populations would maximize the possibility of larvae reaching the nursery grounds.

The tidal circulation, especially the seasonal progression in the phase of the tide, was identified as being potentially important for larval transport. Hence, the barotropic tidal circulation for the Gulf of Carpentaria was obtained from an existing model. This model, however, did not include dynamics to provide the vertical current structure, which can potentially alter the vertical-migration pattern of the penaeid shrimp. Therefore, a vertical current structure was superimposed on the barotropic tidal flow by assuming Ekman dynamics and using observed wind-stress values. No attempt was made to match space and time scales of the barotropic tidal flow and vertical current structure. The biological component of the model con-sisted of several idealized vertical-migration strategies, which were designed to investigate the transport distance and direction that would result from larval residence in different portions of the water column.

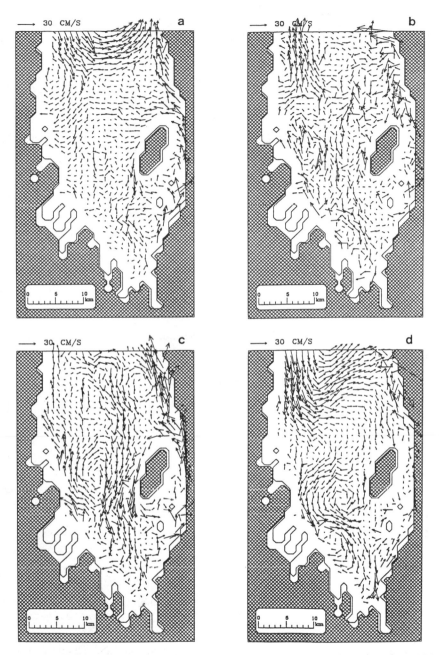

FIGURE 6-4 Current vector fields at 10 m for (a) June 1989, (b) September 1989, (c) October 1989, and (d) June 1990 in Conception Bay determined from the diagnostic steady-state calculation. Figure from deYoung et al. (1994), used with permission.

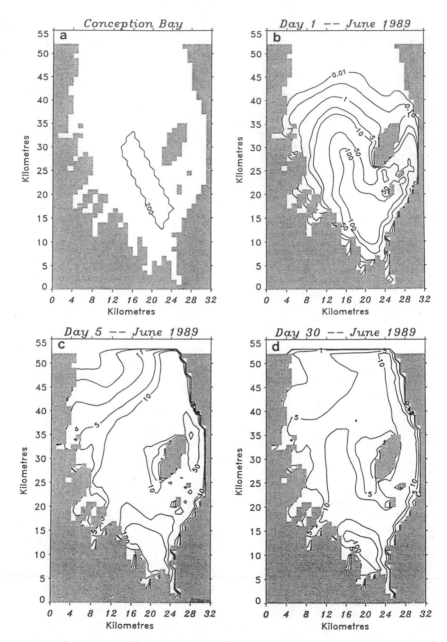

FIGURE 6-5 Vertically integrated capelin larvae concentration obtained using the June 1989 circulation field. Simulated distributions represent (a) day 0, (b) day 1, (c) day 5, and (d) day 30 of the simulation. Larval concentrations are expressed in arbitrary units of numbers. Figure from deYoung et al. (1994), used with permission.

The transport patterns obtained from the many simulations were used to construct seasonal recruitment patterns for penaeid shrimp larvae around the Gulf of Carpentaria. Validation of these was done by comparison with observed patterns. Although comparisons were qualitative, they did show that the simulated patterns reflected observations. Predictions of the general timing of recruitment and spatial dispersal patterns are possible with this model in terms of general regional distributions over seasonal time scales. Moreover, the simulated particle trajectories showed clearly the role of animal behavior in enhancing the transport distance experienced by the larvae (figure 6-6). Particles with no behavior were transported over shorter distances than those with behavior and in some cases even in the opposite direction (figure 6-6).

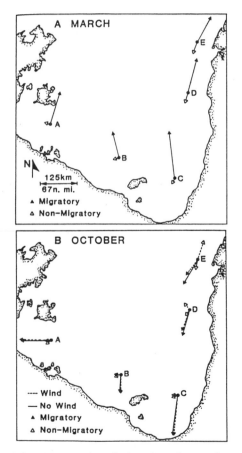

FIGURE 6-6   Spatial variation in larval advection after 20 days of simulation at five release points in the Gulf of Carpentaria, for migratory and nonmigratory larvae in (A) March and (B) October. Wind stress was applied to the model in October. Figure from Rothlisberg et al. (1983).

Discussion and Summary

Understanding the dispersal and structuring of biological distributions in estuarine systems requires that the biological and circulation models be combined. Thus, a critical aspect of interdisciplinary modeling for estuarine systems is the availability of circulation models that resolve scales relevant to the biological/ecological processes. Circulation models have been developed for estuarine systems (Oey 1984; Oey et al. 1985) and these have provided considerable insight as to the processes governing the circulation in these systems. These models could provide a starting point for coupling with biological models, except that none of these was developed from the outset with the intent of application to interdisciplinary modeling studies. For example, the approach used for including tidal circulation in the Narragansett Bay simulation actually misrepresented exchanges of biological material within the bay.

The need for vertical current structure in the Gulf of Carpentaria simulations resulted in the adoption of an ad hoc approach to provide a part of the current structure that was critical to answering questions about patterns of larval recruitment. The approach used for adding the vertical current structure did not necessarily match the scales of motion that were included in the simulated tidal flows and hence the simulated larval transport and direction may be in error. Because the flow field is so important in determining the structure of the biological distributions, every effort needs to be made to ensure that it is the best representation possible. Quantitative measures, such as nondimensional numbers (Froud number, stratification number), exist that can be used to ensure similarities between flows from the natural system and those derived from numerical circulation models. These provide a relatively simple approach for checking the correspondence between simulated and observed circulation.

The approach used to interpolate the velocity from the Eulerian field to the location of the Lagrangian particle includes assumptions about the structure of the subgrid scale flow. Hofmann et al. (1991) show that the degree of error introduced in Lagrangian trajectories is dependent on the interpolation method and the precision of the numerical calculation. In the worst case, the simulated and actual particle trajectory may have little correspondence. The type of model used by Rothlisberg et al. (1983) has enormous potential for improving understanding of the transport pathways and residence times of planktonic estuarine species. Thus, development of circulation models that can be used in this fashion should be a high priority.

In contrast, the oyster-circulation model for Galveston Bay used a detailed circulation model that had good spatial resolution in areas that included oyster reefs, but the grid was not designed around the distribution of oyster reefs. For example, some elements cut across different reef areas. However, even with this caveat, the Galveston Bay study is an example where the circulation model had adequate resolution for biological studies. In this case, however, the oyster models lacked the equivalent fine spatial scale. The post-settlement oyster population dynamics were assumed to apply over the entire area of individual elements independent of the size of the element, even when elements cut across different oyster reefs.

Finite-element models hold promise for modeling circulation in estuarine systems since these are designed for irregular geometry and have the capability of providing resolution where needed. Finite-element circulation models are only now becoming routinely applied to estuarine systems. Thus, these provide a relatively unexplored area for coupling with biological models. The finite-element model used for oyster populations in Galveston Bay provided good spatial resolution in the regions of the bay that included oyster reefs. However, the good spatial resolution could not be used to its fullest advantage because of the nature of the oyster models and the limited amount of environmental input data. Assumptions had to be made about how quantities, such as food supply and predator density, for example, varied over the bay and then used to interpolate a few measurements to the same resolution as the finite-element model.

The use of a diagnostic steady-state circulation field for Conception Bay assumed that the internal density field of the bay is invariant over time. This assumption may be valid if flow in the bay is primarily geostrophic. Ageostrophic motions, such as tidal circulation or wind effects, are not allowed to change the internal density field and thus cannot change the basic circulation structure of the system. Hence, estimated residence times may be too long or too short depending on the ageostrophic contributions to the bay circulation. The diagnostic approach used in the Conception Bay study provides a circulation field that is consistent with hydrographic observations. Therefore, the scales included in the circulation field are determined by the original sampling scheme. These may or may not be appropriate for determining larval residence times.

Inputting data to models does provide an approach for improving the realism of the simulated distributions. Data assimilation techniques allow data to be input to models such that the model and data sets are consistent. Data assimilation has been used with circulation models for some time (see Haidvogel and Robinson 1989), and efforts are now being made to develop data assimilation techniques for biological models (Lawson et al. 1995; Matear 1995; Lawson et al. 1996; Spitz et al. 1998). However, data assimilation requires that data be available. This in turn requires that biological measurements be collected in such a way that they can be used with data-assimilative physical-biological models. Similarly, modeling efforts could be designed around available data sets, but this can less satisfactory, especially if the measurements were not initially collected with the intent of using them with models.

The development of biological instrumentation and measurement techniques that provide good space and time resolution for physical and biological processes is therefore a high priority. Effort has already started in this area (Dickey 1991) and the next generation of sensors will routinely provide data sets that are several orders of magnitude larger than those now available. Already acoustical methods (Brandt et al. 1991; Lou and Brandt 1993) are providing realizations of large-scale distributions of fish and larvae in estuarine systems. Thus, there is a clear need for models that can be used as data synthesis and reduction tools. This in turn will require methods for viewing data and simulated distributions that differ from conventional methods now used. Wheless et al. (1995) provide an example of viewing simulated and/or observed biological and physical distributions in estuarine systems with virtual reality visualization methods. This approach allows the user to interact with and view distributions in ways that previously have not been possible. As such, it should give a better understanding of the processes controlling biological and circulation distributions in estuarine systems. However, no matter the level of complexity of individual models or what techniques are used, modeling for estuarine synthesis requires a series of interlinking circulation, biological, and optical models that are in turn interfaced with a variety of measurement and visualization systems. It is only through the view of an integrated, connected system that advances in understanding of estuarine systems will be made.

Acknowledgments

The facilities and support for completion of this manuscript were provided by the Commonwealth Center for Coastal Physical Oceanography at Old Dominion University.

References

Batchelder, H. P., and C. B. Miller. 1989. Life history and population dynamics of Metridia pacifica: Results from simulation modelling. *Ecological Modelling* 48: 113–36.

Batchelder, H. P., and R. Williams. 1995. Individual-based modelling of the population dynamics of *Metridia lucens* in the North Atlantic. ICES *Journal of Marine Science* 52: 469–82.

Berger, R. C., R. T. McAdory, J. H. Schmidt, and W. D. Martin. 1992. Houston-Galveston navigational channels, Texas. Report 3; Three-dimensional numerical modeling of hydrodynamics and salinity. Report HL-92-7. Vicksburg MS: U.S. Army Engineer Waterways Experiment Station.

Brandt, S. B., E. V. Patrick, D. M. Mason, K. J. Hartman, and M. J. Klebasko. 1991. Bioacoustics and bioenergetics: Trophic supply and demand in the Chesapeake Bay. Pages 105–19 in J. Mihursky (editor), *New perspectives in the Chesapeake Bay system: A science-management partnership.* CRC Publication No. 137. Solomons, MD: Chesapeake Research Consortium.

DeAngelis, D. L., and L. J. Gross. 1992. *Individual-based models and approaches in ecology*. New York: Chapman and Hall.

Dekshenieks, M. M., E. E. Hofmann, and E. N. Powell. 1993. Environmental effects on the growth and development of *Crassostrea virginica* (Gmelin) larvae: A modeling study. *Journal of Shellfish Research* 12: 241–54.

deYoung, B., R. J. Greatbatch, and K. B. Forward. 1993. A diagnostic coastal circulation model with application to Conception Bay, Newfoundland. *Journal of Physical Oceanography* 23: 2617–35.

deYoung, B., J. Anderson, R. J. Greatbatch, and P. Fardy. 1994. Advection-diffusion modelling of larval capelin (*Mallotus villosus*) dispersion in Conception Bay, Newfoundland. *Canadian Journal of Fisheries and Aquatic Sciences* 51: 1297–1306.

Dickey, T. D. 1991. The emergence of concurrent high-resolution physical and bio-optical measurements in the upper ocean and their applications. *Review of Geophysics* 29: 383–413.

Elliott, A. J. 1978. Observations of the meteorologically induced circulation in the Potomac estuary. *Estuarine and Coastal Marine Science* 6: 285–99.

Garvine, R. W. 1985. A simple model of estuarine subtidal fluctuations forced by local and remote wind stress. *Journal of Geophysical Research* 90: 11945–48.

Haidvogel, D. B., and A. R. Robinson (editors). 1989. Special issue: data assimilation. *Dynamics of Atmosphere and Oceans* 13: 171–513.

Hofmann, E. E., K. S. Hedstrom, J. R. Moisan, D. B. Haidvogel, and D. L. Mackas. 1991. Use of simulated drifter tracks to investigate general transport patterns and residence times in the coastal transition zone. *Journal of Geophysical Research* 96: 15041–52.

Hofmann, E. E., J. M. Klinck, E. N. Powell, S. Boyles, and M. Ellis. 1994. Modeling oyster populations. 2. Adult size and reproductive effort. *Journal of Shellfish Research* 13: 165–82.

Hofmann, E. E., E. N. Powell, J. M. Klinck, and E. A. Wilson. 1992. Modeling oyster populations. 3. Critical feeding periods, growth and reproduction. *Journal of Shellfish Research* 11: 399–416.

Hofmann, E. E., E. N. Powell, J. M. Klinck, and G. Saunders. 1995. Modeling diseased oyster populations. 1. Modeling *Perkinsus marinus* infections in oysters. *Journal of Shellfish Research* 14: 121–51.

King, I. P. 1985. Strategies for finite element modeling of three dimensional hydrodynamic systems. *Advanced Water Resources* 8: 69–76.

———. 1988. *A model for three dimensional density stratified flow*. Vicksburg, MS: U.S. Army Engineer Waterways Experiment Station.

Kremer, J. N., and S. W. Nixon. 1978. *A coastal marine ecosystem: Simulation and analysis*. Ecological Studies. Vol. 24. New York: Springer-Verlag.

Lawson, L. M., Y. H. Spitz, E. E. Hofmann, and R. B. Long. 1995. A data assimilation technique applied to a predator-prey model. *Bulletin of Mathematical Biology* 57: 593–617.

Lawson, L. M., E. E. Hofmann, and Y. H. Spitz. 1996. A data assimilation technique applied to a marine ecosystem model. *Deep-Sea Research* 43: 625–51.

Lou, J., and S. B. Brandt. 1993. Bay anchovy *Anchoa mitchilli* production and consumption in mid-Chesapeake Bay based on a bioenergetics model and acoustic measures of fish abundance. *Marine Ecology Progress Series* 98: 223–36.

Madenjian, C. P., and S. R. Carpenter. 1991. Individual-based model for growth of young-of-the-year walleye: A piece of the recruitment puzzle. *Ecological Applications* 1: 268–79.

Malone, T. C., P. J. Neale, and D. Boardman. 1980. Influence of estuarine circulation on the distribution and biomass of phytoplankton sizefractions. *Estuarine Perspectives* 249–62.

Matear, R. J. 1995. Parameter optimization and analysis of ecosystem models using simulated annealing: A case study at Station P. *Journal of Marine Research* 53: 571–607.

McAdory, R. T. Jr., R. C. Berger, and W. D. Martin. 1995. Three-dimensional model and salinity results for use in an oyster model of Galveston Bay. Pages 27–36 in R. Jensen (editor), *Water for Texas.* Proceedings of the 24th Water for Texas Conference.

Oey, L.-Y. 1984. On steady salinity distributions and circulation in partially mixed and well mixed estuaries. *Journal of Physical Oceanography* 14: 629–45.

Oey, L.-Y., G. L. Mellor, and R. I. Hines. 1985. A three dimensional simulation of the Hudson-Raritan estuary. Part 1: Description of the model and model simulations. *Journal of Physical Oceanography* 15: 1676–92.

Powell, E. N., E. E. Hofmann, J. M. Klinck, and S. M. Ray. 1992. Modeling oyster populations 1. A commentary on filtration rate. Is faster always better? *Journal of Shellfish Research* 11: 387–98.

Powell, E. N., E. E. Hofmann, J. M. Klinck, E. A. Wilson-Ormond, and M. S. Ellis. 1995. Modeling oyster populations 5. Declining phytoplankton stocks and the population dynamics of American oyster (*Crassostrea virginica*) populations. *Fisheries Research* 24: 199–222.

Powell, E. N., J. M. Klinck, E. E. Hofmann, and S. M. Ray. 1994. Modeling oyster populations. 4. Rates of mortality, population crashes and management. *U.S. Fish and Wildlife Service Fishery Bulletin* 92: 347–73.

Rothlisberg, P. C., J. A. Church, and A. M. G. Forbes. 1983. Modelling the advection of vertically migrating shrimp larvae. *Journal of Marine Research* 41: 511–38.

Spitz, Y. H., J. R. Moisan, M. R. Abbott, and J. G. Richman. 1998. Data assimilation and a pelagic ecosystem model: Parameterization using time series observations. *Journal of Marine Systems* 16: 51–68.

Weisberg, R. H. 1976. The nontidal flow in the Providence River of Narragansett Bay: A stochastic approach to estuarine circulation. *Journal of Physical Oceanography* 6: 721–34.

Wheless, G. H., A. Valle-Levinson, and W. Sherman. 1995. Virtual reality in oceanography. *Oceanography Magazine* 8(2): 52–58.

CHAPTER 7

An Ecological Perspective on Estuarine Classification

David A. Jay, W. Rockwell Geyer, and David R. Montgomery

Abstract

Estuaries are more numerous and diverse than any other type of marine environment. Scientists and managers currently face a range of issues associated with global alteration of estuarine systems without a thorough understanding of how such changes affect either estuaries or the larger coastal and global oceans to which they are connected. Exhaustive study of all highly altered systems is impractical. It is necessary instead to analyze representative systems characteristic of the spectrum of estuarine types. Identification of representative estuaries requires development of a classification scheme that connects physical processes (circulation and sediment transport) to biogeochemistry and ecology in a predictive manner. Traditional classification efforts focused almost exclusively on the influences of tidal forcing and river inflow on circulation in narrow estuaries. To be useful over the range of estuarine types, a process-based geomorphic classification should include other types of physical forcing; that is, the effects of wind, wind waves, sea ice, and surface heating and cooling. Fluid mechanics and geomorphology can be connected through description of forcing variables for sediment movement in estuaries in terms of hydrodynamic parameters. One important parameter is residence time (T_R). T_R strongly influences larval recruitment and biogeochemical and ecological processes vital to the lower levels of the estuarine food chain. It is a major habitat parameter for higher trophic levels. Knowledge of a mean T_R is, moreover, not sufficient; its temporal variability and spatial heterogeneity should be considered. Finally, we suggest a hypothesis concerning estuarine particle retention processes that, if correct, should allow prediction of residence time and trapping efficiency for at least some estuarine types in the near future.

Introduction

A desire to classify estuaries arises not only out of a tendency toward system-atization inherent in all science, but also from urgent practical necessity. Estuaries are extremely diverse, more so than any other type of marine envi-ronment. Scientists and managers currently face a range of issues associated with global alteration of estuarine systems without a clear understanding of how such changes affect either estuaries or the larger coastal and global oceans to which they are connected. Such an understanding cannot be achieved through exhaustive study because there are too many estuaries in North America alone to carry out comprehensive ecosystem studies of all those undergoing massive alteration. Nor do we as yet have a broad enough view of estuarine processes to choose representative systems for detailed analysis, as has been done, for example, with terrestrial habitats by the LTER program funded by NSF. Development of a scheme for classification of estuarine ecosystems is, therefore, a prerequisite to understanding global change in estuaries. To be useful, such a classification must not be simply retrospective, characterizing well-known systems, but predictive, allowing classification of relatively unknown systems on the basis of a minimum suite of measurements obtainable from climate records, maps, remote sensing, or regional-scale models of the atmosphere and ocean. The necessary input information might include such parameters as latitude, tidal amplitude in the adjacent coastal ocean, presence of seasonal ice cover, freshwater inflow, slope and bedrock type of the tributary river basin(s), and prevailing winds. A successful estuar-ine classification scheme would be one that is useful not just in selection of systems to study but in planning studies of individual systems.

This contribution revisits two physically based approaches to estuarine classification to explore how they can be unified and extended to include ecosystem processes. The two methods considered are classification by nondi-mensional hydrodynamic parameters and by geomorphology. The former approach has in the past focused primarily on a small but vital subset of estu-arine physics, the interaction of tidal and buoyancy forcing in narrow estuar-ies. Buoyancy forcing has traditionally been associated with river flow, but extension to negative and thermal estuaries, where the density field is con-trolled by a balance of evaporation and precipitation, is also possible. Atmos-pheric and wave effects may be included, but the influence of sea ice is more problematic.

A geomorphic approach to estuarine classification as discussed by Pritchard (1952) and Dyer (1973) has an obvious intuitive appeal, but pre-vious attempts have lacked a logical framework allowing detailed elabora-tion. Rather than attempting to broaden earlier estuarine geomorphic classifications, we adapt to estuaries a hierarchical classification scheme from fluvial geomorphology (Montgomery and Buffington 1993, 1997, 1998). Because rivers are simpler and less diverse than estuaries, fluvial geomorphologists have developed a process-based approach with substantial predictive capabilities. We adopt here the structure and style of the Montgomery and Buffington analysis, and the fluvial classifications remain embedded in the larger river-estuary system discussed herein. Multiple types of physical forcing, complex ecosystems, and a variety of geologic settings, however, prevent estuaries from exhibiting as close a relationship between reach type and sediment transport style as can be defined in a fluvial setting. Thus, for example, sediment transport in a bedrock reach may respond differently to external forcing in a fjord than in a river estuary. Such considerations cause the geomorphically based estuarine classification presented below to be more complex than its fluvial counterpart.

The hydrodynamic and geomorphic approaches are superficially very different. But both are based on physical processes, and they can be brought together by defining nondimensional hydrodynamic parameters associated with system geometry and the dominant forcing mode for each geomorphic type, and through use of the concept of residence time. (Residence time or T_R is the average time a parcel of water or other property like salt spends in an estuarine basin. Under steady conditions, it can be estimated as the ratio of the amount of the property in the basin divided by the rate of property exchange through the boundaries of the basin.) Unification of geomorphic and hydrodynamic approaches should in the future allow quantitative predictions to be made concerning physical processes of interest (for example, degree of stratification) and (in some environments) qualitative predictions with regard to biological processes (such as degree of autotrophy versus heterotrophy), from simply measured physical and chemical parameters. Rather than constructing here a comprehensive classification scheme (a lengthy exercise), we show by example how this might be accomplished. The first example used is the concept of T_R and the character of flushing in an estuary. The mean, variance, and spatial distribution of T_R are all important to ecosystem processes, because these processes are influenced by the frequency and intensity of disturbance of the environment. A second example is hydrodynamic trapping of particles and organisms by estuarine circulation, which can be described in terms of two parameters. Using these parameters it should be possible to predict what size of particles are optimally trapped, and the efficiency with which they may be concentrated.

A third estuarine classification approach, enumeration of habitats (see Cowardin et al. 1979; Dethier 1990), is not employed here because it lacks a unified approach to the underlying physical forcing that is vital in structuring estuarine habitats. However, many of the environments enumerated below are similar to those of Dethier (1990). Finally, one might approach classification through analyses of results from coupled numerical biophysical models, such as those discussed by Hofmann (this volume).

Hydrodynamic Classification Approaches

Existing hydrodynamic classification schemes focus on the interaction in narrow estuaries of tidal currents, which provide energy for mixing, with river flow, a source of stratification or buoyancy. Internal mixing is only a secondary sink for circulation energy in most if not all estuaries. Nonetheless, stratification is central to determining the type of mean or residual flow observed and direction and strength of scalar transport. (A scalar is a quantity that, like salinity or concentration, can be expressed as a single number. Velocity is, in contrast, a vector. "Scalar transport" refers to the transport of dissolved and suspended substances by the flow.) Accordingly, the first modern hydrodynamic estuarine classification was based on stratification. Stommel and Farmer (1952) defined four types of estuaries: (1) weakly stratified estuaries; (2) partially mixed systems; (3) salt wedges; and (4) fjords. A more quantitative approach is to define a stratification number G/J, where G is energy dissipation over a defined channel length and J is the rate of gain of potential energy of water moving from land to sea over the same length (Ippen and Harlemann 1961; Prandle 1986; figure 7-1). The inverse of G/J is related to a ratio of buoyancy to shear and can thus be expressed in terms

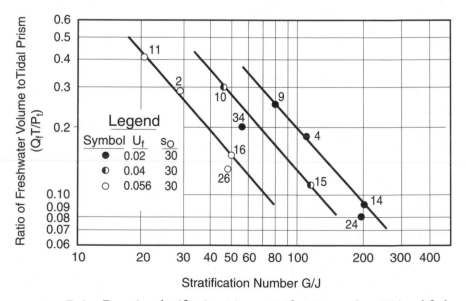

FIGURE 7-1 Estuarine classification using a stratification number G/J (modified from Ippen and Harlemann 1961).

of an estuarine Richardson number as defined by Fischer (1976). Working in a continental shelf context, Simpson and Hunter (1974) define a value of U_T^3/H_C (where U_T is tidal velocity) specifying a critical depth H_C at which shelf fronts separating mixed and stratified regimes should be found. This critical depth is also based on a Richardson number criterion, but the buoyancy flux does not appear explicitly because a particular value is specified. The Simpson and Hunter (1974) approach was subsequently applied to estuarine environments by Nunes Vaz (1991).

G/J and related parameters capture one important aspect of estuarine circulation, but *G/J* alone is insufficient because specific values do not have a one-to-one relationship to actual estuarine types. The ambiguity of a single parameter suggests that at least two hydrodynamic numbers should be employed to describe the interaction of the tides and river flow. The first and best-known two-parameter classification scheme is that of Hansen and Rattray (1966), an outgrowth of estuarine gravitational circulation theory (Hansen and Rattray 1965). Numerous clarifications, modifications, and additions have been made to this basic approach, most focused on the issue of providing a closer connection between the density field and tidal processes (Fischer 1976; Officer 1976; Uncles and Rattray 1983; Oey 1984; Prandle 1986). All of these authors employ some form of two basic classification parameters: (1) stratification $\Delta S/S$, a ratio of top-to-bottom salinity difference to vertical average salinity; and (2) circulation U_S/U_F, a ratio of surface velocity to vertically averaged net outflow.

The Hansen and Rattray (1966) circulation-stratification diagram (figure 7-2) emphasizes that a range of circulation patterns and estuarine types or morphologies are possible for a given degree of stratification. Among the strengths of this approach are the relative simplicity of the parameters employed and its predictive ability with regard to salt-transport mechanisms. Location of an estuary in figure 7-2 suffices to allow calculation of the percentage of diffusive (that is, tidal) salt transport (relative to total landward salt transport) needed to maintain the salt balance (figure 7-3). This ability exemplifies the sorts of predictions that future classification schemes should have with regard to ecological processes. Weaknesses of Hansen and Rattray (1966) include assumption of a steady salt balance, an incomplete parameterization of tidal transport of salt in the Hansen and Rattray (1965) theory (rendering the predictions of figure 7-3 suspect), a tendency for different stations within an estuarine basin to appear in different locations in the diagram, and the absence of a clear connection of U_S/U_F and $\Delta S/S$ to the underlying physical phenomena of tidal forcing and river flow. Prandle (1986) has, however, suggested relationships to remedy this latter defect.

A desire to more directly represent the forcing of residual circulation led Jay and Smith (1988) to suggest a different two-parameter scheme (figure 7-4). This approach employs an external Froude number F_T, the ratio of tidal amplitude ζ to mean depth H (a measure of wave distortion or nonlinearity first employed by Stokes 1847), and an internal Froude number $F_B =$

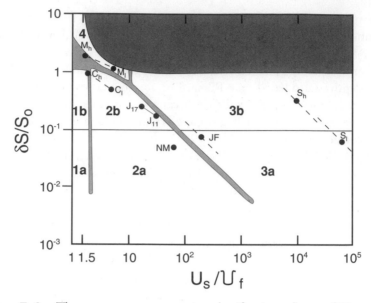

FIGURE 7-2 The two-parameter estuarine classification scheme of Hansen and Rattray (1966). Estuaries are classified by circulation U_S/U_F on the horizontal axis and stratification $\Delta S/S$ on the vertical axis. Type 1 estuaries include those without net upstream bottom flow such that the salt balance must be maintained by tidal transport; type 2 estuaries are partially mixed (for example, the Narrows of the Mersey, NM; the James River estuary, J; and the Columbia River estuary, C); type 3 includes fjords (such as Silver Bay, S, and the Strait of Juan De Fuca, JF); and type 4 includes salt wedges (such as the Mississippi River, M). Subscripts h and l refer to high and low river flow. The shaded area at the top corresponds to a limit of fresh surface outflow over a stagnant bottom layer. Modified from Hansen and Rattray (1966).

$\Delta d/D \, (\Delta \rho_H/\Delta \rho_V)^{1/2}$ (where Δd is the vertical tidal excursion of an isopycnal in the pycnocline, D is its maximum distance off the bottom, and $\Delta \rho_H$ and $\Delta \rho_V$ are the total along-estuary and vertical density differences, respectively) calculated from tidal advection of the salinity field. Systems with weak tides to the lower left of figure 7-4 (for instance, the Mississippi) may be highly stratified, but there is little tidal advection of the density field, and thus only a modest internally generated residual flow. Systems on the far right of figure 7-4 may have very strong tidal advection (for example, Bay of Fundy), but stratification and horizontal density gradients are weak, so again the internal residual flow is not well developed. The strongest residual flow and maximum internal nonlinearity arises in systems like the Columbia and Fraser Rivers with both strong (if variable) stratification and strong advection thereof ($F_B \cong 1$). Strengths of this approach include the direct appearance of tidal parameters in the classification system and its ability to classify entire estuarine basins.

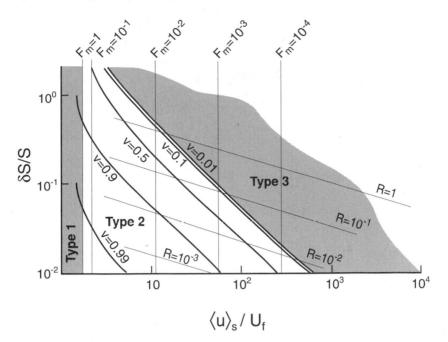

FIGURE 7-3   Contours of ratio (υ) of diffusive salt transport to total salt flux overlain on figure 7-2, given as the total salt transport to tidal wave fluxes over the total landward salt transport. Also shown are contours of tow bulk parameters, a densiometric Froude number $F_m = U_F/(\Delta \rho g h/\rho_0)^{1/2}$ (ratio of freshwater flow speed to internal wave speed), and an estuarine Richardson number $R = \Delta \rho g h/\rho_0 \, U_F/U_T^3$ (ratio of the tendency of freshwater flow to stratify a system relative to the tendency of tidal currents to mix it; R is closely related to G/J). The purpose of the bulk parameters is to relate estuarine classifications and υ to easily measured quantities. Type 1, 2, and 3 estuaries are defined in figure 7-2. Modified from Fischer 1976.

Weaknesses include lack of quantitative predictive ability (stemming from a lack of closed-form solutions in the underlying theory) and difficulty in relating F_B directly to tidal amplitude and river flow.

Friederichs and Madsen (1992) pointed out that the presence of tidal flats can reverse the sense of distortion of a surface tidal wave and suggested a modified version of F_T; $F_{TM} = 5/3 \, \zeta/H - b_T/b$ (where b_T is total mean width including tidal flats and b is channel width). The value of F_{TM} may be either positive (tidal changes in depth predominate) or negative (tidal changes in depth or width predominate), and large absolute values of F_{TM} connote strong nonlinearity.

A recent analysis has considered the effect of surface heating and evaporation on narrow mediterranean estuaries with seasonally weak river inflow (Hearn 1998). In the presence of these surface processes and the virtual

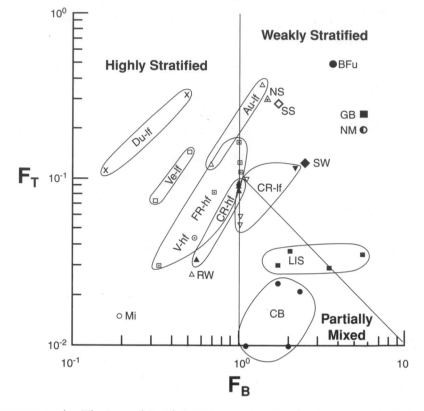

FIGURE 7-4   The Jay and Smith (1988) estuarine classification approach. F_B is an internal Froude number that is a measure of the strength of residual circulation generated by internal processes (with maximum residual occurring for systems with F_B ~1), and F_T (ratio of tidal range to depth) is the barotropic tidal nonlinearity parameter; barotropic nonlinear effects increase with F_T. Systems classified include the Columbia (CR), Fraser (FR), and Mississippi (Mi) River estuaries; Chesapeake Bay (CB); Long Island Sound (LIS); the Narrows of the Mersey (NM); and the Bay of Fundy (Bfu); other systems and data sources are given in Jay and Smith (1988). The symbols *hf* and *lf* indicate high and low river flow, respectively. With changes in seasonal flow or tides, some systems cross boundaries between highly stratified, weakly stratified, and partially mixed classifications. Modified from Jay and Smith 1988.

absence of river inflow, such systems are vertically mixed. A box model suggests that four circulation states should exist. These include the usual two-layer circulation, and strong and weak types of inverse (negative) estuaries, where the surface flow is landward and outflow occurs near the bed. Weak horizontal and vertical gradients lead to slow lateral exchange and long residence times in these systems, especially in the weak inverse state that resembles Tomales Bay in summer. This analysis, while still primarily concerned with buoyancy effects, is a good example of how classical theories based on tides and river flow alone can be generalized.

The above classification schemes need to be generalized to describe a broader range of features and processes relevant to estuarine ecosystems. At a minimum, an improved classification should include as dependent or independent variables parameters descriptive of (1) estuarine geometry; (2) circulatory forcing functions; (3) the idea of residence time; and (4) the role of estuaries in trapping particles. There is a close connection between estuarine geometry and dominant forcing type. Thus, the forcing functions included in Hansen and Rattray (1966) and Jay and Smith (1988) have been restricted to river flow, tidal amplitude, and buoyancy input caused by river flow. The systems encompassed by this system are restricted to narrow embayments, fjords and river estuaries where these factors predominate. The many systems dominated by wind forcing and the influence of wind waves are, in contrast, mostly broad, shallow embayments and plumes. They are either too wide to have a substantial river flow per unit width or, in the case of plumes, lack lateral boundaries. Conversely, narrow estuaries are controlled by atmospheric processes only when river flow and tidal forcing are weak. An enumeration of forcing functions is provided below.

Basic measures of estuarine geometry include those associated with length, width, and depth, but several measures of each may be required by the broad range of estuarine types. Length is described by ratios of salinity intrusion length to total length and total length to tidal wavelength. The former is vital in distinguishing tidal rivers from embayments and fjords, while the latter describes the fit of the tidal wave into an estuarine basin. An important measure of width is the ratio $B/(F_T \lambda)$ of width B to tidal excursion $F_T \lambda$. Wide basins with $B/(F_T \lambda) \gg 1$ often have scalar balances dominated by lateral variations in tidal transport, whereas buoyancy forcing and shear dispersion usually control scalar balances in systems with $B/(F_T \lambda) \ll 1$. An important depth parameter for fjords is a ratio of sill depth to total depth. These examples obviously do not exhaust the list of geometric parameters that may arise in estuarine classification. Other geometric and circulation parameters are introduced below in the discussion of the geomorphic classification approach.

Connection to Ecosystem Processes: Residence Time and Particle Trapping

Residence Time

Determination of residence time T_R is an application of scalar conservation (Jay 1995), and T_R can, under steady-state conditions, be determined from fluxes into (plus internal sources) or fluxes out of (plus internal sinks) a reservoir. The assumption of (tidally averaged) steady content and transport is, however, rarely satisfied. For unsteady situations, alternative approaches involving rates of change of reservoir scalar content may be used (Zimmerman 1988). Because it is a basic measure of the rate of turnover of a substance, T_R is a common element that can be employed to unify many aspects of estuarine ecosystem study. Thus, determination of T_R represents a fundamental connection between hydrodynamics and ecosystem processes.

The physical processes that determine (for conservative scalars) or influence (for biologically active scalars) T_R can be conveniently divided into three groups: (1) flushing, (2) vertical mixing and entrainment, and (3) for particles, trapping mechanisms discussed in the next section. By flushing, we mean the horizontal transport of a scalar into and out of the reservoir in question (usually an estuarine reach or basin, or a layer thereof). Net horizontal scalar transport is often thought of mainly in terms of a two-layer mean or gravitational circulation driven by the density difference between the open ocean and fresh water. But tidal and overtide processes associated with fluctuating density gradients are also vital, especially for particles. Tidal and overtide scalar transports (also known as wave fluxes) arise from correlations between time-varying currents and fluctuations in scalar concentration. These may dominate lateral transport in shallow systems with strong tides. The division between wave and tidal transport of salt can be described in terms of U_S / U_F and $\Delta S/S$ (figure 7-3) or alternatively relative to F_B and F_T. Transports associated with mean two-layer circulation and the tides vary in most narrow systems both with the strength of tidal forcing and with river inflow. Salt-flux calculations (Jay and Smith 1990; Jay et al. 1997) suggest that, for narrow estuaries: (1) the sum of tidal and wave transports is less variable in space and time than either mechanism individually, and (2) spatial variations in transport mechanisms are often responses to particular topographic features. Salt transport in broad, shallow estuaries is often controlled by atmospheric events and lateral tidal dispersion rather than tidal and mean vertical shear transports. The nondimensional width $B/(F_T\lambda)$ described above may be a useful measure in distinguishing the various transport regimes.

Vertical mixing and entrainment mechanisms constitute the second vital collection of processes governing T_R. They control the cross-sectional distribution of scalars, thus in combination with the structure of the velocity field, the spatial distribution of horizontal scalar fluxes into and out of reservoirs. Furthermore, if T_R is to be calculated for several layers, then vertical mixing and entrainment control scalar exchanges between the various layers of the model. Transport of active scalars (those like salinity, temperature, and suspended particulate matter concentration [SPM] that may affect the density distribution) is more complex than that of passive substances (those present in too small a concentration to do so), because their spatial distributions strongly affect flow and vertical mixing, creating a complex feedback between the velocity field and scalar distributions. Vertical mixing and entrainment in positive estuaries can be divided into internal, near-bed, and wind-induced processes. Bottom-boundary mixing is associated with frictional (turbulent) energy loss at the bed; this type of loss is strongly enhanced by sills and constrictions. Winds cause mixing through surface waves and wind-induced shear. Internal mixing is caused by a broad range of processes related to internal waves, fronts, and shear instabilities. Many estuaries, however, are shallow enough that surface, internal, and near-bed mixing cannot be entirely separated. Unlike the atmosphere, convection usually is not dominant in temperate estuaries, except those (negative estuaries) with so little river inflow that their density structure is dominated (seasonally or permanently) by a balance of evaporation and precipitation. It may also be seasonally important during winter in shallow estuaries with small river inflow.

Most calculations of residence time have focused on its average value or its relationship to a specific forcing parameter, usually river flow or tidal range. We here treat T_R as a continuous function of time and consider the importance of its mean, temporal variance, and spatial heterogeneity to ecological processes.

The estuarine circulation theory of Hansen and Rattray (1965) can also be used to define the relationship of mean T_R to steady river flow U_R and tidal velocity (figure 7-5). River flow in a typical estuary varies seasonally by a factor of 5 to 1,000 or more, though large or heavily regulated temperate systems may have a more restricted range of river flow. Few systems exhibit, however, a variability in tidal range (and thus maximum tidal currents) of more than a factor of 2 to 3. Given these typical ranges of variability, residence time is primarily a function of U_R in most temperate estuaries. The strong dependence of T_R on U_R is due to the influence of U_R on (1) estuarine circulation; and (2) estuarine salinity intrusion length L_S and estuarine volume, which decrease as U_R increases. The influence of tides on T_R is more complicated, though usually smaller, than that of river inflow. This complexity appears as a nonmonotonic response of T_R to increasing tidal range for low river-flow levels (figure 7-5). For weak tidal forcing, T_R decreases with increasing tidal forcing because increased mixing decreases L_S and thus the volume of the estuary created by mean estuarine circulation. As tidal forcing further increases, T_R begins to increase again, because greater tidal salt advection creates a larger estuary as tidal range increases. The analysis leading to figure 7-5 provides useful insights, but it has a number of limitations. An important one is that it

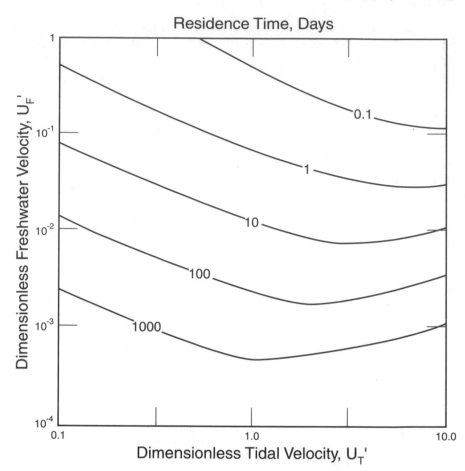

FIGURE 7-5   Analytical estimate of residence time as a function of river flow and tidal range for an idealized estuary, based on the Hansen and Rattray (1965) analytical model. The tidal and river inflow velocities are scaled by a densimetric velocity $(U_d = (\Delta \rho g h / \rho o)^{1/2})$, where h is the water depth. Because river flow is much more variable than tidal range in most estuaries, river flow is responsible for most of the salt residence time variation. The residence time decreases as freshwater flow increases, due to strengthening of the estuarine circulation. For relatively weak tides, increased tidal velocity leads to decreased residence time, due to shortening of the salinity intrusion. For strong tides, tidal dispersion dominates and residence time increases with increased tidal range due to increased salinity intrusion.

assumes a steady salt balance over the residence time calculated. This is a reasonable assumption for rapidly flushed river estuaries but may not be for less dynamic systems. Another limitation is that it does not consider the nature of vertical mixing/entrainment and how this may change with river flow and tidal range.

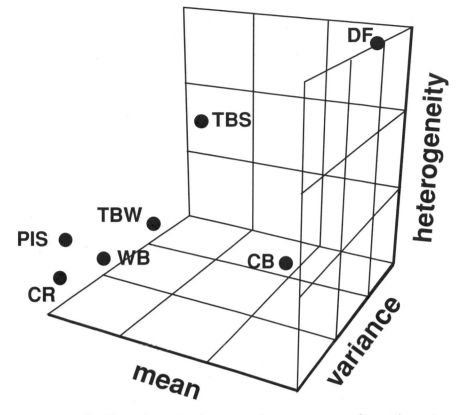

FIGURE 7-6   Three-dimensional conceptual parameter space for residence time T_R, showing mean T_R, T_R temporal variance, and T_R spatial heterogeneity (arbitrary units) for the Columbia River (CR), Tomales Bay winter (TBW), Tomales Bay summer (TBS), Chesapeake Bay (CB), and a hypothetical deep fjord (DF).

For insight into the effects of spatially and temporally varying T_R, it is necessary to resort to more qualitative arguments. Comparison of the estuaries studied by the Land Margin Ecosystem Research program (LMER) suggests that there is an important distinction between systems that are continuously and rapidly flushed by tidal or fluvial processes, and those that are flushed only episodically by storms and oceanic events, as summarized in figure 7-6. Near the short mean residence time extreme are the Columbia River (CR), Tomales Bay in winter (TBW), Plum Island Sound (PIS), and Waquoit Bay (WB), where T_R for each whole system is O (1–5 d). The Columbia is the most extreme case in that flushing is rapid and continuous, with little spatial variability. The T_R for salt in the mid-estuary turbidity maximum (ETM) region is never more than a few days. Other lower estuary areas show even shorter residence times. The strong flows implied by this situation are reflected in the dominance of sand in the bed in all open-water environments and in the absence of long-term deposition of fine materials in channel areas. Nonetheless, the limited spatial heterogeneity present is of considerable importance, because the ETM (where T_R is maximal for the system) is the locus of the most intensive secondary production. Variability, spatial heterogeneity of flushing, or both are somewhat greater in Waquoit Bay and Plum Island Sound than in the Columbia. The clustering of LMER systems at the rapid flushing asymptote reflects both the fact that small systems are more common than large ones, and the greater tractability of sampling in small estuaries.

Infrequently disturbed estuaries are at the other extreme of the space defined in figure 7-6. These systems are characterized by large mean values of T_R and (in most cases) greater T_R variance. A lower layer in a deep, shallow-silled fjord (for instance, some on the west coast of Vancouver Island) may flush only at irregular intervals of one to several years, when major storm events bring unusually dense water over an entrance sill. This leads to a large average T_R for the system, because the lower layer contains most of the water volume. Spatial heterogeneity of T_R is very high, however, because the shallow surface layer of such a system is flushed rapidly by river inflow.

LMER systems other than the Columbia occupy intermediate locations in figure 7-6. Chesapeake Bay is the largest LMER estuary. It is flushed slowly by a two-layer gravitational circulation bringing about a mean T_R of 3 to 9 months, with considerable temporal variability caused by the stochastic influence of strong atmospheric forcing. Winter or spring storms, particularly a hurricane that causes both large river inflow and a major storm surge, may occasionally reduce its T_R to a few days. There are strong seasonal differences in circulation in Tomales Bay that require a distinction in its summer and winter flushing behavior in figure 7-6. During the winter, Tomales Bay is a positive estuary flushed by storms and associated flow events. These provide a mean T_R that may be as small as a few days in some years, but a much larger T_R in those years during which storms are weak or absent. Evaporation exceeds precipitation plus river inflow in Tomales Bay during most summers. The hypersaline inner basin of Tomales Bay becomes a negative, almost neutrally stratified estuary during most summers. This reduces tidal salt transports and brings about a mean T_R of ~100 d (Smith et al. 1991). There is both considerable temporal variability, because of the vagaries of evaporation and precipitation, and spatial heterogeneity, because the circulation of the outer basin remains positive in most years.

In summary, steady-state theory suggests that mean T_R responds more strongly to fluvial than tidal forcing largely because the annual range of river flow is typically much larger than the variability in tidal range. Quantifying transient responses and the spatial variability of T_R will require a combination of numerical modeling, theory, and data analysis, but it is a goal within reach.

Residence Time and Trapping of Particles

Residence time for particles is controlled by flushing and mixing plus particle-trapping mechanisms that, while perhaps not unique to particles, are certainly much more important for particles than for dissolved substances. The most obvious form of particle trapping is convergent near-bed flow, as occurs near the upstream limits of salinity intrusion and two-layer estuarine circulation. That is, upstream bottom flow related to gravitational circulation or a variety of tidal nonlinearities described by Jay and Musiak (1994) depends on the presence of salt, and the landward transport "conveyor belt" must necessarily stop at, or seaward of, the upstream limit of salinity intrusion, leading to trapping of particles in this location. This is, however, far from the only particle-trapping mechanism. Others include along-channel changes in tidal energy and along-channel changes in the balance between overtides and river flow (Allen et al. 1980) and lateral trapping mechanisms associated with asymmetries in vorticity (Zimmerman 1978). The latter include entrapment in lateral embayments, by bay-mouth eddies, and by lateral changes in stratification associated with changes in depth.

   An example of how estuaries might be classified with regard to particle-trapping ability is shown in the trapping-efficiency hypothesis of figure 7-7, where trapping efficiency (ratio of maximum SPM concentration in an ETM to its concentration in oceanic or fluvial source waters) is hypothesized to be a function of F_B and the ratio of settling velocity W_S to the square root of kinematic bed stress, U_* (W_S/U_* is commonly known as the Rouse number). Trapping efficiency is a measure of the ability of a system to concentrate organic matter in the SPM in a situation where it is accessible to food-web processing. Particles of an intermediate size matched to the strength of advection and bed stress in the estuary are most efficiently trapped (Festa and Hansen 1978; Jay and Musiak 1994). Particles that are too coarse (that is, those that settle too rapidly) for conditions prevailing in the system travel, if at all, only as bedload (the bedload limit to the right in figure 7-7). Those that are too fine (settle to slowly) are washload. They cannot be trapped because they have no vertical stratification to allow an enhanced landward transport to develop (the washload limit to the left in figure 7-7). The most effective particle trapping occurs with $F_B \approx 1$, because this is the condition for which the strongest upstream bottom flow occurs. Deep systems (small values of F_B) may be effective sediment traps in the sense that particles are rapidly deposited and rarely disturbed. Because resuspension is infrequent and weak, however, the bed is anoxic below a shallow surface layer and its organic matter cannot support a food web. Very shallow systems cannot concentrate or trap particles in the water column, because even temporary deposition of the relevant materials is absent. The more rapidly particles settle, the more closely F_B must be to one for substantial trapping to occur; that is, larger particles require stronger currents for landward near-bed transport to occur. Note that figure 7-7 suggests a hypothesis to be tested. It is also a two-dimensional slice through a multidimensional parameter space that might include additional factors relating to type or amount of sediment supplied and biological parameters bringing about aggregation. A likely candidate for a third axis is aggregate particle density (relative to water), with dense flocs being more easily trapped than less dense ones.

Suspended Sediment Trapping:[C_{ETM}]/[C_{source}]

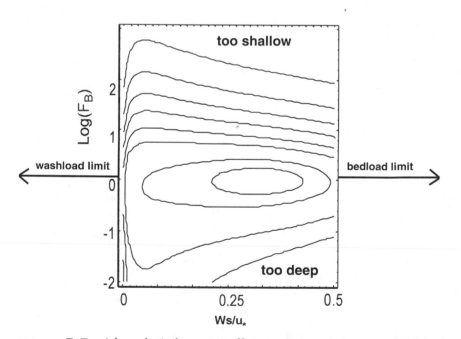

FIGURE 7-7   A hypothetical trapping efficiency diagram—the ratio of SPM concentration in an ETM relative to source-water concentration as a function of W_S/U_* and F_B. The most effective trapping occurs at a sediment size matched to the strength of particle trapping in the estuary.

Combining Geomorphic and Physical Classification Approaches

Along-channel changes in stream ecology, sediment types, and sediment transport processes reflect the physics of river flow down a generally decreasing gradient from steep mountainous terrain to the sea. These processes have a directionality and a simplicity of forcing that has inspired an extensive literature aimed at classification of environments and prediction of sediment transport processes (for example, Hawkes 1975; Schumm 1977; Mosley 1987). We suggest here an extension of the hierarchical structure and predictive capabilities of one such approach (Montgomery and Buffington 1993, 1997, 1998) to estuaries (figure 7-8 and table 7-1). The estuarine portions of the classification scheme generalizes Jay and Smith (1988) to include a broader range of temperate estuarine types, cast in hierarchical form.

One way to classify estuaries is to identify reaches or environments that may occur in each kind of system. It is vital, however, to do this in a structured manner. Thus, the hierarchical approach of Montgomery and Buffington divides the watershed landscape into hillslopes and valleys (figure 7-8). Valleys are made up of three kinds of segments: bedrock, colluvial, and alluvial. Alluvial valley segments are further divided into six kinds of channel reaches that may be distinguished by their slope, type of roughness, hydraulic characteristics, and

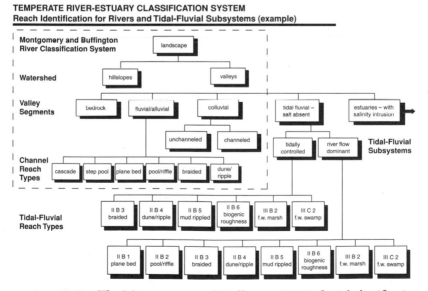

FIGURE 7-8 The Montgomery and Buffington (1993) fluvial classification schema embedded in larger river-estuary system; the numbers and letters in some of the names of the reach types refer to information in table 7-1.

TABLE 7-1 Estuarine reach types.

Estuarine/River Reach Type	Coarse Sediment—Sands and Gravels	Fine Sediment—Silts and Clays	Type of Bed and/or Comments
I. Pelagic nonalluvial			
1. Bedrock/morainal slopes	Supply limited	Supply limited	Bedrock; fjords and straits only
2. Canyons formed by debris flows	Supply limited	Supply limited	Coarse or mixed; fjords and straits only
3. Bedrock sills/channels	Supply limited	Supply limited	Bedrock, with kelp in some cases
4. Relict sediment channels	Supply limited	Supply limited	Fine, mixed (armored), or coarse possibly with kelp
II. Pelagic alluvial			
A. Wave-dominated			
1. Rocky subtidal	Transport limited	Supply limited	Coarse or armored with fine subsurface material, possibly with kelp
2. Estuarine beaches	Transport limited	Supply limited	Coarse or armored with fine subsurface material
B. Rough alluvial			
1. Plane bed	Transport limited	Supply limited	Coarse and armored, fines possible in subsurface
2. Pool/riffle	Transport limited	Supply limited	Coarse
3. Braided	Transport limited	Supply limited	Coarse or mixed
4. Dune ripple	Transport limited	Supply limited	Coarse or mixed
5. Mud rippled	Supply and transport limited	Transport limited	Fine
6. Biogenic roughness	Supply and transport limited	Transport limited	Fine
C. Smooth alluvial			
1. Fluid mud	Supply and transport limited	Transport limited	Fine
2. Anoxic	Supply and transport limited	Transport limited	Fine
III. Flats and marshes			
A. Unvegetated			
1. Rocky intertidal	Transport limited	Supply limited	Coarse or armored with fine subsurface material, possibly with kelp
2. Sand flats	Transport limited	Supply limited	Coarse
3. Mud flats	Supply and transport limited	Transport limited	Fine or mixed
B. With emergent plants			
1. Salt marsh	Supply and transport limited	Transport limited	Fine with plant debris
2. Freshwater marsh	Supply and transport limited	Transport limited	Fine with plant debris
C. Forested			
1. Mangrove/salt swamp	Supply and transport limited	Transport limited	Fine with woody debris
2. Freshwater swamp	Supply and transport limited	Transport limited	Fine with woody debris

sediment transport properties. The structure of and physics behind this schema provides an important predictive capability. Each type of segment occupies a limited range on a spectrum that extends from sediment transport capacity–limited to sediment supply–limited (defined in terms of the materials that structure the bed). Each type responds to changes in sediment supply (relative to river flow) by (among other things) changes in bed configuration.

This classification scheme is able to predict the transport capacity status of a fluvial reach in part because the river itself has created and maintains at least the alluvial reaches in question. This is frequently not the case with estuaries—fjords, for example, occupy troughs previously created by glaciers. Fault-block estuaries like Tomales Bay are a response to local tectonics. Still, estuaries are extensions of rivers. In many cases (such as Atlantic coastal plain estuaries and Puget Sound river deltas) river and estuary are together in the process of filling a depression. Such systems may be either in or out of equilibrium with their surroundings, and the boundary between the two regimes shifts over time as sea-level changes and/or sedimentation occurs. Keeping in mind the significant differences between fluvial and estuarine environments, it is still possible to use a hierarchical classification approach to relate the dominant sediment transport process(es) in each environment to deposition and erosion processes, but cause and effect are more complex.

The six types of fluvial reaches shown in figure 7-8 evolve under the influence of river flow Q_R and sediment supply alone. Classification of estuarine systems is complicated by the presence of at least six distinct transport processes:

- the net motion of river flow (Q_R) through the estuary, which often plays an important role in transporting sediment, despite the presence of other processes;

- oscillatory tidal flow (Q_T), the most important single transport process and source of bed stress in most shallow estuaries;

- internal circulation or buoyancy forcing (Q_I), a collection of processes associated with the presence of water of different densities. Although this type of motion normally requires river inflow for its occurrence, the resulting motion is distinct from and additive to river flow. In fact, it may occur when river flow is very small through an imbalance in evaporation minus precipitation. This category covers several two-layer motions, all having zero net motion over the water column as a whole. These include gravitational circulation forced by the mean density difference between oceanic and fresh water, internal tides and internal motions forced by nonlinearities in the surface tide. The tendency for these motions to concentrate landward flow in the deepest part of a water body can give rise to substantial lateral as well as vertical variability (Fischer 1976);

- atmospherically forced circulation (Q_A) caused both by wind stress at the water's surface and by changes in sea level associated with remote wind-forcing of the adjacent coastal ocean;

- transport and resuspension by wind-waves and swell (Q_W), which are dominant features in some exposed estuarine environments. Typically, net transport due to wave nonlinearities is small, but the enhanced levels of turbulence caused by waves (over and above that associated with currents alone) allows waves and currents together to transport much more material than either acting alone (Grant and Madsen 1979); and

- transport by sea ice (Q_H), a dominant feature in arctic and some temperate systems. While ice may damp tides and prevent movement during the winter when ice is fast to the bed, the spring ice breakup may move large amounts of sediment.

Each of these six types of forcing has a typical time scale and descriptive nondimensional hydrodynamic parameters and is responsible for one or more distinctive styles of sediment transport (table 7-2). The relative importance, for example, of surges associated with remote wind stress over shelf waters in comparison to the barotropic tide might be quantified by $F_A = \varsigma_A/\varsigma_{M_2} \, T_{M_2}/T_A$; that is, the amplitude of the surge relative to the dominant semidiurnal (or M_2) tidal amplitude, multiplied by the ratio of the tidal period to the period of the surge. This parameter can be determined from tidal height records, either on a seasonal, annual average or daily basis. Chesapeake Bay, with hurricane surges of ~10 × the tidal amplitude, is clearly much more strongly forced by storms than the Columbia River estuary or Puget Sound, where the largest storm surges have about the same amplitude as large spring tides. Thus, such hydrodynamic parameters provide a means for connecting this geomorphic classification to the hydrodynamic approaches discussed above.

Sediment transport and geomorphology are more complex in estuaries than in rivers for several reasons in addition to the multiple physical processes responsible for sediment transport. The presence of tides and the prevalence of two-way exchange with the coastal ocean allows more complex ecosystems. These more complex ecosystems bring about a greater role for, and more varieties of, biological roughness elements in estuaries than in rivers. Thus, while vegetated and cohesive streambeds exist, these fluvial environments are less variable and ubiquitous than is the case for their estuarine counterparts. The multiplicity of geomorphic circumstances leading to the formation of estuaries, the varying degree of equilibrium between basin geometry and modern flow processes, and the variety of estuarine types associated with the various possible combinations of the above six basic forcing parameters cause the estuarine classification scheme in figures 7-8 and 7-9a, b, and c to differ from the fluvial subsystem classification

TABLE 7-2
Types of external forcing with associated time scales and nondimensional numbers.

Mechanism and Symbol	Time Scale(s)	Nondimensional Number
River flow—Q_R	Storm and seasonal	Barotropic Froude number: $$F = \frac{U_R}{(gH)^{1/2}}$$
Tidal flow—Q_T	Tidal daily, tidal monthly, seasonal	Tidal Froude number: $$\frac{\zeta}{H} = F_T = \frac{U_T}{(gH)^{1/2}}$$ or modified for tidal flats to: $$F_{TM} = \frac{5}{3}\frac{\zeta}{H} - \frac{\Delta B}{B}$$
Internal/density-driven circulation—Q_I	Tidal monthly and seasonal	Shear Richardson number: $$Ri_I = \frac{(g\Delta\rho_v H)}{(\Delta\rho_0 U_I^2)}$$ Internal Froude number: F_B Internal Froude number: $$G_I = \frac{U_1}{(gH_1)^{1/2}} + \frac{U_2}{(gH_2)^{1/2}}$$
Wind waves—Q_W	5–15 sec, 3–10 days, event and seasonal	Wave height: $\varepsilon_w = \dfrac{\zeta_w}{H}$ Wave steepness: $b = \dfrac{a}{\lambda}$
Atmospheric forcing—Q_A	3–10 days, event and seasonal	
Direct wind stress		$\dfrac{\tau_w}{\tau_B}$
Remote		$\dfrac{\zeta_A}{\zeta_{M2}} \quad \dfrac{T_{M2}}{T_A}$
Sea ice cover	Seasonal	Maximum % seasonal ice cover

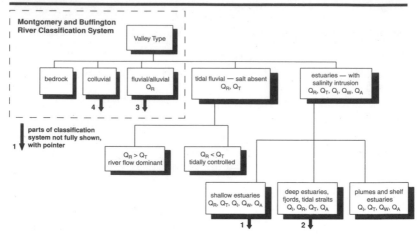

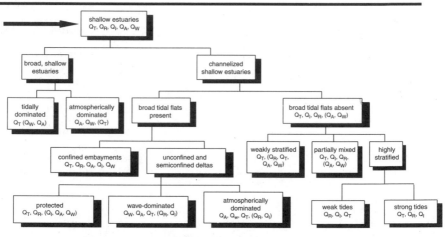

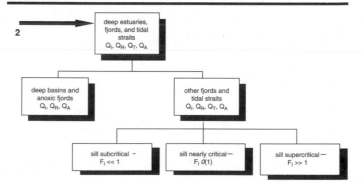

FIGURE 7-9A, B, C River estuary classification scheme showing types of forcing for each subtype.

in figure 7-8. It is necessary to introduce several levels in the hierarchy between the segment and reach or environment level, and similar reach types in different estuary types may have distinctive sediment transport responses. Furthermore, most estuaries will contain several estuarine types between the alluvial channels and the open ocean. Each of these has segments (with diverse small-scale properties), and each may be in or out of equilibrium with present forcing to a different degree. A large estuary may also have several tributary watersheds. The Puget Sound–Straits of Georgia–Strait of Juan de Fuca complex, with its numerous watersheds, fjord basins, sills, straits, deltas, and embayments, is a particularly complicated example. All of the levels of estuarine classification above the reach level will be referred to as "estuarine subsystems," including those (tidal-fluvial and estuaries with salinity intrusion) that are parallel to the valley segments in the Montgomery and Buffington approach.

A major distinction between estuarine and tidal-fluvial subsystem types is the presence in the former of salt and the two-layer internal circulation processes (Q_I in figure 7-9) associated therewith. Because of the narrow, channelized topography of most tidal-fluvial environments, atmospheric and wave forcing (Q_A and Q_W) are usually of minor importance in these subsystems. Most tidal-fluvial channels are ebb dominant (that is, the net transport of sediment and other substances is seaward), because the river flow is larger than asymmetries in the tide that lead to landward transport. Some exceptions to this rule occur in areas of large tides and low topographic relief (for instance, along the English Channel), where flood-dominant sediment transport may be the rule. There are three principal types of estuarine subsystems with salinity intrusion (figures 7-9). These are fjords and deep tidal straits (such as the Strait of Juan De Fuca), plumes and shelf estuaries (such as off the mouth of the Mississippi), and shallow estuaries. Shallow estuaries are in turn divided into those wherein a definite axis of both river flow and tidal propagation can be defined (channelized estuaries) and broad, shallow estuaries and lagoons (for example, the Wadden Sea and some Gulf Coast estuaries); the latter systems often have multiple mouths. Channelized estuarine subsystems are further divided into narrow channelized estuaries (such as the Fraser River) and shallow estuaries with broad tidal flats, including confined embayments (such as Willapa Bay, Washington) and deltas. Deltas may be subdivided in terms of their exposure to atmospheric and wave forcing. Some are protected (for example, at fjord heads), while others are more exposed to wind and wave forcing (such as those in Northern Puget Sound).

It is also necessary to consider for estuaries transport-limitation status (supply versus transport limited) separately for coarse (sands and gravels) and fine sediments (silts, clays, and aggregates formed therefrom; table 7-1). This is vital because there is no fixed relationship between transport-limitation status for the two types of material. Isolation from sources of coarse material may cause estuarine reaches with fine sediments to be simultaneously transport limited for fine material and both supply and transport limited for coarse sediments. That is, no sands and gravels are supplied to such reaches, but if they were supplied, they could not be transported. Conversely, although all coarse-bedded estuarine segments are supply limited for fine materials, they may be either supply or transport limited for coarse material.

A final distinction between fluvial and estuarine systems is the greater importance of biogenic roughness in the latter. Biogenic roughness occurs in many estuarine environments but is particularly important in estuarine reaches relatively remote from direct fluvial influence, wherein coarse sediments are absent because they have been trapped in some more landward, transport-limited reach. There are also smooth-bedded estuarine environments where sediments are so fine that grain roughness does not affect the flow, and even biogenic roughness is absent. This occurs principally in anoxic and fluid-mud bedded reaches.

It is desirable to connect the sediment transport information embedded in this geomorphic scheme to the particle-trapping processes discussed above. A first step in this direction would be to identify the estuarine subsystem types to which the trapping-efficiency diagram (figure 7-7) applies. This identification would assist in determining whether additional parameters are needed, (so that figure 7-7 can be viewed as a slice through a higher-dimensional space). Placement of a substantial number of systems displaying ETM behavior in figure 7-7 might then inspire modeling or theoretical work leading to a predictive ability with regard to particle trapping.

Disturbance frequency, magnitude, and intensity are important processes structuring ecosystem and community dynamics in a variety of environments (Wiens 1976; Sousa 1984). Lotic and riparian ecosystems in particular are strongly structured by spatial and temporal variance of geomorphological disturbance processes (Swanson et al. 1988; Montgomery 1999). Extension of these concepts to estuarine ecosystems is illustrated by considering the mean, variance and spatial heterogeneity of the residence time T_R (figure 7-6). While other types of physical forcing (as in table 7-2) and numerous biological factors not present in fluvial environments contribute to the complexity of estuarine ecosystems, this approach may provide a framework for connecting the dynamics of physical and biological systems and for further developing process-based estuarine classification schemes.

In summary, a geomorphic estuarine classification schema with a hierarchical structure provides a convenient means to identify environments or reach types found in the various types of estuarine subsystems, relates estuarine types to dominant sediment transport processes, and allows a prediction as to whether sediment transport in each reach type is limited by transport capacity or sediment supply for coarse and fine sediments. The principal connection to past or possible future hydrodynamic classification schemes arises through the nondimensional hydrodynamic parameters associated with each sediment transport-forcing mode. Additional connections to biologically relevant processes such as particle trapping and residence time are possible.

Conclusions and Recommendations

A need to understand and respond to the rapid global alteration of estuarine systems requires that representative systems be studied in detail, and that better means be developed to avoid and ameliorate future impacts and compensate for past ecosystem losses. The urgency of these issues suggests that a process-based estuarine ecosystem classification scheme should be developed in the near future. Traditional hydrodynamic estuarine classification systems have focused on the interaction of buoyancy and tidal forcing in narrow estuaries. Although existing systems are too restricted with regard to types of systems and the range of forcing processes included, they may be generalized to create a broader, more useful framework. As a more immediate goal, we have suggested relationships that would (once quantified) allow estimates of residence time and particle-trapping efficiency from bathymetric data and a few measurements. Geyer et al. (this volume) takes a comparative approach to analysis of estuarine ecosystems and sets longer-term goals for estuarine classification.

A geomorphic classification schema with a hierarchical structure has also been presented. This approach identifies the major sediment-transport mechanisms for a spectrum of estuarine types, identifies the types of reaches or environments found in each, and allows determination of transport capacity versus supply limitation separately for coarse and fine sediments. A connection to hydrodynamic classification approaches is provided through nondimensional parameters associated with each of six major types of estuarine forcing. The same hydrodynamic numbers are likely to appear as governing parameters for estimation of particle-trapping efficiency and residence time. Thus, it should also be possible to connect this geomorphic approach to these and other parameters of ecosystem relevance.

Acknowledgment

The work of David Jay and W. Rockwell Geyer was supported by the NSF's Land-Margin Ecosystem Research program projects for the Columbia River estuary (OCE8907118 and extension) and Waquoit Bay. Additional support for David Jay was provided by the Skokomish Tribe. Support for David R. Montgomery was provided by the EPA's Pacific Northwest Ecosystem Research Consortium, through cooperative agreement CR824682.

References

Allen, G. P., J. C. Salomon, P. Bassoullet, Y. Du Penhoat, and C. DeGrandpré. 1980. Effects of tides on mixing in macrotidal estuaries. *Sedimentary Geology* 26: 69–80.

Cowardin, L. M., V. Carter, F. C. Golet, and E. T. LaRoe. 1979. *Classification of wetlands and deepwater habitats of the United States.* FWS.OBS-79/31, Washington, DC: U.S. Fish and Wildlife Service.

Dethier, M. N. 1990. *A marine and estuarine habitat classification system for Washington state.* Olympia, WA: Dept. of Natural Resources.

Dyer, K. R. 1973. *Estuaries, a physical introduction.* New York: Wiley-Interscience.

Festa, J. F., and D. V. Hansen. 1978. Turbidity maxima in partially mixed estuaries—a two-dimensional numerical model. *Estuarine and Coastal Marine Science* 7: 347–59.

Fischer, H. P. 1976. Mixing and dispersion in estuaries. *Annual Review of Fluid Mechanics* 8: 107–33.

Friederichs, C. T., and O. S. Madsen. 1992. Nonlinear diffusion of the tidal signal in frictionally dominated embayments. *Journal of Geophysical Research* 97: 5637–50.

Grant, W., and Madsen, O. S. 1979. Combined wave and current interaction with a rough bottom. *Journal of Geophysical Research* 84: 1797–1808.

Hansen, D. V., and M. Rattray Jr. 1965. Gravitational circulation in straits and estuaries. *Journal of Marine Research* 23: 104–22.

———. 1966. New dimensions in estuary classification. *Limnology and Oceanography* 11: 319–26.

Hawkes, H. A. 1975. River zonation and classification. Pages 312–74 in B. A. Whitton (editor), *River ecology.* Berkeley: University of California Press.

Hearn, C. J. 1998. Application of the Stommel model to shallow mediterranean estuaries and their characterization. *Journal of Geophysical Research* 103: 10391–10405.

Ippen, A. T., and D. R. F. Harlemann. 1961. Technical Bulletin 5. Committee for Tidal Hydraulics, Vicksburg, MS: U.S. Army Corps of Engineers Waterways Experiment Station.

Jay, D. A. 1995. Residence time, box models and shear fluxes in tidal channel flows. Pages 3–12 in K. R. Dyer and R. J. Orth (editors), *Changes in fluxes in estuaries.* Fredensborg: Olsen and Olsen.

Jay, D. A., and J. D. Smith. 1988. Circulation in and classification of shallow, stratified estuaries. Pages 21–41 in J. Dronkers and W. van Leussen (editors), *Physical processes in estuaries.* Heidelberg, Germany: Springer-Verlag.

———. 1990. Circulation, density distribution and neap-spring transitions in the Columbia River Estuary. *Progress in Oceanography* 25: 81–112.

Jay, D. A., and J. D. Musiak. 1994. Particle trapping in estuarine turbidity maxima. *Journal of Geophysical Research* 99: 20, 446–61.

Jay, D. A., R. J. Uncles, J. Largier, W. R. Geyer, J. Vallino, and W. R. Boynton (LMER Scalar Transport Working Group). 1997. A review of recent developments in estuarine scalar flux estimation. *Estuaries* 20: 262–80.

Montgomery, D. R. 1999. Process domains and the river continuum. *Journal of the American Water Resources Association* (in press).

Montgomery, D. R., and J. M. Buffington. 1993. *Channel classification, prediction of channel response, and assessment of channel condition.* Report TFW-SH10-93-002 for the SHAMW Committee of the Washington State Timber/Fish/Wildlife Agreement. Washington State Department of Natural Resources.

———. 1997. Channel reach morphology in mountain drainage basins. *Geological Society of America Bulletin* 109: 596–611.

———. 1998. Channel processes, classification, and response potential. Pages 13–42 in R. J. Naiman and R. E. Bilby (editors), *River ecology and management.* New York: Springer-Verlag.

Mosely, M. P. 1987. The classification and characterization of rivers. Pages 194–320 in K. Richards (editor), *River channels: Environment and process.* Oxford, England: Blackwell.

Nunes Vaz, R. A., and G. W. Lennon. 1991. Modulation of estuarine stratification and mass transport at tidal frequencies. Pages 505–20 in B. B. Parker (editor), *Progress in tidal hydrodynamics.* New York: John Wiley and Sons.

Oey, L.-Y. 1984. On the steady salinity distribution and circulation in partially mixed and well mixed estuaries. *Journal of Physical Oceanography* 14: 629–45.

Officer, C. B. 1976. *Physical oceanography of estuaries (and associated coastal waters).* New York: Wiley-Interscience.

Prandle, D. 1986. Generalized theory of estuarine dynamics. Pages 42–57 in J. van de Kreeke (editor), *Physics of shallow estuaries and bays.* Berlin, Germany: Springer-Verlag.

Pritchard, D. W. 1952. Estuarine hydrography. *Advances in Geophysics* 1: 243–80.

Rattray, M., Jr. and R. J. Uncles. 1983. On the predictability of the ^{137}Cs distribution in the Severn Estuary. *Estuarine, Coastal and Shelf Science* 16: 475–87.

Schumm, S. A. 1977. *The fluvial system.* New York: John Wiley & Sons.

Simpson, J. H., and J. R. Hunter. 1974. Fronts in the Irish Sea. *Nature* 250: 404–6.

Smith, S. V., J. T. Hollibaugh, S. J. Dollar, and S. Vink. 1991. Tomales Bay metabolism: C-N-P stoichiometry and ecosystem heterotrophy at the land-sea interface. *Estuarine, Coastal and Shelf Science* 33: 223–57.

Sousa, W. P. 1984. The role of disturbance in natural communities. *Annual Review of Ecology and Systematics* 15: 353–91.

Stokes, G. G. 1847. On the theory of oscillatory waves. *Transactions of the Cambridge Philosophical Society* 8: 441–55.

Stommel H., and H. G. Farmer. 1952. *On the nature of estuarine circulation.* Woods Hole Oceanographic Institution Reference Notes 52–51, 52–63, and 52–88.

Swanson, F. J., T. K. Katz, N. Caine, and R. G. Woodmansee. 1988. Landform effects on ecosystem patterns and processes. *BioScience* 38: 92–98.

Wiens, J. A. 1976. Population responses to patchy environments. *Annual Review of Ecology and Systematics* 7: 81–120.

Zimmerman, J. F. T. 1978. Topographic generation of residual circulation by oscillatory (tidal) currents. *Geophysical and Astrophysical Fluid Dynamics* 11: 35–47.

————. 1988. Estuarine residence times. Pages 75–84 in B. J. Kjerfve (editor), *Hydrodyamics of estuaries.* Vol. 1. Columbia, SC: University of South Carolina Press.

CHAPTER 8

Interaction between Physical Processes and Ecosystem Structure:

A Comparative Approach

W. Rockwell Geyer, James T. Morris, Frederick G. Prahl, and David A. Jay

Abstract

This review assesses our current understanding of the role of physical processes on estuarine ecology and advocates a comparative approach in future efforts to understand the linkage between the physics and biology. Physical processes affect the ecology at a broad range of spatial and temporal scales. These interactions also occur simultaneously at different trophic levels and may include multiple, interacting variables. Due to the breadth of scales and the complexity of these interactions, our understanding of the relationship between the physics and ecology of estuaries is still seriously limited. One important step toward improving our understanding of these processes is the systematic comparison of physical and ecological processes among different estuarine systems. Such a comparative approach provides much more dynamic range than can be accomplished by studies within an individual system. More effective use of numerical models will also be of considerable value, both for process studies and for realistic simulations of particular estuaries. Both simple and complex numerical models should be used in conjunction with comparative studies among estuaries to quantify the relationships between physical forcing and ecological processes.

Introduction

Estuaries are already used, abused, and managed heavily for resource harvests, habitation, and recreation. The pressures on estuarine resources from population redistribution and growth will increase significantly in the next few decades. The resulting changes in nutrient loading and hydrology at the regional scale, as well as accelerated sea-level rise and climate change at a global scale, will lead to significant alterations of estuarine habitats. Already

77% of the total water discharge of the 139 largest river systems in North America, Europe, and former Soviet republics is significantly modified by dams or water diversions (Dynesius and Nilsson 1994). Human activities such as channelization and damming of rivers have altered the timing, magnitude, and nature of inputs of materials to estuaries (Hopkinson and Vallino 1995; Rozengurt et al. 1987). Freshwater diversions have also significantly altered the salinity and sediment budgets of estuaries (Kjerfve and Magill 1990; Simenstad et al. 1992; Jay and Simenstad 1996). Nutrient inputs to estuaries are increased by conversion of land to agricultural purposes (Correll et al. 1992) or increasing urbanization of coastal watersheds (Valiela et al. 1990, 1992; LMER Coordinating Committee 1992). The history of coastal eutrophication due to anthropogenic activities is well documented (Turner and Rabalais 1994; Balls et al. 1995; Officer et al. 1984), though the effects differ depending largely on physical processes. We cannot stop changes from happening, nor is that the role of the research community. Our challenge is to provide a quantitative, predictive understanding of the responses of estuaries to changes in forcing variables, at the temporal and spatial scales, that are relevant to estuarine resources and management concerns. A goal of estuarine research should be to identify and understand the cause-and-effect relationships among the physical and biological processes that define estuarine structure and function. We must understand the basic interactions between physical and ecological processes before we can reliably predict the responses of estuaries to changes in physical processes brought about by human activities and in order to distinguish natural environmental variation from anthropogenic effects.

We have only a rudimentary knowledge of how the physical regimes of estuaries affect ecology and vice versa. The variability in the physics and of the chemical and biological components of estuaries is large. Estuaries differ, for instance, in trophic efficiency, utilization of primary production, and even in the domination of energy flow by detrital or herbivore pathways (Baird and Ulanowicz 1993; Smith and Hollibaugh 1993). This same variability provides a large signal with which to investigate the linkages between physics and ecology among the great diversity of estuarine regimes. We recommend a comparative approach that exploits the large differences in forcing and response between different estuaries in order to better understand and quantify the response of estuaries to changes in forcing variables. This cross-system comparison can then be used to develop predictive models for assessing the response of individual estuaries to anticipated changes in forcing conditions. Such an approach is closest to being realized in the context of physical responses to physical forcing variables, such as changes in sea level or freshwater inflow. However, even the purely physical response characteristics of estuaries are not known adequately to provide confident, a priori predictions, and far less is known about the ecological responses to physical forcing and the possible feedback mechanisms that may occur.

In spite of these difficulties, the time is right to clearly define our goals for attaining this predictive capability, and for defining research priorities that will achieve this goal efficiently. We suggest a shift in emphasis from focused studies of individual estuaries to a cross-cutting, integrative approach that will permit the gains in understanding within one system to benefit our understanding of other environments that have not been studied in as much detail. This does not mean that focused studies of individual estuaries should be abandoned, but rather that tools must be developed to generalize the results of these studies, with respect to the physical mechanisms as well as the biogeochemistry and ecology.

The Coupling of Physics, Biogeochemistry, and Ecology

At virtually every spatial scale, within every component of estuarine ecosystems, physical processes influence the distribution and fate of chemicals and organisms (figure 8-1). The delivery of nutrients to the biota depends not only on the sources of nutrients to the estuary, but also fundamentally on the advective and dispersive processes that transport and retain the dissolved and particulate material (Rudek et al. 1991; Eyre 1993). The recruitment of estuarine fauna is also dependent on the physical regime, as exemplified by the influence of estuarine circulation on the recruitment of blue crab larvae in Delaware and Chesapeake Bays (Epifanio et al. 1984). The primary productivity of salt-marsh ecosystems involves a sensitive feedback loop between variations in physical variables such as sea level, tidal flooding, and pore-water salinity (Morris 1995), which in turn influences the physical regime (Leonard and Luther 1995).

At the nearly microscopic scales that influence the ecology of small benthic invertebrates, the feedback between physics and biology controls not only ecology but also near-bottom turbulence characteristics, sediment transport, and sediment-water exchange (Nowell et al. 1981). Microbes play a key role in construction of large aggregates that in turn play a large role in support of epibenthic zooplankton in estuarine turbidity maxima. Along with turbulence, zooplankton consuming the bacteria play a strong role in aggregate disruption (Crump and Baross 1996). The synchrony of biological and physical events and the relative time scales at which different processes operate are critically important. Physical processes such as diurnal and semidiurnal tidal height variation and wind stress that vary periodically on time scales equivalent to biological reaction rates can have significant effects on ecological processes. For example, Litaker et al. (1993) found that most of the variance in biologically important variables in a shallow mid-Atlantic estuary occurred at time scales of 12 to 96 hours because of physical events, including runoff,

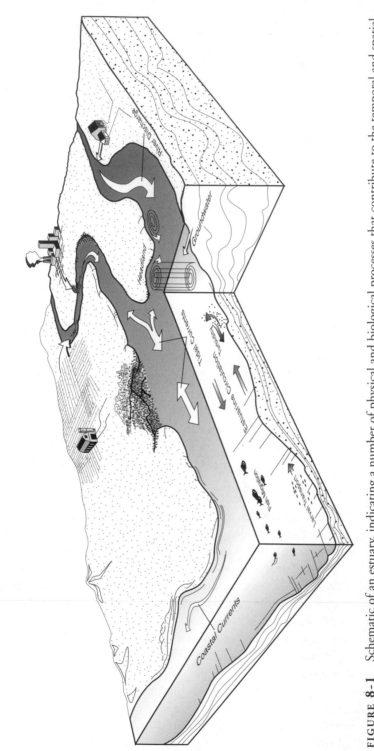

FIGURE 8-1 Schematic of an estuary, indicating a number of physical and biological processes that contribute to the temporal and spatial variability of the estuarine regime. Tidal currents (including tide-induced residual circulations at headlands), estuarine circulation, and river discharge contribute to the water motion. Anthropogenic inputs and groundwater flow provide geochemical sources. Physical processes modulate exchange with fringing marshes and the adjoining coastal ocean and affect the turbidity distribution. Biological transport can be passive, in the case of larval transport, or active, in the case of fish migration.

insolation, wind, tides, and temperature change. Dominant epibenthic zoo-plankton in river estuaries with strong tidal currents synchronize their vertical migration with the tides (rather than responding to light, which is essentially absent in their environment) to optimize their retention in the estuary (Morgan et al. 1997). Stratification events in the Delaware estuary are controlled by river discharge, and when stratification occurs, productivity is stimulated because of the formation of a shallow surface mixed layer (Pennock 1985; Sharp et al. 1986). The occurrence of anoxic bottom water in the upper Chesapeake Bay is also associated with stratification, caused by high discharge, but the timing and duration of the high-flow period has a major impact on the initiation of hypoxic conditions (Schubel and Pritchard 1986). Interannual variation in peak algal biomass in the Chesapeake is associated with changes in nutrient loading (Boynton et al. 1982) and with variation in vertical stratification, which influences the vertical flux of regenerated ammonium from sediments to the euphotic zone (Malone et al. 1988). Physical processes that affect productivity can have an enhancing effect on other trophic groups such as bacteria (Sharp et al. 1986; Hoch and Kirchman 1993). However, not only are there resource or bottom-up effects, but at every trophic level there are direct effects of the timing and magnitude of physical processes, like the onshore movement of Atlantic cod in the Gulf of St. Lawrence, which depends on nearshore sea temperatures and currents that are linked to wind-forced upwellings and downwellings (Rose and Leggett 1988a, 1988b). Another example of the sensitivity of a key component of a food web to variation in a physical condition is provided by Lagoa Munda, Brazil, where the harvest of benthic mollusks is strongly dependent on changes in salinity and interannual variation in rainfall. Sudden drops in salinity produce massive mortality of the mollusks, and the recovery of their populations takes several years (Oliveira and Kjerfve 1993).

The spatial and temporal distributions of ecologically significant variables respond to different forcing functions on different time scales, such as the response of suspended sediment and salinity distribution to tidal circulation, wind, and river discharge, and the variation in the strength of these variables over the spring-neap cycle (Powell et al. 1989). Variability on short time scales in such ecologically important factors as nutrients and suspended sediments can be as great as the seasonal variability (Caffrey and Day 1986). Physical processes operate on multiple time scales and adjust to perturbations at different rates. For example, the low frequency response of the velocity field of the Eastmain estuary, James Bay, Canada, was much faster to equilibrate to a rapid change in discharge than was the distribution of salinity (Lepage and Ingram 1988). Wind forcing assumes a greater role as a determinant of estuarine salinity distributions as river discharge decreases (Litaker et al. 1987). Thus, the relative degree of control among physical processes of the temporal and spatial patterns within estuaries varies among estuaries. Furthermore, examples are known where the degree of variation in a physical parameter like salinity, rather than the mean, controls key ecological parameters such as population density and community structure (Montague and Ley 1993). The variability in the physics and in the chemical and biological components of estuaries is large, but the same variability also poses a means of investigating and defining fundamental processes and interactions.

Probably no physical process has a greater effect on estuarine ecology than temperature regulation. Inasmuch as temperature is a driver of evapotranspiration, the hydrology of intertidal sediments varies seasonally with temperature (Howes and Goehringer 1994). Furthermore, the density and viscosity of water are temperature sensitive. For example, as temperature rises from 5 to 30°C hydraulic conductivity increases by a factor of 2 (Dingman 1994). Also the equilibria of volatile organics and other gases between air and water change significantly with temperature (Dewulf et al. 1995). Thus, physical transport processes that are critically important to ecosystem structure and function are highly sensitive to temperature.

Typically, biological processes have a greater sensitivity to temperature than do physical processes (Hartenstein 1972), and a number of physical processes have underlying biological roots that are temperature sensitive. For example, the strong coherence often observed between sedimentation rates, concentrations of suspended sediments, and water temperature has been attributed to temperature-sensitive bioturbation (Wolaver et al. 1988; Gardner et al. 1989; Hutchinson et al. 1996). The literature abounds with examples of the temperature dependence of microbial processes ranging from denitrification (Nowicki 1994) to the decomposition of specific compounds like DMSP (Kiene and Service 1991). Bacterial metabolism generally increases three to six times for a 10°C rise in temperature (Sander and Kalff 1993). Estuaries with seasonally variable temperature may undergo shifts in the regulation of their food webs. For instance, Shiah and Ducklow (1994) have suggested that bacterial production in the Chesapeake Bay is limited by temperature during winter rather than by substrate supply. Seasonally variable water temperatures appear to drive cycles of nutrient regeneration and consumption in sediments that can change the direction of nutrient flux (Boynton et al. 1980). Temperature-dependent microbial metabolism in sediments can also affect nutrient flux by altering the depth of the surface oxidized layer. This is thought to explain the seasonal variation in release of sediment phosphorus from Danish lakes (Jensen and Andersen 1992).

An understanding of climate effects on estuaries is only beginning to emerge. Temperature, river inflow, and budgets of dissolved and particulate substances respond strongly to climate fluctuations. Thus, river flow and sediment input to major West Coast estuaries from San Francisco Bay to the

Fraser River respond to the El Niño–Southern Oscillation (ENSO) and lower frequency climate cycles (Redmond and Koch 1991; Schwing et al. in review). The response of fluvial sediment and organic matter input to estuaries is stronger than that of the river flow Q_R alone, because fluvial suspended-sediment transport generally follows a power law; sediment input $\sim Q_R{}^n$, where $n = 1.5$ to 3. Estuarine water-temperature behavior is complicated by its response to air temperature and upwelling, both of which follow ENSO cycles but are not necessarily in phase. These climate influences extend up the food chain to salmon (Mantua et al. 1997), but not all salmon populations have the response to increased flow. High flows favor some Columbia River salmonids, while several Fraser River runs show better survival of juveniles under low-flow conditions (Pearcy 1992; Beamish et al. 1994).

Temperature affects every biological process at every trophic level including the migration (Hettler and Chester 1982) and reproductive success (Secor and Houde 1995) of marine organisms. However, temperature does not affect the metabolic rates of all trophic groups equally. For instance, bacterial metabolism is inhibited in cold waters to a greater degree than photosynthesis, which shifts the balance of energy flow between microbes and herbivores and results in fundamental differences in the trophic dynamics of near-arctic and tropical marine food webs (Pomeroy and Wiebe 1988; Pomeroy and Deibel 1986). More comparative work of this nature is needed before estuarine responses to climate can be understood. Questions about the distribution of estuarine metabolism across climate gradients and the ecological responses to climate are unresolved. Is the change in ecosystem metabolism across a latitudinal temperature gradient as sensitive to temperature as the individual, in situ biological rate processes, or does temperature compensation occur at the scale of the ecosystem?

The discharge of dissolved and particulate materials from watersheds is controlled by complex interactions between climate, lithology, sea level, and geomorphology. On a broad scale, the interaction of climate, particularly annual precipitation, and the underlying lithology determine the weathering rate and dissolved yield of minerals that are carried in the discharge from watersheds to estuaries (Bluth and Kump 1994). The discharge of materials is modified by important interactions between the intertidal zone sediments and surface water (Vörösmarty and Loder 1994). These interactions are determined in part by the local geomorphology that determines the hydraulic gradient and the turnover of water and nutrients in intertidal sediments (Harvey et al. 1987; Howes and Goehringer 1994). Groundwater turnover and the advection of solutes may be modified also by the accumulation of sediments of low hydraulic conductivity (Millham and Howes 1994). The discharge of regional groundwater into coastal-zone sediment is important also and is an example of the significance of regional-scale geomorphology, because it exerts a major control over the flux of solutes between intertidal sediments and flood waters (Nuttle and Harvey 1995). Finally, changes in relative mean sea level modify the balance between surface runoff and groundwater discharge by altering the hydraulic gradient of regional groundwater (Nuttle and Portnoy 1992).

One of the physical features of estuaries that sets them apart from other ecosystems is the regular hydrologic pulsing by tides that allows for regular periodic interactions between intertidal sediments and surface water. In some estuaries the discharge of shallow groundwater that drains from the marshes into the tidal creeks provides a nutrient subsidy to the surface water (Whiting and Childers 1989; Morris, this volume). These advective movements of shallow groundwater evidently change with tidal amplitude and between neap and spring tides (Chapman 1938; Carr and Blackley 1986). Within the vegetated intertidal zone there are spring-neap cycles of sediment oxidation and reduction, depending on subsurface drainage and the elevation of the marsh surface (Armstrong et al. 1985), and these oxidation-reduction cycles affect the immobilization and solubilization of trace metals (Gambrell et al. 1991). Prahl et al. (1997) indicate that the intertidal inputs of solutes may control their concentrations in the main channel of the Columbia River estuary. The hydroperiod or frequency of inundation of tidal marshes has a significant effect on plant community composition (Olff et al. 1988) and controls access to marsh surface habitats by nekton (Rozas 1995). It follows that the nature of these interactions must vary among estuaries as a function of tidal harmonics.

There are important effects on ecological processes of longer tidal periods as well. These include interannual variations in the solar annual tidal cycle that modify the size of intertidal habitat and affect the recruitment of commercially important fishes (Morris et al. 1990). Changes in tidal amplitude associated with the 18.6-year nodal cycle affect the recruitment of fish larvae by altering the retention and turnover of water within estuaries (Cabilio et al. 1987). The amplitude of this cycle in the Pensacola, Florida, area is about 9 cm (Marmer 1954), which is highly significant given the low relief of intertidal area along much of the coast and the volume of tidal exchange that a change in amplitude of this magnitude represents. This 18.6-year tidal cycle is thought also to account for the periodic stabilization of mudflats by colonization by mangroves during periods of abnormal exposure and diminished tidal amplitude (Wells and Coleman 1981). Finally, there is a strong relationship of the very high tides at the peak of the annual and nodal cycles with catastrophic coastal erosion events that may breach sand spits protecting estuaries and generally cause major changes in estuarine habitats (Wood 1978).

Estuarine circulation is a unique feature of estuaries that affects the spatial and temporal distributions of suspended particles, dissolved substances, and organisms. Tidal currents moving up-estuary affect these distributions by stirring and suspending bottom sediments, and the strength of these currents varies with tidal amplitude. This can affect the formation and location of the estuarine turbidity maximum (Vale and Sundby 1987; Uncles and Stephens 1993), as well as the temporal and longitudinal distributions of dissolved oxygen, due to the high oxygen demand of resuspended sediments (Nelson et al. 1994). In the mixing zone between fresh and salt water, cation exchange processes result in the desorption of ammonium from suspended sediments, and in this region of the estuary the sorption-desorption processes are more important to ammonium transport than diffusive fluxes from the sediment (Simon 1989). Different nutrients are affected by different physical processes. For example, Eyre (1993) suggested that phosphorus was controlled primarily by sedimentary processes that operate on a time scale of hours (such as tidal currents), and that differential sorting of particle sizes had different effects on the distributions of particulate nitrogen and phosphorus.

These processes that are controlled by estuarine circulation have important effects on the distributions and activity of marine organisms. For example, the mixing zone is often an area of high nutrient concentrations, as discussed above, and it is often here that a chlorophyll maximum occurs. Herbivore populations, such as menhaden, are often associated with the chlorophyll maximum in the mixing zone (Friedland et al. 1996). The location of the chlorophyll maximum and its attendant food web will shift with movements of the mixing zone. In addition to their effect on resuspension of nutrients and sediments in the mixing zone, tidal currents also can destratify the water column, depending on their strength, and mix oxygenated surface water with nutrient-rich bottom water leading to bloom conditions (Webb and D'Elia 1980).

The physical process of mixing of fresh and salt water along estuarine salinity gradients also affects a number of chemical and biological processes independently of circulation effects. Along these salinity gradients the removal of dissolved humic substances, metals, and phosphorus by flocculation constitutes a significant part of the chemical mass balance between rivers and the ocean. Most of this removal occurs in the mixing zone between 0 and 5 ppt (Sholkovitz 1976; Sholkovitz et al. 1978). River-borne iron, which consists largely of mixed iron oxide-organic matter colloids, is rapidly removed from the water column during estuarine mixing as seawater cations neutralize the colloids and allow for their flocculation (Boyle et al. 1977). Turner and Milward (1994) indicate that about 13% of iron was removed in this manner. Cation exchange processes that are sensitive to salinity include the reversible sorption of metal cations onto suspended particulate matter (Paalman et al. 1994). In general, partitioning of trace metals between dissolved and solid phases is salt sensitive and will respond to both the concentration and quality of particle populations induced by tidal resuspension, though trace metals do not all behave qualitatively in the same way (Turner and Milward 1994). Even the partitioning of gases between the atmosphere and the dissolved phase is affected by salinity due to changes in solubility (Yurteri et al. 1987).

Low salinity regions of estuaries are areas of rapid chemical and biological reactions. In the Tamar estuary there is a rapid drop in dissolved oxygen in the mixing zone, which was thought to be a result of plasmolysis of freshwater phytoplankton and rapid microbial consumption of O_2 (Morris et al. 1978). This explanation appears to be consistent with what is known about the distribution of bacteria. For example, in the St. Lawrence estuary the abundance of free bacteria was greatest at the freshwater end and declined exponentially with increasing salinity, while attached bacteria had highest mean concentrations at salinities between 0.5 and 5 ppt (Painchaud et al. 1995). The dominance of various forms of microbial metabolism varies along salinity gradients as well. This includes methanogenesis and sulfate reduction, which dominate at the freshwater and saltwater ends, respectively (Bartlett et al. 1987). High salinity (ca. 28 ppt) was shown to inhibit ammonium- and nitrite-oxidizing bacteria, and variation in the response of nitrifying bacteria to salinity may affect the distribution of nitrogen species along estuarine salinity gradients (Furumai et al. 1988).

In spite of the wealth of examples of physical-biological interactions in estuaries, there are only a handful of cases in which these coupling processes are well understood. Moreover, there have been few attempts to compare these processes in different estuaries. The extensive correlative study by Nixon et al. (1986) of the influence of nutrient loading on primary productivity provides an important first step toward developing a mechanistic model of the influence of external forcing factors on ecological response. However, Boynton et al. (1995) point out that we still lack an understanding of how the variations in flushing characteristics of different estuaries influence the response to loading. Vollenweider (1975) showed that the response of lakes to nutrient loading varies with residence time, and a similar dependence would be expected for estuaries although it has not been verified. Monbet (1992) used a correlative approach to show that tidal conditions influence estuarine primary productivity, with macrotidal estuaries showing a diminished response to nutrient loading. Monbet's analysis did not distinguish the mechanics of this response. There are a number of ways that tides may influence primary productivity, for example, changing the turbidity and hence the light penetration, changing the stratification, or changing the flushing characteristics. Without a more definitive analysis of the physical response to the variation in tidal range, the next step of predicting the ecological response to tidal variations is seriously compromised.

The above example indicates that we need to go beyond correlative comparisons to mechanistic comparisons between systems. In order to develop predictive, mechanistic models of the ecological responses of estuaries to changing forcing conditions, we must integrate process-oriented research with comparative studies of estuaries. It is not adequate to understand the physical-biological coupling processes in a few, well-studied estuaries; we also require a systematic means of extending the results from one estuary to others that have not been the subject of detailed, interdisciplinary studies.

The synthesis of our understanding of estuarine processes among different estuaries should be addressed at two levels: (1) improved estuarine classification; and (2) interdisciplinary, numerical modeling. These two approaches appear at first glance as polar opposites, the first being a descriptive, reductionist approach, and the second being a quantitative, detailed representation of the estuarine regime. Rather than viewing these as opposites, it is more constructive to view estuarine science as continuum of approaches whose goal is twofold: understanding and prediction. Estuarine classification provides a framework for addressing similarities and differences between different systems, which in order to be useful should contribute to our understanding of estuarine processes and allow us to make predictions. For example, the identification of an estuary as partially mixed allows the knowledge gained from other systems of that type to be applied to that system, including the prediction of response to changes in forcing conditions. A simulation model would be expected to provide a more quantitative prediction, but it would suffer from lack of generality. Yet simulation models have great promise as tools for gaining understanding and investigating the differences between different systems, much in the same spirit as estuarine classification but with a more rigorous, quantitative approach.

The remainder of this chapter charts a course for the future of physical oceanographic research in estuaries as it relates to estuarine ecology. First, the complementary roles of estuarine classification and numerical simulations are discussed. These sections are followed by discussion of specific issues of particular relevance to physical-biological interactions, including residence time, the transport of passive and active particles, time dependence, spatial variability, and interactions between multiple systems.

Estuarine Classification: A Framework for Comparison

Estuarine classification is a basic tool for the comparison of different systems. At the minimum, estuarine classification provides a language with which to describe the similarities and differences between systems. Classification of estuaries by physical parameters and/or geomorphologic characteristics has a long history. Stommel and Farmer's (1952) division of estuaries into well mixed (type 1), partially mixed (type 2), fjord-like (type 3), and salt wedge (type 4) still provides a useful descriptive basis for physical classification. Hansen and Rattray (1966) provided the first parametric classification, using stratification and estuarine circulation as the two parameters that quantitatively classify the estuarine regime. The classification system was based on an analytical model of estuarine circulation, providing a mechanistic connection between the two parameters and the estuarine transport processes. It is compromised, however, by overly restrictive assumptions about a steady-state salt balance, inadequate parameterization of tide-induced mixing, and neglect of transverse circulation and spatial variations in estuarine geometry. More recent classification approaches (Prandle 1985; Jay and Smith 1990) have improved on the parameterization of tidal processes; however, the predictive power of these approaches is still limited by incomplete knowledge of mixing processes and neglect of transient processes and spatial variability of estuarine morphology.

In any attempt to classify estuaries, we need to recognize the inherent variability, both temporal and spatial, within individual estuaries (Jay et al., this volume). The Columbia River varies between highly stratified and well mixed within the fortnightly tidal cycle (Jay and Smith 1990). Fjords typically have very different physical regimes in the deep basins and sills (Ebbesmeyer and Barnes 1980; Geyer and Cannon 1982). Upon closer inspection, most estuaries reveal characteristics of different "classes" at different times and places, which cannot be effectively described by the existing framework of classification. Further developments in estuarine classification should incorporate the variability as well as the mean conditions in the framework of classification. The spring-neap variability of stratification is a clear choice as a defining parameter for estuarine hydrography; similarly the sensitivity of the estuarine regime to variability of freshwater inflow should be parameterized. New approaches have to be devised to characterize spatial variability. One important index may be the ratio of cross-estuary to along-estuary salinity gradients; another might be the tortuosity of estuarine channels. These considerations of estuarine variability as part of estuarine classification will better articulate the connections between estuarine structure and estuarine processes, which is the ultimate goal of classification.

As estuarine classification is refined, it begins to represent a model for estuarine processes. For example, the Hansen-Rattray classification scheme is based on an underlying model of the estuarine circulation. Although that model does not encompass our current understanding of the controlling variables, it exemplifies the continuum between estuarine classification and estuarine modeling. Classification of estuaries starts out at a descriptive level, but models emerge from an understanding of the underlying mechanisms that determine the differences defined by the classification scheme. Physical processes are readily represented by models because the constitutive relations are known a priori. For particles, chemicals, and biota, the situation is more complex, and processes are not as readily represented by simple, parametric models. These processes are beginning to be treated in large, complex numerical models. Now the task is to learn what these complex models are telling us about the underlying mechanisms.

Numerical Modeling: A Tool for Synthesis

The greatly increased power of computers in the last decade increases the potential for numerical simulations to capture the complex processes that occur in estuaries. This power is a blessing and a curse, because although the processes can be represented in ever finer detail, our understanding of the mechanisms and our ability to compare the processes occurring in different systems have not increased commensurately with increased complexity of models. In fact, numerical models of estuaries are much more commonly used for simulation of a particular environment than for process studies and intersite comparative studies, due in part to the urgent management needs with respect to estuarine water quality (Johnson et al. 1993; Cerco and Cole 1993). Yet numerical models are particularly well suited to process studies and intersite comparisons, providing a valuable complement to field observations and theoretical investigations. Increased use of numerical models for these purposes will provide a much needed synthesis of the processes that influence the dynamics and ecology of estuaries, addressing issues that are beyond the capabilities of correlative studies or simple, parametric analyses.

Numerical models are frequently used for process studies in estuaries, but the strategy is typically to reduce the dimensionality of the system (Festa and Hansen 1978; Nunes Vaz et al. 1989) or to isolate a particular physical process (Chao 1988; Chao 1990). These exercises are informative, but they do not address the full suite of processes that may influence a particular estuary, nor do they guide the investigator toward the dominant process in a certain setting. Processes such as secondary flows are inherently three-dimensional (Geyer 1993; Geyer et al. 1998), and they will not be manifest in two-dimensional simulations. Likewise, a model study of the relative importance of different forcing functions on the response of a particular system requires that the model include a realistic rendition of the full suite of forcing variables. Thus, there are certain questions that require simulations, rather than idealizations of an estuarine regime.

The challenge, which is yet to be fully realized in our field, is to use complex, three-dimensional simulations as tools for developing insights into mechanisms and for comparing processes among estuaries. The output of three-dimensional models should be regarded analogously to a hypothetical observational program in which measurements were obtained everywhere. Not that the model output should be accepted as reality; there are too many unresolved questions about turbulence closure (Simpson et al. 1996; Stacey et al. 1999), grid resolution (Luettich and Westerink 1995), and differencing schemes (Baptista et al. 1995) to put too much faith in a particular numerical result. However, modelers can learn from observationalists with respect to identifying the underlying mechanisms. Many observational studies of estuaries have addressed the question of salt flux in estuaries (Hunkins 1981), with the objective of understanding the mechanisms and how they relate to the estuarine geometry and forcing conditions, that is, to relate the salt flux to estuarine classification. This reductionist approach is the surest way to develop an understanding of a particular system that can then be generalized across estuarine regimes. The same methodology can be applied to numerical simulations, with the satisfying difference that all of the terms in the salt flux can be quantified. Bowen (1999) took advantage of this attribute of a three-dimensional model to look in detail at the mechanism of oscillatory shear dispersion, which could only be effectively quantified by resolving the variations across the estuarine cross section and vertically through the boundary layer. Geyer et al. (1998) used a realistic simulation of the Hudson River estuary to investigate the mechanisms of particle trapping. They found that the trapping process is inherently three-dimensional, involving the generation of secondary flows and fronts. These examples illustrate that three-dimensional simulations provide an effective means of investigating estuarine processes, particularly those affected by the temporal or spatial variability of the estuarine regime.

Estuarine simulation models would also provide a valuable tool for comparative studies; however, this capability has rarely been exploited. Model investigations of parameter sensitivity (for example, Fong 1998) can quantify the influence of different forcing functions and geometries across a meaningful range of estuarine conditions. Sensitivity studies are typically based on idealizations of estuaries, which are not intended to produce quantitatively accurate representations of a particular environment. Simulation models have the advantage of providing a quantitative comparison with observations, against which the quality of the simulation can be verified. The use of simulation models for comparisons of the processes between estuaries would provide more quantitative comparisons than idealizations and would more closely relate to observational comparisons. The idealized sensitivity studies are essential to developing an understanding of the controlling variables, but the simulations provide a more focused look at the processes in particular environments. Just as future field work in estuaries should emphasize intersite comparative studies, comparative model simulations would provide deeper insights into the mechanisms, as well as disseminating advanced modeling approaches to multiple sites.

Quantifying Residence Time

A pervasive problem in interdisciplinary studies is finding the common ground between the disciplines. There are certain aspects of the physics of estuaries that are particularly relevant to ecology, and others that are less so. One of the most important quantities relating the physics to the ecology of estuaries is residence time. A widely cited example is the work of Vollenweider (1975), who demonstrated that in lakes it is not just the nutrient loading, but rather the product of nutrient loading and residence time that determines the impact on phytoplankton production. Unfortunately, estuarine physicists have been rather unenthusiastic about attempting to quantify residence time, due in part to how easily misinterpreted a single number would be in characterizing the complex exchange processes that influence an estuary.

Residence time is defined for steady-state conditions as the ratio of the mass of a scalar in a reservoir to the rate of renewal of the scalar. The concept is most useful in systems in which the mixing time scale within the reservoir is rapid in comparison with the residence time, which leads to nearly homogeneous scalar concentrations (for example, salt in the ocean). In estuaries this constraint is rarely met; furthermore the residence time is complicated by spatial variability and time dependence of the flushing processes (Oliveira and Baptista 1997; Zimmerman 1976) as well as nonconservative behavior of many scalars of interest. In spite of these difficulties, the concept is still worth pursuing because of its relevance to the fate of waterborne material.

Measurements of bioreactive compounds in estuaries support the idea that the speed of a transport process relative to the rates of biogeochemical processes is an important determinant of the temporal and spatial distributions of biochemicals in estuaries (Wofsy et al. 1981; Imberger et al. 1983; Muller et al. 1994) as well as phytoplankton (Doering et al. 1994; Balls et al. 1995). However, the variation in residence time may be dominated in some estuaries by freshwater inputs and in others by tidal exchange (Knoppers et al. 1991), and the relative effects of these alternative exchange mechanisms on estuarine structure and function are not the same. Residence time is decreased by freshwater discharge and tidal exchange alike, but unlike freshwater discharge, the tidal prism results in a characteristic dilution of nutrients and suspended material in estuaries by coastal water and periodic cycles of abundance of bacteria (Kirchman et al. 1984) and phytoplankton (Litaker et al. 1993). In spite of the potential hazards, residence time provides a valuable tool for determining the importance of physical exchange vis-à-vis in-situ transformations of ecological quantities, and more effort should be placed in developing more accurate and sophisticated approaches to estimating residence time.

There are several avenues for future research that offer promise for refining our understanding and ability to quantify residence time. Numerical water-mass tracking by Awaji et al. (1980), Signell and Geyer (1990), Signell and Butman (1992), and Oliveira and Baptista (1997) have provided key insights into mechanisms of tidal exchange. As numerical models improve and become more thoroughly verified, this approach will become an important means of assessing flushing rates in estuarine systems. Advances in drifter-tracking technology (Hitchcock et al. 1996) will provide unprecedented opportunities to document Lagrangian motions and exchange in estuaries. Deliberate tracers such as Rhodamine dye and SF_6 are providing increasingly precise information about fluid-exchange processes as sampling approaches are refined (Clark et al. 1996). Future applications of these approaches should be combined with ecological investigations, so as to provide direct examination of the relative influence of physical and biogeochemical processes in controlling distributions of ecological variables.

Non-Lagrangian Quantities: Particles and Swimmers

Many of the biogeochemical variables that are most important to water quality and ecology do not behave as the water behaves. The most obvious example is sediment, which is moved by the water but does not follow water parcels exactly, due to its settling velocity. A dramatic outcome of this non-Lagrangian behavior of sediment is the estuarine turbidity maximum, the zone of high concentrations of suspended sediment that occurs as a result of the settling of suspended sediment close to the landward end of estuaries (Postma 1967; Uncles and Stephens 1993).

The transport and fate of sediment is amenable to analysis and modeling in a similar manner to the fluid motion, as long as the settling velocity and the bottom flux condition can be specified. The chapter by Jay et al. (this volume) illustrates the trapping characteristics of sediment under different parameter regimes, based on simple, hydrodynamic principles. Unfortunately, both settling velocity and the bottom boundary condition for sediment flux are extremely challenging problems that involve not just physics but also the biochemical properties of the sediment particles that influence their adhesion to the bed and to each other. Virtually all particles in estuaries form aggregates, which increases their settling velocity relative to the individual sediment grains (Kranck and Milligan 1992). To make matters worse, these aggregates change size as they pass through the estuary or even as the turbulence varies through the tidal cycle (van Leussen 1989). The bottom boundary condition for sediment flux is perhaps even more difficult, because in addition to the complex particle interactions occurring at the sediment-water interface, there is usually an active benthic community that influences the bottom stress and may contribute directly to the resuspension or trapping of sediment (Nowell et al. 1981).

Our understanding of sediment transport will benefit from rigorous comparisons of the transport processes in different estuaries, as well as in linking field observations with the numerous laboratory studies of sediment transport (van Leussen 1989; Mehta 1989). Even the basic question of what controls the position of the estuarine turbidity maximum cannot be answered in general, even though it has been resolved for individual estuaries. However, detailed comparison of the conditions in different estuarine systems should help resolve the relative roles of tidal, estuarine, and bathymetric factors in determining the spatial distribution of suspended sediment. Likewise, aggregation processes and biological effects on particles would be better understood with more detailed comparisons among estuaries.

Plankton are similar to inanimate particles in that only their vertical velocity differs significantly from that of the water. Epifanio et al. (1984) showed how the recruitment of blue crab larvae relied on the interaction of the vertical motion of the larvae with the horizontal motion of the water because of estuarine circulation. Tyler and Seliger (1978) similarly related the distribution of toxic dinoflagellates to advection by the estuarine circulation. While these examples provide encouragement, there remain large obstacles to a broader understanding of the influence of the flow on distributions of living organisms. Uncertainty in initial distributions and behavior present formidable challenges to the modeling of the distributions of organisms. Much of the variability cannot be attributed to the physics, regardless of how sophisticated the transport model. Yet we need to more clearly identify the role of transport in different systems, and in particular to identify the environments in which transport is a dominant variable and those in which it is relatively unimportant. The Quashnet River estuary in Cape Cod has a residence time of about half a day (Geyer 1997), compared to Chesapeake Bay's residence time of 3 to 6 months. The rapid flushing rate of the Quashnet River exceeds the growth time scale of virtually every planktonic organism, while the relatively slow flushing of the Chesapeake should have a relatively minor influence on planktonic organisms, allowing a more fully developed planktonic food web. More attention to the ecological implications of variations of physical parameters is likely to yield considerable insights into the causes of variations between different estuarine systems.

Time Scales of Variability

Temporal variations are often large enough to totally alter the physical and ecological conditions of estuaries over the whole spectrum of time scales from tidal to interannual. Thus the variability itself must be recognized as a key defining element of an estuary. Virtually all estuaries vary seasonally due to variations in river flow, sometimes with dramatic changes in flushing characteristics between high and low discharge (Largier and Slinger 1991). Tidal variations impose a number of time scales of variation including the semi-daily and daily oscillations as well as monthly and fortnightly (Jay and Smith 1990) and even annual and interannual fluctuations (see Morris et al. 1990; Morris, this volume). Variability in atmospheric forcing is increasingly recognized as an important agent in estuarine variability (Elliott 1982; Weisberg 1976; Goodrich et al. 1987), often rivaling the estuarine circulation in its influence on flushing. These various modes of variability in estuaries must be considered as intrinsic characteristics of estuaries, both for estuarine classification and for the ultimate goal of understanding the influence of physical forcing on the ecology.

Understanding temporal variability is particularly important for identifying long-term trends in estuarine function, which may be the result of anthropogenic alterations but may also be forced by natural variations of long period or even the aliasing of large-amplitude, short-term fluctuations. Long time series are of tremendous value in assessing trends, but the sampling strategy must either resolve or filter out the short-term variability, and extreme care must be exerted in attributing cause to long-term variations, for which the statistics are very poorly constrained.

Comparisons between systems provide useful complements to long-term records at single sites for addressing the influence of anthropogenic stresses on ecosystems. Valiela et al. (1992) compared three estuaries with different levels of development as a means of predicting the temporal progression of impact of a single estuary to increased anthropogenic stress. This approach depends critically on one's ability to identify and separate the factors contributing to variations between sites, both natural and anthropogenic. At our present level of understanding of these factors, this "space-for-time substitution" is qualitative rather than quantitative, although it should be regarded as an important tool to develop for the purpose of predicting changes in estuarine systems.

Spatial Variability

Just as temporal variability must be viewed as an intrinsic characteristic of estuarine systems, so must spatial variability. By their very definition, estuaries are the sites of strong gradients, thus the conditions within one part of an estuary will differ greatly from those in another part. The processes dominating the tidal-fluvial part of a river estuary are vastly different from those in the salt wedge; likewise the flow through intertidal marshes bears little resemblance to the swift motion in estuarine channels. In the effort to simplify and parameterize estuarine characteristics, the variability within individual estuaries cannot be overlooked. Estuaries cannot be regarded in general as homogeneous reactor vessels, but rather as networks of reactors with varying rates of transport between the different components.

Estuarine classification would appear to be confounded by the variability within estuaries. For example, an estuary may be well mixed in some reaches and highly stratified in others, depending on its geometry and the freshwater delivery. On the one hand, the classification approach must be flexible enough to embrace this variability, but on the other hand it must retain enough specificity to be useful for distinguishing different estuaries. In part, the problem is solved by incorporating variability within the definitions of different types of estuaries. For instance, most fjords have regions of strong spatial gradients and fronts (sills) as well as zones of weak horizontal gradients (basins). Lagoons have channels and intertidal marshes. All estuaries have benthic and pelagic zones. Thus spatial variability is intrinsic to the definitions of individual classes of estuaries.

Numerical models provide an effective means of examining the influence of spatial variability, given adequate resolution to represent the interactions between different zones. Finite element approaches (Luettich and Westerink 1995) have the advantage of adaptive grid generation, which can provide high resolution in complex or critical regions of the domain. Numerical approaches to linking benthic and pelagic processes show promise in estuarine models (see Cerco and Cole 1993). Efforts to date should be viewed as the first step—there are many layers of complexity on both sides of the sediment-water interface that will challenge the modelers and observationalists for decades. Marsh-estuary interactions present a challenging modeling problem that involves groundwater transport, wetting and drying of the marsh surface, and complex tidal dispersion processes. Numerical models have the potential to address these problems, albeit with considerable attention to the details of the numerous exchange processes.

Multisystem Interactions: A Regional View

The estuaries of the East Coast of the United States are generally close enough to each other that the outflow of one estuary coalesces with the outflow of adjacent estuaries. The plume of the Hudson River impinges on Delaware Bay, and the Delaware impinges on the Chesapeake, and so on. Chesapeake Bay is in fact a large network of estuaries that interact with each other and with the main stem. Thus the dynamics of these systems are coupled, and a complete understanding of physical, biogeochemical, or ecological processes within one estuary requires an understanding of the coupling. An example of the complexity of this coupling is the variation in stratification of the subestuaries of Chesapeake Bay. In an often cited study, Haas (1977) observed that the stratification of the York River varied inversely with the spring-neap cycle of tidal amplitude variation, which he logically attributed to variations in vertical mixing rate. Then Hayward et al. (1982) discovered that these variations in stratification were actually due to spring-neap changes in tidal advection within the main stem of the Chesapeake that altered the stratification within the subestuaries. Biogeochemical and ecological conditions are even more sensitive to remote forcing, due to the longer time scales of these processes compared to physical phenomena. For example, the species diversity within any given estuarine habitat will be strongly influenced by the speciation in adjacent and even more distant habitats, as long as there is a pathway for recruitment.

The interactions between estuaries argues for a regional approach to analyzing estuarine processes. Particularly in problems where the receiving waters are relevant, estuaries cannot be treated in isolation. Modeling of these interacting systems could be pursued with a network of simple models of individual estuaries, whose boundary conditions mutually interact. However, in order to effectively quantify the transport between systems, the coastal waters have to be resolved explicitly. Finite-element models can provide multiple scales of resolution within a single domain. For finite-difference models, nested grids can be used to preserve the high resolution within estuaries and adequate spatial resolution to encompass the coastal waters. As always the case with coastal modeling, the specification of the open boundaries is problematical, both from the standpoint of mathematical well-posedness and with respect to external forcing. Nesting coastal models with ocean-basin models is becoming a practical means of obtaining meaningful boundary conditions for coastal simulations (Oey and Chen 1992; Shankar et al. 1997).

Recommendations

Advances in our understanding of the coupling between physical, biogeochemical, and ecological processes over the next decade will be fostered by the formal integration of research results among different estuarine systems. This can be achieved by improving our approaches to estuarine classification and making more effective use of numerical simulations.

- Estuarine classification has to focus on the physical processes that are most closely coupled to biogeochemical and ecological processes, such as residence time and exchange rates between different environments within estuaries.

- Numerical model simulations need to be used specifically for the purpose of better understanding processes, particularly the complex, time-dependent, and three-dimensional processes that affect the biology and geochemical transport processes in estuaries.

- Numerical models should be used hand in hand with detailed field measurements for intersite comparative studies, to provide a formal approach to examining the differences in mechanisms between sites.

- Temporal variability of estuaries at a broad range of scales must be recognized as an intrinsic characteristic of estuarine systems. We must describe and predict this variability at time scales from tidal to interannual. Extreme care must be exerted to distinguish between anthropogenically induced trends and natural variations of long period. Long time series are essential, but they can be complemented by systematic comparisons of similar sites to ascertain the response of estuaries to anthropogenic stresses.

- Spatial variations within individual environments also must be understood and effectively modeled. The transport of material between zones within estuaries is critical to the biogeochemical and ecological functions of the entire systems. We must better understand the transport and storage of sediments, nutrients, organic carbon, and organisms between these zones in order to effectively model the ecological processes within estuaries.

- Most estuaries are dynamically coupled to neighboring estuarine regimes. Thus a predictive understanding of estuarine processes ultimately requires a regional-scale approach.

In conclusion, we recommend that estuarine research over the next decade emphasize quantitative comparisons among estuaries, combining observations with advanced modeling approaches to address the complex and highly variable processes within and among interacting estuarine systems.

References

Armstrong, W., E. J. Wright, S. Lythe, and T. J. Gaynard. 1985. Plant zonation and the effects of the springneap tidal cycle on soil aeration in a Humber salt marsh. *Journal of Ecology* 73: 323–39.

Awaji, T., N. Imasato, and H. Kunishi, 1980. Tidal exchange through a strait: A numerical experiment using a simple model basin. *Journal of Physical Oceanography* 10: 1499–1508.

Baird, D., and R. E. Ulanowicz. 1993. Comparative study on the trophic structure, cycling and ecosystem properties of four intertidal estuaries. *Marine Ecology Progress Series* 99: 221–37.

Balls, P. W., A. Macdonald, K. Pugh, and A. C. Edwards. 1995. Long-term nutrient enrichment of an estuarine system: Ythan, Scotland (1958–1993). *Environmental Pollution* 90: 311–21.

Baptista, A. M., E. E. Adams, and P. Gresho. 1995. Benchmarks for the transport equation: The convection-diffusion equation and beyond. Pages 241–68 in D. R. Lynch and A. M. Davies (editors), *Quantitative skill assessment for coastal ocean models.* Washington, DC: American Geophysical Union.

Bartlett, K. B., D. S. Bartlett, R. C. Harriss, and D. I. Sebacher. 1987. Methane emissions along a salt marsh salinity gradient. *Biogeochemistry* 4: 183–202.

Beamish, R. J., C-E. M. Neville, and B. L. Thompson. 1994. A relationship between Fraser River discharge and interannual production of Pacific salmon (*Oncohynchus* spp.) and Pacific herring (*Clupea pallasi*) in the Strait of Georgia. *Canadian Journal of Fisheries and Aquatic Sciences* 51: 2843–55.

Bluth, G. J. S., and L. R. Kump. 1994. Lithologic and climatologic controls of river chemistry. *Geochimica et Cosmochimica Acta* 58: 2341–59.

Bowen, M. M. 1999. Salt transport in the Hudson Estuary. Ph.D. thesis, Woods Hole Oceanographic Institution, Massachusetts Institute of Technology Joint Program in Physical Oceanography, Woods Hole, MA.

Boyle, E. A., J. M. Edmond, and E. R. Sholkovitz. 1977. The mechanism of iron removal in estuaries. *Geochimica et Cosmochimica Acta* 41: 1313–24.

Boynton, W. R., J. H. Garber, R. Summers, and W. M. Kemp. 1995. Inputs, transformations, and transport of nitrogen and phosphorus in Chesapeake Bay and selected tributaries. *Estuaries* 18: 285–314.

Boynton, W. R., W. M. Kemp, and C. G. Osborne. 1980. Nutrient fluxes across the sediment-water interface in the turbid zone of a coastal plain estuary. Pages 93–109 in V. S. Kennedy (editor), *Estuarine perspectives.* New York: Academic Press.

Boynton, W. R., W. M. Kemp, and C. W. Keefe. 1982. A comparative analysis of nutrients and other factors influencing estuarine phytoplankton production. Pages 69–90 in V. S. Kennedy (editor), *Estuarine comparisons.* New York: Academic Press.

Cabilio, P., D. L. Dewolfe, and G. R. Daborn. 1987. Fish catches and long-term tidal cycles in northwest Atlantic fisheries: A nonlinear regression approach. *Canadian Journal of Fisheries and Aquatic Sciences* 44: 1890–97.

Caffrey, J. M., and J. W. Day Jr. 1986. Control of the variability of nutrients and suspended sediments in a Gulf Coast estuary by climatic forcing and spring discharge of the Atchafalaya River. *Estuaries* 9: 295–300.

Carr, A. P., and W. L. Blackley. 1986. The effects and implication of tides and rainfall on the circulation of water within salt marsh sediments. *Limnology and Oceanography* 31: 266–76

Cerco, C. F., and T. Cole. 1993. Three-dimensional eutrophication model of Chesapeake Bay. *Journal of Environmental Engineering* 119: 1006–25.

Chao, Shenn-Yu. 1988. Wind-driven motion of estuarine plumes. *Journal of Physical Oceanography* 18: 1144–66.

———.1990. Tidal modulation of estuarine plumes. *Journal of Physical Oceanography* 20: 1115–23.

Chapman, V. J. 1938. Studies in salt-marsh ecology. *Journal of Ecology* 26: 144–79.

Clark, J. F., P. Schlosser, M. Stute, and H. J. Simpson, 1996. SF_6-^3He tracer release experiment: A new method of determining longitudinal dispersion coefficients in large rivers. *Environmental Science and Technology* 30: 1527–32.

Correll, D. L., T. E. Jordan, and D. E. Weller. 1992. Nutrient flux in a landscape: Effects of coastal land use and terrestrial community mosaic on nutrient transport to coastal waters. *Estuaries* 15: 431–42.

Crump, B., and J. A. Baross. 1996. Particle-attached bacteria and heterotrophic plankton associated with the Columbia River estuarine turbidity maxima. *Marine Ecology Progress Series* 138: 265–73.

Dewulf, J., D. Drijvers, and H. van Langenhove. 1995. Measurement of Henry's Law constant as function of temperature and salinity for the low temperature range. *Atmospheric Environment* 29: 323–31.

Dingman, S. L. 1994. *Physical hydrology*, pages 1–575. New York: Macmillan.

Doering, P. H., C. A. Oviatt, J. H. McKenna, and L. W. Reed. 1994. Mixing behavior of dissolved organic carbon and its potential biological significance in the Pawcatuck River estuary. *Estuaries* 17: 521–36.

Dynesius, M., and C. Nilsson. 1994. Fragmentation and flow regulation of river systems in the northern third of the world. *Science* 266: 753–62.

Ebbesmeyer, C. C., and C. A. Barnes. 1980. Control of a fjord basin's dynamics by tidal mixing in embracing sill zones. *Estuarine and Coastal Marine Science* 11: 311–30.

Elliott, A. J. 1982. Wind-driven flow in a shallow estuary. *Oceanol. Acta* 5: 7–10.

Epifanio, C. E., C. C. Velenti, and A. E. Pembroke. 1984. Dispersal and recruitment of blue crab larvae in Delaware Bay, USA. *Estuarine, Coastal and Shelf Science* 18: 1–12.

Eyre, B. 1993. Nutrients in the sediments of a tropical north-eastern Australian estuary, catchment and nearshore coastal zone. *Australian Journal of Marine and Freshwater Research* 44: 845–66.

Festa, J. F., and D. V. Hansen. 1978. Turbidity maxima in partially mixed estuaries: A two-dimensional numerical model. *Estuarine and Coastal Marine Science* 7: 347–59.

Fong, D. A. 1998. Dynamics of freshwater plumes: Observations and numerical modeling of the wind-forced response and alongshore freshwater transport. Ph.D. thesis, Woods Hole Oceanographic Institution, Massachusetts Institute of Technology Joint Program in Physical Oceanography, Woods Hole, MA.

Friedland, K. D., D. W. Ahrenholz, and J. F. Guthrie. 1996. Formation and seasonal evolution of Atlantic menhaden juvenile nurseries in coastal estuaries. *Estuaries* 19: 105–14.

Furumai, H., T. Kawasaki, T. Futawatari, and T. Kusuda. 1988. Effect of salinity on nitrification in a tidal river. *Water Science and Technology* 20: 165–74.

Gambrell, R. P., J. B. Weisepape, W. H. Patrick Jr., and M. C. Duff. 1991. The effects of pH, redox, and salinity on metal release from a contaminated sediment. *Water, Air, and Soil Pollution* 57–58: 359–67.

Gardner, L. R., L. Thombs, D. Edwards, and D. Nelson. 1989. Time series analyses of suspended sediment concentrations at North Inlet, South Carolina. *Estuaries* 12: 211–21.

Geyer, W. R. 1993. Three-dimensional tidal flow around headlands. *Journal of Geophysical Research* 98: 955–66.

———. 1997. Influence of wind on dynamics and flushing of shallow estuaries. *Estuarine Coastal and Shelf Science* 44: 713–22.

Geyer, W. R., and G. A. Cannon. 1982. Sill processes related to deep water renewal in a fjord. *Journal of Geophysical Research* 87: 7985–96.

Geyer, W. R., R. P. Signell, and G. C. Kineke. 1998. Lateral trapping of sediment in a partially mixed estuary. Pages 115–26 in *8th International Biennial Conference on Physics of Estuaries and Coastal Seas, 1996*. Rotterdam, The Netherlands: A. A. Balkema.

Goodrich, D. M., W. C. Boicourt, P. Hamilton, and D. W. Pritchard. 1987. Wind-induced destratification in the Chesapeake Bay. *Journal of Physical Oceanography* 17: 2232–40.

Haas, L. W. 1977. The effect of the spring-neap tidal cycle on the vertical salinity structure of the James, York and Rappahannock Rivers, Virginia, U.S.A. *Estuarine and Coastal Marine Science* 5: 485–96.

Hansen, D. V., and M. Rattray Jr. 1966. New dimensions in estuary classification. *Limnology and Oceanography* 11: 319–25.

Hartenstein, R. 1972. *Principles of physiology*. New York: Van Nostrand Reinhold Co.

Harvey, J. W., P. F. Germann, and W. E. Odum. 1987. Geomorphological control of subsurface hydrology in the creekbank zone of tidal marshes. *Estuarine Coastal and Shelf Science* 25: 677–91.

Hayward, D., C. S. Welch, and L. W. Haas. 1982. York River destratification: An estuary-subestuary interaction. *Science* 216: 1413–14.

Hettler, W. F., and A. J. Chester. 1982. The relationship of winter temperature and spring landings of pink shrimp, *Penaeus duorarum*, in North Carolina. *Fishery Bulletin* 80: 761–68.

Hitchcock, G. L., D. B. Olson, S. L. Cavendish, E. C. Kanitz. 1996. A GPS-tracked surface drifter with cellular telemetry capabilities. *Marine Technology Society Journal* 30: 44–49.

Hoch, M. P., and D. L. Kirchman. 1993. Seasonal and inter-annual variability in bacterial production and biomass in a temperate estuary. *Marine Ecology Progress Series* 98: 283–95.

Hopkinson, C., and J. J. Vallino. 1995. The relationships among man's activities in watersheds and estuaries: A model of runoff effects on patterns of estuarine community metabolism. *Estuaries* 18: 598–621.

Howes, B. L., and D. D. Goehringer. 1994. Porewater drainage and dissolved organic carbon and nutrient losses through the intertidal creekbanks of a New England salt marsh. *Marine Ecology Progress Series* 114: 289–301.

Hunkins, K. 1981. Salt dispersion in the Hudson Estuary. *Journal of Physical Oceanography* 11: 729–38.

Hutchinson, S. E., F. H. Sklar, and C. Roberts. 1996. Short term sediment dynamics in a southeastern U.S.A. Spartina marsh. *Journal of Coastal Research* 11: 370–80.

Imberger, J., T. Berman, R. R. Christian, E. B. Sherr, D. E. Whitney, L. R. Pomeroy, R. G. Wiegert, and W. J. Wiebe. 1983. The influence of water motion on the distribution and transport of materials in a salt marsh estuary. *Limnology and Oceanography* 28: 201–14.

Jay, D., and J. D. Smith. 1990. Circulation, density structure and neap-spring transitions in the Columbia River estuary. *Progress in Oceanography* 25: 81–112.

Jay, D. A., and C. A. Simenstad. 1996. Downstream effects of water withdrawal in a small, high-gradient basin: Erosion and deposition on the Skokomish River Delta. *Estuaries* 19: 501–17.

Jay, D. A., and J. D. Smith. 1989. Residual circulation in and classification of shallow, stratified estuaries. Pages 21–41 in Dronkers and van Leussen (editors), *Physical processes in estuaries*. Berlin, Germany: Springer-Verlag.

Jensen, H. S., and F. O. Andersen. 1992. Importance of temperature, nitrate, and pH for phosphate release from anaerobic sediments of four shallow, eutrophic lakes. *Limnology and Oceanography* 37: 577–89.

Johnson, B. H., K. W. Kim, R. E. Heath, B. B. Hsieh, and H. L. Butler. 1993. Validation of three-dimensional hydrodynamic model of Chesapeake Bay. *Journal of Hydraulic Engineering* 119: 2–20.

Kiene, R. P., and S. K. Service. 1991. Decomposition of dissolved DMSP and DMS in estuarine waters: Dependence on temperature and substrate concentration. *Marine Ecology Progress Series* 76: 1–11.

Kirchman, D., B. Peterson, and D. Juers. 1984. Bacterial growth and tidal variation in bacterial abundance. *Marine Ecology Progress Series* 19: 247–59.

Kjerfve, B., and K. E. Magill. 1990. Salinity changes in Charleston Harbor 1922–1987. *Journal of Waterway, Port, Coastal, and Ocean Engineering* 116: 153–68.

Knoppers, B., B. Kjerfve, and J. P. Carmouze. 1991. Trophic state and water turnover time in six choked coastal lagoons in Brazil. *Biogeochemistry* 14: 149–66.

Kranck, K., and T. G. Milligan. 1992. Characteristics of suspended particles at an 11-hour anchor station in San Francisco Bay, California. *Journal of Geophysical Research* 97: 11373–82.

Largier, J. L., and J. H. Slinger. 1991. Circulation in highly stratified southern African estuaries. *South African Journal of Aquatic Science* 1,2: 103–15.

Leonard, L. A., and M. E. Luther. 1995. Flow hydrodynamics in tidal marsh canopies. *Limnology and Oceanography* 40: 1474–84.

Lepage, S., and R. G. Ingram. 1988. Estuarine response to a freshwater pulse. *Estuarine Coastal and Shelf Science* 26: 657–67.

Litaker, W., C. S. Duke, B. E. Kenney, and J. Ramus. 1987. Short-term environmental variability and phytoplankton abundance in a shallow tidal estuary. *Marine Biology* 96: 115–21.

———. 1993. Short-term environmental variability and phytoplankton abundance in a shallow tidal estuary. 2. Spring and fall. *Marine Ecology Progress Series* 94: 141–54.

LMER Coordinating Committee (W. Boynton, J. T. Hollibaugh, D. Jay, M. Kemp, J. Kremer, C. Simenstad, S. V. Smith, and I. Valiela). 1992. Understanding changes in coastal environments: The Land Margin Ecosystems Research Program. *EOS* 73: 481–85.

Luettich, R. A., and J. J. Westerink. 1995. Continental shelf scale convergence studies with a barotropic tidal model. Pages 349–72 in D. R. Lynch and A. M. Davies (editors), *Quantitative skill assessment for coastal ocean model.* Washington, DC: American Geophysical Union.

Malone, T. C., L. H. Crocker, S. E. Pike, and B. W. Wendler. 1988. Influences of river flow on the dynamics of phytoplankton production in a partially stratified estuary. *Marine Ecology Progress Series* 48: 235–49.

Mantua, N.J., S. R. Hare, Y. Zhang, J. M. Wallace, and R. C. Francis. 1997. A Pacific interdecadal climate oscillation with impacts on salmon production. *Bulletin of the American Meteorological Society* 78: 1069–79.

Marmer, H. A. 1954. Tides and sea level in the Gulf of Mexico. *Fishery Bulletin* 55: 101–18.

Mehta, A. J. 1989. On estuarine cohesive sediment suspension behavior. *Journal of Geophysical Research* 94: 14303–14.

Millham, N. P., and B. L. Howes. 1994. Nutrient balance of a shallow coastal embayment: 1. Patterns of ground water discharge. *Marine Ecology Progress Series* 112: 155–67.

Monbet, Y. 1992. Control of phytoplankton biomass in estuaries: A comparative analysis of microtidal and macrotidal estuaries. *Estuaries* 15: 563–71.

Montague, C. L., and J. A. Ley. 1993. A possible effect of salinity fluctuation on abundance of benthic vegetation and associated fauna in northeastern Florida Bay. *Estuaries* 16: 703–17.

Morgan, C. A., J. R. Cordell, and C. A. Simenstad. 1997. Sink or swim? Copepod population maintenance in the Columbia River estuarine turbidity maxima region. *Marine Biology* 129: 309–17.

Morris, A. W., R. F. C. Mantoura, A. J. Bale, and R. J. M. Howland. 1978. Very low salinity regions of estuaries: Important sites for chemical and biological reactions. *Nature* 274: 678–80.

Morris, J. T. 1995. The mass balance of salt and water in intertidal sediments: Results from North Inlet, South Carolina. *Estuaries* 18: 556–67.

Morris, J. T., B. Kjerfve, and J. M. Dean. 1990. Dependence of estuarine productivity on anomalies in mean sea level. *Limnology and Oceanography* 35: 926–30.

Muller, F. L. L., M. Tranter, and P. W. Balls. 1994. Distribution and transport of chemical constituents in the Clyde Estuary. *Estuarine Coastal and Shelf Science* 39: 105–26.

Nelson, B. W., A. Sasekumar, and Z. Z. Ibrahim. 1994. Neap-spring tidal effects on dissolved oxygen in two Malaysian estuaries. *Hydrobiologia* 285: 7–17.

Nixon, S. W., C. A. Oviatt, J. Frithsen, and B. Sullivan. 1986. Nutrients and the productivity of estuarine and coastal marine ecosystems. *Journal of the Limnological Society of South Africa* 12: 43–71.

Nowell, A. R. M., P. A. Jumars, and J. E. Eckman. 1981. Effects of biological activity on the entrainment of marine sediments. *Marine Geology* 42: 133–53.

Nowicki, B. L. 1994. The effect of temperature, oxygen, salinity, and nutrient enrichment on estuarine denitrification rates measured with a modified nitrogen gas flux technique. *Estuarine Coastal and Shelf Science* 38: 137–56.

Nunes Vaz, R. A., G. W. Lennon, and J. R. de Silva Samarasinghe. 1989. The negative role of turbulence in estuarine mass transport. *Estuarine Coastal and Shelf Science* 28: 361–77.

Nuttle, W. K., and J. W. Harvey. 1995. Fluxes of water and solute in a coastal wetland sediment. 1. The contribution of regional groundwater discharge. *Journal of Hydrology* 164: 89–107.

Nuttle, W. K., and J. W. Portnoy. 1992. Effect of rising sea level on runoff and groundwater discharge to coastal ecosystems. *Estuarine Coastal and Shelf Science* 34: 203–12.

Oey, L-Y, and P. Chen. 1992. A nested-grid ocean model: With application to the simulation of meanders and eddies in the Norwegian Coastal Current. *Journal of Geophysical Research* 97: 20063–86.

Officer, C. B., R. B. Biggs, J. L. Taft, L. E. Cronin, M. Tyler, and W. R. Boynton. 1984. Chesapeake Bay anoxia: Origin, development and significance. *Science* 223: 22–27.

Olff, H., J. P. Bakker, and L. F. M. Fresco. 1988. The effect of fluctuations in tidal inundation frequency on a salt-marsh vegetation. *Vegetatio* 78: 13–19.

Oliveira, A., and A. M. Baptista. 1997. Diagnostic modeling of residence times in estuaries. *Water Resources Research* 33: 1935–46.

Oliveira, A. M., and B. Kjerfve. 1993. Environmental responses of a tropical coastal lagoon system to hydrological variability: Mundau-Manguaba, Brazil. *E.C.S.S.* 37: 575–91.

Paalman, M. A. A., C. H. van der Weifden, and J. P. G. Loch. 1994. Sorption of cadmium on suspended matter under estuarine conditions: Competition and complexation with major sea-water ions. *Water, Air and Soil Pollution* 73: 49–60.

Painchaud, J., D. Lefaivre, J. C. Therriault, and L. Legendre. 1995. Physical processes controlling bacterial distribution and variability in the upper St. Lawrence estuary. *Estuaries* 18: 433–44.

Pearcy, W. G. 1992. *Ocean ecology of north Pacific salmon.* Washington Sea Grant Program. Seattle, WA: University of Washington Press.

Pennock, J. R. 1985. Chlorophyll distribution in the Delaware estuary: Regulation by light-limitation. *Estuarine Coastal and Shelf Science* 21: 711–25.

Pomeroy, L. R., and D. Deibel. 1986. Temperature regulation of bacterial activity during the spring bloom in Newfoundland coastal waters. *Science* 233: 359–61.

Pomeroy, L. R., and W. J. Wiebe. 1988. Energetics of microbial food webs. *Hydrobiologia* 159: 7–18.

Postma, H. 1967. Sediment transport and sedimentation in the estuarine environment. Pages 158–79 in G. H. Lauff (editor), *Estuaries.* Washington, DC: American Association for the Advancement of Science.

Powell, T. M., J. E. Cloern, and L. M. Huzzey. 1989. Spatial and temporal variability in South San Francisco Bay (USA). 1. Horizontal distributions of salinity, suspended sediments, and phytoplankton biomass and productivity. *Estuarine Coastal and Shelf Science* 28: 583–97.

Prahl, F. G., Small L. F., and Eversmeyer B. 1997. Biogeochemical characterization of suspended particulate matter in the Columbia River estuary. *Marine Ecology Progress Series* 160: 173–84.

Prandle, D. 1985. On salinity regimes and vertical structure of residual flows in narrow tidal estuaries. *Estuarine Coastal and Shelf Science* 20: 615–35.

Redmond, K. T., and R. W. Koch. 1991. Surface climate and streamflow variability in the western United States and their relationship to large-scale circulation indices. *Water Resources Research* 27: 2381–99.

Rose, G. A., and W. C. Leggett. 1988a. Atmosphere-ocean coupling and Atlantic cod migrations: Effects of wind-forced variations in sea temperatures and currents on nearshore distributions and catch rates of *Gadus morhua*. *Canadian Journal of Fisheries and Aquatic Sciences* 45: 1234–43.

————. 1988b. Atmosphere-ocean coupling in the northern Gulf of St. Lawrence. *Canadian Journal of Fisheries and Aquatic Sciences* 45: 1222–33.

Rozas, L. P. 1995. Hydroperiod and its influence on nekton use of the salt marsh: A pulsing ecosystem. *Estuaries* 18: 579–90.

Rozengurt, M., M. J. Herz, and M. Josselyn. 1987. The impact of water diversions on the river-delta-estuary-sea ecosystems of San Francisco Bay and the Sea of Azov. In D. M. Goodrich (editor), *San Francisco Bay: Issues, resources, status and management.* NOAA Estuary of the Month Seminar Series 6. Washington, DC: NOAA.

Rudek, J., H. W. Paerl, M. A. Mallin, and P. W. Bates. 1991. Seasonal and hydrological control of phytoplankton nutrient limitation in the lower Neuse River Estuary, North Carolina. *Marine Ecology Progress Series* 75: 133–42.

Sander, B. C., and J. Kalff. 1993. Factors controlling bacterial production in marine and freshwater sediments. *Microbial Ecology* 26: 79–99.

Schubel, J. R., and D. W. Pritchard. 1986. Responses of upper Chesapeake Bay to variations in discharge of the Susquehanna River. *Estuaries* 9: 236–49.

Schwing, F. B., P. Orton, D. A. Jay, H. Batchelder, and L. K. Rosenfeld (in review). The 1998 East Pacific Ocean Conference—the 1997–98 El Niño: Did its bite match its hype? *EOS.*

Secor, D. H., and E. D. Houde. 1995. Temperature effects on the timing of striped bass egg production, larval viability, and recruitment potential in the Patuxent River (Chesapeake Bay). *Estuaries* 18: 527–44.

Shankar, J, H. F. Cheong, and C-T Chan. 1997. Boundary fitted grid models for tidal motions in Singapore coastal waters. *Journal of Hydraulic Research* 35: 3–20.

Sharp, J. H., L. A. Cifuentes, R. B. Coffin, and J. R. Pennock. 1986. The influence of river variability on the circulation, chemistry, and microbiology of the Delaware estuary. *Estuaries* 9: 261–69.

Sherwood, C. R., D. A. Jay, R. B. Harvey, P. Hamilton, and C. A. Simenstad, 1990. Historical changes in the Columbia River estuary. *Progress in Oceanography* 25: 271–97.

Shiah, F. K., and H. W. Ducklow. 1994. Temperature regulation of heterotrophic bacterioplankton abundance, production and specific growth rate in Chesapeake Bay. *Limnology and Oceanography* 39: 1243–58.

Sholkovitz, E. R. 1976. Flocculation of dissolved organic and inorganic matter during the mixing of river water and seawater. *Geochimica et Cosmochimica Acta* 40: 831–45.

Sholkovitz, E. R., E. A. Boyle, and N. B. Price. 1978. The removal of dissolved humic acids and iron during estuarine mixing. *Earth and Planetary Science Letters* 40: 130–36.

Signell, R. P., and W. R. Geyer. 1990. Numerical simulation of tidal dispersion around a coastal headland. Pages 210–22 in R. T. Cheng (editor), *Residual currents and long-term transport.* Coastal and Estuarine Series. New York: Springer-Verlag.

Signell, R. P., and B. Butman. 1992. Modeling tidal exchange and dispersion in Boston Harbor. *Journal of Geophysical Research* 97: 15591–606.

Simenstad, C. A., D. A. Jay, and C. R. Sherwood. 1992. Impacts of watershed management on land-margin ecosystems: the Columbia River Estuary as a case study. Pages 266–306 in R. Naimen (editor), *New perspectives for watershed management—balancing long-term sustainability with cumulative environmental change.* New York: Springer-Verlag.

Simon, N. S. 1989. Nitrogen cycling between sediment and the shallow-water column in the transition zone of the Potomac River and estuary. 2. The role of wind-driven resuspension and adsorbed ammonium. *Estuarine Coastal and Shelf Science* 28: 531–47.

Simpson, J. H., W. R. Crawford, T. P. Rippeth, A.R. Campbell, and J. V. S. Cheok. 1996. The vertical structure of turbulent dissipation in shelf seas. *Journal of Physical Oceanography* 26: 1579–90.

Smith, S. V., and J. T. Hollibaugh. 1993. Coastal metabolism and the oceanic carbon balance. *Reviews in Geophysics* 31: 75–89.

Stacey, M. T., S. G. Monismith, and J. R. Burau. Observations of turbulence in a partially stratified estuary. *Journal of Physical Oceanography* (in press).

Stommel, H., and H. G. Farmer. 1952. *On the nature of estuarine circulation.* Part 1. Reference 52–51. Woods Hole, MA: Woods Hole Oceanographic Institution.

Turner, A., and G. E. Milward. 1994. Partitioning of trace metals in a macrotidal estuary. Implications for contaminant transport models. *Estuarine Coastal and Shelf Science* 39: 45–58.

Turner, R. E., and N. N. Rabalais. 1994. Coastal eutrophication near the Mississippi River delta. *Nature* 368: 619–21.

Tyler, M. A., and H. H. Seliger. 1978. Annual subsurface transport of a red tide dinoflagellate to its bloom area: Water circulation patterns and organism distributions in the Chesapeake Bay. *Limnology and Oceanography* 23: 227–46.

Uncles, R. J., and J. A. Stephens. 1993. The freshwater-saltwater interface and its relationship to the turbidity maximum in the Tamar Estuary, United Kingdom. *Estuaries* 16: 126–41.

Vale, C., and B. Sundby. 1987. Suspended sediment fluctuations in the Tagus Estuary on semi-diurnal and fortnightly time scales. *Estuarine Coastal and Shelf Science* 25: 495–508.

Valiela, I., J. Costa, K. Foreman, J. M. Teal, B. Howes, and D. Aubrey. 1990. Transport of groundwater-borne nutrients from watersheds and their effects on coastal waters. *Biogeochemistry* 10: 177–97.

Valiela, I., K. Foreman, M. LaMontagne, D. Hersh, J. Costa, P. Peckol, B. DeMeo-Andreson, C. D'Avanzo, M. Babione, C. H. Sham, J. Brawley, and K. Lajtha. 1992. Couplings of watersheds and coastal waters: Sources and consequences of nutrient enrichment in Waquoit Bay, Massachusetts. *Estuaries* 15: 443–57.

van Leussen, W. 1989. Aggregation of particles, settling velocity of mud flocs: A review. Pages 347–403 in J. Dronkers and W. van Leussen (editors), *Physical processes in estuaries*. Berlin, Germany: Springer-Verlag.

Vollenweider, R. A. 1975. Input-output models with special reference to the phosphorus loading concept in limnology. *Schweizerische Zeitschrift fur Hydrologie* 37: 53–84.

Vörösmarty, C. J., and T. C. Loder III. 1994. Spring-neap tidal contrasts and nutrient dynamics in a marsh-dominated estuary. *Estuaries* 17: 537–51.

Webb, K. L., and C. F. D'Elia. 1980. Nutrient and oxygen redistribution during a spring neap tidal cycle in a temperate estuary. *Science* 207: 983–84.

Weisberg, Robert H. 1976. The nontidal flow in the Providence River of Narragansett Bay: A stochastic approach to estuarine circulation. *Journal of Physical Oceanography* 6: 721–34.

Wells, J. T., and J. M. Coleman. 1981. Periodic mudflat progradation, northeastern coast of South America: A hypothesis. *Journal of Sedimentary Petrology* 51: 1069–75.

Whiting, G. J., and D. L. Childers. 1989. Subtidal advective water flux as a potentially important nutrient input to southeastern U.S.A. saltmarsh estuaries. *Estuarine Coastal and Shelf Science* 28: 417–31.

Wofsy, S. C., M. B. McElroy, and J. W. Elkins. 1981. Transformations of nitrogen in a polluted estuary: Nonlinearities in the demand for oxygen at low flow. *Science* 213: 754–57.

Wolaver, T. G., R. F. Dame, J. D. Spurrier, and A. B. Miller. 1988. Sediment exchange between a euhaline salt marsh in South Carolina and the adjacent tidal creek. *Journal of Coastal Research* 4: 17–126.

Wood, F. J. 1978. *The strategic role of Perigean spring tides in nautical history and North American coastal flooding, 1635–1976.* Washington, DC: NOAA.

Yurteri, C., D. F. Ryan, J. J. Callow, and M. D. Gurol. 1987. The effect of chemical composition of water on Henry's Law constant. *Journal Water Pollution Control Federation* 59: 950–56.

Zimmerman, J. T. F. 1976. Mixing and flushing of tidal embayments in the Western Dutch Wadden Sea. Part 1: Distribution of salinity and calculation of mixing time scales. *Netherlands Journal of Sea Research* 10: 397–439.

PART THREE

Linking Biogeochemical Processes and Food Webs

ALTERATIONS IN THE RATE AND TYPE of biogeochemical and biological processes in estuaries bring about changes in estuarine food webs that are relevant to management concerns. An example is the reduction in fish habitat caused by low-oxygen conditions and excessive algal blooms at the surface of the water. Some of the important biogeochemical transformations occurring in estuaries are sulfate reduction, which produces toxic hydrogen sulfide; the mineralization of organic matter, which produces soluble nitrogen compounds for growing algae; and denitrification, which removes excess nitrogen from the water column and transforms it to inert nitrogen gas. Although all of these transformations take place through bacterial metabolism, we can measure the rates and study the controls on these processes without looking at the species or biomass of bacteria. We have developed an approach that treats the microbes and the process in question as a single unit, a "black box." Thus we can examine the controls on processes without opening the black box to look at the bacteria inside.

Measurements of biogeochemical processes are often made on the very small scale. We measure oxygen utilization rates, for example, in several hundred milliliters of sediment confined in plastic tubes. How can we apply the results to whole estuaries, and how can we scale the estimates for a whole estuary up to apply to a whole coastal region? The examples in this section on the problem of moving from the small scale to larger scales rely mostly on a simple empirical approach, such as regression analyses of the influence of amounts of nutrients entering an estuary on rates of biological processes.

In chapter 9, Seitzinger discusses denitrification, the biogeochemical process by which fixed inorganic nitrogen in the form of ammonia, nitrite, and nitrate is transformed into dinitrogen gas. This process, which requires aerobic and anaerobic microzones in close proximity, is responsible for most of the loss of fixed nitrogen from the oceans. Denitrification also removes inorganic nitrogen from coastal zones, taking away a nutrient that promotes the process of eutrophication. This chapter illustrates how measurement of rates at specific sites builds knowledge of controls that can be used to estimate annual denitrification for an estuary, a region, and the entire ocean.

Seitzinger began by gathering information on the rate of denitrification, first at one location in an estuary and then at two more. From the correlations observed, she hypothesized that denitrification was highest near the site of the maximum amount of input of nitrogen to the estuary. She next conducted experiments in the mesocosms at the University of Rhode Island's Marine Experimental Research Laboratory. From the relationship between the loading rate, the rate of input of nitrogen to a mesocosm, and the denitrification rate she estimated a rate for the whole estuary. To develop a generalized view of a whole region, she carried out a comparative study of rates in four estuaries and analyzed the data by correlation analysis. Finally, she examined the effects of additional inputs, particulate and dissolved organic nitrogen, different residence times of the water, and differences in denitrification in fresh and salt water.

In chapter 10, Rabalais et al. demonstrate the value of large data sets collected over several decades in determining the causes of changes in coastal waters. The question under study was the response of the biological systems of the Gulf of Mexico to nutrients from the Mississippi River. The author's approach was to compare the concentrations of nutrients and oxygen in the waters over space and time. In their presentation, changes such as the doubling of the concentration of nitrate in the river over the last 40 years are linked to nutrient changes in the ocean, to changes in phytoplankton species, and to the formation of hypoxic (low-oxygen) bottom water.

Another example is drawn from the well-studied Chesapeake Bay. In chapter 11, Boynton and Kemp present the results of an empirical synthesis, in which the analysis is based solely on observations, not on information from experiments or mechanistic studies. Long-term studies in this bay have produced a remarkable time-series data set for key physical forces such as hourly to daily observations on river flow, water level, and wind velocity, and weekly to monthly observations on water quality variables such as nutrient, oxygen, and phytoplankton concentrations. The authors report that year-to-year variations in river flow and nutrient loading correlate with mean concentrations of phytoplankton, dissolved oxygen, and rates of benthic nutrient recycling.

Chapter 12 contains a report of the workshop for this section. Its participants took up a particularly challenging question: How well are we able to synthesize current knowledge about how processes and controls at the bottom of the food web affect higher trophic levels? That is, can biogeochemical processes be usefully related to growth and survival of fish, birds, and mammals? The report discusses synthetic goals and models needed to explain long-term trends found in data sets, cross-system comparisons, mass balances, and experiments.

CHAPTER 9

Scaling Up: Site-specific Measurements to Global-scale Estimates of Denitrification

Sybil P. Seitzinger

Abstract

Site-specific process studies of denitrification are used in correlation analyses and in existing regional and global models to explore controlling factors and the spatial pattern of denitrification at a number of scales. Initial studies of denitrification in four estuaries had suggested that, at the ecosystem scale, external dissolved inorganic nitrogen (DIN) loading was an important factor controlling denitrification, with approximately 40–60% of the DIN loading removed by denitrification. More recent analyses using data from 14 coastal marine systems demonstrate that the relationship between denitrification and DIN loading is much more variable (3–100%) than earlier studies suggested. Expanding that analyses to include loading of organic-N as well as DIN results in two relatively good relationships; one between total N (TN) loading and denitrification rates ($r^2 = 0.7$) and another between the percentage of TN loading that is removed by denitrification and water residence time ($r^2 = 0.75$). The combination of water residence time and N loading also provides a good prediction of the percentage of N inputs removed by denitrification in rivers ($r^2 = 0.8$). Estimates of the regional distribution of denitrificaiton in estuaries of the North Atlantic suggest that while the amount of TN transport by rivers is highest in the tropics and lowest in the high latitudes, denitrification in tropical estuaries is predicted to show the opposite pattern. That is, denitrification in tropical estuaries will be low because almost 70% of the TN export by rivers in tropical latitudes is discharged directly onto the shelf by large rivers, thus bypassing estuaries. Estimates of denitrification for the continental shelf of the North Atlantic suggest that ~80% occurs in the mid to high latitude areas, with only 20% in tropical shelf sediments. Results of a global model of DIN transport by world rivers indicate that almost 90% of the denitrification in rivers and estuaries is in the northern hemisphere, with almost half of this occurring in China, India, and southeast Asia. While there have been considerable advances in our understanding of denitrification rates and controlling factors in the last 20 years, there are still large uncertainties in estimating current or future patterns of denitrification in rivers, estuaries, and continental shelves at ecosystem, regional, or global scales.

Introduction

The coastal zone, sandwiched between the terrestrial and oceanic region, plays a pivotal role in the biogeochemical cycles of carbon, nutrients, trace metals, and other elements. Estuaries, bays, and continental shelves can be sinks or transformers of inputs originating from both the terrestrial and oceanic regions. A comprehensive understanding of processes within the coastal zone, their response to human impacts, as well as the linkages between the coastal zone and adjacent systems, requires site-specific studies, together with regional- and global-scale approaches (Gordon et al. 1996). At all scales, synthesis can be facilitated by process studies, correlation analyses, and models (Hobbie, chapter 1, this volume). In the current chapter I use site-specific process studies of denitrification in combination with correlation analyses and models to provide insights into nitrogen biogeochemistry in the coastal zone at a number of spatial scales.

The current analysis is not intended to be a comprehensive review of denitrification in coastal ecosystems, but rather to provide some examples of different approaches to estuarine synthesis using a single process, denitrification. The overall structure begins with denitrification studies, which were initially conducted in one estuary, Narragansett Bay, Rhode Island (USA) (figure 9-1). Subsequently, denitrification measurements were made across a range of estuaries with different characteristics. The results are used to develop correlations between denitrification and ecosystem-level properties. Regional-scale estimates of denitrification in estuaries and continental shelves of the North Atlantic Basin are presented based on a model developed from the site-specific process studies and a series of correlation analyses. Finally, a global-scale approach is discussed in which information from all the above was used to develop a spatially explicit model that provides estimates of denitrification in freshwater and coastal marine ecosystems globally.

Site-specific Process Studies in One Estuary

Denitrification is carried out by many heterotrophic, generally facultative, anaerobic bacteria. These bacteria utilize nitrite or nitrate as the terminal electron acceptor during the oxidation of organic matter and produce the gaseous

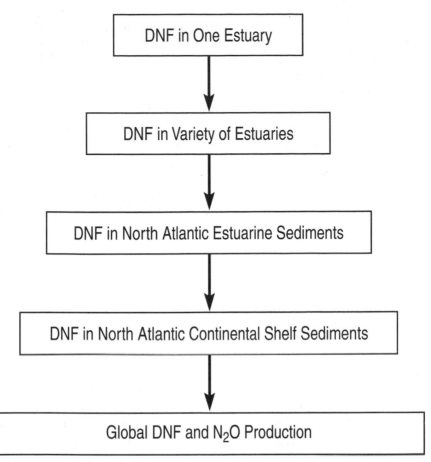

FIGURE 9-1   Scaling up denitrification (DNF) from process-based measurements to regional and global scales.

end products N_2, NO, or N_2O. The N gases diffuse into the atmosphere and are lost from the ecosystem; thus denitrification results in the permanent removal of fixed N from an ecosystem. Three major requirements for denitrification are (1) low-oxygen or anoxic conditions, (2) organic matter, and (3) nitrate or nitrite. Sediments are ideal sites for denitrification since they generally are anaerobic below the surface few mm and have an ample supply of organic matter from the overlying water. Nitrate and nitrite can be supplied to the sediments either directly from the overlying water, through nitrification in the sediments of ammonia released during decomposition of organic matter, or from nitrate and nitrite advected through sediments with groundwater intrusion. Many coastal marine ecosystems are receiving increased nutrients from a variety of anthropogenic sources. Because many coastal ecosystems are nitrogen limited, denitrification plays a particularly important role. It not only permanently removes nitrogen but decreases the N:P ratio of an ecosystem and the amount of N transported to downstream systems.

Some of the first evidence that denitrification was important in estuarine sediments came from benthic nutrient flux and metabolism measurements in Narragansett Bay, Rhode Island (USA), which indicated that less inorganic N was recycled from the sediments to the overlying water than expected based on Redfield (1934) stoichiometry (Nixon et al. 1976). The ratio of oxygen consumed to inorganic N released (inorganic N = DIN = ammonia, nitrate plus nitrite) from Narragansett Bay sediments averaged 13 over an annual cycle, compared to a theoretical ratio of 6.25 to 7.6 (6.25 if NH_4 is the end product, and 7.6 if half of the ammonia released is oxidized to nitrate; table 9-1). Subsequent denitrification measurements over an annual cycle demonstrated that removal of inorganic N as N_2 by denitrification accounted for the missing N in benthic fluxes in Narragansett Bay (Seitzinger et al. 1984). Approximately 520 mmol N m^{-2} y^{-1} was released from the sediments as N_2, resulting in an O_2:N ratio of 8:1, which was very close to the theoretical ratio of 7.6, assuming that half of the NH_4 mineralized was nitrified to nitrate (table 9-1).

Insight into the importance of denitrification at the ecosystem scale was gained by placing the denitrification rates in the context of the N budget for Narragansett Bay. Phytoplankton in the bay are primarily N limited, with benthic N recycling supplying approximately 25% of the annual phytoplankton N demand (Nixon 1981). Denitrification contributes to that N limitation, reducing by approximately 40% the amount of N that is recycled from the sediments in a form available to phytoplankton. In terms of the overall N budget for Narragansett Bay, denitrification was identified as a major sink for external N in the ecosystem. External sources of N to the bay from atmospheric deposition, rivers, and sewage amounted to 258×10^6 mol DIN y^{-1} (table 9-2). Removal of N by denitrification annually accounted for 136×10^6 mol DIN y^{-1}, which was equivalent to approximately 50% of the external DIN inputs to the estuary.

Integration of Denitrification with Ecosystem Scale Properties

Effect of N Loading: Experimental Manipulations and Correlation Analyses

Experimental Manipulations. The Narragansett Bay denitrification studies suggested that areas of the bay with higher nutrient loading had higher

TABLE 9-1

Initial evidence that denitrification was important in estuarine sediments came from a comparison of measured O:N ratios in benthic nutrient fluxes from Narragansett Bay, Rhode Island, and theoretical ratios based on Redfield stoichiometry (Nixon et al. 1976). Subsequent measurements of denitrification in Narragansett Bay sediments indicated that denitrification could account for the "missing" N (Seitzinger et al. 1984).

Redfield Equation for Decomposition of Organic Matter:

$$\left(CH_2O\right)_{106}\left(NH_3\right)_{16}H_3PO_4 + 106O_2 => 106CO_2 + 16NH_4 + 1PO_4$$

$$8NH_4 + 16O_2 => 8NO_3$$

Theoretical[a] O_2:N 7.6:1

Narragansett Bay Benthic Fluxes:

	O_2 Uptake (mmol O_2 m^{-2}y^{-1})	DIN Release (mmol O_2 m^{-2}y^{-1})	Denitrification (mmol O_2 m^{-2}y^{-1})	O_2:N
Measured	11,440	870		13:1
Denitrification	11,440	870	520	8:1

[a] Assuming half of ammonia mineralized is nitrified.

TABLE 9-2

DIN budget for Narragansett Bay placing N removal by denitrification in context of ecosystem N budget (from Seitzinger et al. 1984).

	× 10⁶ mol N yr⁻¹
Sources	
Precipitation	8
Rivers	147
Sewage	103
Total	258
Sinks	
Loss Offshore	?
Sedimentation[a]	35
Denitrification	136
Total	171

[a] Particulate N.

denitrification rates (Seitzinger et al. 1984). Recognition of the magnitude of denitrification as a sink for external N loading to the bay suggested that denitrification might "buffer" nitrogen levels in coastal systems as nutrient loading and eutrophication increased. While studies at various locations within the bay indicated that denitrification rates would increase as nutrient loading increased, the results were only correlative; data were available from only three sites, and a number of other factors also varied among the sites. To experimentally examine the effect of external nutrient-loading rates on N removal by denitrification, denitrification studies were conducted during an ongoing eutrophication experiment using the mesocosms at the Marine Ecosystem Research Laboratory (MERL) at the University of Rhode Island. The mesocosms are cylindrical fiberglass tanks that contained coupled benthic and pelagic components. A 5 m deep, 13 m^3 mixed water column overlay a 40 cm deep layer of sediment; both unfiltered water and vertically intact sediments were from Narragansett Bay. Extensive descriptions of the MERL mesocosms, and details of that specific eutrophication experiment, can be found in Nixon et al. (1984). Measurements of benthic denitrification were made in mesocosms with nutrient loadings ranging from 1.4 up to 94 mmol DIN m^{-2} d^{-1}. The lower end of the range typified DIN-loading rates to Narragansett Bay, while the highest loading rates were similar to those in New York Harbor (Nixon et al. 1984). The response of denitrification to increased N loading occurred relatively quickly, within 2 months. Denitrification rates were positively correlated with DIN loading rates (figure 9-2A; Seitzinger and Nixon 1985). The results indicated that as N-loading rates to an estuary increase, denitrification rates would also increase, and that the response would be quite rapid. The increase in denitrification rates was a linear function of the N-loading rates ($r^2 = 0.98$). However, while the absolute rate of denitrification increased at higher nutrient-loading rates, the percentage of N removed appeared to decrease with increased loading; denitrification was equal to 23% of the DIN inputs at the highest loading rates. Nitrous oxide fluxes also increased, but more rapidly than N$_2$ fluxes, suggesting that coastal eutrophication could lead to a nonlinear increase in emissions of this radiatively important trace gas (figure 9-2B).

Correlation Analyses. Development of a more generalized view of the importance of denitrification in estuaries and the response to human impacts necessitated comparative studies across a range of estuaries with different characteristics, including N-loading rates, extent of intertidal area, flushing rates, and anthropogenic influence. Denitrification studies were expanded to three additional estuaries: Delaware Bay, Ochlockonee Bay, Florida, and the Tejo estuary, Portugal. These estuaries were chosen first because external N loadings were known, and because they differed in a number of other

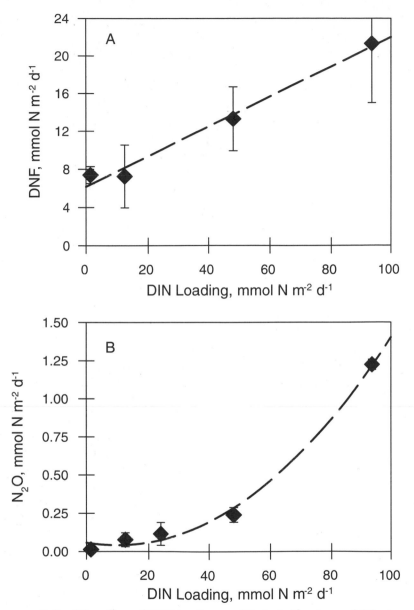

FIGURE 9-2 The effect of DIN loading on (A) denitrification and (B) nitrous-oxide production in sediments (Seitzinger and Nixon 1985). The experiments were carried out on mesocosms at the Marine Ecosystem Research Laboratory.

characteristics thought potentially to be important in controlling denitrification rates. Delaware Bay has relatively high nitrate concentrations in the upper estuary and relatively coarse-grained sediments, compared to Narragansett Bay. The Tejo estuary in Portugal has high N loading from urban and agricultural sources in the watershed; about half of the estuary is intertidal. In contrast, Ochlockonee Bay, Florida, is a relatively pristine estuary with few anthropogenic sources other than from atmospheric deposition; in addition, it is shallow (mean depth 1 m) and has a short residence time (average 3 days). The results of those four studies were then used to explore possible correlations between denitrification rates and various ecosystem-level characteristics. A relatively strong correlation was found between denitrification rates and external DIN loading, with denitrification removing an amount of N equivalent to 40–60% of the external DIN loading (Seitzinger 1988a).

There are now at least fourteen coastal marine systems for which both denitrification and DIN-loading estimates are available (table 9-3). Using this larger database it is clear that the relationship between denitrification and DIN loading is not constant. Across the fourteen coastal marine systems studied, denitrification removes an amount of N ranging from a low of 3% to more than 100% of the DIN loading, with an average of 48% $\pm$ 39 (S.D.). There is no statistically significant relationship between denitrification and DIN loading across coastal marine systems. However, the frequency distribution of these data shows that in half of the systems studied, N removal by denitrification was equal to between 30 and 60% of the DIN inputs (figure 9-3). The results from the remaining studies were about equally distributed below 30% (four systems) and above 60% (three systems).

The above analyses were based on comparisons with DIN loading. However, both dissolved organic and particulate N forms contribute to N loading as well and in many cases comprise 25% to more than 80% of the total N (TN) inputs (Seitzinger and Sanders 1997). A recent compilation of TN loading and denitrification rates for nine coastal systems (Nixon et al. 1996) plus data from three additional systems has permitted further analysis of the relationship between denitrification and ecosystem-level characteristics. A linear regression analysis of the data from twelve coastal systems indicates that denitrification removes an amount of N equivalent to ~30% of the TN inputs ($r^2 = 0.69$), with the percentage of removal ranging from 7 to 89% (figure 9-4). The relatively good relationship between denitrification and TN inputs, but not necessarily with just DIN inputs, is consistent with the changing view of the biological availability of dissolved and particulate organic N inputs to coastal systems. While traditionally dissolved and particulate organic N inputs often were considered refractory, recent evidence suggests that they are significant contributors to biologically available nitrogen (see

TABLE 9-3
N loading and denitrification in a variety of coastal marine systems. Units: mmol N m^{-2} yr^{-1}.

Coastal System	DIN	Total N	Denitrification	Percentage of DIN Removed by Denitrification	Percentage of Total N Removed by Denitrification	Residence Time[1] (months)	Note
Delaware Bay	1,368	1,900	832	61	44	3.3	2
Boston Harbor	4,320	9,095	666	15	7	0.33	3
Galveston Bay	2,350		321	14		2.3	4
Guadalupe Estuary	629	949	356	57	38	3	5
Norsminde Fjord	11,541		320	3		0.17	6
Ochlockonee Bay	1,577	5,991	700	44	12	0.15	7
Baltic Sea	95	191	88–161	93–169	46–84	240	8
Narragansett Bay	1,442	1,980	384–517	27–36	19–26	0.85	9
Potomac Estuary	1,299	2,095	330	25	16	5	10
Patuxent Estuary	695	902	282	41	31		11
Choptank Estuary	207	305	271	131	89		12
Chesapeake Bay (MD portion)	1,027	1467	351	34	24		13
Chesapeake Bay (all)	657	938	243	37	26	7	14
Tejo Estuary	4,468		2,059	46			15
Scheldt Estuary		13,400	5,420		40	3	16

[1] Residence time from Nixon et al. 1996 unless noted.

[2] TN compiled in Nixon et al. 1996; DIN estimated assuming equals 72% of TN based on Culberson et al. 1987; denitrification from Seitzinger 1988b.

[3] Nowicki 1994.

[4] Zimmerman and Benner 1994.

[5] Yoon and Benner 1992; residence time from Zimmerman and Benner 1994.

[6] Nielsen et al. 1995.

[7] Seitzinger 1987; TN Seitzinger unpublished measurements.

[8] TN compiled in Nixon et al. 1996; DIN estimated assuming DIN = 50% TN (Granéli et al. 1990); denitrification range from Shaffer and Rönner 1984 (measurements of deep Baltic water) and Nixon et al. 1996 (from TN mass-balance calculations).

[9] DIN and TN from Nixon et al. 1995 (DIN revised from earlier estimates of Nixon 1981); denitrification from range of Nowicki 1994 (384) and Seitzinger et al. 1984 (517).

[10] Boynton et al. 1995; DIN calculated assuming equal to 62% TN (USGS unpublished data).

[11] Boynton et al. 1995; DIN calculated assuming equal to 77% TN (USGS unpublished data).

[12] Boynton et al. 1995; DIN calculated assuming equal to 68% TN (USGS unpublished data).

[13] Boynton et al. 1995; DIN calculated assuming equal to 70% TN (average Susquehanna, Potomac, Choptank, and Patuxent Rivers using USGS unpublished data).

[14] Nixon et al. 1996; see 13 for DIN calculation.

[15] Seitzinger 1988a.

[16] Nixon et al. 1996.

Carlsson et al. 1993, 1995; Seitzinger and Sanders 1997), which would be subject to denitrification once the N has entered the biological N cycle.

Water residence time is an important factor determining variation in the percentage of the TN inputs that are denitrified (Nielsen et al. 1995; Nixon et al. 1996). Using data from ten systems, a linear relationship was developed

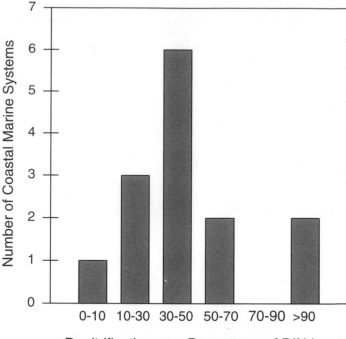

FIGURE 9-3 The number of coastal marine systems with various ranges of the percentage of DIN input removed by denitrification. When there was more than one estimate for a system, the average was used. See table 9-3 for sources of the data.

between the percentage of the external inputs of total N (TN) that are removed by denitrification and the log of the mean water residence time ($r^2 = 0.75$; Nixon et al. 1996; figure 9-5). The correlation between the percentage of the external TN inputs that are removed by denitrification and estuarine water residence time ($r^2 = 0.75$) and between denitrification rates and external TN loading ($r^2 = 0.69$) are similar, indicating that both relationships provide an equally good estimate of estuarine denitrification. While correlations do not demonstrate a cause-and-effect relationship, the two correlations noted above are consistent with each other. As nutrient inputs increase, phytoplankton production rates increase (Boynton et al. 1982; Nixon 1992), thus increasing organic matter deposition to sediments and benthic metabolism (Nixon and Pilson 1983). Benthic metabolism, as measured by sediment oxygen consumption, correlates with denitrification rates, thus connecting N loading with benthic denitrification (Seitzinger and Giblin 1996). A longer water residence time in a coastal system would result in an N molecule passing through the phytoplankton/benthic mineralization/phytoplankton cycle more times, and thus increase the overall percentage of the N input that is denitrified. (A multiple regression using both

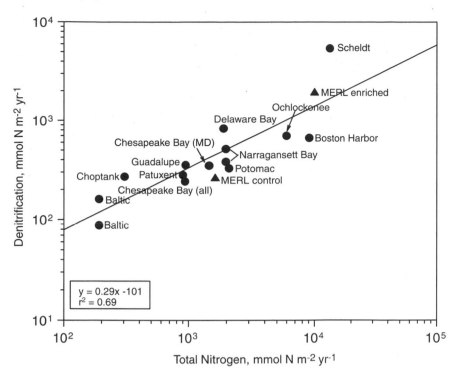

FIGURE 9-4   Relationship between denitrification and external TN loading to a range of coastal marine ecosystems. See table 9-3 for sources of the data.

water residence time and TN loading did not substantially improve the prediction of denitrification; $r^2 = 0.76$.)

The above analyses provide insight into ecosystem-level processes affecting denitrification rates. On smaller scales, a number of factors have been demonstrated to affect denitrification rates including, but not limited to, benthic community composition and infaunal densities (Kristensen et al. 1991; Pelegri et al. 1994), bottom-water oxygen (Nowicki 1994), and nitrate (Nielsen et al. 1995) concentrations, temperature (Nowicki 1994), and benthic algae (Andersen et al. 1984; Jørgensen and Sørensen 1988; Nielsen and Sloth 1994). Further investigations into factors at all scales will likely increase our ability to predict the effects of human disturbance in coastal systems.

Effect of Denitrification on Ecosystem N:P Ratios: Comparisons across Estuaries

Denitrification decreases the amount of N but not phosphorus (P) that is recycled from the sediments, thus decreasing N for phytoplankton and

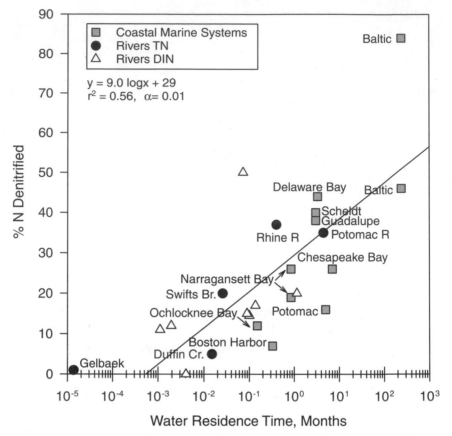

FIGURE 9-5   Relationship between percentage of TN loading removed by denitrification and water residence time in a range of coastal marine ecosystems and rivers.
Data sources: coastal marine systems (see table 9-3); Duffin Creek (Hill 1979, 1981,
1983); Potomac River (Seitzinger 1988a); Rhine (Billen et al. 1991; Flintrop et al.
1996; Friedrich and Müller 1984); Swifts Brook (Robinson et al. 1979; Kaushik and
Robinson 1976); Gelbæk (Pind et al. 1997; Christensen et al. 1990). Denitrification
in streams and rivers where only DIN inputs were quantified are also shown (not included in regression): Purukohukohu, New Zealand (Cooper and Cooke 1984); Bear
Brook (Richey et al. 1985); River Dorn (Cooke and White 1987); Delaware
(Seitzinger 1988b); South Platte (Sjodin et al. 1997); Neversink (Burns 1998).

resulting in low N:P ratios of benthic fluxes. Sediment-water fluxes of oxygen, ammonia, nitrate, and phosphate were measured in many of the same
systems where denitrification rates were measured. Those measurements have
shown that denitrification often can account for the low N:P ratio of inorganic nutrients recycled from sediments (often ~3:1 to ~10:1). However,
quantitative analysis of the effect of denitrification on system-level N:P ratios
is generally lacking. The inorganic as well as total N:P ratios in coastal

marine ecosystems often are considerably lower than N:P ratios of external inputs, which has in part been attributed to denitrification. An analysis of the quantitative effect of denitrification on system N:P ratios requires comprehensive budgets for external inputs of both N and P. The recent summary of total N and P budgets for eight estuaries and coastal systems compiled by Nixon et al. (1996) permitted an initial analysis of the contribution of denitrification to a reduction in the N:P ratio of coastal systems (Seitzinger 1998). The basic assumption in that analysis was that the N:P ratio of an ecosystem would be the same as the N:P ratio of the external inputs if no processes were preferentially adding or removing N or P. External inputs of total N and total P used in that analysis included rivers, direct atmospheric deposition, N_2 fixation, direct anthropogenic discharges, and the coastal ocean (data compiled in Nixon et al. 1996).

The N:P ratio of external inputs varied among the eight systems examined and ranged from 7 (Guadalupe estuary) to 48 (Potomac River estuary; figure 9-6). The extent to which the reported denitrification rates could decrease N:P at the ecosystem scale varied considerably among these eight ecosystems (figure 9-6). Denitrification alone could decrease the N:P ratio of the system by approximately half or more, in four out of eight systems. In the remaining four systems, denitrification could decrease the N:P ratio by 25% or less. The

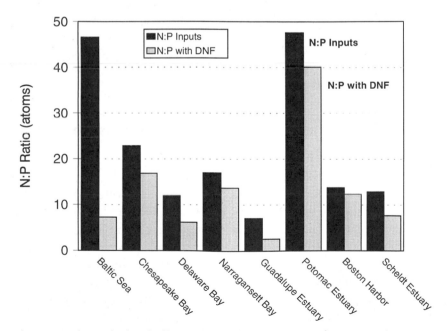

FIGURE 9-6   Calculated effect of denitrification in coastal marine ecosystems on N:P ratio of external inputs. Figure from Seitzinger 1998.

magnitude of the effect is not correlated with the N:P of the external inputs. For example, in the Baltic where the N:P of external inputs is high (~47) denitrification could decrease the N:P to less than 8. The N:P of external inputs to the Potomac (48) is similar to the Baltic, although denitrification could account for a decrease in N:P to only 40. In systems with N:P ratios of external inputs less than 16, denitrification can decrease the ratio to as low as 2.5 (Guadalupe estuary). The above analyses were done using annual average data and do not take into account seasonal or spatial differences, which ecologically can be very important.

In addition to denitrification, N_2 fixation and net N and P burial can affect the N:P ratio. The N:P ratio of net sediment burial in these eight coastal ecosystems is often considerably greater than 16:1 and thus would tend to increase N:P ratios. N_2 fixation also increases N:P ratios in a system. However, N_2 fixation generally is a minor source of N in coastal ecosystems and therefore would increase the N:P ratios only slightly. The quantitative effect of N_2 fixation and burial on N:P ratios in coastal systems are evaluated in Seitzinger (1998). The above processes add new N or P or permanently remove N and P in a system. However, N versus P limitation of primary production in a system can also be affected by differential rates of N and P recycling that affect the N:P ratio of inorganic nutrients available for uptake. With respect to this, differences in P recycling between marine and freshwater systems is an important factor contributing to N versus P limitation (Caraco et al. 1989). Further detailed analyses, including analyses of spatial and temporal variation, are warranted to more fully understand the integrated effects on nutrient limitation in aquatic ecosystems by processes that add new N or P, that permanently remove N or P, and that result in differential recycling of N and P.

Cross-system Comparison: Freshwater versus Estuarine

In the above analyses, comparisons were made across a range of similar types of systems (that is, coastal marine). Comparisons among different ecosystem types can provide additional insights, which are not readily apparent from comparisons of similar systems; on the other hand, they can broaden the generality of concepts developed in similar systems. For example, a comparison of denitrification in estuarine and freshwater sediments revealed some unexpected differences. In estuarine sediments denitrification generally removes 35–40% of the mineralized N. In contrast, in freshwater sediments denitrification often appears to remove a much larger percentage (80–100%) of the mineralized N (Seitzinger 1988a). This is unexpected given that lakes tend to be more P limited and estuaries tend to be N limited. As noted above, denitrification reduces the N:P ratio of recycled nutrients, thus leading a system toward N limitation. The percentage of the overall N removed by denitrification in lakes and estuaries is not obviously different, given the fairly large range in both (Seitzinger 1988a; Howarth et al. 1996). A synthesis of information for lakes and coastal marine ecosystems on N and P inputs, denitrification, N and P burial, and N_2 fixation could clarify the relative contribution of these processes in controlling N:P ratios, and thus nutrient limitation in freshwater and coastal marine ecosystems.

While a number of comparisons of biogeochemical cycles in estuaries and lakes have been made, rivers and estuaries generally have not been compared. However, there is not only a physical continuum from small headwater (first-order) streams to larger rivers that enter into estuaries, but also a biological continuum, for example, in energy utilization and organic matter storage (the River Continuum Concept based on stream size in a single drainage; Vannote et al. 1980, see figure 3-7), which theoretically could be extended to include estuaries. A comparison of denitrification across a range of stream and river sizes and in estuaries suggests that similar factors are controlling the amount of TN inputs that are removed. As noted in a previous section, the percentage of TN inputs that are removed by denitrification in estuaries shows a good correlation with the log of the water residence time (Nixon et al. 1996). While there are only a few studies of denitrification in streams and rivers where water residence time and TN inputs are quantified, the data suggest a similar relationship as that for estuaries (figure 9-5). A linear regression using data from both estuary (all data in table 9-3) and stream and river studies returns the following equation:

$$\% \text{ total N denitrified} = 9 * \log \text{residence time} + 29$$

($r^2 = 0.56$, which is significant at the 0.01 level; residence time in months). The rivers included in this regression range from small headwater streams with very short residence times to tidal freshwater portions of major rivers (Potomac and Delaware) at the heads of estuaries.

Regional-scale Approaches

Denitrification in Estuaries of the North Atlantic Basin

Denitrification in estuaries and rivers can decrease the transport of N from terrestrial to continental shelf regions. A perspective on the potential significance of estuarine denitrification in decreasing N inputs to continental shelf regions requires information on N loading to estuaries within a region of shelf, N removal within the respective estuaries, and N inputs to the shelf region from other sources. Such a perspective recently became possible using a synthesis of information on N inputs and transport in the North Atlantic

Ocean and its watersheds (Howarth et al. 1996). In that study, the watershed of the North Atlantic Basin was divided into nine regions. Nitrogen-(TN-) loading rates from terrestrial sources for each of those regions were developed (Howarth et al. 1996), as were atmospheric inorganic N inputs to the associated continental shelf areas (Prospero et al. 1996). Rivers in the North Atlantic Basin were estimated to export 935 Gmol TN/y (Howarth et al. 1996). A considerable portion of that N (about 44% or 415 Gmol N/y), as noted by Nixon et al. (1996), is debouched directly onto the continental shelf by large rivers (Mississippi, Orinoco, Grijalva–Usumacinta, Magdalena, and Amazon) and thus does not pass through an estuary (*sensu* Prichard "semi-enclosed body of sea water measurably diluted with freshwater"). Therefore, almost half of the TN export by rivers in the North Atlantic Basin is not subject to removal by estuarine denitrification, or burial in estuaries. Total N loading to estuaries in the North Atlantic Basin is estimated to be 521 Gmol N/y. Assuming that most estuaries have a water residence time of 0.5 to 12 months, one would predict that approximately 10 to 50% of the total N input to estuaries from rivers would be removed by denitrification (figure 9-5; Nixon et al. 1996). Accordingly, denitrification within estuaries would remove 52–260 Gmol N/y or decrease the total N export from terrestrial to continental shelves in the North Atlantic by 6 to 28%.

We can obtain some insight into the latitudinal distribution of estuarine denitrification using information on the regional distribution of TN inputs to estuaries in the North Atlantic Basin (figure 9-7a). Almost half of the N removed by denitrification in estuaries of the North Atlantic Basin is predicted to occur in high-latitude regions (>45°N), with approximately 30% in mid-latitude (20–45°N) and 20% in tropical (0–20°N) estuaries (figure 9-7b). This is the reverse of the pattern of N export from the watersheds that drain into these latitudinal belts where the largest N export is from watersheds in tropical latitudes (~40%) and the lowest is from high-latitude regions (~25%; figure 9-7a). The difference in the pattern of river N export and estuarine denitrification is because 70% of the TN in tropical latitudes bypasses estuaries (for example, is debouched directly onto the shelf by large rivers) as does 45% in mid-latitudes. Approximately equal amounts of N are estimated to be removed in estuaries in the eastern (55%) and western (45%) half of the basin. In the above analysis, moving from site-specific studies to a regional approach necessitated loss of small-scale detail, generalization of assumptions, and an increase in the level of uncertainty of the conclusions. This was due to both insufficient understanding of the process(es) being modeled to extrapolate site-specific studies to all other locations and insufficient information on all parameters at the smaller scale throughout the region. However, the strength of such approaches is that it points out potential patterns that are not obvious from smaller-scale studies and can lead to additional productive investigations.

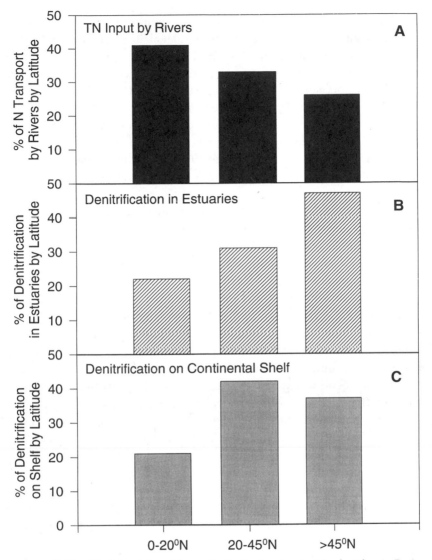

FIGURE 9-7 Model-predicted latitudinal distribution in North Atlantic Basin of (A) TN transport by rivers, (B) denitrification in estuaries, and (C) denitrification in shelf sediments. Each category expressed as latitudinal distribution of the total of each category.

Denitrification in North Atlantic Continental Shelves

What is the fate of the N that is transported to continental shelves either through export from estuaries or direct discharge to the shelf by large rivers? Denitrification in continental shelves has been estimated to remove more than 50% of the total nitrogen inputs to the oceans (Christensen et al.

1987). However, there have been very few direct measurements of denitrification in shelf sediments (Devol 1991; Devol and Christensen 1993), although a number of indirect methods have been used (Billen 1978; Christensen 1989; Raaphorst et al. 1992; Haines et al. 1981; Law and Owens 1990; Lohse et al. 1993; Koike and Hattori 1979; Florek and Rowe 1983; Gardner et al. 1993). Global-scale estimates of denitrification in shelf sediments have generally been based on application of a single rate to all shelf areas. For example, by applying an average nitrate consumption rate for continental shelf and slope sediments to the global shelf area, Christensen et al. (1987) estimated that 3.6×10^{12} mol N/y are removed by denitrification.

Recently, a model was developed to estimate the spatial distribution of denitrification coupled to nitrification in continental shelves of the North Atlantic (Seitzinger and Giblin 1996). That model was built on the limited number of denitrification measurements in continental shelf sediments and on previously identified relationships in estuaries between denitrification and benthic oxygen consumption (Seitzinger 1990) and between benthic oxygen consumption and phytoplankton production (Nixon and Pilson 1983). Using regression analysis, a relationship was developed between continental shelf phytoplankton production and benthic oxygen consumption. That relationship was then linked to a regression between benthic oxygen consumption and denitrification in shelf sediments (figure 9-8). Using the relatively extensive data on primary production in shelf regions, the regional variation in denitrification in shelf sediments throughout the North Atlantic Basin was estimated.

A summary of the model output by latitudinal region indicates that the majority (approximately 80%) of the denitrification in North Atlantic shelf sediments occurs in the mid (20°–45°N) to high-latitude (>45°N) areas, with only 20% in tropical shelf sediments (0°–20°N; figure 9-7c). Similar amounts of N are removed by denitrification in the mid- and high-latitude areas.

For the North Atlantic Basin as a whole, model-predicted N removal by denitrification in continental shelf sediments exceeds N inputs from terrestrial and atmospheric sources combined by about a factor of 2 (figure 9-9). Terrestrial inputs from rivers and estuarine export plus atmospheric deposition account for $59–76 \times 10^{10}$ mol N y^{-1}, while denitrification is estimated to remove at least twice that amount (143×10^{10} mol N y^{-1}). Burial removes an additional small amount of N. The denitrification estimate for shelf sediments by Seitzinger and Giblin (1996) likely underestimates total denitrification as only denitrification coupled to sediment nitrification was included, and not denitrification associated with nitrate diffusing into the sediments from the overlying water.

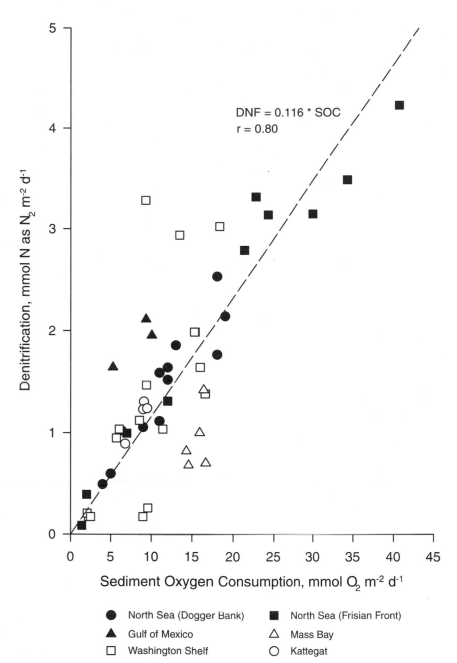

FIGURE 9-8 Relationship between continental shelf denitrification rates (coupled to sediment nitrification) and sediment oxygen consumption (SOC). Regression line is indicated by dashed line. Washington shelf (Devol and Christensen 1993, adjusted for N_2 production due to nitrate from overlying water) and Massachusetts Bay (Kelly and Nowicki 1993; Giblin et al. 1994) data are from direct N_2 flux measurements; North Sea (Raaphorst et al. 1990 and 1992), Gulf of Mexico (Gardner et al. 1993), and Katteggat region of the Baltic (Enoksson et al. 1990) data are from stoichiometric calculations (see Seitzinger and Giblin 1996 for details). Figure from Seitzinger and Giblin 1996. Reprinted with kind permission from Kluwer Academic Publishers.

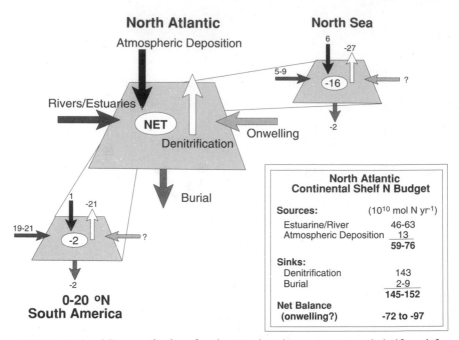

FIGURE 9-9 Nitrogen budget for the North Atlantic continental shelf, and for subregions of the North Atlantic shelf including the North Sea and 0–20° N off South America (based on denitrification model predictions of Seitzinger and Giblin 1996, estuarine export and burial from Nixon et al. 1996, and atmospheric deposition from Prospero et al. 1996).

The above analysis indicates that N removal by denitrification exceeds N inputs from rivers and the atmosphere for the entire North Atlantic shelf. However, N mass balances for subregions differ. For example, in the North Sea/Norwegian Sea region, N inputs from rivers and atmospheric deposition can account for only about half of the N sink from denitrification plus burial (figure 9-9). In contrast, the model suggests that in the tropical latitudes of the western North Atlantic, river inputs plus atmospheric deposition are similar to the estimated N removal by denitrification (figure 9-9). The additional N needed to support the estimated denitrification sink in North Atlantic shelf sediments as a whole is postulated to come from oceanic sources (such as onwelling or advection) (Nixon et al. 1996; Seitzinger and Giblin 1996).

Transport of N between the shelf and oceanic regions in the North Atlantic Basin generally is not well documented. However, there are estimates for the

South Atlantic Bight region (Cape Hatteras to South Florida) and they are consistent with the "missing" N suggested by the above mass-balance calculations. For example, model-estimated N removal by denitrification and burial in the South Atlantic Bight are approximately balanced by the combined N input due to export from estuaries (Nixon et al. 1996), atmospheric deposition (Prospero et al. 1996), plus across-slope transport of nitrate (Lee et al. 1991; figure 9-10).

While there are considerable uncertainties in the predictions from the North Atlantic shelf model, the exercise begins to elucidate the pivotal role that the continental shelf plays not only in the fate of terrestrially derived N, but in the oceanic N budget. Clearly, future studies of denitrification in continental shelf regions are required to verify the model results.

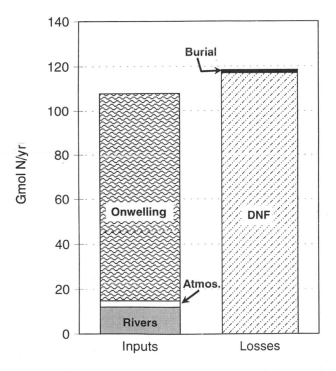

FIGURE 9-10   South Atlantic Bight continental shelf nitrogen budget, using model-estimated denitrification rates (Seitzinger and Giblin 1996), burial (Nixon et al. 1996), atmospheric deposition (Prospero et al. 1996), and onwelling of N across the slope/shelf boundary (Lee et al. 1991). Gmol = 10^9 mol.

Global-scale Model of Denitrification in Freshwater and Coastal Marine Ecosystems

The final example comes from a spatially explicit global model of denitrification in freshwater and coastal marine ecosystems. That model builds on previous studies, a synthesis of additional literature, and global databases (Seitzinger and Kroeze 1998). This model, which we call the N-model, was originally developed to estimate nitrous oxide production in freshwater and coastal marine ecosystems globally, particularly as they are affected by human activities in the surrounding watersheds. The global distribution of denitrification in rivers and estuaries was related to N inputs to those systems, which in turn were related to human activities in the surrounding watersheds. The four watershed parameters used to estimate DIN export by rivers to estuaries were fertilizer N use, human population density, atmospheric N deposition, and water runoff; gridded (1° latitude by 1° longitude) global databases for each of these parameters were used in the N-model. The watershed parameters (land use and N inputs) are fairly well known, and the regression between these parameters and DIN export by rivers had a high correlation coefficient ($r^2 > 0.8$) both for the original formulation of Caraco and Cole (1999) and when we applied global databases to the modified version (Seitzinger and Kroeze 1998). Therefore, we have reasonable confidence in the estimate of DIN export by rivers to estuaries on a regional basis. Denitrification in estuaries was assumed to have removed 50% of the DIN transported to estuaries by rivers based on earlier analyses of data from five estuaries (Seitzinger 1988a). In the current analysis of data from fourteen estuaries it is clear that the percentage of DIN inputs that are removed by denitrification can vary considerably among estuaries (table 9-3). However, in half of those studies, the amount of N removed by denitrification fell between 30 and 60% of the DIN inputs (figure 9-3), suggesting that 50% removal is not an unreasonable average for global estuaries. In the global N model we assumed that all DIN exported by rivers enters estuaries, although, as noted previously, some large rivers discharge directly to shelf areas.

Fewer studies were available to develop predictive relationships for denitrification in rivers. As for estuaries, denitrification was modeled as a function of DIN transport by rivers; it was assumed that an amount of N equivalent to half of the DIN transport by rivers was removed by denitrification before the river reached the estuary (Seitzinger and Kroeze 1998). The effect of varying the percentage of denitrification in rivers from 25 to 50% removal of DIN was also explored (Kroeze and Seitzinger 1998). A recent analysis of denitrification and N input data for eleven rivers suggests that a reasonably good relationship exists between either percentage of DIN or percentage of TN inputs removed by denitrification and mean water

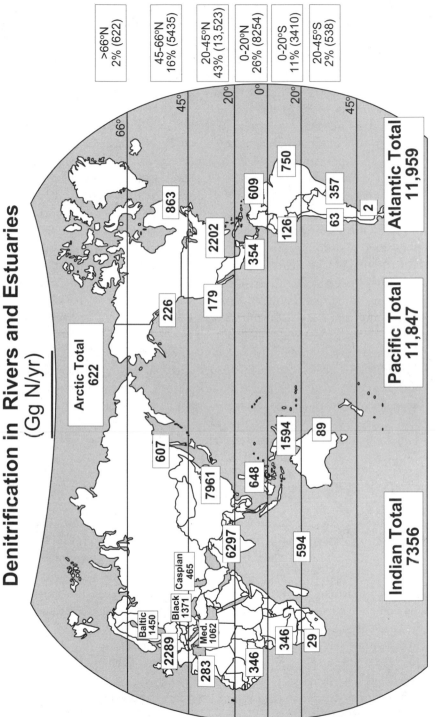

FIGURE 9-11 Global geographical distribution of N-model-predicted denitrification in rivers plus estuaries by different latitudinal zones for oceanic regions. (Note that denitrification in rivers was assigned to the latitudinal zone where the river discharges to the coast.) Gg = 10⁹ g.

residence time in rivers (figure 9-5). However, river-specific dynamics, such as water residence time, were not available for rivers globally and thus could not be explicitly used in the model formulation.

The model results suggest that denitrification in rivers and estuaries is a significant sink for N relative to global anthropogenic N inputs. Denitrification in rivers and estuaries is estimated to remove about 32 Tg N y^{-1} globally, which is equivalent to ~30% of the total N inputs (103 Tg N y^{-1}) to the exoreic terrestrial environment (exorheic watersheds are drained by rivers to the ocean; Tg = 10^{12} g). Synthetic fertilizer (73.6 Tg N) accounts for more than 70% of the N input, atmospheric DIN deposition (22.5 Tg N) for more than 20%, and human sewage for only 7%. Additional N is removed by denitrification in continental shelf regions as discussed above. According to the N-model, almost 90% of the denitrification in rivers plus estuaries is in the Northern Hemisphere, with approximately 70% in rivers and estuaries located between 0 and 45° N (figure 9-11). This distribution is consistent with the distribution of land masses and subsequently anthropogenic activity. Almost half of the denitrification in rivers and estuaries globally is predicted to occur in China, India, and Southeast Asia. In contrast, less than 10% of the global denitrification in freshwater and estuarine ecosystems is predicted to occur in Africa or South America.

While there are considerable uncertainties in a global model such as this, it provides insight into the broad-scale spatial distributions and identifies regions of the world that are likely to be major sinks for anthropogenic nitrogen. Most recently, this model has been used to make projections for DIN export, denitrification, and nitrous oxide production in the year 2050 using estimates of population distributions, fertilizer use, and atmospheric deposition for 2050 (Kroeze and Seitzinger 1998).

Summary/Conclusions

The above examples of site-specific process studies of denitrification combined with correlation analyses and models provide insights into nitrogen biogeochemistry in the coastal zone at a number of spatial scales. The considerable amount of research on coastal denitrification, of which only a small portion was included in the current chapter, has substantially increased our knowledge of the rates, controlling factors, and ecological significance of denitrification in coastal marine ecosystems compared to what it was in the early 1980s. The above syntheses point out some of those advances in our understanding. However, just as important, such syntheses help to identify the large uncertainties in our knowledge. This becomes particularly evident as we attempt to predict current or future denitrification rates in specific estuaries and when developing spatially explicit regional or global-scale estimates.

This chapter also illustrates, in a more general sense, what is gained by doing these types of synthesis efforts and what some of the pitfalls are. One message is that, almost always, there is not as much data as we would like to have. However, we need to be willing to venture into the analyses and extrapolate with caution. Following that, we need to continue to evaluate the relationships as more data become available. This was illustrated by the initial relationship between DIN and denitrification that was developed using data from four estuaries. The four estuaries were chosen because they differed in many ways that were considered potentially important for denitrification. The initial relationship looked very promising. However, adding more sites revealed the importance of using TN rather than DIN and also showed that water residence time is a factor. We need to be careful of limited data sets or bias in the available data sets but must push on nevertheless until more information becomes available. What we gain by doing this is identification of fundamental gaps in our knowledge and a focus for future research.

Acknowledgments

Preparation of this chapter was sponsored, in part, by the NOAA Office of Sea Grant (NJSG-97-370), the NOAA Coastal Ocean Program, and the NOAA Mid-Atlantic Bight National Undersea Research Program. I would like to thank Anne Giblin for reviewing the manuscript.

References

Andersen, T. K., M. H. Jensen, and J. Sorensen. 1984. Diurnal variation of nitrogen cycling in coastal, marine sediments. 1. Denitrification. *Marine Biology* 83:171–76.

Billen, G. 1978. A budget of nitrogen recycling in North Sea sediments off the Belgian coast. *Estuarine, Coastal and Shelf Science* 7:127–46.

Billen, G., C. Lancelot, and M. Meybeck. 1991. N, P, and Si retention along the aquatic continuum from land to ocean. Pages 19–44 in R. F. C. Mantoura, J.-M. Martin, and R. Wollast (editors.), *Ocean margin processes in global change*. New York: John Wiley & Sons Ltd.

Boynton, W. R., W. M. Kemp, and C. W. Keefe. 1982. A comparative analysis of nutrients and other factors influencing estuarine phytoplankton production. Pages 69–90 in V. S. Kennedy (editor), *Estuarine comparisons*. New York: Academic Press.

Boynton, W. R., J. H. Garber, R. Summers, and W. M. Kemp. 1995. Inputs, transformations, and transport of nitrogen and phosphorus in Chesapeake Bay and selected tributaries. *Estuaries* 18(1B): 285–314.

Burns, D. A. 1998. Retention of NO_3^- in an upland stream environment: A mass balance approach. *Biogeochemistry* 40: 733–96.

Caraco, N. F., J. J. Cole, and G. E. Likens. 1989. Evidence for sulphate-controlled phosphorus release from sediments of aquatic systems. *Nature* 341: 316–18.

Caraco, N. F., and J. J. Cole. 1999. Human impact on aquatic nitrogen loads: A regional scale study using large river basins. *Ambio* 28: 167–70.

Carlsson, P., A. Z. Segatto, and E. Granéli. 1993. Nitrogen bound to humic matter of terrestrial origin—a nitrogen pool for coastal phytoplankton? *Marine Ecology Progress Series* 97: 105–16.

Carlsson, P., E. Granéli, P. Tester, and L. Boni. 1995. Influences of riverine humic substances on bacteria, protozoa, phytoplankton, and copepods in a coastal plankton community. *Marine Ecology Progress Series* 127: 213–21.

Christensen, J. P. 1989. Sulfate reduction and carbon oxidation rates in continental shelf sediments, an examination of offshelf carbon transport. *Continental Shelf Research* 9(3): 223–46.

Christensen, J. P., J. W. Murray, A. H. Devol, and L. A. Codispoti. 1987. Denitrification in continental shelf sediments has major impact on the oceanic nitrogen budget. *Global Biogeochemical Cycles* 1: 97–116.

Christensen, P. B., L. P. Nielsen, J. Sorensen, and N. P. Revsbech. 1990. Denitrification in nitrate-rich streams: Diurnal and seasonal variation related to benthic oxygen metabolism. *Limnology and Oceanography* 35(3): 640–51.

Cooke, J. G., and R. E. White. 1987. The effect of nitrate in stream water on the relationship between denitrification and nitrification in a stream-sediment microcosm. *Freshwater Biology* 18: 213–26.

Cooper, A. B., and J. G. Cooke. 1984. Nitrate loss and transformation in 2 vegetated headwater streams. *New Zealand Journal of Marine and Freshwater Research* 18: 441–50.

Culberson, C. H., J. R. Pennock, B. W. Lee, R. B. Biggs, T. M. Church, and J. H. Sharp. 1987. Data from the YABLED cruises September 1981–July 1984. Oceanographic Data Report No. 4. Newark: University of Delaware.

Devol, A. H. 1991. Direct measurement of nitrogen gas fluxes from continental shelf sediments. *Nature* 349: 319–21.

Devol, A. H., and J. P. Christensen. 1993. Benthic fluxes and nitrogen cycling in sediments of the continental margin of the eastern North Pacific. *Journal of Marine Research* 51: 345–72.

Enoksson, V., F. Sorensson, and W. Graneli. 1990. Nitrogen transformations in the Kattegat. *Ambio* 19: 159–66.

Flintrop, C., B. Hohlmann, T. Jasper, C. Korte, O. G. Podlaha, S. Scheele, and J. Viezer. 1996. Anatomy of pollution: Rivers of North Rhine–Westphalia, Germany. *American Journal of Science* 296: 58–98.

Florek, R. J., and G. T. Rowe. 1983. Oxygen consumption and dissolved inorganic nutrient production in marine coastal and shelf sediments of the Middle Atlantic Bight. *Int. Rev. Ges. Hydrobiol.* 68(1): 73–112.

Friedrich, G., and D. Müller. 1984. Rhine. Pages 265–315 in B. A. Whitton (editor), *Ecology of European Rivers*. New York: Blackwell Scientific Publications.

Gardner, W. S., E. E. Briones, E. C. Kaegi, and G. T. Rowe. 1993. Ammonium excretion by benthic invertebrates and sediment-water nitrogen flux in the Gulf of Mexico near the Mississippi River outflow. *Estuaries* 16(4): 799–808.

Giblin, A. E., C. S. Hopkinson, J. Tucker, B. Nowicki, and J. R. Kelly. 1994. Metabolism, nutrient cycling and denitrification in Boston Harbor and Massachusetts Bay sediments in 1993. MWRA Environmental Quality Department. Technical Report Series No. 94. Boston, MA: Massachusetts Water Resources Authority.

Gordon, Jr., D. C., P. R. Boudreau, K. H. Mann, J. E. Ong, W. L. Silvert, S. V. Smith, G. Wattayakorn, F. Wulff, and T. Yanagi. 1996. Land-ocean interactions in the coastal zone (LOICZ). *Biogeochemical modeling guidelines*. LOICZ/R&S/95-5. Texel, The Netherlands: LOICZ Core Project, Netherlands Institute for Sea Research.

Granéli, E., K. Wallström, U. Larsson, W. Granéli, and R. Elmgren. 1990. Nutrient limitation of primary production in the Baltic Sea area. *Ambio* 19(3): 142–51.

Haines, J. R., R. M. Atlas, R. P. Griffiths, and R. Y. Morita. 1981. Denitrification and nitrogen fixation in Alaskan continental shelf sediments. *Applied and Environmental Microbiology* 41(2): 412–21.

Hill, A. R. 1979. Denitrification in the nitrogen budget of a river ecosystem. *Nature* 27: 291–92.

———. 1981. Nitrate-nitrogen flux and utilization in a stream ecosystem during low summer flows. *Canadian Geographer* 25: 225–39.

———. 1983. Nitrate-nitrogen mass balances for two Ontario Rivers. Pages 457–77 in T.D. Fontaine III and S. M. Bartell (editors), *Dynamics of lotic ecosystems*. Ann Arbor, MI: Ann Arbor Science.

Hobbie, J. E. 2000. Estuarine science: The key to progress in coastal ecological research. Pages 1–12 in J.E. Hobbie (editor), *Estuarine science: A synthetic approach to research and practice*. Washington, DC: Island Press.

Howarth, R. W., G. Billen, D. Swaney, A. Townsend, N. Jaworski, K. Lajtha, J. A. Downing, R. Elmgren, N. Caraco, T. Jordan, F. Berendse, J. Freney, V. Kudeyarov, P. Murdoch, and Z. Zhao-Liang. 1996. Regional nitrogen budgets and riverine N & P fluxes for the drainages to the North Atlantic Ocean: Natural and human influences. *Biogeochemistry* 35: 75–139.

Jorgensen, K. S., and J. Sorensen. 1988. Two annual maxima of nitrate reduction and denitrification in estuarine sediment (Norsminde Fjord, Denmark). *Marine Ecology Progress Series* 48: 147–54.

Kaushik, N. K., and J. B. Robinson. 1976. Preliminary observations on nitrogen transport during summer in a small spring-fed Ontario stream. *Hydrobiologia* 49(1): 59–63.

Kelly, J. R., and B. L. Nowicki. 1993. Direct measurements of denitrification in Boston Harbor and Massachusetts Bay sediments. Environmental Quality Technical Report. Series No. 93-3. Boston, MA: Massachusetts Water Resources Authority.

Koike, I., and A. Hattori. 1979. Estimates of denitrification in sediments of the Bering Sea shelf. *Deep-Sea Research* 26A: 409–15.

Kristensen, E., M. H. Jensen, and R. C. Aller. 1991. Direct measurements of dissolved inorganic nitrogen exchange and denitrification in individual polychaete *Nereis virens* burrows. *Journal of Marine Research* 49: 355–77.

Kroeze, C., and S. P. Seitzinger. 1998. Nitrogen inputs to rivers, estuaries and continental shelves and related nitrous oxide emissions in 1990 and 2050: A global model. *Nutrient Cycling in Agroecosystems* 52: 195–212.

Law, C. S., and N. J. P. Owens. 1990. Denitrification and nitrous oxide in the North Sea. *Netherlands Journal of Sea Research* 25(1/2): 65–74.

Lee, T. N., J. A. Yoder, and L. P. Atkinson. 1991. Gulf Stream frontal eddy influence on productivity of the southeast U.S. continental shelf. *Journal of Geophysical Research* 96(C12): 22, 191–22, 205.

Lohse, L., J. F. P. Malschaert, C. P. Slomp, W. Helder, and W. van Raaphorst. 1993. Nitrogen cycling in North Sea sediments: Interaction of denitrification and nitrification in offshore and coastal areas. *Marine Ecology Progress Series* 101: 283–96.

Nielsen, K., L. P. Nielsen, and P. Rasmussen. 1995. Estuarine nitrogen retention independently estimated by the denitrification rate and mass balance methods: A study of Norsminde Fjord, Denmark. *Marine Ecology Progress Series* 119: 275–83.

Nielsen, L. P., and N. P. Sloth. 1994. Denitrification, nitrification and nitrogen assimilation in photosynthetic microbial mats. Pages 319–24 in L. J. Stal and P. Caumette (editors), *Microbial mats. Structure, development and environmental significance*. NATO ASI Series.

Nixon, S. W. 1981. Remineralization and nutrient cycling in coastal marine ecosystems. Pages 111–138 in B. J. Neilson and L. E. Cronin (editors), *Estuaries and nutrients*. Clifton, NJ: Humana Press.

———. 1992. *Quantifying the relationship between nitrogen input and the productivity of marine ecosystems. Ecological Management of the marine environment*, 57–83. Proceedings of the Eighth Marine Technology Conference 5: 1992.

Nixon, S. W., and M. E. Q. Pilson. 1983. Nitrogen in estuarine and coastal marine ecosystems. Pages 565–648 in E. J. Carpenter and D. G. Capone (editors), *Nitrogen in the marine environment*. New York: Academic Press.

Nixon, S. W., C. A. Oviatt, and S. S. Hale. 1976. Nitrogen regeneration and the metabolism of coastal marine bottom communities. Pages 269–83 in J. M. Anderson and A. Macfadyen (editors), *The role of terrestrial and aquatic organisms in decomposition processes*. Blackwell Scientific Publications.

Nixon, S. W., M. E. Q. Pilson, C. A. Oviatt, P. Donaghay, B. Sullivan, S. P. Seitzinger, D. Rudnick, and J. Frithsen. 1984. Eutrophication of a coastal marine ecosystem—an experimental study using the MERL microcosms. Pages 105–35 in *Flows of energy and material in marine ecosystems: Theory and practice*. New York: Plenum Press.

Nixon, S. W., S. L. Granger, and B. L. Nowicki. 1995. An assessment of the annual mass balance of carbon, nitrogen, and phosphorus in Narragansett Bay. *Biogeochemistry* 31: 15–61.

Nixon, S. W., J. Ammerman, L. Atkinson, V. Berounsky, G. Billen, W. Boicourt, W. Boynton, T. Church, D. DiToro, R. Elmgren, J. Garber, A. Giblin, R. Jahnke, N. Owens, M. E. Q. Pilson, and S. Seitzinger. 1996. The fate of nitrogen and phosphorus at the land-sea margin of the North Atlantic Ocean. *Biogeochemistry* 35: 141–80.

Nowicki, B. L. 1994. The effect of temperature, oxygen, salinity, and nutrient enrichment on estuarine denitrification rates measured with a modified nitrogen gas flux technique. *Estuarine, Coastal and Shelf Science* 38: 137–56.

Pelegri, S. P., L. P. Nielsen, and T. H. Blackburn. 1994. Denitrification in estuarine sediment stimulated by the irrigation activity of the amphipod *Corophium volutator*. *Marine Ecology Progress Series* 105: 285–90.

Pind, A., N. Risgaard-Petersen, and N. P. Revsbech. 1997. Denitrification and microphytobenthic NO_3^- consumption in a Danish lowland stream: Diurnal and seasonal variation. *Aquatic Microbial Ecology* 12: 275–84.

Prospero, J. M., K. Barrett, T. Church, F. Dentener, R. A. Duce, N. J. Galloway, H. Levy II, J. Moody, and P. Quinn. 1996. Atmospheric deposition of nutrients to the North Atlantic Basin. *Biogeochemistry* 35: 27–73.

Redfield, A. C. 1934. On the proportions of organic derivatives in sea water and their relation to the composition of plankton. Pages 176–92 in *James Johnstone Memorial* volume. Liverpool, England: University Press.

Richey, J. S., W. H. McDowell, and G. E. Likens. 1985. Nitrogen transformations in a small mountain stream. *Hydrobiologia* 124: 129–39.

Robinson, J. B., H. R. Whiteley, W. Stammers, N. K. Kaushik, and P. Sain. 1979. The fate of nitrate in small streams and its management implications. Pages 247–59 in R. C. Lohr et al. (editors), *Best management practices for agriculture & silviculture*. Ann Arbor, MI: Ann Arbor Science.

Seitzinger, S. P. 1987. Nitrogen biogeochemistry in an unpolluted estuary: The importance of benthic denitrification. *Marine Ecology Progress Series* 41: 177–86.

———. 1988a. Denitrification in freshwater and coastal marine ecosystems: Ecological and geochemical importance. *Limnology and Oceanography* 33: 702–24.

———. 1988b. Benthic nutrient cycling and oxygen consumption in the Delaware Estuary. Pages 132–47 in S. K. Majumdar, E. W. Miller, and L. E. Sage (editors), *The Ecology and restoration of the Delaware River Basin*, Easton: Pennsylvania Academy of Science.

———. 1990. Denitrification in aquatic sediments. Pages 301–22 in N. P. Revsbech, and J. Sorensen (editors), *Denitrification in Soil and Sediment*. New York: Plenum Press.

———. 1998. An analysis of processes controlling N:P ratios in coastal marine ecosystems. Pages 65–83 in *Effects of nitrogen in the aquatic environment*. Stockholm: Swedish Royal Academy of Sciences.

Seitzinger, S. P., and A. E. Giblin. 1996. Estimating denitrification in North Atlantic continental shelf sediments. *Biogeochemistry* 35: 235–59.

Seitzinger, S. P., and C. Kroeze. 1998. Global distribution of nitrous oxide production and N inputs in freshwater and coastal marine ecosystems. *Global Biogeochemical Cycles* 12(1): 93–113.

Seitzinger, S. P., and S. W. Nixon. 1985. Eutrophication and the rate of denitrification and N_2O production in coastal marine sediments. *Limnology and Oceanography* 30: 1332–9.

Seitzinger, S. P., and R. W. Sanders. 1997. Contribution of dissolved organic nitrogen from rivers to estuarine eutrophication. *Marine Ecology Progress Series* 159: 1–12.

Seitzinger, S. P., S. W. Nixon, and M. E. Q. Pilson. 1984. Denitrification and nitrous oxide production in a coastal marine ecosystem. *Limnology and Oceanography* 29: 73–83.

Shaffer, G., and U. Rönner. 1984. Denitrification in the Baltic proper deep water. *Deep-Sea Research* 31(3): 197–220.

Sjodin, A. L., W. M. Lewis Jr., and J. F. Saunders III. 1997. Denitrification as a component of the nitrogen budget for a large plains river. *Biogeochemistry* 39: 327–42.

van Raaphorst, W., H. T. Kloosterhuis, A. Cramer, and K. J. M. Bakker. 1990. Nutrient early diagenesis in the sandy sediments of the Dogger Bank area, North Sea: Pore water results. *Netherlands Journal of Sea Research* 26(1): 25–52.

van Raaphorst, W., H. T. Kloosterhuis, E. M. Berghuis, A. J. M. Gieles, J. F. P. Malschaert, and G. J. van Noort. 1992. Nitrogen cycling in two types of sediments of the southern North Sea (Frisian front, broad fourteens): Field data and mesocosm results. *Netherlands Journal of Sea Research* 28(4): 293–316.

Vannote, R. L., G. W. Minshall, K. W. Cummins, J. R. Sedell, and C. E. Cushing. 1980. The river continuum concept. *Canadian Journal of Fisheries and Aquatic Science* 37: 130–37.

Yoon, W. B., and R. Benner. 1992. Denitrification and oxygen consumption in sediments of two south Texas estuaries. *Marine Ecology Progress Series* 90: 157–67.

Zimmerman, A. R., and R. Benner. 1994. Denitrification, nutrient regeneration and carbon mineralization in sediments of Galveston Bay, Texas, USA. *Marine Ecology Progress Series* 114: 275–88.

CHAPTER 10

Gulf of Mexico Biological System Responses to Nutrient Changes in the Mississippi River

Nancy N. Rabalais, R. Eugene Turner, Dubravko Justić, Quay Dortch,
William J. Wiseman Jr., and Barun K. Sen Gupta

Abstract

Freshwater inflow from the Mississippi River system is a major feature of the Louisiana continental shelf. This freshwater flow correlates closely with nutrient flux, surface-water net productivity, and bottom-water oxygen deficiency. Thus, changes in riverine inputs to the coastal ocean will likely affect biological systems on the adjacent shelf. Some changes are already happening as Mississippi River nutrient concentrations and loadings to the Gulf of Mexico have dramatically accelerated. Since the 1950s, the concentrations of dissolved N and P have doubled and Si has decreased by 50%, the dissolved Si:N ratio dropped from 4:1 to 1:1, and seasonal trends have changed. The resulting nutrient composition in the receiving coastal waters, on the average, shifted towards stoichiometric nutrient ratios closer to the Redfield ratio and more balanced than previously. Now, N and P are less limiting for phytoplankton growth, while some increase in Si limitation is probable. In spite of a probable decrease in coastal Si availability, the overall productivity of diatoms appears to have increased as evidenced by (1) equal or greater net phytoplankton community uptake of silicate in the mixing zone, compared to the 1950s; and (2) greater accumulation rates of biologically bound silica (BSi) in sediments beneath the plume. The increased percentage of BSi in Mississippi River bight sediments that parallels increased N loading to the system is direct evidence for the effects of eutrophication on the shelf adjacent to the Mississippi River. Composition shifts of individual phytoplankton species (heavily silicified diatoms → lightly silicified diatoms; diatom → non-diatom) indicates some responses to reduced Si supplies or changes in nutrient ratios or both. Finally, analysis of benthic foraminifera indicates an increase in oxygen-deficiency stress this century, with a dramatic increase since the 1950s. Increased bottom-water hypoxia could result from increased organic loading to the seabed or shifts in material flux (quantity and quality) to the lower water column or both.

Introduction

The eutrophication of estuaries and enclosed coastal seas has increased over the last several decades, particularly in river-dominated ecosystems. Other evidence suggests a long-term increase in frequency of phytoplankton blooms, including noxious forms, for example, in the Baltic Sea, Kattegat, Skagerrak, and Dutch Wadden Sea (Smayda 1990). Also, an increase in the areal extent or severity of hypoxia or both was observed, for example, in Chesapeake Bay (Officer et al. 1984), the northern Adriatic Sea (Justić et al. 1987), and some areas of the Baltic Sea (see Andersson and Rydberg 1988). Hypoxia is also present in the northern Gulf of Mexico (figure 10-1), which

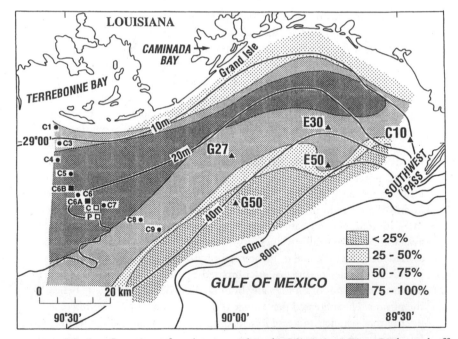

FIGURE 10-1   Location of study area within the Mississippi River Bight and off Terrebonne Bay for hypoxia monitoring; stippled intensity corresponds to frequency of occurrence of midsummer hypoxia at monitoring stations (1985–1987, 1990–1993, N. N. Rabalais, R. E. Turner, and W. J. Wiseman Jr., unpublished data). Transect C stations (closed circles), mooring locations, C6A and C6B (closed squares), LaSER stations (closed triangles), and "platform" and "control" stations of Fucik (1974) and Ward et al. (1979; open squares marked C and P).

receives the outflows of the Mississippi and Atchafalaya Rivers. In this presentation, we investigate the biotic responses of this ecosystem to changes in the delivery of nutrients and sediments by the Mississippi River.

The Mississippi River ranks among the world's top ten rivers in discharge and sediment yields to the coastal ocean (Milliman and Meade 1983). Freshwater inflow from the Mississippi and Atchafalaya (which carries one-third of the flow) is a major feature of the Louisiana shelf. Discharge from the Mississippi and Atchafalaya Rivers rapidly forms the Louisiana Coastal Current, a highly stratified coastal current that flows, on average, westward along the Louisiana coast and then southward along the Texas coast. Coastal winds are from the southeast at the beginning of the flood season. As spring progresses, the winds along the south Texas coast become favorable for upwelling, and the local flow reverses and returns back towards the north and east (Cochrane and Kelly 1986). As a result, density stratification, due primarily to salinity, persists for much of the year.

Fresh water attributed to the Mississippi and Atchafalaya Rivers may be traced as far west as Port Aransas on the Texas coast (Smith 1980) and as far east as the Atlantic seaboard, under combinations of major floods and appropriate oceanographic conditions (Atkinson and Wallace 1975; Tester and Atkinson 1994; Walker et al. 1994). Within the northern Gulf of Mexico, the influence of the Mississippi and Atchafalaya outflows is evident in plots of surface salinity accumulated from numerous hydrographic cruises between 85°W (just east of Cape San Blas, Florida) to 95°W (just west of Galveston Bay, Texas) (figure 10-2). The influence of the freshwater input is more obvious to the west of the Mississippi delta (approximately 89°W) than to the east, with additional inputs near 91°W (the Atchafalaya delta). Plots of nutrient values (example given for silicate in figure 10-2) show a more pronounced decrease in concentration over an equal distance from the deltas when compared to the salinity plot. This indicates a nonconservative mixing due to biological uptake.

High biological productivity in the immediate (320 g C m^{-2} yr^{-1}) and extended plume (290 g C m^{-2} yr^{-1}) of the Mississippi River (Lohrenz et al. 1990; Sklar and Turner 1981; respectively) is mediated by high nutrient inputs and regeneration, and favorable light conditions. Small-scale and short-term variability in productivity are the consequence of various factors, such as nutrient concentrations, temperature, and salinity (Lohrenz et al. 1990, 1997), but on a seasonal time scale they are most influenced by Mississippi River flow and nutrient flux to the system (Justić et al. 1993). In this system, "new" nutrients become depleted along the river-to-ocean mixing gradient through dilution and biological uptake, and regenerated nutrients support primary production for great distances from the river mouth (Dortch et al. 1992a; Bode and Dortch 1996).

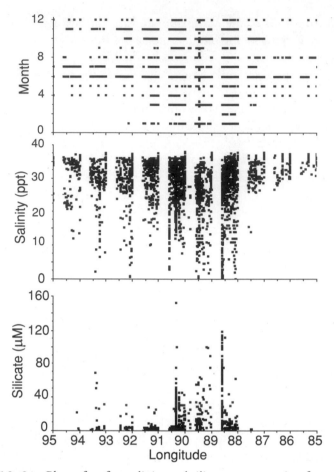

FIGURE 10-2 Plots of surface salinity and silicate concentration from a series of hydrographic cruises in 10–100 m water depth for 1972–1991, for months and longitudes indicated (modified from Rabalais et al. 1996).

Oxygen-depleted bottom waters are seasonally dominant features of the Louisiana continental shelf adjacent to the deltas of the Mississippi and Atchafalaya Rivers (Rabalais et al. 1991, 1992a, 1994a, 1994b, 1998). The areal extent of bottom-water hypoxia (< 2 mg l^{-1} dissolved oxygen) in mid-summer may cover up to 20,000 km^2, with the spatial configuration varying interannually (figure 10-1). Samples along a transect on the southeastern shelf and continuous time series data off Terrebonne Bay document hypoxic bottom waters as early as February and as late as October, with widespread, persistent, and severe hypoxia or anoxia from mid-May to mid-September. Spatial and temporal variability in the distribution of hypoxia exists and is at least partially related to the amplitude and phasing of the Mississippi River

discharge (Rabalais et al. 1994a; Wiseman et al. 1997). This oxygen deficiency is also related to the amplitude and phasing of nutrient flux to the
coastal waters and subsequent production and flux of carbon from surface
waters to the lower water column and seabed. The development and persistence of bottom-water hypoxia requires a strong, persistent pycnocline that
prevents effective reoxygenation of deep waters from the upper layers. The fall
destruction of the salinity-dominated pycnocline by wind mixing and thermal
cooling of surface waters heralds the termination of hypoxic conditions.

The ability to detect changes in an estuary or coastal system is often ham-
pered by the lack of data that are temporally and spatially complete enough to
document adequately the variability of the system and deviations from the
mean condition. This chapter draws on several syntheses of monitoring and
hydrographic cruise data, comparisons with published data from prior
decades, and historical records preserved in the sediments to document how
changes in riverine nutrient fluxes have likely altered biological systems in the
adjacent Gulf of Mexico. We present data from a variety of published syntheses that demonstrate in aggregate a close coupling between riverborne nutrients, net productivity, and hypoxia, as well as elucidate the effects of
anthropogenic nutrient loads on a coastal marine ecosystem. First, we identify
the links between riverine nutrient fluxes and biological response in a river-
dominated coastal system. Second, we document long-term nutrient changes
in riverine nitrogen, phosphorus, and silica fluxes to the continental shelf and
probable changes in the receiving waters. We then address biological responses
to these nutrient alterations, including shifts in probable nutrient limitations
for phytoplankton growth, shifts in phytoplankton community structure,
phytoplankton production, and indicators of eutrophication and increased
oxygen stress. A similar presentation of results has been given previously in
Rabalais et al. (1996).

Methods

While the influence of the discharge of the Mississippi and Atchafalaya
Rivers can be measured at great distances from the deltas, we limit our discussion of consequences of riverine nutrient changes to the areas influenced
by the immediate and extended plumes of the current birdfoot delta. In the
Mississippi River bight these areas extend west to about 90°30'W or the
entrance to Terrebonne Bay (figure 10-1). The oceanographic data synthe-
sized include the following: (1) monitoring cruises conducted during the
period 1985–1993 across the width of the Louisiana shelf, primarily in midsummer (stations covered by stippled areas in figure 10-1); (2) data from the
southeastern shelf off Terrebonne Bay on a biweekly to monthly basis in
1985–1986 and 1990–1993 (transect C in figure 10-1); (3) more frequent

sampling at an instrument mooring (stations C6A and C6B in figure 10-1); (4) six cruises covering 10 to 80 m water depth within the Mississippi River bight between 1987 and 1990 (that is, LaSER stations in figure 10-1); and (5) miscellaneous other cruises within the study area. Comparative data from the literature were available for periods in the 1950s and 1970s. Long-term river nutrient and flow data were available from the USGS and the U.S. Army Corps of Engineers. We also analyzed ^{210}Pb-dated sediment cores for historical biological records of overlying water productivity and oxygen stress. Research and statistical methods are provided in the primary literature as cited.

Synthesis Results

River-shelf Couplings

There is great daily and weekly variability in current flow and stratification on the shelf and, therefore, no simple description of the couplings between carbon production in surface waters and delivery and recycling in bottom waters at these time scales. However, there is evidence of an ecological "signal" (couplings) amid the "noise" (the variability) when more extensive data sets and longer time periods are examined.

Hydrographic data from the southeastern shelf in midsummer from 1985–1991 and off Terrebonne Bay on a biweekly to monthly basis in 1985–1986 and 1990–1993 (transect C in figure 10-1) were averaged by month and compared to long-term (1954–1988) Mississippi River flow at Tarbert Landing (Justić et al. 1993; figure 10-3). The surface layer (0 to 0.5 m) shows an oxygen surplus relative to the saturation values during February–July; the maximum occurs during April and May and coincides with the maximum flow of the Mississippi River. The bottom layer (approximately 20 m), on the contrary, exhibits an oxygen deficit throughout the year with the maximum deficit in July. Bottom hypoxia in the northern gulf is most pronounced during periods of high water-column stability when surface-to-bottom density differences are greatest (Rabalais et al. 1991; Wiseman et al. 1997).

The correlation between Mississippi River flow and surface oxygen surplus peaks at a time lag of 1 month, and the highest correlation for bottom oxygen deficit is for a time lag of 2 months (Justić et al. 1993). These findings suggest that the oxygen surplus in the surface layer following high flow depends on nutrients ultimately coming from the river but regenerated many times. Annual mass balance calculations (Turner and Rabalais 1991; Dortch et al. 1992a) and N uptake measurements in the fall suggest that every N atom is recycled approximately four times, although recycling may be less important

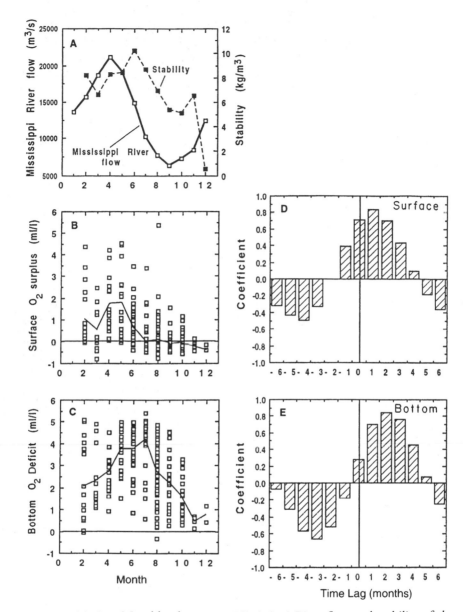

FIGURE 10-3 Monthly changes in Mississippi River flow and stability of the water column ($\delta\sigma t$) (A), surface O_2 surplus (B), and bottom O_2 deficit (C) for stations within stippled area of figure 10-1 in 15 to 30 m water depth. The surface O_2 surplus and the bottom O_2 deficit were calculated as (O_2-O_2') and $-(O_2-O_2')$, respectively, where O_2 was the measured oxygen concentration and O_2' was the oxygen concentration at 100% saturation. Cross-correlation analysis of the data on river flow, surface O_2 surplus (D), and bottom O_2 deficit (E). Modified from Justić et al. 1995.

in the spring (Dortch et al. 1992a; Bode and Dortch 1996). This is an important finding, since a surplus of oxygen relative to the saturation value is a good indicator of net productivity in the surface waters. An oxygen surplus also means that there is an excess of organic matter derived from primary production that can be redistributed within the system; some of this will eventually reach the sediments. The development of summer hypoxia in the northern Gulf of Mexico (as exemplified by stations C6A and C6B, figure 10-1) is associated with the decay of organic matter accumulated during spring phytoplankton blooms (Qureshi 1995).

Changes in Lower Mississippi River Nutrients

Major alterations in the morphology of the main river channel, widespread landscape alterations in the watershed, and anthropogenic additions of nitrogen and phosphorus have resulted in dramatic water-quality changes this century (Turner and Rabalais 1991). Water-quality data for the lower Mississippi River was collected from stations at St. Francisville, Luling, New Orleans, and Venice (Turner and Rabalais 1991, 1994a; Rabalais et al. 1996). The mean annual concentration of nitrate was approximately the same in 1905–1906 and 1933–1934 as in the 1950s, but it has doubled in the last 40 years (results for St. Francisville for 1954–1994 shown in figure 10-4). The mean annual concentration of silicate was approximately the same in 1905–1906 as in the early 1950s, then it declined by 50%. Concentrations

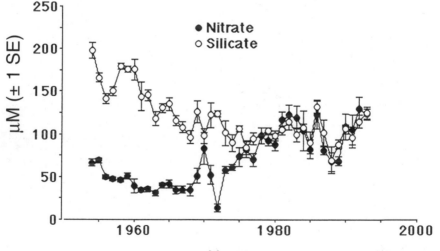

FIGURE 10-4   Average annual concentrations of nitrate and silicate from 1955 through 1994 in lower Mississippi River for the station at St. Francisville. Modified and updated from Turner and Rabalais (1991, 1994a); Rabalais et al. (1996).

of nitrate and silicate appear to have stabilized, but trends are masked by increased variability in the 1980s and early 1990s data. We found no substantial records of total phosphorus concentrations in the lower Mississippi River before 1972. Although the concentration of total phosphorus appears to have increased since 1972, variations among years are large (Turner and Rabalais 1991).

The rise in nitrate since 1960 was coincidental with an increase in nitrogen fertilizer application in the Mississippi River watershed; fertilizer was applied in sufficient amounts to account for the changes in water quality (Turner and Rabalais 1991). The decrease in silicate since 1960 appears to be a consequence of upstream phosphorus additions that stimulated freshwater diatom production and an eventual burial in freshwater sediments of silica in diatom remains (see Schelske and Stoermer 1971; Schelske et al. 1986), thus reducing the annual supply of riverine silicate to coastal waters.

The silicate:nitrate ratios have also changed with the changes in nutrients (in Turner and Rabalais 1991, 1994a; Rabalais et al. 1996, and derivable from figure 10-4) from approximately 4:1 to 1:1. The ratio appears stable at 1:1 through 1994 with little variation. The seasonal patterns in nitrate and silicate concentration have also changed during this century (Turner and Rabalais 1991; Rabalais et al. 1996). There was no pronounced peak in nitrate concentration earlier this century, whereas there was a spring peak from 1975 to 1985, presumably related to seasonal agricultural activities. A seasonal summer-fall maximum in silicate concentration, in contrast, is no longer evident. The seasonal shifts in nutrient concentrations and ratios become increasingly relevant in light of the close temporal coupling of river flow to surface-water net productivity (1-month lag) and subsequent bottom-water oxygen deficiency (2-month lag) described above.

The proportions of dissolved Si, N, and P in the lower Mississippi River have changed historically and now closely approximate the Redfield ratio (Si:N:P = 16:16:1; Redfield 1958; Justić et al. 1995a, 1995b). We compared the data for two periods: 1960–1962 and 1981–1987 (table 10-1; data from Turner and Rabalais 1991). Substantial increases in N (300%) and P (200%) concentrations occurred over several decades, and Si decreased (50%). No data on total P concentration in the Mississippi River were reported prior to 1973; however, total P in the river showed a moderate increase between 1973 and 1987. By applying a linear least-squares regression on the 1973–1987 data, we estimated ($p < 0.01$) that the total P concentration increased twofold between 1960–1962 and 1981–1987. Accordingly, the Si:N ratio decreased from 4.2 to 0.9, the Si:P ratio decreased from 39.8 to 14, and the N:P ratio increased from 9 to 15. By applying the Redfield ratio as a criterion for stoichiometric nutrient balance, one can distinguish between P-deficient, N-deficient, and Si-deficient rivers, and those having a well-balanced nutrient composition. The nutrient ratios for the Mississippi River (1981–1987 database) show an almost perfect coincidence with the Redfield ratio (Justić et al. 1995b). The proportions of Si, N, and P have changed over time in such a way that they now suggest a balanced nutrient composition.

Nutrient Changes on the Adjacent Continental Shelf

We analyzed extensive nutrient data sets from the northern Gulf of Mexico to examine how the coastal nutrient structure may reflect long-term changes in the proportions of dissolved Si, N, and P in riverine loads (Justić et al. 1994, 1995a, 1995b). Fully reliable long-term data sets to examine the nutrient composition 30 years ago, however, were not available. Accordingly, we reconstructed the past coastal nutrient composition (figure 10-5) by assuming that the relative proportion of nutrients in the river-dominated coastal waters reflect changing composition of riverine nutrients (table 10-1). This assumption takes into account that the Mississippi River is the most important nutrient source to the northern Gulf of Mexico. Also, a similar reconstruction technique for the northern Adriatic Sea produced results that closely paralleled the real data (Justić et al. 1995a, 1995b). The detailed reconstruction procedure is given in Justić et al. 1995a). By calculating the specific rates of change for Si:N, Si:P, and N:P ratios in the Mississippi River, we obtained a reasonable estimation of coastal nutrient composition 30 years ago. Comparison of the reconstructed data with the available historical nutrient data (Thomas and Simmons 1960; Turner and Rabalais 1994a) showed a reasonable agreement between the measured and the reconstructed nutrient ratios.

Comparison of measured and reconstructed nutrient ratios for the northern gulf adjacent to the Mississippi River outflow reveals long-term changes in proportions of nutrients in the surface waters (Justić et al. 1995a, 1995b; figure 10-5). The reconstructed nutrient ratios for 1960, on average, scatter further from the center of the grid (that is, the Redfield ratio) than the recent data. By applying the Redfield ratio as the criterion for balanced nutrient composition, P and N deficiencies have decreased while Si deficiency has increased. Equally important, recent nutrient ratios approximate the Redfield ratio, suggesting an almost perfectly balanced nutrient composition.

Probable nutrient limitation (Dortch and Whitledge 1992) was also assessed by comparing the ambient nutrient concentrations with the k_s for nutrient uptake and, in the case of Si, a threshold value for uptake. Plots of relative frequencies (Justić et al. 1994) show that dissolved N concentrations in the surface layer of the northern Gulf of Mexico during the period 1985–1992 were lower than 1 μM in about 13% of the cases. Reactive P was below 0.1 μM in 17% of the cases, while reactive Si concentrations lower than 2 μM occurred in 25% of the cases. In contrast, the corresponding frequencies

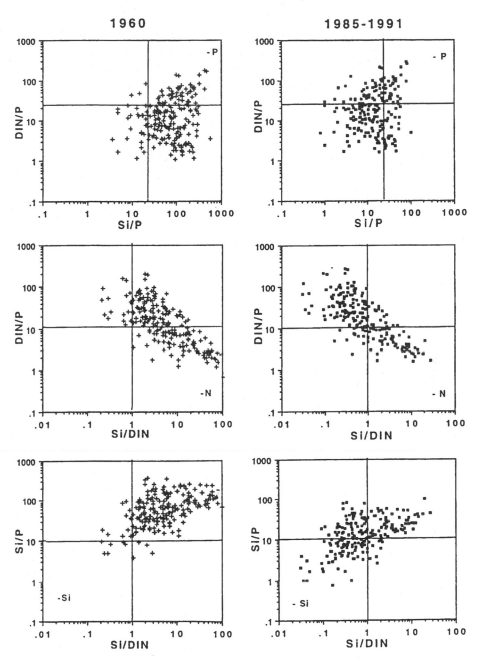

FIGURE 10-5 Scatter diagrams of atomic ratios of dissolved inorganic nitrogen (DIN), reactive phosphorus (P), and reactive silica (Si) in surface water of the northern Gulf of Mexico for (left) reconstructed data for 1960 and (right) ambient nutrient ratios during 1985–1991. From Justić et al. 1995a.

TABLE **10-1**

Concentrations (μM) and atomic ratios of nitrogen (N), phosphorus (P), and silica (Si) in the lower Mississippi River and the northern Gulf of Mexico for 1960–1962 and 1981–1987; x = mean value, n = number of data, S = standard error, $p < 0.001$ = highly significant difference in nutrient concentrations between the two periods, based on a two-sample t-test. (Modified from Justić et al. 1995a, presented in Rabalais et al. 1996.)

		Mississippi River		Northern Gulf of Mexico	
		1960–1962	1981–1987	1960[d]	1985–1991
Nutrient concentration (μM):					
	x	36.5	114	2.23	8.13
N[a]	n	72	200	219	219
	S	2.9	6.0	0.16	0.60
		($p < 0.001$)			
	x	3.9	7.7	0.14	0.34
P[b]	n	—	234	231	231
	S	—	0.4	0.01	0.02
		($p < 0.001$)			
	x	155.1	108	8.97	5.34
Si[c]	n	72	71	235	235
	S	7.5	4.3	0.55	0.33
		($p < 0.001$)			
Average atomic ratios:					
Si:N		4.2	0.9	4.0	0.7
N:P		9	15	16	24
Si:P		39.8	14	64	16

[a] N-NO_3 for the Mississippi River, dissolved inorganic nitrogen (DIN = NO_3 + NH_4 + NO_2) for the northern Gulf of Mexico.
[b] Total P for the Mississippi River, reactive P for the northern Gulf of Mexico.
[c] Reactive Si.
[d] Reconstructed data.

were 39%, 41%, and 10%, respectively, in 1960. These findings are important because studies of nutrient uptake kinetics (Rhee 1973; Harrison et al. 1977; Goldman and Glibert 1983; Nelson and Brzezinski 1990) indicate that concentrations of 1 μM, 0.1 μM, and 2 μM may be considered as threshold values for N, P, and Si uptake, respectively. Thus, it appears that overall nutrient limitation has decreased.

Changes in Phytoplankton Species Composition

The changes in riverine and coastal nutrient concentrations and ratios over time should lead to observable changes in phytoplankton species composition. Published reports of phytoplankton species composition for 1955–1957 near the delta (Simmons and Thomas 1962) and for 1972–1973 approximately 80 km west of the delta (Fucik 1974; Ward et al. 1979) were compared with recently obtained data (1990–1993) from near the delta and at stations C6A and C6B in 20 m water depth off Terrebonne Bay (figure 10-1; Dortch et al. unpubl. data; presented in Rabalais et al. 1996; tables 10-2 and 10-3). Dominant species for each area and time were picked by ranking the phytoplankton by concentration, summing the ranks, and picking the top-ranked groups. This comparison is qualitative because of differences in locations, seasons sampled, and methodology. The methodology used in earlier reports may have missed common non-diatoms, such as small coccoid cyanobacteria and phytoflagellates, which now often dominate. Consequently, it was not possible to determine whether the dominance of these groups has increased. A conservative approach was taken in this comparative analysis. Data were compiled by matching season and location as closely as possible and including in the recent data only phytoplankton types that would have been observed in the earlier studies. In addition, because of uncertainties of taxonomy, differences were only noted where organisms were identified unambiguously.

Demonstrable changes have occurred in the diatom and non-diatom species composition from the 1950s and 1970s to present. Some heavily silicified diatom species are either not observed at all in recent samples (boldface in tables 10-2 and 10-3) or are much less dominant. For example, no *Melosira* species, which appear heavily silicified, were observed in 1990–1993, but these species were present in both 1955–1957 and 1972–1973. *Asterionella japonica* (=*Asterionellopsis glacialis*, Round et al. 1990) was observed recently at low salinities in the spring but is not a dominant species as it was from 1955–1957.

Similarly, more lightly silicified diatoms are documented for the 1970s and are present now, especially at higher salinities. *Rhizosolenia fragilissima* and *Ceratulina pelagica*, which are so lightly silicified they are sometimes difficult to see, were not reported for 1955–1957. Two species of *Leptocylindricus* spp. were frequently dominant in 1990–1993 but were a minor constituent in 1955–1957. During 1972–1973 lightly silicified diatoms were reported, including *Rhizosolenia delicatula, Leptocylindricus danicus,* and *Ceratulina pelagica,* but a more quantitative analysis would be required to determine if their abundance was less than at present.

The data suggest that the shift in dominant diatom composition toward less silicified species occurred between 1955–1957 and 1972–1973, but methodological differences preclude conclusions about changes in non-diatoms. Since Si availability continued to decrease after the early 1970s (Turner and Rabalais 1991), a continued shift in species composition to non-diatoms would be expected. The phytoplankton at C6A and C6B in 1990–1993 were often numerically dominated by small flagellates and cyanobacteria (not shown in

TABLE 10-2

Comparison of dominant diatoms near the river mouth and in the plume between 1955–1957 (Simmons and Thomas 1962) and 1990–1992 (Dortch et al. unpublished data) for spring and summer. Species in bold indicate those previously observed but not currently found or much less dominant, or vice versa. Where taxonomic names have changed, the original published name is retained. (Modified from Rabalais et al. 1996.)

	Spring			Summer	
	Near-River	Plume		Near-River	Plume
May 1957[a]	**Melosira sp.** **Melosira distans** Cyclotella sp. Cyclotella meneghiniana Asterionella japonica Chaetoceros spp.	Chaetoceros sp. Chaetoceros vanheurkii Chaetoceros affinis Thalassiosira nitzschioides Asterionella japonica Nitzschia sp. Skeletonema costatum **Eunotia spp.**	June 1955[a]	Cyclotella spp. **Melosira spp.** Skeletonema costatum Chaetoceros affinis	Cyclotella spp. Skeletonema costatum Chaetoceros affinis
May 1991[b]	Cyclotella spp. < 8μm **Extubocellulus spinifer?** Thalassiosira/Cyclotella	**Dactyliosolen fragilissimus** Pseudo-nitzschia spp. Leptocylindrus minimus	July 1990[b]	Skeletonema costatum Cyclotella spp. < & > 8μm Thalassiosira spp. Thalassiosira/Cyclotella Unident. centric > 8μm Thalassionema nitzschioides	Skeletonema costatum Cyclotella spp. < & > 8μm **Dactyliosolen fragilissimus** **Ceratulina pelagica** Thalassiosira/Cyclotella Thalassionema nitzschioides
			July 1991[b]		Cyclotella spp. > 8μm Chaetoceros spp. **Ceratulina sp. (not pelagica)** **Dactyliosolen fragilissimus**
			July 1992[b]	Thalassiosira/Cyclotella Skeletonema costatum Cyclotella spp. < 8μm Unident. pennate < 8μm **Dactyliosolen fragilissimus**	Sigmoid nitzschiod pennate < 8μm Skeletonema costatum Cyclotella spp. < 8μm

[a] Data for 1955–1957 are for four areas east of the Mississippi River delta (two with salinity < 18 ppt combined under "Near-River" and two with salinity of 18–32 ppt combined under "Plume." Most of the 1990–1992 data are from west of the delta bur are matched for salinity and distance from the river (no stations west of 89.92°W). Methods used by Simmons and Thomas (1962) were reliable for diatoms > 0.6 μm.

[b] Water samples for 1990–1992 were preserved with glutaraldehyde and filtered onto 0.2, 3, and 8 μm polycarbonate filters (Dortch et al. 1992b). Organisms on the 3 and 8 μm filters were frozen and counted later. The 0.2 and 3 μm filters were counted immediately by epifluorescence microscopy; the 8 μm filters were stained with proflavin.

TABLE 10-3

Comparison of dominant phytoplankton species mid-shelf along the Louisiana coast during the periods 1972–1973[a] and 1990–1993.[b] High river flow in spring 1990 and 1993 make these periods most comparable. Species in bold indicate those previously observed but not currently or much less dominant or vice versa. Where taxonomic names have changed, the original published name is retained. (Modified from Rabalais et al. 1996.)

Year	Location	September	July
1972/1973	Platform	*Skeletonema costatum*	*Cyclotella striata* *Thalassionema nitzschioides* *Ceratulina pelagica* **Melosira sp.** *Coscinodiscus radiatus*
1972/1973	Control	*Rhizosolenia deliculata* *Gonyaulax monilata* *Gonyaulax turbynei* *Chaetoceros pelagicus* *Skeletonema costatum*	*Cyclotella striata* **Melosira sp.** *Coscinodiscus radiatus* *Pleurosigma navicalaceum* *Rhizosolenia schrubsolei*
1990	C6A	*Pseudo-nitzschia* spp. *Rhizosolenia delicatula* *Ceratulina pelagica* *Dactyliosolen fragilissimus* *Chaetoceros* spp.	*Dactliosolen fragilissimus* *Leptocylindrus minimus* *Pseudo-nitzschia* spp. *Cylindrotheca closterium*
1991	C6B	*Pseudo-nitzschia* spp. *Cyclotella spp.* < 8μm *Chaetoceros* spp. Unident. pennate < 8μm Nitzschioid-linear pennate < 8μm	*Cyclotella* spp. < 8μm *Cyclotella* spp. > 8μm Unident. auto.dino > 8μm *Thalassiosira/Cyclotella* *Scrippsiella* sp.
1992	C6B	*Chaetoceros curvisetus* *Chaetoceros socialis* *Chaetoceros* spp. *Cyclotella* spp. < 8μm *Ceratulina pelagica*	**Trichodesmium sp.** *Leptocylindrus minimus* *Pseudo-nitzschia* spp.
1993	C6B	*Pseudo-nitzschia* spp. *Chaetoceros* spp. *Skeletonema costatum* Unident. centric > 8μm *Dactyliosolen fragilissimus*	*Chaetoceros* spp. *Thalassionema nitzschioides* **Dinophysis caudata** *Thalassiosera/Cyclotella* **Ceratium furca**

[a] The data of Fucik (1974) and Ward et al. (1979) were obtained from "Control" and "Platform" (fig. 10-1). Their data indicate they used the standard Utermohl method on net tow samples; consequently, this table includes only diatoms, dinoflagellates, and large filamentous cyanobacteria, which they would surely have observed if they had been present.

[b] Data for 1990–1993 (Dortch et al. unpublished data) from nearby station C6A and C6B (fig. 10-1), methods in table 10-2.

table 10-3). They were not considered in this comparison, because it is not clear whether they would have been observed in the 1972–1973 study, even if present in large numbers. It is also tempting to hypothesize that the presence of *Trichodesmium* sp. in 1990–1993, but not in the earlier studies, indicates decreased Si availability. *Trichodesmium* sp., however, is a bloom-forming species, which has been previously reported for this region (Eleuterius et al. 1981).

Several species with importance to human health are now present but were either absent before or have increased in dominance. The dominance of *Pseudo-nitzschia* spp. on the Louisiana shelf appears to have increased dramatically since the 1950s, and concentrations now frequently exceed 1×10^6 cells 1^{-1} (Dortch et al. 1997). Several species known to produce domoic acid, which causes amnesic shellfish poisoning, have been identified in samples from Louisiana waters (Parsons et al. 1998) and high cellular domoic acid concentrations have recently been measured (Dortch, Doucette, Parsons, pers. comm.) Since this species is heavily silicified for a marine diatom (Conley et al. 1989), it is difficult to attribute its increase to decreasing Si:N ratios, although it could be responding to increasing N availability. *Dinophysis caudata*, a dinoflagellate that may be associated with diarrhetic shellfish poisoning (Dickey et al. 1992), was not recorded in the earlier studies. Recently, it was often present at low to moderate concentrations but was sometimes dominant and reached concentrations as high as 1×10^5 cells 1^{-1}.

The increasing N availability and decreasing Si:N ratios appear to have led to increases in dominance of lightly silicified diatoms and non-diatoms. The indicated shift from heavily to lightly silicified forms may have altered carbon flux via directly sinking phytoplankton cells, if silica has become limiting. Subsequently, organic loading to the seabed and, possibly, oxygen depletion may have been affected. On the other hand, an increase in non-diatom forms that are less preferentially grazed than diatoms may have increased the flux of these cells into the lower water column and altered oxygen consumption in a different manner.

Silicate-based Phytoplankton Community Response

Bien et al. (1958) first documented the dilution and nonconservative uptake of silicate in the Mississippi River plume by sampling from the river mouth seaward in 1953 and 1955. A notable characteristic of the mixing diagram is that the concentration of silicate often falls below the conservative mixing line, thus indicating biological uptake (figure 10-6). Uptake can be statistically modeled as a deviation from this mixing line, which we did for thirty-one adequately sampled data sets (Turner and Rabalais 1994a). We found that the concentration of silicate at the 20 ppt mixing point declined in the last several decades during the winter-spring (January–April) and summer months

(June–August); however, there was no discernible change during the fall-winter months (October–December). We normalized for the effects of varying concentrations in the riverine end-member (see Loder and Reichard 1981) and compared the estimated net silicate uptake at 30 ppt as a function of silicate riverine end-member concentration (figure 10-6). Nonconservative uptake of silicate was indicated in all data collections. The net uptake (at 30 ppt) above dilution ranged from 1 to 19% of the intercept concentration, and the data groups for before and after 1979 were remarkably similar. Further, the net silicate uptake appears to be even higher after 1979 than before 1979 (figure 10-6). The results from this analysis suggests that net silicate uptake in the dilution gradient from river to sea has remained the same, or even increased, as the riverine concentration of silicate decreased.

Biologically Bound Silica and Carbon Accumulation

As noted earlier, nearly coincidental trends in silicate (decreasing) and nitrogen (increasing) concentrations in the Mississippi River caused the average annual dissolved silicate:nitrate-N atomic ratio (Si:N) to decline from about 4:1 earlier this century to 1:1 this decade. The present Si:N ratio (1:1) is thought to be a significant limiting threshold for diatom growth, intraspecific competition, and production (Officer and Ryther 1980; Smayda 1990;

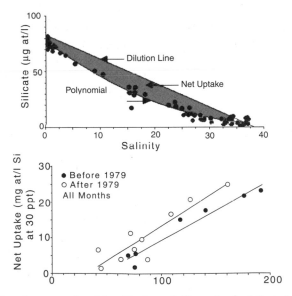

FIGURE 10-6   An example of the dilution of silicate in the Mississippi River (for June 1953) upon mixing with sea water (upper panel). The shaded area is the net uptake of silicate not due to dilution. A polynomial line of best fit was used to estimate the silicate concentration at 20 and 30 ppt. The estimated net uptake of silicate above dilution losses at 30 ppt, for all months of data (lower panel). Modified from Turner and Rabalais (1994a).

Dortch and Whitledge 1992; Turner and Rabalais 1994a). Thus, two contrasting hypotheses predict changes in phytoplankton on this coast since the 1950s. The first is that the coastal phytoplankton are nitrogen, not silica, limited, and higher nitrogen loading will result in proportionally higher phytoplankton production rates. A competing hypothesis is that the combination of lower silica fluxes and a Si:N ratio near 1:1 will result in lower production rates through limits on diatom production along with species-composition shifts. This is a nontrivial issue for managers, because diatoms are an ecologically important constituent of phytoplankton and contribute significantly to the organic loading of bottom waters and sediments and the subsequent oxygen depletion.

We documented that surficial sediments, directly downstream and beneath the surface riverine dilution plume, reflected the in situ primary production and subsequent transport of organic carbon from surface to bottom waters within the Mississippi River bight (Rabalais et al. 1992b; Turner and Rabalais 1994b). We further quantified the silica in the skeletal remains of diatoms sequestered as biologically bound silica (BSi) in dated sediment cores from the same region. The highest concentrations of BSi were in sediments deposited in 25 to 50 m water depth in the middle of the sampling area. The percent BSi in sediments from deeper waters (110 and 200 m) were generally stable through time but rose in the shallower stations (10 and 20 m) around the beginning of this century. At the intermediate depths (27 to 50 m), where both the percent BSi concentration and accumulation rates were highest, coincidental changes in the percent BSi with time were evident, especially in the 1955 to 1965 period (a rise and fall) and a post-1975 (1980?) rise that was sustained to the sampling date (1989; Turner and Rabalais 1984b; figure 10-7). The general pattern that emerges is a constant accumulation of BSi from 1800 to 1900, then a slow rise, followed by a more dramatic rise in the past two decades. Diagenesis of the BSi undoubtedly occurs in these cores but will be relatively low because the sedimentation rate is high (> 0.5 cm yr^{-1}). Furthermore, others have found the record of BSi to be a good indicator of in situ production. Conley et al. (1993) summarized for freshwater lakes that, in general, accumulation of BSi in sediments mimics overlying water column productivity, and that the more diatoms produced by nutrient-enhanced growth, the more BSi will be deposited. Additional information is in Turner and Rabalais (1994b).

If the assumption is made that the BSi:C ratio at the time of deposition remained constant during this century, then the increased BSi deposition represents a significant change in carbon deposition rates (up to 43% higher in core sediments dated after 1980 than those dated between 1900 and 1960). The pattern in percentage of BSi changes parallels the documented increases in nitrogen loading in the lower Mississippi River, over the same period during which the silicate concentrations have been decreasing (figure 10-7). We

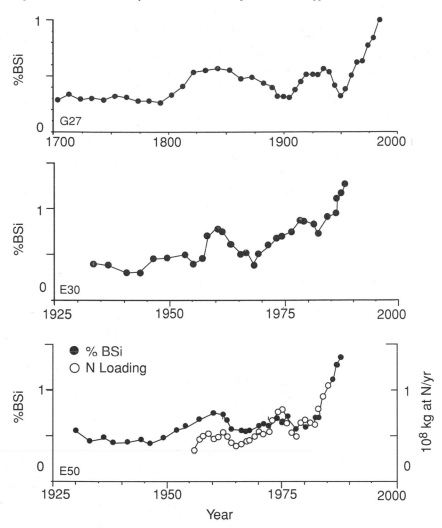

FIGURE 10-7   Average percentage of biologically bound silica (BSi) concentration of sediments in each section of ^{210}Pb-dated sediment cores from stations G27, E30, and E50. A 3-year running average for each sampling date is shown. The percentage of BSi data for station E50 are superimposed with a 3-year running average of the nitrogen loading from the Mississippi River through the delta passes. Modified from Turner and Rabalais (1994b).

conclude from our analyses that the flux of diatoms from surface to bottom waters, beneath the Mississippi River plume, increased this century. These changes were coincident with changes in riverine nitrogen loadings and resulted in higher organic sedimentation to bottom-water layers. The depletion of bottom-water oxygen, its persistence and areal coverage on this shelf, is thus indicated to have been altered this century.

Consequences to Hypoxic Bottom-water Formation and Severity

Long-term changes in the severity and extent of hypoxia cannot be assessed directly, because systematic sampling of bottom-water dissolved oxygen concentrations did not begin until 1985. Prior records of hypoxia, dating to 1973, were obtained sporadically as components of other studies; previous events were drawn from anecdotal relationships with shrimp trawl yields. Therefore, biological, mineral, or chemical indicators of hypoxia preserved in sediments, where accumulation rates record historical changes, provide clues to prior hydrographic and biological conditions.

Dominance trends of benthic foraminifera serve as indicators of reduced oxygen levels or carbon-enriched sediments or both (Sen Gupta et al. 1981; Sen Gupta and Machain-Castillo 1993). The same series of ^{210}Pb-dated sediment cores used for BSi analyses (Turner and Rabalais 1994b, figure 10-7) were used for determination of benthic foraminifera. Some downcore shifts in species abundance at station G27 (figure 10-7) in the Mississippi River bight were interpreted as foraminiferal responses to increasing oxygen stress (Sen Gupta et al. 1993, 1996). Benthic foraminiferal density and diversity are generally low in this environment, but a comparison of assemblages in surficial sediments from areas differentially affected by oxygen depletion indicates that the dominance of *Ammonia parkinsoniana* over *Elphidium* spp. is much more pronounced under hypoxia than in well-oxygenated waters. The relative abundance of *A. parkinsoniana* is also correlated with percentage of BSi (food source indicator) in sediments. In the context of modern hypoxia, species distribution in dated sediment cores reveals stratigraphic trends in the *Ammonia/Elphidium* ratio that indicate an overall increase in oxygen stress (in intensity or duration) in the last 100 years (figure 10-8). In particular, the stress seems especially severe since the 1950s. For this time period, both *Ammonia* and *Elphidium* become less important components of the assemblage, while *Buliminella morgani* shows an unusual dominance as also shown by Blackwelder et al. (1996). *Quinqueloculina* (a significant component of the modern assemblage only in well-oxygenated waters) has been absent from the record of this core since the 1870s but was a conspicuous element of the fauna in the previous 100 years. Thus, there are indications that oxygen-deficiency stress increased as nutrient loads and carbon flux to the seabed increased.

Predictions

One reason to synthesize data in estuarine and coastal studies is to understand processes and likely responses of the system to a variety of human impacts. Once a series of coherent patterns and interactions is outlined, as we have done for the continental shelf adjacent to the Mississippi River effluent,

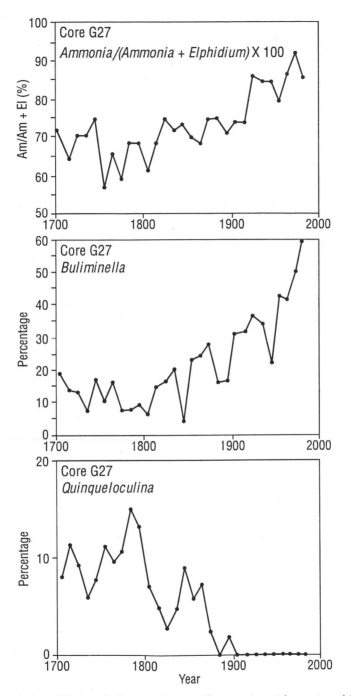

FIGURE 10-8 Changes in benthic foraminifera species with stratigraphic depth in ^{210}Pb-dated sediment core from station G27 in the Mississippi River Bight. From Rabalais et al. (1996).

it may be used to predict the response of the coastal system to alterations in riverine nutrient fluxes. Policies for nutrient control in freshwater systems seldom take into account the impact, or lack of impact, on coastal systems. Controls in freshwater systems often target phosphorus, based on the numerous laboratory and field studies of the stimulatory effect of phosphorus on freshwater ecosystems. Coastal systems, however, are usually thought to be nitrogen limited, at least part of the time, and this includes the shelf adjacent to the Mississippi River.

If freshwater phosphorus loading is reduced, we predict that the rate of BSi burial in freshwater sediments will decrease because freshwater diatom production will be lower (similar to the decreased percentage of BSi in more recent Lake Michigan sediments; shown in Conley et al. 1993). The release of dissolved Si from the sediments in the watershed will exceed the uptake of dissolved Si in the water column. The result will be an increase in dissolved Si within the Mississippi River and in the adjacent coastal waters (a return to the 1950s and turn-of-the-century levels). At the same time, controls on N use and loadings within the watershed may or may not be affected.

Based on the biological consequences of documented historical riverine changes, we foresee three possible scenarios for future system responses as riverine fluxes of silicate increase (figure 10-9). First, if N concentration remains the same, overall N limitations to phytoplankton productivity will be similar to present, but Si will no longer be limiting. The result would be increased BSi and carbon accumulation in the sediments, and an increase in the extent and severity of hypoxia. Second, if N concentrations increase, there will be no N or Si limitations and the Si:N ratio will be balanced. The result would be greatly increased BSi and carbon accumulations as well as substantial increase in the severity of hypoxia. Third, if N concentrations are reduced (for example, to 1950s values), then N would again return to the limiting nutrient status; although Si would be in abundant supply, the system would be restricted by N supplies and hypoxia would decrease.

If water quality improves, the close coupling between riverine nutrient loading and phytoplankton production should enable reversal of the current effects of nutrient increases and nutrient ratio changes. However, the management of one nutrient alone (Si or N) may not be sufficient to reduce eutrophication to an acceptable level if the compensatory qualitative adaptations of species lead to new phytoplankton communities, including those with noxious or toxic species.

Acknowledgments

Data were collected during programs funded by NOAA's Ocean Assessments Division, Louisiana Board of Regents LaSER Award 86-LUM(1)-083-13 and LEQSF

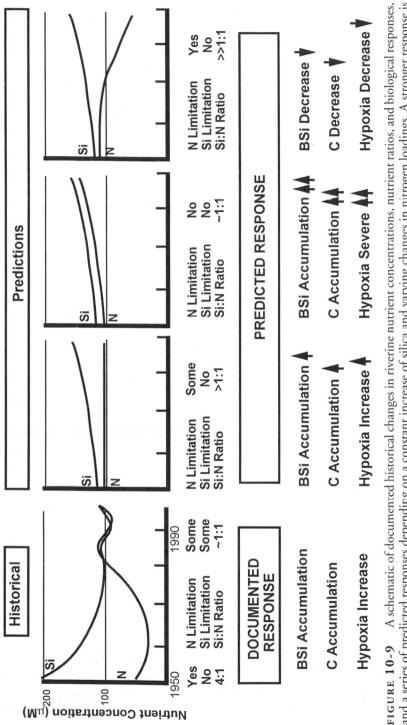

FIGURE 10-9 A schematic of documented historical changes in riverine nutrient concentrations, nutrient ratios, and biological responses, and a series of predicted responses depending on a constant increase of silica and varying changes in nitrogen loadings. A stronger response is indicated by double arrows. (See text for discussion; modified from Rabalais et al. 1996).

Award (1987–90)-RD-A-15, the Louisiana Sea Grant College Program, NOAA's National Undersea Research Center, Louisiana State University, the Louisiana Universities Marine Consortium, and the NOAA Coastal Ocean Program Office, Nutrient Enhanced Coastal Ocean Productivity (NECOP) Study Grant No. NA90AA-D-SG691 to the Louisiana Sea Grant College Program, Awards No. MAR31, MAR24 and MAR92-02. We thank the many research assistants and associates who helped with the collection and analyses of samples as well as captains and crew of the R/V *R. J. Russell*, R/V *Acadiana,* and R/V *Pelican*. Funding for this synthesis and manuscript preparation was provided by the NOAA NECOP program.

References

Andersson, L., and L. Rydberg. 1988. Trends in nutrient and oxygen conditions within the Kattegat: Effects on local nutrient supply. *Estuarine Coastal and Shelf Science* 26: 559–79.

Atkinson, L. P., and D. Wallace. 1975. The source of unusually low surface salinities in the Gulf Stream off Georgia. *Deep-Sea Research* 23: 913–16.

Bien, G. S., D. E. Contois, and W. H. Thomas. 1958. The removal of soluble silica from fresh water entering the sea. *Geochimica et Cosmochimica Acta* 14: 35–54.

Blackwelder, P., T. Hood, C. Alvarez-Zarikian, T. Nelsen, and B. McKee. 1996. Benthic foraminifera from the NECOP study area impacted by the Mississippi River plume and seasonal hypoxia. *Quarternary International* 31: 19–36.

Bode, A., and Q. Dortch. 1996. Uptake and regeneration of inorganic nitrogen in coastal waters influenced by the Mississippi River: Spatial and seasonal variations. *Journal of Plankton Research* 18: 2251–68.

Cochrane, J. D., and F. J. Kelly. 1986. Low-frequency circulation on the Texas-Louisiana shelf. *Journal of Geophysical Research* 91: 10645–59.

Conley, D. J., S. S. Kolham, and E. Theriot. 1989. Differences in silica content between marine and freshwater diatoms. *Limnology and Oceanography* 34: 205–13.

Conley, D. J., C. L. Schelske, and E. F. Stoermer. 1993. Modification of the biogeochemical cycle of silica with eutrophication. *Marine Ecology Progress Series* 101: 179–92.

Dickey, R. W., G. A. Fryxell, H. R. Granade, and D. Roelke. 1992. Detection of the marine toxins okadaic acid and domoic acid in shellfish and phytoplankton in the Gulf of Mexico. *Toxicon* 30: 355–59.

Dortch, Q., and T. E. Whitledge. 1992. Does nitrogen or silicon limit phytoplankton production in the Mississippi River plume and nearby regions? *Continental Shelf Research* 12: 1293–1309.

Dortch, Q., A. Bode, and R. R. Twilley. 1992a. Nitrogen uptake and regeneration in surface waters of the Louisiana continental shelf influenced by the Mississippi River. Pages 52–56 in *Proceedings, nutrient enhanced coastal ocean productivity workshop.* Publication No. TAMU-SG-92-109. Texas Sea Grant College Program. College Station, TX: Texas A&M University.

Dortch, Q., D. Milsted, N. N. Rabalais, S. E. Lohrenz, D. G. Redalje, M. J. Dagg, R. E. Turner, and T. E. Whitledge. 1992b. Role of silicate availability in phytoplankton species composition and the fate of carbon. Pages 76–83 in *Proceedings, nutrient enhanced coastal ocean productivity workshop*. Publication No. TAMU-SG-92-109. Texas Sea Grant College Program. College Station, TX: Texas A&M University.

Dortch, Q., R. J. Robichaux, S. Pool, D. Milsted, G. Mire, N.N. Rabalais, T. Soniat, G. Fryxell, R. E. Turner, and M. L. Parsons. 1997. Abundance and flux of *Pseudonitzschia* in the northern Gulf of Mexico. *Marine Ecology Progress Series* 146: 249–64.

Eleuterius, L., H. Perry, C. Eleuterius, J. Warren, and J. Caldwell. 1981. Causative analysis on a nearshore bloom of *Oscillatoria erythraea* (*Trichodesmium*) in the northern Gulf of Mexico. *Northeast Gulf Science* 5: 1–11.

Fucik, K. W. 1974. The effect of petroleum operations on the phytoplankton ecology of the Louisiana coastal waters. Master's thesis, Texas A&M University.

Goldman, J. C., and P. M. Glibert. 1983. Kinetics of inorganic nitrogen uptake by phytoplankton. Pages 223–74 in E. J. Carpenter and D. G. Capone (editors), *Nitrogen in the marine environment*. New York: Academic Press.

Harrison, P. J., H. L. Conway, R. W. Holmes, and C. O. Davis. 1977. Marine diatoms in chemostats under silicate or ammonium limitation. 3. Cellular chemical composition and morphology of three diatoms. *Marine Biology* 43: 19–31.

Justić, D., T. Legović, and L. Rottini-Sandrini. 1987. Trends in the oxygen content 1911–1984 and occurrence of benthic mortality in the northern Adriatic Sea. *Estuarine Coastal and Shelf Science* 25: 435–45.

Justić, D., N. N. Rabalais, R. E. Turner, and W. J. Wiseman Jr. 1993. Seasonal coupling between riverborne nutrients, net productivity and hypoxia. *Marine Pollution Bulletin* 26: 184–89.

Justić, D., N. N. Rabalais, and R. E. Turner. 1994. Riverborne nutrients, hypoxia and coastal ecosystem evolution: Biological responses to long-term changes in nutrient loads carried by the Po and Mississippi Rivers. Pages 161–67 in K. R. Dyer and R. J. Orth (editors), *Changes in fluxes in estuaries: Implications from science to management*. Proceedings of ECSA22/ERF Symposium, International Symposium Series. Fredensborg, Denmark: Olsen & Olsen.

Justić, D., N. N. Rabalais, R. E. Turner, and Q. Dortch. 1995a. Changes in nutrient structure of river-dominated coastal waters: Stoichiometric nutrient balance and its consequences. *Estuarine Coastal and Shelf Science* 40: 339–56.

Justić, D., N. N. Rabalais, and R. E. Turner. 1995b. Stoichiometric nutrient balance and origin of coastal eutrophication. *Marine Pollution Bulletin* 30: 41–46.

Loder, T. C., and R. P. Reichard. 1981. The dynamics of conservative mixing in estuaries. *Estuaries* 4: 64–69.

Lohrenz, S. E., M. J. Dagg, and T. E. Whitledge. 1990. Enhanced primary production at the plume/oceanic interface of the Mississippi River. *Continental Shelf Research* 10: 639–64.

Lohrenz, S.E., G.L. Fahnenstiel, D. G. Redalje, G. A. Lang, X.-G., Chen, and M. J. Dagg. 1997. Variations in primary production of northern Gulf of Mexico continental shelf waters linked to nutrient inputs from the Mississippi River. *Marine Ecology Progress Series* 155: 45–54.

Milliman, J. D., and R. H. Meade. 1983. World-wide delivery of river sediment to the oceans. *Journal of Geology* 91: 1–21.

Nelson, D. M., and M. A. Brzezinski. 1990. Kinetics of silicic acid uptake by natural diatom assemblages in two Gulf Stream warm-core rings. *Marine Ecology Progress Series* 62: 283–92.

Officer, C. B., and J. H. Ryther. 1980. The possible importance of silicon in marine eutrophication. *Marine Ecology Progress Series* 3: 83–91.

Officer, C. B., R. B. Biggs, J. L. Taft, L. E. Cronin, M. Tyler, and W. R. Boynton. 1984. Chesapeake Bay anoxia: Origin, development, and significance. *Science* 223: 22–27.

Parsons, M. L., Q. Dortch, and G. A. Fryxell. 1998. A multi-year study of the presence of potential domoic acid-producing *Pseudo-nitzschia* species in the coastal and estuarine waters of Louisiana, USA. Pages 184–87 in B. Reguera, J. Blanco, Ma L. Fernandez, and T. Wyatt (editors), *Harmful algae.* Santiago de Compostela, Spain: Xunta de Galicia and Intergovernmental Oceanographic Commission of UNESCO, GRAFISANT.

Qureshi, N. A. 1995. The role of fecal pellets in the flux of carbon to the sea floor on a river-influenced continental shelf subject to hypoxia. Ph.D. diss., Louisiana State University.

Rabalais, N. N., R. E. Turner, W. J. Wiseman Jr., and D. F. Boesch. 1991. A brief summary of hypoxia on the northern Gulf of Mexico continental shelf: 1985–1988. Pages 35–46 in R. V. Tyson and T. H. Pearson (editors), *Modern and ancient continental shelf anoxia.* Geological Society Special Publication No. 58. London: The Geological Society.

Rabalais, N. N., R. E. Turner, and W. J. Wiseman Jr. 1992a. Distribution and characteristics of hypoxia on the Louisiana shelf in 1990 and 1991. Pages 15–20 in *Proceedings, nutrient enhanced coastal ocean productivity workshop.* Publication No. TAMU-SG-92-109. Texas Sea Grant College Program. College Station, TX: Texas A&M University.

Rabalais, N. N., R. E. Turner, and Q. Dortch. 1992b. Louisiana continental shelf sediments: Indicators of riverine influence. Pages 131–35 in *Proceedings, nutrient enhanced coastal ocean productivity workshop.* Publication No. TAMU-SG-92-109. Texas Sea Grant College Program. College Station, TX: Texas A&M University.

Rabalais, N. N., W. J. Wiseman Jr., and R. E. Turner. 1994a. Hypoxic conditions in bottom waters on the Louisiana-Texas shelf. Pages 50–54 in M. J. Dowgiallo (editor), *Coastal oceanographic effects of summer 1993 mississippi river flooding. Special NOAA Report.* Silver Spring, MD: U.S. Dept. of Commerce, National Oceanic and Atmospheric Administration, Coastal Ocean Program.

———. 1994b. Comparison of continuous records of near-bottom dissolved oxygen from the hypoxia zone of Louisiana. *Estuaries* 17: 850–61.

Rabalais, N. N., R. E. Turner, D. Justić, Q. Dortch, W. J. Wiseman Jr. and B. K. Sen Gupta. 1996. Nutrient changes in the Mississippi River and system responses on the adjacent continental shelf. *Estuaries* 19: 386–407.

Rabalais, N. N., R. E. Turner, W. J. Wiseman Jr., and Q. Dortch. 1998. Consequences of the 1993 Mississippi River flood in the Gulf of Mexico. *Regulated Rivers: Research & Management* 14: 161–77.

Redfield, A. C. 1958. The biological control of chemical factors in the environment. *American Scientist* 46: 205–22.

Rhee, G. Y. 1973. A continuous culture study of phosphate uptake, growth rate and polyphosphate in *Scenedesmus* sp. *Journal of Phycology* 9: 495–506.

Round, F. E., R. M. Crawford, and D. G. Mann. 1990. *The diatoms. Biology and morphology of the genera.* Cambridge, England: Cambridge University Press.

Schelske, C. L., and E. F. Stoermer. 1971. Eutrophication, silica depletion, and predicted changes in algal quality in Lake Michigan. *Science* 173: 423–24.

Schelske, C. L., D. J. Conley, E. F. Stoermer, T. L. Newberry, and C. D. Campbell. 1986. Biogenic silica and phosphorus accumulation in sediments as indices of eutrophication in the Laurentian Great Lakes. *Hydrobiologia* 143: 79–86.

Sen Gupta, B. K., R. F. Lee, and M. S. May. 1981. Upwelling and an unusual assemblage of benthic foraminifera on the northern Florida continental slope. *Journal of Paleontology* 55: 853–57.

Sen Gupta, B. K., and M. L. Machain-Castillo. 1993. Benthic foraminifera in oxygen-poor habitats. *Marine Micropaelontology* 20: 183–201.

Sen Gupta, B. K., R. E. Turner, and N. N. Rabalais. 1993. Oxygen stress in shelf waters of northern Gulf of Mexico: 200-year stratigraphic record of benthic foraminifera. Page A138 Abstract in *Geological Society of America, 1993 Annual Meeting.* Boulder, Colo.: Geological Society of America.

————. 1996. Seasonal oxygen depletion in continental-shelf waters of Louisiana: Historical record of benthic foraminifers. *Geology* 24: 227–30.

Simmons, E. G., and W. H. Thomas. 1962. Phytoplankton of the eastern Mississippi delta. *Publications of the Institute of Marine Science, University of Texas* 8: 269–98.

Sklar, F. H., and R. E. Turner. 1981. Characteristics of phytoplankton production off Barataria Bay in an area influenced by the Mississippi River. *Contributions in Marine Science* 24: 93–106.

Smayda, T. J. 1990. Novel and nuisance phytoplankton blooms in the sea: Evidence for global epidemic. Pages 29–40 in E. Graneli, B. Sundstrom, R. Edler, and D. M. Anderson (editors), *Toxic marine phytoplankton.* New York: Elsevier Science Publishing Co.

Smith, N. P. 1980. On the hydrography of shelf waters off the central Texas gulf coast. *Journal of Physical Oceanography* 10: 806–13.

Tester, P. A., and L. P. Atkinson. 1994. Low salinity water in the Gulf Stream off North Carolina. Pages 72–76 in M. J. Dowgiallo (editor), *Coastal oceanographic effects of summer 1993 Mississippi River flooding.* Special NOAA Report. Silver Spring, MD: National Oceanic and Atmospheric Administration, Coastal Ocean Program.

Thomas, W. H., and E. G. Simmons. 1960. Phytoplankton production in the Mississippi River Delta. Pages 103–16 in F. P. Shepard (editor), *Recent sediments, northwest Gulf of Mexico.* Tulsa, OK: American Association of Petroleum Geologists.

Turner, R. E., and N. N. Rabalais. 1991. Changes in Mississippi River water quality this century. Implications for coastal food webs. *BioScience* 41: 140–47.

———. 1994a. Changes in the Mississippi River nutrient supply and offshore silicate-based phytoplankton community responses. Pages 147–50 in K. R. Dyer and R. J. Orth (editors), *Changes in fluxes in estuaries: Implications from science to management.* Proceedings of ECSA22/ERF Symposium, International Symposium Series. Fredensborg, Denmark: Olsen & Olsen.

———. 1994b. Coastal eutrophication near the Mississippi River delta. *Nature* 368: 619–21.

Walker, N. D., G. S. Fargion, L. J. Rouse, and D. C. Biggs. 1994. The great flood of summer 1993: Mississippi River discharge studied. *Eos, Transactions, American Geophysical Union* 75: 409, 414–15.

Ward, C. H., M. E. Bender, and D. J. Reish (editors). 1979. The Offshore Ecology Investigation. Effects of oil drilling and production in a coastal environment. *Rice University Studies* 65: 1–589.

Wiseman, W. J. Jr., N. N. Rabalais, R. E. Turner, S. P. Dinnel, and A. MacNaughton. 1997. Seasonal and interannual variability within the Louisiana coastal current: Stratification and hypoxia. *Journal of Marine Systems* 12: 237–48.

CHAPTER 11

Influence of River Flow and Nutrient Loads on Selected Ecosystem Processes
A Synthesis of Chesapeake Bay Data

Walter R. Boynton and W. Michael Kemp

Abstract

In this chapter we assembled and analyzed two data sets, one a discontinuous 22-year time series (1972–1977, 1985–1993) of observations from a single mesohaline site in Chesapeake Bay, and the other, a much shorter time series from that site plus similar sites in four bay tributaries. For all locations, the data set includes measurements of river flow, nutrient-loading rate, phytoplankton primary production rates and biomass, water-column nutrient concentrations, and sediment-water exchanges of ammonium. In addition, data on sedimentation rates of chlorophyll *a* and bottom-water dissolved oxygen concentrations were analyzed at one site.

We examined a series of hypotheses concerning the influence of river flow and nutrient loading on these variables toward the goal of understanding underlying mechanisms. Significant relationships to flow and associated nutrient loads were found for all variables, some being stronger than others. In most cases, the influence of flow was found to extend over relatively short time periods (months to 2 years) and there were temporal lags between flow events and ecosystem responses on time scales of weeks to months. Results of analyses based on the time series from one location and on comparative analyses of data from five different sites were qualitatively similar; in this system it was not necessary to invoke comparative analyses to capture a large enough signal in forcing and response to observe interpretable patterns. Analyses generally indicated that relationships proximal to flow or nutrient loading rate were stronger (for example, nutrient load versus water-column nutrient mass) than those more removed from the direct influence of flow or nutrient load (for example, flow versus sediment nutrient releases).

These analyses indicate the importance of freshwater flow and associated nutrients in shaping chemical and biological responses in this estuary. Analyses are continuing and the next step will be to examine the effects of flow and nutrient loads on submersed vascular plant distributions and zooplankton and benthic communities.

Introduction

During the last decade there has been an increasing number of environmental measurements taken in coastal and estuarine systems, and this trend seems destined to continue for the foreseeable future. In part, this activity has been stimulated by increased awareness of natural resource deterioration in these environments due to human activities in the drainage basin as well as in the receiving water bodies. Common now are reports of declining or collapsed fisheries, toxic algal blooms, development of hypoxia and anoxia in deeper waters, and loss of submersed aquatic vegetation communities (Nixon 1990).

Despite much larger databases for many of these systems, we are still unable to confidently answer many fundamental questions concerning how these systems work and, from a practical viewpoint, what resource managers need to do to reverse declines in water quality and abundance of living resources. One reason for this state of affairs is that analyses and interpretations of these data sets have been limited. This is particularly true for many data sets collected in monitoring programs and ad hoc field surveys. In addition, scientific data collected in research programs are often interpreted within relatively narrow areas of scientific interest having little value at the larger scales of organization relevant for resource management (Malone et al. 1993). To be useful, these data need to be pulled together into some sort of synthesis that focuses on time, space, and organizational scales appropriate to the questions being asked.

In recent years, some very ambitious numerical efforts have been initiated and serve as one type of data synthesis. For example, sophisticated hydrodynamic models have been developed for a number of estuarine systems (for instance, Long Island Sound, Tampa Bay, Chesapeake Bay) and act as a framework for synthesis of large data sets as well as forecasting tools. In other cases, these models have been coupled with water-quality models and used as diagnostic tools in water-quality management programs, as is the case in Chesapeake Bay (Cerco and Cole 1992). While these tools have obvious advantages, they are expensive and time consuming to construct, analyze, and maintain.

The purpose of this chapter is to present the results of a direct, empirical type of synthesis whereby variations in key properties of coastal ecosystems are related to changes in riverine nutrient loading (Rigler 1982; Peters 1991). Specifically, we describe here the results of regression modeling based on data collected in Chesapeake Bay. In this work we have primarily focused on examining the influence of freshwater inputs (and associated nutrient loads) on several ecological processes. The importance of freshwater inputs is obvious; it is a central feature in the definition of estuarine systems, it influences physical dynamics (Boicourt 1992), is well correlated with nutrient inputs (Summers 1993), and has been implicated in regulating either directly or indirectly estuarine processes ranging from primary production (Boynton et al. 1982; Cloern et al. 1983) to benthic secondary production (Flint 1985) to fish recruitment (Stevens 1977) and catch (Sutcliffe 1973; Sutcliffe 1977; Ennis 1986). The emphasis here is the exploration of data sets for patterns that conform to expected relationships or suggest new relationships (see Meeuwig et al. 1998) rather than statistical testing for significant differences or temporal trends. We wish to examine environmental data for relationships and to use these as clues to suggest underlying mechanisms.

Approach and Methods

Conceptual Model

The focus of these analyses is to investigate the influence of river flow and associated nutrient inputs on selected ecological processes in Chesapeake Bay. Most, if not all, of these hypothesized direct or indirect effects of river flow on ecological processes have been documented in other systems. For example, phytoplankton biomass and community composition have been shown to be regulated by river discharge in San Francisco Bay (Cloern et al. 1983) and Texas estuaries (Flint 1985), while buoyancy effects of fresh water have been extensively investigated in various estuaries (Boicourt 1992), and responses of benthic respiration and nutrient regeneration to variations in phytoplankton production and deposition have also been examined (Flint 1985; Cowan et al. 1996). Here we consider the extent to which these effects of river flow are manifest in Chesapeake Bay and we have organized this analysis around a simple conceptual model (figure 11-1). In this model, river flow adds directly to the nutrient pools (1) and influences buoyancy of the water-column. River flow also determines the geographic positioning of water-column events (that is, events such as plankton blooms tend to shift seaward in high-flow periods and landward in low-flow periods) and the location of water-column deposition of organic matter to the benthos. Phytoplankton production (2) and biomass (3) are responsive to nutrient pools and phytoplankton biomass is lost to the benthic community via sinking (4). The benthic community recycles nutrients to the water column (6). Finally, deep-water dissolved-oxygen depletion (5) is influenced by stratification of the water column, organic matter derived from

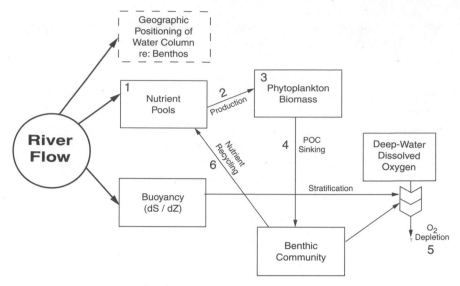

FIGURE 11-1   A simple schematic diagram showing the influences of river flow on ecosystem stocks and processes examined in this study. The mechanistic relationships between river flow and the stocks and processes shown in the diagram are explained in the text.

phytoplankton, and respiration of this organic matter by the benthic community.

Obviously, the perspective in this conceptual model is heavily biased toward bottom-up (as opposed to top-down) control of ecological interactions. We recognize that top-down effects can be important, and dominant, in some estuarine situations. For example, Alpine and Cloern (1992) found dramatic changes in the temporal pattern of phytoplankton production and biomass in San Francisco Bay following the introduction of a suspension-feeding clam. Meeuwig et al. (1998) found that herbivory by mussels was a strong modifier of algal biomass-nutrient relations in some Canadian estuarine systems. There are also more numerous and well-known examples from limnology (Carpenter and Kitchell 1988). In this preliminary analysis, we chose to emphasize the bottom-up perspective for simplicity and elegance.

Ecological Inference

Our approach to ecological synthesis consists of several steps, the first of which involves developing empirical models to specify the relationships of interest and to aid in selection of appropriate variables. Rigler (1982) differentiated ecological research into empirical and explanatory categories, the

former with the goal of prediction and the latter with mechanistic explanations of predictions. In regression modeling the mechanistic relationships
between variables are not specified while in explanatory approaches (such as
simulation modeling) every interaction is exactly specified. In a sense,
empirical approaches are more holistic while explanatory schemes are more
reductionistic.

As an example of empirical model development, interannual variability in
algal biomass may be of interest and appropriate variables might be chlorophyll a
concentrations or some other measure of algal stock. The next step is to select a
group of probable causative variables and in the above example these might be
river flow, nutrient-supply rate, light availability, or others. Much of the above
obviously requires previous knowledge in establishing relationship between
variables so there is a natural interaction between reductionist and holistic
approaches. We also recognize the need to establish alternative hypotheses in
empirical approaches as suggested by Peters (1991). The idea here is to explore
all reasonable explanations, rejecting most because they do not support the
hypothesis and leaving us with one (or more) that can be supported and further
explored. It is this step that largely differentiates this approach from a simple
statistical examination of a data set for statistically significant relationships.

We have also employed comparative approaches in a portion of the analyses presented in this chapter wherein similar data from a variety of systems
are used in the analysis. This technique has the advantage of increasing the
signal range for both independent and dependent variables and hence
increasing the chance of interpretable patterns emerging from what is admit-
tedly a complex set of interactions (Vollenweider 1976; Nixon 1988).
However, comparative approaches generally require "scaling" of variables in a
fashion that makes them comparable among systems and this in itself can be
a complex and interesting problem (Schnieder 1994).

In this chapter we present two groups of empirical analyses; the first is
based on a data set collected at one location in Chesapeake Bay for a number
of years (~13 years) while the second examines similar issues but uses a comparative approach based on data collected at multiple locations in Chesapeake
Bay for shorter periods of time (1 to 4 years). We take advantage of a long data
record in the former and inherent differences among systems in the latter; in a
sense this can be thought of as a time-space substitution with both approaches
being useful in testing ideas about ecosystem behavior (Pickett 1991).

Study Area

Chesapeake Bay is the largest estuary in the United States, having an area of
6,500 km^2, a length > 300 km, a width of 5–30 km, and mean depth of 8.4 m;
it is closely embraced by the land (drainage basin surface area: bay surface
area = 28:1). The surface area of the bay system is equally divided between the

mainstem bay and the numerous (approximately thirty) tributary rivers and bays; however, about 66% of the volume is contained in the deeper mainstem (figure 11-2).

The hydraulic fill time (volume divided by freshwater inputs) is approximately 1 year and water residence times range from 3 to 6 months. The mainstem bay is stratified from late winter through early fall; stratification in tributaries is generally weaker and less persistent (Boicourt 1992). Water-column stratification is in part responsible for chronic hypoxic and anoxic conditions in deeper regions of the system (Kemp et al. 1992). It appears that the volume of hypoxic water has increased since colonial times (Cooper and Brush 1991), much of it in the last three to four decades (Boicourt 1992).

The bay and its watershed lie in the coastal corridor of dense human population between New York and Virginia; population in the watershed is now 13.6 million and is projected to soon be 16.2 million (Magnien et al. 1995). Current nitrogen- and phosphorus-loading rates averaged for the entire bay are about 13 gN m^{-2} yr $^{-1}$ and 1 gP m^{-2} yr $^{-1}$, respectively; however, loading rates to distinct portions of the bay system range from both a factor of 5 higher and lower than these and thus provide a good opportunity for comparative analyses. Since European settlement, bay-wide loading rates of nitrogen and phosphorus have increased about six- and seventeenfold, respectively (Boynton et al. 1995).

One of the important characteristics of estuarine systems such as Chesapeake Bay is temporal variations in inputs such as freshwater flow (figure 11-3). During the past several decades, the magnitude of annual average freshwater input to the head of Chesapeake Bay has varied by a factor of 2.4; average annual flows from the Susquehanna River are about 1,200 m^3 sec $^{-1}$ and represent about 50% of the freshwater flow to the entire Chesapeake system. However, seasonal patterns of flow are even more variable, especially during the "spring freshet." This important hydrological event has occurred between January and May in recent decades, though typically during March or April, and has varied in magnitude by a factor of 5 (figure 11-3).

Data Sources

Chesapeake Bay and associated tributaries is one of the most studied estuarine systems in the United States and a tremendous amount of data are available, especially from the last decade. Much of this information has been collected as part of the EPA's Chesapeake Bay Program and closely related monitoring programs conducted by the states of Maryland and Virginia.

To provide some indication of the intensity of this program, about 40,000 measurements of such variables as chlorophyll *a* have been made in the mainstem bay alone during the last decade. During the last decade the challenge

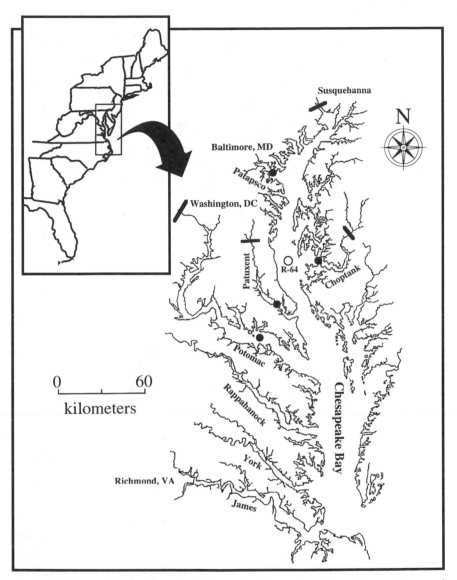

FIGURE 11-2 A map showing Chesapeake Bay and major tributary rivers and location of this estuarine system on the East Coast of the United States. Bold lines indicate locations at the fall line where freshwater inflows and nutrient loads are monitored. Bold dots indicate locations where water-quality, phytoplankton, and sediment-water flux measurements were made; these measurements as well as sedimentation rates were made at the site indicated by the open circle (R-64).

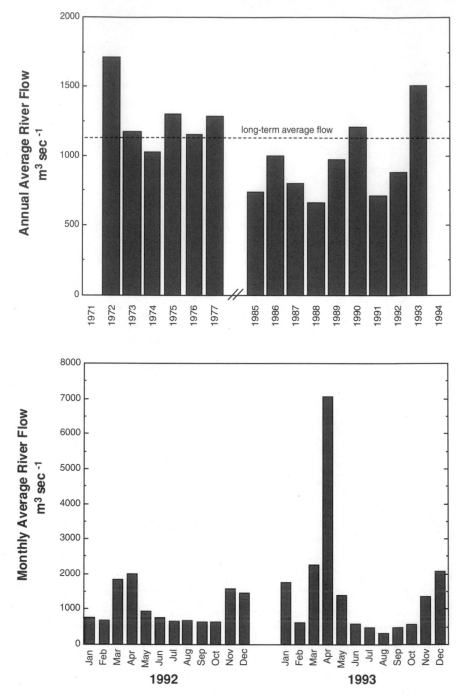

FIGURE 11-3 Estimates of annual average river flow (1972–1977 and 1985–1993) from the Susquehanna River entering the head of Chesapeake Bay. Flows are for the years used in regression analyses. Also shown are monthly average flows for the years 1992 and 1993 from the same location to indicate the variation in magnitude and seasonal patterns of freshwater inputs (USGS 1993).

of simply finding enough data to see if interpretable seasonal patterns exist has changed to one of managing large databases in such a way that any patterns present can be found. A brief description of the data collection program is provided in table 11-1 for the variables used in the analyses presented here; details concerning collection and analytical methodologies have been described in detail elsewhere (see sources listed in table 11-1).

Statistical Methods

Regression techniques have been applied to ecological problems for quite some time, often yielding useful results. In part, the attraction of this approach lies in its simplicity. As opposed to water quality, fisheries, and ecosystem models, data requirements can be relatively small, the time required to explore many possible relationships short, and there are few, if any, assumptions to be made concerning the form of mechanistic relationships. The relative ease of using the technique makes it, therefore, very attractive as a tool for scanning data in search of suspected relationships and as a tool that often suggests new interpretations.

While regression can be straightforward there are, of course, limitations that are both practical and conceptual. As with any statistical technique, strong correlation between variables does not, however tempting, indicate causal relationships. In addition, covariance among variables can lead to spurious conclusions. Assumptions concerning the distribution of data and other criteria for strict application of parametric techniques are often difficult or impossible to check.

In spite of these problems, and the apparent simplicity of the approach, there have been many useful applications of regression techniques to ecological problems. During the 1960s and early 1970s, Vollenweider and his colleagues developed a series of "mixed reactor regression models" relating algal standing stock (used as an indicator of trophic condition) to nutrient loading (primarily phosphorus loading to lakes). They found significant relationships that were useful in classifying lakes according to trophic status and in suggesting the amount of nutrient loading needed to change the trophic status of a lake (Vollenweider 1976). Similar, but less inclusive, efforts have been made relative to estuarine systems (Boynton et al. 1982; Meeuwig et al. 1998). It has long been taken as a fundamental tenet of ecology that there is some relationship, probably complex, between rates of primary production and fishery yields. Such a relationship was documented by Oglesby (1977) for lakes and later by Nixon (1988) for estuarine, coastal, and marine systems. More complex relationships between standing stock size, growth rates, and production in marine food webs have also been determined using regression modeling approaches (Sheldon et al. 1977; Ennis 1986).

TABLE **11-1**

Brief description of data sources used in development of regression models presented in this chapter. Each program component is a part of the Chesapeake Bay Water Monitoring Program, which was initiated in 1984 and continues through the present time (Magnien et al. 1995). Phytoplanktonic production and chlorophyll *a* data from the 1972–1977 period are from Mihurshi et al. (1977).

Program Component	Variables Measured	Number of Stations	Field Technique	Sampling Frequency and Duration	Reference
Water-quality variables	T, S, DO, pH, chl *a*, dissolved and particulate N, P and Si concentration (vertical profiles)	50	One to five water column samples depending on total depth.Standard oceanographic analytical techniques.	16–20/yr 1984–present	Magnien et al. 1994
Freshwater and nutrient-loading rates	T, DO, conductivity, pH, chl *a*, *E. coli,* BOD, COD, TSS, total and dissolved N,	Fall line of all major rivers	Standard river gauges estimating daily flow.Regular parameter sampling and statistical modeling of flow-concentration relationships.	1–4/month (daily flow) 1972–present	Summers 1993
Sedimentation rates	Sedimentation rates of C, N, P, Si, chl *a*, and seston	One site in middle Chesapeake	One fixed vertical array. Collecting cups positioned in upper mixed layer, just beneath the pycnocline, and 1 m above the bottom.	Spr, sum, fall (~1/week) 1984–1993	Boynton et al. 1994
Sediment-water exchange rates	Net sediment exchanges of O2, NO2, NO3, DIP, Si, CO2	8	Estimated from shipboard incubation of intact sediment cores. Incubations were under ambient conditions.	Spr, sum, fall 1/month 1984–1996	Boynton et al. 1994
Phytoplankton component	Primary production rates, chl *a* concentration, and species composition	34	Short-term (3-hr), constant light ^{14}C incubations. Fluorometric chl *a* determinations.	16–20/year 1984–present	Sellner 1993

Results and Discussion

Time-series Observations at a Single Site

In this section, we examine the influence of river flow on phytoplanktonic production and biomass, deposition rate of spring-bloom phytoplankton, deep-water dissolved-oxygen declines, and recycling of ammonium from estuarine sediments. General pathways of the influence of river flow on these processes are summarized in figure 11-1. All of these analyses are based on time-series data collected at one location in the central portion of Chesapeake Bay (R-64, figure 11-2). We used seasonally or annually averaged data (6–8 or 16–20 observations, respectively) in these analyses rather than single, instantaneous values because these were the time scales of interest and because we wanted to avoid short-term variablity related to organism response times, changes in water residence times, and the like. These results were selected to serve as examples of the utility of synthesis as a framework to think about interrelationships of estuarine processes; there is a great deal of additional analysis that could be conducted on these and other data sets.

Algal Production and Biomass

The starting point for these investigations was suggested by previous analyses from lakes (for example, Vollenweider 1976) and coastal and estuarine systems (for example, Boynton et al. 1982; Nixon 1988) where statistically significant relationships were found between nutrient-loading rates and algal production and algal biomass. In our case, we used river flow as the independent variable because in Chesapeake Bay it is strongly correlated with nutrient-loading rates (Summers 1993) and provides most of the buoyancy that results in seasonal water-column stratification and hence definition of the upper mixed layer (Boicourt 1992).

We were initially doubtful about the possibility of finding strong relationships between flow and algal parameters. Previous investigators had adopted comparative approaches to obtain a sufficiently large range in loads and phytoplanktonic responses to observe significant relationships (Nixon 1988). Our initial concept was that there were so many factors controlling algal parameters that any one variable, even one like river flow that has multiple influences on the system, would explain only a small portion of the observed variability.

It appears that this is not the case. Results indicate strong relationships of river flow to biomass and, to a lesser extent, production (figure 11-4). In both, a large percentage (59% and 78%) of interannual variability was explained by river flow alone. This result reinforces the general conclusion that river flow is a dominant factor regulating some basic ecosystem processes in systems like Chesapeake Bay.

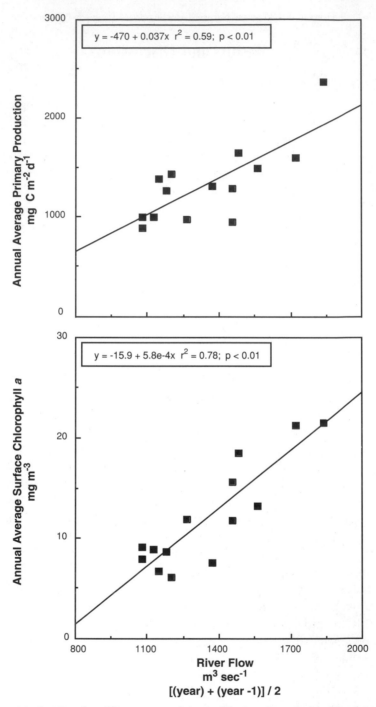

FIGURE 11-4 Results of linear regression models showing relationships between annual average phtoplankton primary production and surface chlorophyll *a* concentrations and freshwater flows from the Susquehanna River. River flow was calculated as the average of flow in the present and preceding year. Production and chlorophyll *a* data are from the R-64 site during the periods 1972–1977 and 1985–1993.

In this analysis several variations of independent (river flow) and dependent (phytoplankton production and biomass) variables were also tested, each being a variation representing a modified hypothesis concerning river influence on algal parameters (table 11-2). For example, in some trials just the spring freshet was used as the flow variable to explore the idea that this short but high period of flow was a key event in the annual or summer portion of the phytoplankton cycle. Many were statistically significant indicating the general importance of river flow. However, the flow variable that explained the most variability was an average of annual flow from the current year and from the previous year. This combination was suggested by examination of field data where it was noticed that production and biomass in years of average flow were higher than expected if they were preceded by a year of exceptionally high flow. This, in turn, suggests some nutrient retention or "nutrient memory" over time scales of a year rather than seasonal periods as suggested by bay water residence times (Boynton et al. 1990). Given the shallow depths of the bay, interannual retention of nutrients in the water column is not likely. The only likely multiyear nutrient storage site is sediments (Boynton et al. 1995). We suggest that in years of especially high flow, above-normal algal biomass is generated during the spring bloom. Recycling of this material supports high production through summer, which serves to conserve nutrients in the bay and make possible a large fall bloom. The deposition of the fall bloom to sediments, coupled with falling water temperatures, preserves nutrients through winter and they become available the next spring to support production and algal biomass at higher than expected levels. Kemp and Boynton (1984) proposed a similar sequence of

TABLE 11-2

A summary of results from linear regression analyses examining data sets for relationships between river flow and phytoplankton production and biomass. Entries in the table are r^2 values. Single and double asterisks indicate significance at the 0.05 and 0.01 probability levels, respectively. The number of observations in each analysis was fourteen. Data are from Mihurski et al. (1977), USGS (1993), and Magnien et al. (1994).

Phytoplankton Variables	River Flow Averages					
	Jan.–Mar.	Jan.–May	Jan.–Jun.	Jan.–Jul.	Avg. Annual	Two-Year Avg.
Annual average chlorophyll *a*	0.22	0.50**	0.67**	0.47**	0.64**	0.78**
Annual average production	0.14	0.20	0.15	0.13	0.22	0.58**
Summer average chlorophyll *a*	0.17	0.47**	0.74**	0.74**	0.62**	0.56**
Summer average production	0.18	0.31*	0.25	0.23	0.26	0.51**

events based on observations made in the Patuxent River estuary, but in that case the sequence did not include more than one annual cycle.

Spring-bloom Deposition

In most years, the annual cycle of phytoplankton biomass accumulation exhibits a distinct maxima associated with the spring bloom in the mesohaline regions of the Chesapeake Bay and there is considerable interannual variability in the magnitude of this spring peak as a consequence of interannual differences in nutrient input from the watershed (Malone et al. 1988). Studies by Sellner (1993) and White and Roman (1992) indicated that the spring bloom was not extensively grazed by zooplankton. If it was deposited to deep waters, as seems likely, it would become available to support a host of processes including macrofaunal growth, microbial respiration, and associated oxygen consumption and sediment nutrient releases.

Deposition rates of total chlorophyll *a* were measured using fixed sediment traps (weekly or biweekly measurement periods) from 1985 to 1992 (Kemp and Boynton 1992; Boynton et al. 1994; Roden et al. 1995). Deposition from spring blooms (integrated from day 50 through day 150 in all years) ranged from 541 mg m^{-2} in 1989 to 1,190 mg m^{-2} in 1990. Estimates of spring-bloom deposition rates followed qualitative trends in algal biomass for some years but not others (Magnien et al. 1994).

Inspection of spring-bloom deposition rate and river flow data suggested that there might be a more consistent relationship between deposition and flow than to algal biomass during the spring bloom and deposition. A series of analyses was performed (Boynton at al. 1993) wherein the period of time during which flow was averaged was different (figure 11-5).

These analyses indicated that river flow that occurred just prior to the spring bloom had the most influence on the magnitude of subsequent deposition rates. Low river flow from December through February was always associated with small spring deposition rates as were freshets that occurred late in the spring after the time of normal spring-bloom development (for example, May 1989); the largest deposition events (1987, 1988, and 1990) were all associated with river flow patterns that featured a distinct above-normal pulse in flow from December through February. In this case, analyses suggested a close temporal coupling between flow and an ecosystem response. Spring-bloom deposition appears to be responsive to relatively recent river flow events, with integrated effects of flows from previous seasons not being evident.

A strong departure from the expected pattern was observed in 1991 wherein a relatively small deposition event was associated with very high and sustained river flows that began in fall and continued throughout the winter. This

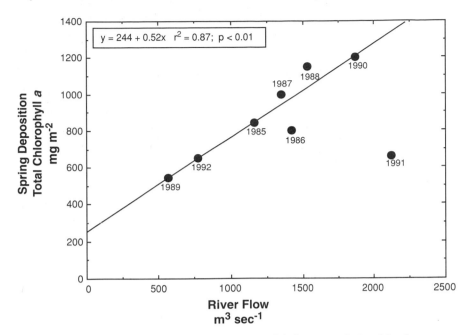

$$y = 244 + 0.52x \quad r^2 = 0.87; \quad p < 0.01$$

FIGURE 11-5 Results of linear regression model showing relationships between the magnitude of spring phytoplankton bloom deposition and freshwater inflows from the Susquehanna River. Spring-bloom deposition of total chlorophyll *a* was estimated from moored sediment trap arrays (duplicate weekly or biweekly measurements from collecting cups positioned beneath the pycnocline). Spring-bloom deposition was calculated as the total chlorophyll *a* mass collected from the initiation to termination of bloom deposition during each year. River flow was averaged for the months of December through March for each year preceding the spring bloom. All data are from the R-64 site. The 1991 data were not included in the regression model.

suggested that either the bloom did not develop or that deposition did not occur as usual. Water-column chlorophyll data suggested a strong bloom in 1991 (Magnien et al. 1994), which weakens the former hypothesis. It appears probable that the 1991 bloom deposited farther downstream than usual and thus was not measured at our fixed station. Maps of chlorophyll concentration in surficial sediments made immediately after bloom deposition in 1993 (another year with a strong freshet) indicated that most of the bloom deposited 20–30 km downstream of the sediment trap location. This analysis suggests that these systems are responsive to forcing events of relatively short duration and that the spatial location of the spring-bloom epicenter can be shifted seaward in years of high sustained river flow. These observations generally, but not always, conformed to the simple "bottom-up" control model presented earlier. The outlier in this data set was useful, as is often the case, leading us to consider additional explanations that eventually resulted in a better understanding of these systems.

Seasonal Declines in Deep-water Oxygen

In many coastal areas, including Chesapeake Bay, there is strong interest in the scientific and management communities to better understand processes regulating dissolved-oxygen declines. In Chesapeake Bay low dissolved-oxygen conditions occur primarily in those areas where the water column is stratified and where there is substantial deposition of organic matter from the upper water column to deep waters and sediments. Establishing relationships between myriad environmental variables and oxygen conditions has not been a simple task in the bay. For example, Seliger and Boggs (1988) pointed out that low dissolved-oxygen conditions in the bay could be explained by river flow and water-column stratification and concluded that nutrient inputs (and the organic matter produced) may have little to do with anoxia. Kemp et al. (1992) argued that physical and biological processes are coupled, with freshwater flow providing buoyancy for stratification as well as nutrients for organic matter production. Boicourt (1992) has suggested the possibility of a phase shift in freshwater flow versus anoxia relationships.

Our data are not adequate to entirely resolve this problem but it is possible to test for relationships between deep-water oxygen characteristics and organic matter deposition rates for one region of the mainstem bay where seasonal oxygen problems are chronic. Several anecdotal observations indicated that such relationships might exist. First, hypoxic or anoxic conditions developed in deep waters for some period of time during each year since intensive monitoring began in 1984. Low oxygen concentrations in deep waters were associated with even the lowest flow conditions observed. Second, in 1989, the spring freshet (and associated nutrient load) did not enter the bay until mid-May. The spring phytoplankton bloom did not develop to any significant extent and deep-water oxygen depletion was delayed for about a month.

Finally, in 1992, the spring freshet was very small. Spring chlorophyll *a* concentration in the water column and chlorophyll *a* deposition rates were among the lowest on record and dissolved-oxygen concentrations declined slowly, not reaching mg l^{-1} until early July. These results suggest that deep-water oxygen conditions are regulated, at least in part, by the amount of organic matter deposited during spring.

Bottom-water oxygen concentrations were routinely measured (weekly or biweekly) at the R-64 station from 1985 to 1992 (figure 11-2). Water depth at this site is about 17 m and vertical water-column stratification is generally strong in this region of the bay. The daily rate of change of oxygen concentration (d DO/dt) was calculated using spring measurements from 1985 through 1992. The time period over which rates of change were calculated varied slightly among years but in most cases included the period from the beginning of March through the middle of May. The criterion used to determine the starting point was that the first observation should not be followed by any oxygen measurements of higher concentrations. Typically, during late winter and early spring, deep-water oxygen concentrations exhibit both small increases and decreases over time but are usually close to saturation. The final oxygen measurement used was the last measurement made before oxygen concentration declined below 1 mg l^{-1}. The rates of oxygen decline for the years 1985 through 1992 calculated from these data were linear, statistically significant ($p < 0.01$; $r^2 > 0.90$), and differed appreciably among years (by more than twofold).

The fact that dissolved-oxygen concentrations began declining during early spring suggested that these declines were caused by respiration of spring deposition events rather than later summer events. Accordingly, average spring deposition rates of total chlorophyll *a* were calculated for each year using deposition data collected between early February and the beginning of May. Chlorophyll *a* deposition rates were regressed against the rate of dissolved-oxygen decline derived from regressions of time versus dissolved-oxygen concentration (figure 11-6). These results suggest a strong influence of organic matter availability on the rate of oxygen decline. However, at least two alternative explanations exist. First, it can be hypothesized that different spring rates of oxygen decline are caused by interannual differences in temperature regimes. Oxygen decline would be more rapid in warm years than in cold years because of the influence of temperature on respiration rates (Sampou and Kemp 1994). This explanation seems unlikely to be the prime cause because interannual temperature differences have been small over the period of record. Additionally, warm and cool springs were not correlated with high and low rates of oxygen decline. The second hypothesis is that the cause is related to interannual differences in the strength of water-column stratification. In years when the water column is highly stratified, less mixing of oxygen from surface to deep-water occurs and rates of oxygen decline would be greater. Stratification certainly plays a major role in determining deep-water oxygen characteristics. However, the case for stratification being the dominant cause of interannual differences in oxygen decline rates is weak because years of high and low stratification do not correspond well to years of high and low rates of oxygen decline.

Sediment Ammonium Fluxes

The final example in this sequence concerns possible relationships between river flow and sediment nutrient recycling. It is hypothesized that variations in river flow and associated nutrient inputs regulate spring-bloom size and

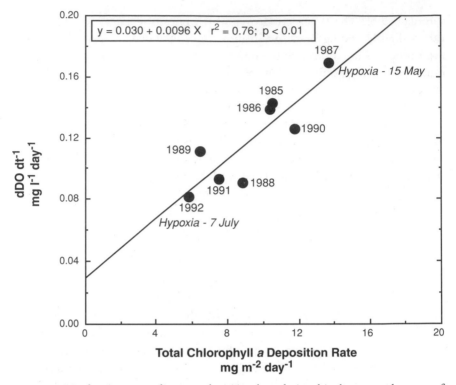

FIGURE 11-6 A scatter diagram showing the relationship between the rate of decline in dissolved-oxygen concentrations in deep water (dDO dt^{-1}) and average deposition rates of total chlorophyll a during the spring-bloom period. Data are from the 1985–1992 period and were collected at the R-64 site. The date on which hypoxia (DO concentration <1 mg l^{-1}) was first encountered during highest (1987) and lowest (1992) deposition years is also indicated.

organic matter deposition rates. Deposited organic matter, in turn, serves as a substrate for decomposers, which eventually regulates nutrient releases from sediments. We attempted again a series of regressions using different time averaging of flow and benthic nutrient recycling rates. Again, most combinations indicated a positive relationship; the strongest relationship between river flow and sediment ammonium flux was found using winter (December to February) flow rates, as in the deposition-versus-flow relationship, and summer (June to September) average benthic ammonium fluxes (figure 11-7). This implies a time delay between nutrient input and benthic nutrient recycling. In this estuary, springtime respiration rates remain relatively low at temperatures below 10°C for both benthic (Boynton et al. 1990) and water-column communities (Smith and Kemp 1995); rates increase exponentially with vernal warming beyond these thresholds. Deposition of organic matter to sediments derived from the spring bloom starts in late February and ends by

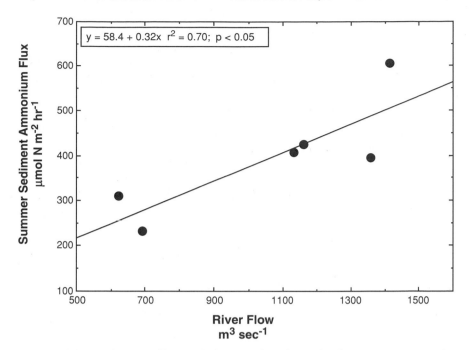

FIGURE 11-7   A scatter diagram indicating the relationship between summer sediment ammonium flux (June–September) and winter flow from the Susquehanna River. Sediment-water flux data are from the R-64 site and were collected during the 1988–1993 period.

mid-May. However, large sediment fluxes of ammonium are not evident until June when bottom waters are above 15°C and coupled rates of nitrification and denitrification begin to decline with oxygen depletion (Kemp et al. 1990). Relative to the other relationships presented here, the river flow-nutrient cycling relationship was the weakest. In part, this may be due to a more limited data set. It may also be because this process is the farthest removed from the influence of flow, at least as conceptualized here. In this view, other factors have more of a chance to come into play (for instance, infaunal community activities, sediment redox conditions, nitrification-denitrification, focusing of organic matter from shoal areas to deeper waters) modifying or fundamentally changing the nature of the flow-recycling relationship.

Intersite Comparative Analyses

In this section, we present additional examinations of estuarine features as they relate to variations in river flow and attendant nutrient-loading rates. The approach here is comparative wherein data from five sites with different nutrient-loading rates were used rather than a time series from a single site.

Here we use space (different sites) to examine the possible causes of temporal variability just as we used a time series of observations in the previous analyses (Pickett 1991). There are several distinct advantages to a comparative approach to synthesis. The range in both independent and dependent variables can be expanded if sites are chosen with this in mind, thereby increasing the possibility of observing patterns, if they exist. Additionally, the inclusion of multiple sites or different systems inherently increases the generality of conclusions; the possibility that observed relationships are only unique to a single site is diminished. However, difficulties present themselves with comparative analyses as well, the most prominent being the differences in the characteristic scales (such as volume, depth, residence time) among sites. Hence, there is a need to analyze data in a way that accounts for scaling differences so that ecological variables of interest are comparable among different systems. In fact, the use of nutrient-loading rate was adopted here because nutrient loads were known to be different among the systems we studied and could be scaled to the respective estuarine areas. The scaling of variables (such as nutrient-loading rate to estuarine area) is, in itself, an issue that could benefit from some serious consideration.

Nutrient Stocks

Perhaps the most direct relationship to consider is that between nutrient loads and water-column stocks (figure 11-8). Our analyses indicated a very strong relationship between annual average total nitrogen (TN) load and average annual TN mass in the water column; to a lesser extent the same was found for total phosphorus (TP). The weaker TP relationships may have resulted from the fact that a large percentage of the TP load is in the form of inorganic particulate phosphorus and hence not as prone to remain in the water column (Boynton et al. 1995). However, all results were poor when specific chemical species (such as nitrate) rather than totals were used. Apparently, specific nutrient species are transformed rapidly enough so that simple relationships to load are not apparent at that level of detail.

Another feature of these relationships is the attenuated response of concentrations to loads. For example, TN loads varied by a factor of about 10 while water-column mass varied by only a factor of 3. Similar attenuated responses were found for a variety of variables examined in a series of estuarine nutrient budgets (Boynton et al. 1995) and in a set of marine mesocosms exposed to a range of nutrient-enrichment rates (Nixon et al. 1986). This suggests that either internal sinks (such as sediment burial and denitrification) are quite active or that these nutrients are efficiently transported out of the system. In the case of Chesapeake Bay, both seem to be involved (Boynton et al. 1995). Finally, in some regression models the intercept values contain information of

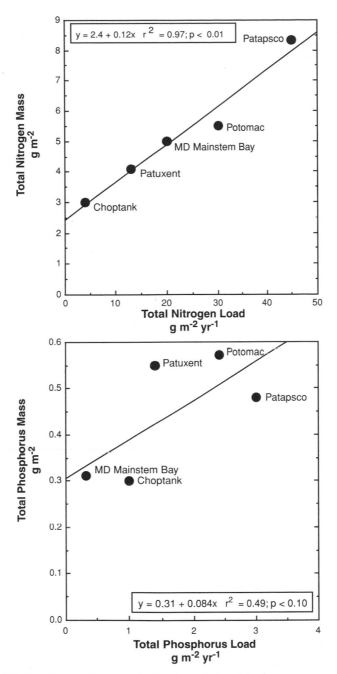

FIGURE 11-8 Scatter diagrams indicating relationships between annual average total nitrogen (TN) and total phosphorus (TP) mass in the water-column and average annual TN and TP loads to five locations in the Chesapeake Bay system. Loads and water-column masses were measured at the fall-line sites and estuarine stations, respectively, indicated in figure 11-2. Data were averaged for the 1985–1996 periods.

ecological interest. In this case, TN and TP values at zero loading rates are still substantial (~17 μM N and 1 μM P for a 10 m water column) and suggest potentially high productivity even under pristine conditions.

Phytoplankton Chlorophyll Stocks

In the mid-1970s limnologists developed a series of useful statistical models relating nutrient-loading rates and algal biomass for a large sampling of lakes (Vollenweider 1976). These relationships were used to estimate the degree to which nutrient-loading rates would need to be decreased to move a particular lake from one trophic state (as defined by chlorophyll *a* concentration) to another. Surprisingly, few comparable relationships have been developed for coastal and marine ecosystems (Nixon et al. 1986; Nixon 1988). We initially attempted a direct duplication of the Vollenweider (1976) model using average annual (or ice-free periods) surface-water chlorophyll *a* concentration (μg l^{-1}) as the dependent variable and annual average phosphorus loading rate (adjusted for the freshwater fill time and mean depth of the receiving water body) as the independent variable. This selection of variables did not produce either predictive or significant statistical results ($r^2 < 0.10$; $p > 0.10$) We then reasoned that, because algal blooms often develop in deep waters, particularly in spring in Chesapeake Bay, vertically integrated water-column chlorophyll *a* (mg m^{-2}) would be a better estimate of algal biomass; however, results were only marginally better. We then substituted nitrogen for phosphorus and results improved to the degree shown in figure 11-9. We have also obtained sufficient data to add results of the MERL eutrophication experiment (Nixon et al. 1986) and portions of Hillsborough Bay, Florida, data (Johannson 1991) to this analysis.

The results support the concept that, for some estuarine systems, phytoplankton biomass levels respond in positive linear relation to nutrient-loading rates. Further, there is some indication that different systems respond in a similar fashion when loading rates are scaled for local conditions of depth and flushing rates. This sort of analysis could be expanded to include other systems to explore the robustness of the relationship; a successful test would increase confidence in the conceptual model on which it is based. However, the conceptual model used here explicitly favors bottom-up control. It is almost certain that such a model would not work in instances where top-down controls become dominant, as in cases where intensive benthic grazing by introduced species (Alpine and Cloern 1992) or aquacultural activities are important (Meeuwig et al. 1998). Finally, the scaling of the nutrient load for estuarine flushing characteristics used in this example is primitive and would not be appropriate for estuarine systems with larger tides or limited freshwater inflows (Monbet 1992). More realistic formulations are needed. However, this is an example of where a synthesis

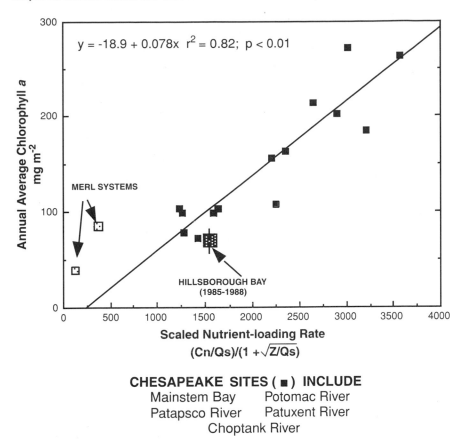

$y = -18.9 + 0.078x$ $r^2 = 0.82$; $p < 0.01$

CHESAPEAKE SITES (■) INCLUDE

| Mainstem Bay | Potomac River |
| Patapsco River | Patuxent River |
| Choptank River |

FIGURE 11-9 A scatter diagram relating annual average total chlorophyll *a* mass to nitrogen-loading rate. Data are from the five estuarine sites indicated in figure 11-3 and were collected during the 1985–1987 period. Nitrogen-loading rates were scaled following the method used by Vollenweider (1976) where: C_n = nitrogen-loading rate (mg N m^{-2} yr^{-1}); Q_s = hydraulic fill time (years); Z = mean depth (m). Hillsborough Bay, Florida, data are from Johansson (1991) and the MERL data are from Oviatt et al. (1986).

activity clearly suggests some additional lines of inquiry; in this case, the scaling of important characteristics of ecosystems.

Sediment Nutrient Releases

In an earlier example, we related river flow to sediment nitrogen releases (figure 11-7) using time-series data. We considered the same processes again but used a comparative approach with data from several sites that encompassed a large range in total nitrogen-loading and sediment ammonium-recycling rates. As in the previous case, the conceptual model linking nutrient loading to the ecosystem from external sources and sediment nutrient recycling involved load-related algal biomass, which was deposited to sediments and eventually served as substrate supporting sediment nutrient releases. Results from one set of analyses are shown in figure 11-10.

The load-recycling relationship suggests several interesting insights. First, the slope of the regression indicates that for every unit reduction in TN load there would be about an equivalent reduction in sediment ammonium recycling. However, flux data are from summer when values are high; typical values from the remaining months are only 10–30% of these values. Overall, there still appears to be a strong linkage between load and flux. The intercept value of 120 μmol N m^{-2} hr^{-1} is sufficient to support relatively low rates of phytoplanktonic primary production ($\sim$0.3 g C m^{-2} day^{-1}, assuming Redfield C:N proportions). The intercept value would be lower if data from the Choptank River were excluded, as possibly they should be, because of problems with estimating the TN loads. At low nutrient-loading

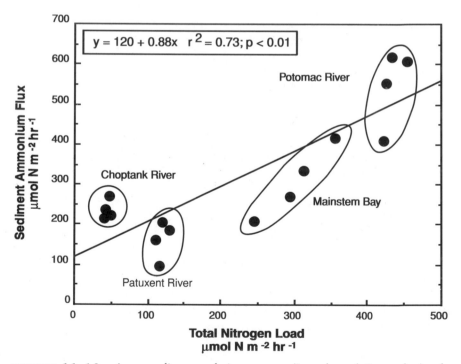

FIGURE 11-10   A scatter diagram relating summer (June through September) sediment ammonium flux to average annual total nitrogen load in four locations in Chesapeake Bay. Loads and sediment fluxes were measured at the fall line and estuarine sites, respectively, indicated in figure 11-2. Data were from the 1985–1988 period.

rates the amount of nitrogen recycled from sediments is small, but this could still be important in more oligotrophic environments.

Choptank River fluxes are higher than expected for a given level of nutrient loading. This discrepancy may be more apparent than real. Results based on nutrient budget calculations indicate that the lower Choptank River receives substantial nutrient additions from the mainstem bay (Boynton et al. 1995). Groundwater discharges directly to tidal waters are also important. If this is the case, nutrient loading to the Choptank River is higher than shown in figure 11-10 and would have the effect of displacing Choptank River fluxes to the right, more in line with those of other systems. The fact that Choptank fluxes diverged so strongly prompted us, and others, to reexamine loads to this system, not an uncommon benefit of synthetic analyses such as these.

While only TN versus ammonium flux is shown in figure 11-10, other load-recycling relationships were examined. In general, sediment-water fluxes (for instance, sediment oxygen consumption, silica) were consistently better correlated with TN loading than with TP loading (Boynton et al. 1994). Even sediment recycling of PO_4^{3-} exhibited a stronger relationship with TN loads than with TP loads. In part, this may result from the fact that there is a considerably broader range in TN loads than TP loads (Boynton et al. 1995). It may not be possible to resolve TP influences on sediment nutrient recycling over this relatively narrow loading range. Alternatively, the poor correspondence with TP loads may indicate that most of the phytoplankton debris that reaches sediments, and eventually supports sediment-water nutrient recyling, was produced more in response to N than P availability in the water column. (D'Elia et al. 1986; Fisher et al. 1992).

There may be additional inferences to be drawn from this comparative analysis of TN loading versus sediment ammonium recycling (figure 11-10). Specifically, even in this limited (4 year) data set, there appear to be qualitative relationships of recycling to loading to each system; the pattern is most obvious for the mainstem bay site, but there is a hint apparent for most sites. The provocative observation here is that the slope of each cluster of points for 4 years tends to increase as ecosystem system size decreases. Thus, the fraction of TN loading that appears in summer benthic ammonium recycling is larger with small systems. This implies that smaller systems retain and recycle nutrients more efficiently, or that a larger fraction of primary production is deposited to and recycled in smaller (shallower) systems (Kemp and Boynton 1992; Boynton et al. 1995). Are there some rules for estuarine scaling to be gleaned from these types of observations? We intend to add more observations to this analysis in the future; it may well be that continued empiricism may provide answers to these questions.

Concluding Remarks

Examination of time-series and comparative data at sites in Chesapeake Bay have revealed surprisingly strong and linear relationships of primary production, benthic-pelagic coupling, and nutrient recycling to both freshwater flow and nutrient-loading rates. It appears that in well-sampled systems comparative analyses are not necessary to obtain sufficient range in variables for a pattern to emerge; in the case of Chesapeake Bay, clear signals were seen when river flow varied by about a factor of 2. However, comparative analyses increase the generality of results.

There are extensive data on water quality, physical forces, and ecological processes for Chesapeake Bay; we have only scratched the surface for inferences that could be drawn from these data. Data are available to explore the relations between freshwater flow (and nutrient loading) and distributions of seagrass, zooplankton, and benthic invertebrate communities. In addition, spatial sampling in many regions of the Chesapeake system is sufficient to develop volume-weighted estimates of processes, biomasses, and pool sizes; these variables would presumably be more representative of estuarine conditions than estimates based on samples from a single station.

The approach used here started with a conceptual model of how freshwater flow or nutrient loads influenced key processes and properties of an estuarine ecosystem. We have found these conceptualizations to be far more profitable than approaches that start with a "blind" search for correlations. However, most of our work has utilized simple linear regression techniques. More sophisticated techniques such as multiple linear and nonlinear regression, multivariate analyses, classification, and regression-tree and time-series approaches appear attractive because of the complexity of estuarine processes.

Based on our initial effort comparing ecological responses to variations in physical forces among Chesapeake Bay subsystems, we are optimistic about the utility of applying comparative analysis methods with time-series data for multiple ecosystems. We suggest that contrasting similar time-series analysis among different systems may help identify key scaling relationships needed to generate fundamental scientific understanding that is not site-specific.

Acknowledgments

We acknowledge the following colleagues and institutions for their assistance in completing this work: Michael Haire and Robert Summers (Maryland Department of the Environment) provided nutrient-loading-rate estimates; Robert Magnien and Bruce Michaels (Maryland Department of Natural Resources) provided water-column nutrient data, and Kevin Sellner (Academy of Natural Sciences of Philadelphia) supplied phytoplankton stock and primary-production-rate data. We also thank Jonathan Garber, Jean Cowan, Janet Barnes, Frances Rohland, Lisa Matteson,

David Jasinski, and James Hagy for their assistance in conducting fieldwork and organizing data files.

This work was supported by the Maryland Department of the Natural Resources Chesapeake Bay Water Quality Monitoring Program (Grant No. RAT 7/98-045); NSF Land Margin Ecosystem Research Program (Grant No. DEB9412113); U.S. EPA Exploratory Research Program, Multiscale Experimental Ecosystem Research Center Project (Grant No. R819640); U.S. EPA Chesapeake Bay Program (Grant No. CB993586-01-1); and Maryland Sea Grant College (Project No. R/P-32).

References

Alpine, A. E., and J. E. Cloern. 1992. Trophic interactions and direct physical effects control phytoplankton biomass and production in an estuary. *Limnology and Oceanography* 37(5): 946–55.

Boicourt, W. C. 1992. Influences of circulation processes on dissolved oxygen in the Chesapeake Bay. Pages 7–59 in D. E. Smith, M. Leffler, and G. Mackiernan (editors), *Oxygen dynamics in the Chesapeake Bay—a synthesis of recent results.* College Park, MD: Maryland Sea Grant Program.

Boynton, W. R., W. M. Kemp, and C. W. Keefe. 1982. A comparative analysis of nutrients and other factors influencing estuarine phytoplankton production. Pages 69–90 in V. S. Kennedy, (editor), *Estuarine comparisons.* New York: Academic Press.

Boynton, W. R., W. M. Kemp, J. M. Barnes, J. L. Cowan, S. E. Stammerjohn, L. L. Matterson, F. M. Rohland, M. Marvin, and J. H. Garber. 1990. Long-term characteristics and trends of benthic oxygen and nutrient fluxes in the Maryland portion of Chesapeake Bay. Pages 339–54 in *New perspectives in the Chesapeake System: A research and management partnership.* Chesapeake Research Consortium Publication No. 137. Solomons, MD: CRC.

Boynton, W. R., W. M. Kemp, J. M. Barnes, L. L. Matteson, F. M. Rohland, D. A. Jasinski, and H. L. Kimble. 1993. Ecosystem Processes Component Level 1. Interpretive Report. [UMCEES] CBL Ref. No. 93-030a. Chesapeake Biological Laboratory. Solomons: University of Maryland.

————. 1994. Ecosystem Processes Component Level 1, No. 11. Interpretive Report. [UMCEES] CBL Ref. No. 94-031a. Chesapeake Biological Laboratory. Solomons: University of Maryland.

Boynton, W. R., J. H. Garber, R. Summers, and W. M. Kemp. 1995. Inputs, transformations and transport of nitrogen and phosphorus in Chesapeake Bay and selected tributaries. *Estuaries* 18 (1B): 285–314.

Carpenter, S. R., and J. F. Kitchell. 1988. Consumer control of lake productivity. *BioScience* 38(11): 764–69.

Cerco, C., and T. Cole. 1992. *Application of the three-dimensional eutrophication model CE-QUAL-ICM to Chesapeake Bay.* Draft Technical Report. Vicksburg, MS: U.S. Army Engineer Waterways Experiment Station.

Cooper, S. R., and G. S. Brush. 1991. Long-term history of Chesapeake Bay anoxia. *Science* 254: 992–96.

Cloern, J. E., A. E. Alpine, B. E. Cole, R. I. J. Wong, J. F. Arthur, and M. D. Ball.

1983. River discharge controls phytoplankton dynamics in the northern San Francisco Bay estuary. *Estuarine, Coastal and Shelf Science* 16: 415–529.

Cowan, J. L., J. R. Pennock, and W. R. Boynton. 1996. Seasonal and interannual patterns of sediment-water nutrient and oxygen fluxes in Mobile Bay, Alabama (USA): Regulating factors and ecological significance. *Marine Ecology Progress Series* 141: 229–45.

D'Elia, C. F., J. G. Sanders, and W. R. Boynton. 1986. Nutrient enrichment studies in a coastal plain estuary: Phytoplankton growth in large-scale, continuous cultures. *Canadian Journal of Fisheries and Aquatic Sciences* 43: 397–406.

Ennis, G. P. 1986. Stock definition, recruitment variability, and larval recruitment processes in the American lobster, *Homarus americanus:* A review. *Canadian Journal of Fisheries and Aquatic Sciences* 43: 2072–84.

Fisher, T. R., E. R. Peele, J. W. Ammerman, and L. W. Harding. 1992. Nutrient limitation of phytoplankton in Chesapeake Bay. *Marine Ecology Progress Series* 82: 51–63.

Flint, R. W. 1985. Long-term estuarine variability and associated biological response. *Estuaries* 8(2A): 158–69.

Johannson, J. O. R. 1991. Long-term trends of nitrogen loading, water quality, and biological indicators in Hillsborough Bay, Florida. Pages 157–76 in S. F. Treat and P. A. Clark (editors), *Proceedings, Tampa Bay Area Scientific Information Symposium 2.* Tampa, FL: Bay Area Scientific Information Symposium.

Kemp, W. M., and W. R. Boynton. 1984. Spatial and temporal coupling of nutrient inputs to estuarine primary production: The role of particulate transport and deposition. *Bulletin of Marine Science* 35(3): 522–35.

———. 1992. Benthic-pelagic interactions: Nutrients and oxygen dynamics. Pages 149–209 in D. E. Smith, M. Leffler, and G. Mackiernan (editors), *Oxygen dynamics in the Chesapeake Bay—a synthesis of recent results.* College Park, MD: Maryland Sea Grant Program.

Kemp, W. M., P. Sampou, J. Caffrey, M. Mayer, K. Henriksen, and W. R. Boynton. 1990. Ammonium recycling versus denitrification in Chesapeake Bay sediments. *Limnology and Oceanography* 35: 1545–63.

Kemp, W. M., P. A. Sampou, J. Garber, J. Tuttle, and W. R. Boynton. 1992. Seasonal depletion of oxygen from bottom waters of Chesapeake Bay: Roles of benthic and planktonic respiration and physical exchange processes. *Marine Ecology Progress Series* 85: 137–52.

Magnien, R. E., D. K. Austin, and B. D. Michael. 1994. *Chemical/physical properties component: Level I data report.* Chesapeake Bay Projects Division, Chesapeake Bay and Special Projects Program, Water Management Administration. Baltimore, MD: Maryland Department of Enviroment.

Magnien, R., D. Boward, and S. Bieber (editors). 1995. *The state of the Chesapeake Bay,* 1–45. Chesapeake Bay Program. Annapolis, MD: Environmental Protection Agency.

Malone, T. C., L. H. Crocker, S. E. Pile, and B. W. Wendler. 1988. Influences of river flow on the dynamics of phytoplankton production in a partially stratified estuary. *Marine Ecology Progress Series* 48: 235–49.

Malone, T. C., W. Boynton, T. Horton, and C. Stevenson. 1993. Nutrient loading to surface waters: Chesapeake Bay case study. Pages 8–38 in Uman, M. F. (editor), *Keeping pace with science and engineering*. Washington, DC: National Academy Press.

Meeuwig, J. J., J. Rasmussen, and R. H. Peters. 1998. Turbid waters and clarifying mussels: Their moderation of empirical chl:nutrient relations in estuaries in Price Edward Island, Canada. *Marine Ecology Progress Series* 171: 139–50.

Mihurski, J. A., D. R. Heinle, and W. R. Boynton. 1977. *Ecological effects of nuclear steam electric station operations on estuarine systems*. UMCEES Ref., No. 77-28-CBL. Solomons, MD: Chesapeake Biological Laboratory.

Monbet, Y. 1992. Control of phytoplankton biomass in estuaries: A comparative analysis of microtidal and macrotidal estuaries. *Estuaries* 15(4): 563–71.

Nixon, S. W. 1988. Physical energy inputs and the comparative ecology of lake and marine ecosystems. *Limnology and Oceanography* 33: 1005–25.

———. Marine eutrophication: A growing international problem. *Ambio* 19: 101.

Nixon, S. W., C. A. Oviatt, J. Frithsen, and B. Sullivan. 1986. Nutrients and the productivity of estuarine and coastal marine systems. *Journal of the Limnological Society of South Africa* 12(1/2): 43–71.

Oglesby, R. T. 1977. Relationship of fish yield to lake phytoplankton standing crop, production, and morphoedaphic factors. *Journal of the Fisheries Research Board of Canada* 34: 2271–79.

Oviatt, C. A., A. A. Keller, P. A. Sampou, and L. L. Beatty. 1986. Patterns of productivity during eutrophication: A mesocosm experiment. *Marine Ecology Progress Series* 28: 69–80.

Peters, R. H. 1991. *A critique for ecology*. Cambridge, England: Cambridge University Press.

Pickett, S. T. A. 1991. Space for time substitution as an alternative to long-term studies. Pages 110–35 in G. E. Likens (editor), *Long-term studies in ecology: Approaches and alternatives*. New York: Springer-Verlag.

Rigler, F. H. 1982. Recognition of the possible: An advantage of empiricism in ecology. *Canadian Journal of Fisheries and Aquatic Sciences* 39: 1323–31.

Roden, E. E., J. H. Tuttle, W. R. Boynton, and W. M. Kemp. 1995. Carbon cycling in mesohaline Chesapeake Bay sediments. 1: POC deposition rates and mineralization pathways. *Journal of Marine Research* 53: 799–819.

Sampou, P. A., and W. M. Kemp. 1994. Factors regulating plankton community respiration in Chesapeake Bay. *Marine Ecology Progress Series* 110: 249–258.

Schneider, D. C. 1994. Quantitative ecology: Spatial and temporal scaling. New York: Academic Press.

Seliger, H. H., and J. A. Boggs. 1988. Long-term pattern of anoxia in the Chesapeake Bay, Pages 570–83 in M. P. Lynch and E. C. Krome (editors) *Understanding the estuary: Advances in Chesapeake Bay research*. Publication No. 129 CBP/TRS 24/88. Edgewater, MD: Chesapeake Research Consortium.

Sellner, K. 1993. *Long term phytoplankton monitoring and assessment program*. Maryland Chesapeake Bay Water Quality Monitoring Program. Benedict, MD: Benedict Estuarine Research Laboratory.

Sheldon, R. W., W. H. Sutcliffe Jr., and M. A. Paranjape. 1977. Structure of pelagic food chain and relationship between plankton and fish production. *Journal of the Fisheries Research Board of Canada* 34: 2344–53.

Smith, E. M., and W. M. Kemp. 1995. Seasonal and regional variations in plankton community production and respiration for Chesapeake Bay. *Marine Ecology Progress Series* 116: 21–731.

Stevens, D. E. 1977. Striped bass (*Morone saxatilis*) year class strength in relation to river flow in the Sacramento-San Joaquin estuary, California. *Transactions of the American Fisheries Society* 106: 34–42.

Summers, R. M. 1993. *Point and non-point source nitrogen and phosphorus loading to the northern Chesapeake Bay.* Technical Report. Baltimore, MD: Maryland Department of the Environment, Water Management Administration, Chesapeake Bay and Special Projects Program.

Sutcliffe, W. H. Jr. 1973. Correlations between seasonal river discharge and local landings of American lobster (*Homarus americanus*) and Atlantic halibut (*Hippoglossus hippoglossus*) in the Gulf of St. Lawrence. *Journal of the Fisheries Research Board of Canada* 30: 856–59.

Sutcliffe, W. H. Jr., K. Drinkwater, and B. S. Muir. 1977. Correlations of fish catch and environmental factors in the Gulf of Maine. *Journal of the Fisheries Research Board of Canada* 34: 19–30.

United States Geological Survey. 1993. *Water resources data, Maryland and Delaware water year 1993.* Water-Data Report MD-DE-93-1. Towson, MD: U.S. Geological Survey.

Vollenweider, R. A. 1976. Advances in defining critical loading levels of phosphorus in lake eutrophication. *Memorie-Istituto Italiano de Idrobilogia* 33: 53–83.

White, J. R., and M. R. Roman. 1992. Seasonal study of grazing by metazoan zooplankton in the mesohaline Chesapeake Bay. *Marine Ecology Progress Series* 86: 251–61.

CHAPTER 12

Linking Biogeochemical Processes to Higher Trophic Levels

James N. Kremer, W. Michael Kemp, Anne E. Giblin, Ivan Valiela,
Sybil P. Seitzinger, and Eileen E. Hofmann

Abstract

How well are we able to synthesize current understanding of how bottom-up
controls affect higher trophic levels? We consider this question with respect
to five general approaches: Long-term data collection, cross-ecosystem comparisons, computational models, nutrient and energy budgets, and experiments. The current status of each approach is assessed, published examples
are identified, strengths and weaknesses are summarized, and recommendations are offered with specific attention to their utility for management. The
foundation of estuarine synthesis is the formulation of models, both conceptual and mathematical. To the extent that models permeate this work, we
emphasize the need to repeatedly question, revise, and test with data the fundamental assumptions of the models implicitly or explicitly underlying each
approach.

Introduction

Synthesis: the assembly of parts into a unified whole. In estuarine ecology, this
term has two consistent yet slightly different connotations depending at least
partly on scale. First, information may be gathered in order to identify incon-
sistencies and to assess the relative importance of contributing parts. Most
often, such synthesis applies to a single ecosystem such as a nutrient budget
for an estuary. Second, comparable data may be assembled to identify patterns across time or space that may be of predictive value, or may suggest
hypotheses about relationships appropriate for further study. A synopsis of
seasonal chlorophyll levels in one estuary or a plot of production versus N
loading in different systems develops a pattern that may predict behavior at
different times or of other unstudied systems. These generalizations may be

useful within a single ecosystem but are especially powerful when applied across various diverse systems.

An additional distinction deserves mention. Some syntheses may yield predictive power yet provide no direct information about cause and effect. Others based on mechanistic premises explicitly explore causality but may be marginally useful as predictive tools. Ideally, an ecologically complete synthesis would provide a general perspective based firmly on sound causal understanding allowing prediction within and among systems. Practically, we usually obtain a partial understanding, an incomplete mix of causal insight and predictive power.

This working group considered a range of methodological approaches to a particularly challenging issue: How well are we able to synthesize current understanding of how bottom-up controls affect higher trophic levels? Can biogeochemical processes be usefully related to fish, birds, and the like? Our strategy was to consider five general approaches. For each case, we have identified some published examples that serve to demonstrate the approach and assess the extent to which it may have been useful so far in relating nutrients to higher trophic levels. Each approach has inherent strengths and weaknesses related to the type of synthesis it promotes and to the extent to which it relates to causality or prediction. Based on our general assessment of these factors, we offer recommendations on how useful each approach may be, the kind of specific research questions that may be addressed productively, and the promise of this approach for application to management issues over the next decade.

The foundation of estuarine synthesis is formulation of models, both conceptual and mathematical. They are absolutely implicit in all the synthetic approaches we consider. In this chapter we consider synthesis and modeling in a number of contexts. These include modeling to explain trends and patterns in long-term data collections, modeling to explain linkages suggested by correlations noted in cross-system comparisons, conceptual models that are the basis of mass balances, and models used for the design and interpretation of experiments. In all cases, the fundamental assumptions of these models must be repeatedly questioned, revised, and tested.

Long-term, Continuous Data Collection

Status of Approach

It is widely accepted among ecologists that obtaining continuous, long-term records of physical, chemical, and biological processes and properties will lead to improved understanding of how ecosystems function. Indeed, there are numerous examples of how data records collected in such Sustained Ecological Research (SER) and monitoring programs have led to important scientific insights (see Likens 1992). Satellite remote sensing also offers the possibility of long time series of a limited number of variables over large space and time scales. Such data sets are inherently a synthesis over time, but mainly they stimulate a large-scale view and thus stimulate conceptual and quantitative synthesis. The existence of statistically significant covariance or regressions between two or more properties of ecosystems helps to form and test hypotheses concerning factors regulating ecological processes. If data records involve sampling frequencies that are consistent and if they extend for long enough to reveal their periodic nature, spectral analysis can be used to characterize frequency distributions of variance (Platt and Denman 1975). The degree of coherence between two such periodic time series can be tested statistically, as can the existence of time lags between recurring, and apparently related, events. Similar analyses can be conducted for data collected from spatial series collected from a moving vehicle such as a ship or airplane. Time and space scales of contemporaneous observations must, however, be consistent for measurements of the properties to be subjected to regression or spectral analyses. Because causality cannot be inferred from these statistical relations derived from time-series estuarine data, clearly conceived conceptual models must accompany analyses and interpretations.

Examples

Many of the successful examples of SER have been for terrestrial ecosystems where interannual covariations between biogeochemical processes and physical forces have been identified. There are also relatively long-term (10- to 20-year) data sets for many coastal ecosystems for key physical forces such as hourly-to-daily observations on river flow, water level, and wind velocity, and weekly-to-monthly data for water quality variables such as nutrient, oxygen, and phytoplankton concentrations. In an example from Chesapeake Bay, given in detail in Boynton and Kemp (chapter 11, this volume), year-to-year variations in river flow and nutrient loading correlate with mean concentrations of phytoplankton, dissolved oxygen, and rates of benthic nutrient recycling (Boynton et al. 1991; Kemp and Boynton 1992). There are also SER data that allow analysis of the response of marsh vegetation to nutrient additions (Valiela 1989). The correlations found in these studies can be positive, as in the relationship between nutrient loading and algal biomass and productivity (figure 12-1A), or negative, as in the relationship between river flow and algae when salinity changes control the amount of benthic grazing (figure 12-1B).

When a time series of observations is not available, ecological samples can be spatially segregated by age to produce a space-for-time-substitution (Picket 1989). This creates a pseudo-time series of observations. In estuarine sediments, for example, strata in estuarine sediment cores can be aged by ^{210}Pb or

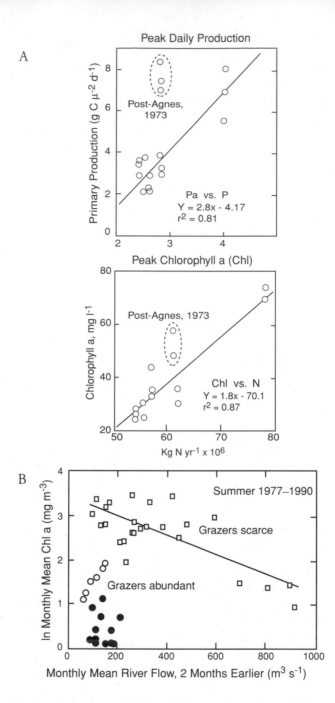

FIGURE 12-1 Long-term algal data: (A) regressions of peak daily production and chlorophyll *a* concentration vs. annual nitrogen loading from 1972–1977 in Chesapeake Bay (Boynton et al. 1982, used with permission); (B) log of monthly mean chlorophyll *a* concentration vs. mean monthly river flow 2 months earlier for the 3 months of each year centered around the annual chlorophyll *a* maximum from 1977–1990 in San Francisco Bay (Alpine and Cloern 1992, used with permission).

pollen analysis and analyzed for various ecologically relevant data such as algal composition, chemical residues, geochemical end-products, and fish scales. In this way, changes in burial rates for organic carbon, inorganic sulfur, and diatom species in estuarine sediments can be directly related to changes in inputs from the coastal ocean and from continental watersheds (Cooper and Brush 1991; Eadie et al. 1994).

Several estuarine studies have attempted to relate long-term trends in fishery yields (compiled annually) to interannual variations in major climatological forces. Sutcliffe (1972) found significant positive correlations between variations in river flow and phytoplankton chlorophyll and between river flow and harvest of five coastal fisheries in two Canadian embayments between 1936 and 1968 (figure 12-2A), indicating the importance of nutrient loading for stimulating coastal food chains. Similar analyses indicate the importance of both habitat and food availability for coho salmon (river flow and upwelling strength) along the Oregon coast (figure 12-2B, Scarnecchia 1981). Van Winkle et al. (1979) used another technique, spectral analysis, to examine coherence in frequency distributions of variance between climatological and coastal fisheries yield. Livingston and Duncan (1979) have used these methods to relate river-flow, water-quality, and fish-trawl data for a coastal environment in the Florida panhandle. A unique 25-year continuous record of relative abundance of phytoplankton and zooplankton for the North Sea and North Atlantic revealed long-term trends that were related to changes in climatology and associated variations in timing of the spring bloom (Glover et al. 1972; Colebrook 1979; Williams 1984). For estuarine fish populations, most analyses of long-term data use harvest information as indices of fish abundance; explanations for observed population changes are typically related to climatic variations or changes in fishing pressure (see Steele and Henderson 1984; Houde and Rutherford 1993; Hofmann and Powell 1998; Fogarty and Murawski 1998). While these examples are provocative, it is clear that there are only a few appropriate long-term time-series data sets that allow examination of relations between biogeochemical processes and production at upper trophic levels.

In several coastal environments, higher-frequency time-series (seconds-to-minutes frequency, days-to-weeks duration) or continuous spatial data have been obtained for physical and biogeochemical properties. Platt and Denman (1975) carried out spectral analyses of data on water temperature, salinity, and chlorophyll distributions in time or space and found that at most scales phytoplankton distributions follow the same Kolmogorov turbulent energy dissipation model that describes conservative physical properties. In a few cases, novel acoustical and optical sampling methods have generated similar continuous data sets for higher trophic-level plankton populations in addition to, temperature, salinity, and chlorophyll a (copepods, Mackas and Boyd 1979;

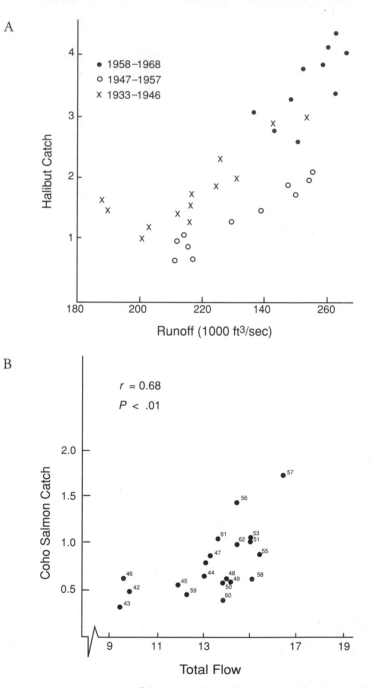

FIGURE 12-2 Long-term fish catch relationships: (A) halibut catch (10^5 lb) of Quebec vs. discharge of the St. Lawrence River (1,000 ft^3 s^{-1}) for 1933–1962 ($r = 0.8$), with a 10-year lag and 3-year running averages (Sutcliffe 1972, used with permission); (B) coho salmon catch (10^6 kg) in the Oregon commercial fishery vs. river flows from five coastal rivers (10^9 m^3 s^{-1}) for 1942–1962 (Scarnecchia 1981, used with permission).

krill, Weber et al. 1986). Analyses of these data indicate that these animal communities are more patchily distributed than their food and habitat at fine spatial scales, perhaps as a result of their particular swimming behaviors. Fish distributions have also been sampled acoustically in marine and estuarine habitats (Horwood and Cushing 1978) to analyze spatial and temporal covariance of predator and prey populations (Rose and Leggett 1990; Horne and Schneider 1997). However, there are virtually no sets of observations that allow examination of covariability and spatial/temporal scaling of organisms at both lower and upper trophic levels. Recent advances in methods, such as the undulating towed-instrument systems, optical plankton counters, and acoustical methods for large plankton and fish, will greatly facilitate future analyses of physical-biological coupling at multiple trophic levels. The need to combine data from a number of disparate sampling methods will remain an important consideration.

Natural or human-induced changes often provide opportunities to obtain insights into how coastal and estuarine systems respond to change. These opportunities include extreme events like Hurricane Andrew and changes due to dredging, certain fishing activities, or local climate alterations. To take full advantage of the opportunity, experimental or monitoring protocols should be in place and a long time series available.

Strengths and Weaknesses

Estuarine monitoring, and research programs that systematically collect long-term, continuous data, provide a baseline against which to measure ecosystem changes in response to both chronic and drastic changes in forcing functions. Scientifically, this provides a basis for understanding how individual properties and processes of the ecosystem change with changes in physical forces and how the ensemble of ecosystem interactions modulates these responses. As data sets expand with increasing years of sampling and increasing numbers of variables, the resulting increases in degrees of freedom add statistical power for drawing meaningful inferences. This is certainly one important way in which an integrated understanding of ecosystem dynamics is generated. Empirically based models also can make use of results from long-term field data collection programs to produce independent predictions of ecological dynamics. Such data-rich models are among the most robust predictive tools available to ecologists (Peters 1991). Another type of model, the mechanistic model constructed from data from experimental studies, also makes use of field observations as validation. This combination of mechanistic numerical modeling with a long-term continuous program of data collection has tremendous potential for predictions in space and time.

From a resource management perspective, long-term monitoring programs are essential for measuring effects of perturbations and of corrective policies intended to remediate harmful impacts. Without a baseline and a long-term data set, it is impossible to assess the effects of any remedial actions to manage estuarine resources. When such data sets are sufficiently long to have covered a wide range of climatological conditions, they can also help managers and scientists to distinguish ecological changes related to natural variations from those related to anthropogenic forces.

It is often the case, however, that long-term data collection programs include only water-quality variables and fisheries statistics; what is worse is that the two data collection efforts are rarely coordinated. Without contemporaneous data on organisms at intermediate trophic levels and conditions of their habitats, there is little chance for such programs to relate changes in fish populations to trophic interactions or alterations in habitat. Long-term monitoring efforts may also be compromised by inadequate attention to relevant time and space scales. A low number of samples in an estuary may be inadequate to characterize spatial variability within the basin; infrequent or poorly timed visits to the site may not provide data capable of establishing the very trends that the long-term program seeks to identify.

Recommendations

Improved understanding of how biogeochemical processes are coupled to production at higher trophic levels in estuaries can be facilitated with more and better continuous records of temporal (and spatial) data on key ecological variables. Future developments in estuarine synthesis will require stronger commitments to long-term measurements to elucidate interannual variations of ecologically important properties and processes. Long-term data series will be most amenable to ecological synthesis if a broad range of variables can be measured concurrently. For this approach to flourish, new technologies (such as, moored platforms with data transmission, acoustical and optical methods for plankton and fish, remotely operated incubation techniques for rate measurements, innovations for sampling a wider range of variables) need to be developed and applied. In addition, similar suites of techniques are needed for shorter-duration measurements at finer time and space scales. Programs established to obtain fine- or coarse-scale continuous data in time or space domains need to start with clear research and management objectives that include conceptual and computation models.

Measurements need to be made in the context of a model to identify linkages and interrelationships. Sampling frequencies for these measurements need to consider inherent periodicity and scales of the variables being sampled; considerable information exists on the characteristic time and space scales of various organisms, processes, and properties. In addition, useful guidelines can be obtained from sampling theory (for example, sample at half the Nyquist frequencies). Finally, the full value of continuous long-term data sets can be best realized when combined with other approaches to synthesis. The implementation of long-term sampling should be based on a substantive conceptual model, and ideally coupled, initially and during sampling, with numerical models. For example, time-series data are needed to calibrate and validate computational models of estuarine ecosystem dynamics.

Utility for Management

Long-term records are perhaps the most useful information for the evaluation of management decisions. Such data sets characterize the state of an ecosystem prior to implementation of management actions, including information on the scales of natural variability for the specific site. Interannual variability may confound inferences drawn from relatively short-term time series, misleading managers as to the need for action or the evaluation of results from actions taken. Adequate long-term ecological data are, however, rarely available. They are perceived as routine monitoring and, therefore, less exciting than cutting-edge research; yet without such data, management is likely to be inadequately informed.

Cross-ecosystem Comparisons

Status of the Approach

Aquatic scientists use cross-ecosystem comparisons to examine how ecological variables may be related to one another. The goal is to infer how internal processes or external driving forces influence other aspects of ecosystem structure or function. Downing (1992) distinguished three types of cross-ecosystem comparisons: (1) composite comparisons where many variables are simultaneously compared using statistical techniques such as principal components analysis; (2) variable-focused comparisons where only one or a few variables are considered usually visually or by regression analysis; and (3) connection-focused comparisons that examine the responses of ecosystems to experiments where one or a few variables have been manipulated. This discussion will emphasize variable-focused comparisons but will also point out a few examples where composite comparisons have been used fruitfully. Connection-focused comparisons will be discussed here in a limited context.

Although cross-system comparisons are one of the oldest approaches in ecology (Heal and Grime 1992), their value has been criticized because "correlation does not imply causality." Box et al. (1978) have also pointed out difficulties inherent in using some statistical techniques, such as regression analysis, on nonexperimental data. In spite of the potential problems, there has recently been a great deal of interest in using cross-ecosystem comparisons for data synthesis. An issue of *Limnology and Oceanography* (1988, vol. 33, no. 4, part 2) was completely devoted to the comparative ecology of fresh and marine waters, and Cole et al. (1992) have edited a book on the patterns, mechanisms, and theories of the comparative analysis of ecosystems. The advantages and disadvantages of cross-system comparisons for synthesizing our understanding of how ecosystems function have been well debated in the literature (Peters 1988; Lehman 1986; Pace 1992). A number of thoughtful articles have pointed out how the pitfalls of the comparative approach can be minimized and how the value of the approach can be maximized (Peters et al. 1992; Carpenter et al. 1992; Downing 1992). Here we outline how cross-system comparisons have been used to link biogeochemical processes to higher trophic levels and explore some of the strengths and weakness of a comparative approach for synthesizing what we know about this topic.

Examples

Some of the most successful cross-system comparisons examined the factors controlling primary production in lakes. Vollenweider (1968) demonstrated that the trophic status was related to depth and phosphorous loading. Dillon and Rigler (1975) found a very high correlation between spring phosphorous concentration and mean summer chlorophyll concentration. These relationships were then used to construct simple models of lake chlorophyll concentration based on phosphorous loading, water budget, and morphometry (Dillon and Rigler 1975; Vollenweider and Dillon 1974). For a variety of lakes around the globe, Schindler (1978) found that phosphorus input and water renewal time (figure 12-3A) explained a high proportion of the variance in both annual phytoplankton production and mean chlorophyll. The relationship between phosphorus inputs and lake productivity is robust enough that it is used as a predictive management tool.

In estuaries and coastal marine ecosystems, a strong relationship between nutrient loading and chlorophyll concentrations as well as primary production also emerged (Boynton et al. 1982; Nixon and Pilson 1983; Nixon 1992). In marine systems, primary production was closely related to nitrogen inputs rather than phosphorus (figure 12-3B). Again, the relationship is considered to be strong enough that reduction of nitrogen inputs is part of management strategies to alleviate eutrophication.

Cross-system comparisons examining the links between biogeochemical processes and higher trophic levels first focused on fish production in lakes. Limnologists began to look for relationships between fish yield and lake characteristics about the same time that relationships between phosphorous and lake chlorophyll were being developed. The first index was the lake morphoedaphic index (MEI), which integrated information on morphometry

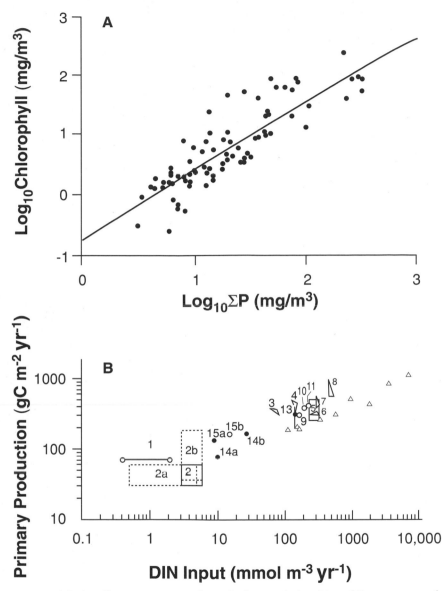

FIGURE 12-3 Cross-ecosystem phytoplankton relationships: (A) mean annual chlorophyll *a* concentration vs. mean annual concentration of total phosphorus for various lakes (Schindler 1978, used with permission); (B) primary production by ¹⁴C vs. annual input of dissolved inorganic nitrogen for various marine systems (data sources and sites identified in Nixon 1992, used with permission).

and trophic status to describe fisheries yield from Canadian lakes (Ryder 1965). The MEI used lake total dissolved solids as a proxy of lake nutrient content and mean depth as an index of lake morphometry. Good relationships between the MEI and fish yield were then found for other locations although the same regression line could not describe all regions. The next step was to look for correlations between fish yield and chemical or biological measurements. With this approach, Hanson and Leggett (1982) found a strong correlation between total phosphorous and fish yield in twenty-one Ontario lakes. Even better correlations were noted between primary production and fish yield in temperate (Oglesby 1977) (figure 12-4A) and tropical (Melack 1976) lakes.

Many investigators continued to work with the original morphoedaphic index because information on total dissolved solids (TDS) was available for a large number of lakes where primary production or total phosphorus data were lacking. By the early 1980s more than 100 papers had been published on the relationship of the MEI to fish yield, some praising and expanding upon the method, others criticizing the approach (reviewed in Ryder 1982). Larger-scale analyses of the MEI produced insights into possible differences in the local versus regional controls of fish yield. For example, Schlesinger and Regier (1982) examined data from forty-three intensively fished lakes from 62°N to 15°S to determine the controls of fish yields on a global scale. They concluded that climate (mean air temperature) accounted for 74% of the variability while the morphoedaphic index accounted for an additional 7–9%. This finding contrasts with regional data where the MEI usually accounted for a much greater proportion of the variability than climatic variables. It also contrasts with the relationship Schindler (1978) found between phosphorus loading and primary production where loading, not latitude, explained most of the variance.

There have been numerous other attempts to link biogeochemical processes to production at higher trophic levels. Peters (1988) reviewed a number of studies in lakes where correlations were found between total phosphorus (TP) concentrations and biomass of higher trophic levels. High correlations (r^2 = 0.72) between TP and crustacean zooplankton biomass were found in lakes from around the world (figure 12-4B), and less significant correlations between TP and benthic biomass were also found (r^2 = 0.48). Hanson and Peters (1984) observed that the correlations of TP and higher trophic level biomass are often much better than correlations with phytoplankton biomass.

While many studies have simply looked for correlations between variables, cross-systems comparisons have also been used to test hypotheses. Pace (1986) used a cross-system comparison to test hypotheses on the effects of nutrient loading on zooplankton size structure. In a comparison of thirteen lakes, both microzooplankton and macrozooplankton exhibited positive and highly

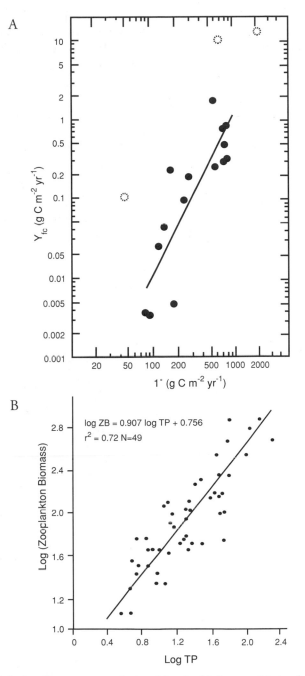

FIGURE 12-4 Cross-system relationships for higher trophic levels: (A) fish yield vs. annual primary productivity for fifteen lakes; area less than 10 km² (•), smaller lakes (°) (Oglesby 1977, used with permission); (B) log zooplankton biomass vs. log total phosphorus concentration for forty-seven lakes worldwide (Hanson and Peters 1984, used with permission).

significant correlations with lake TP, and the size structure (expressed as microzooplankton biomass per unit of total zooplankton) did not change with increased nutrients. This was in contrast to theory that suggested that microzooplankton should dominate in more productive lakes.

More complex statistical models have been extensively used in food-web analysis but rarely for cross-ecosystem studies. In one example, where Carpenter et al. (1992) examined both the interannual variance of variables within lakes and the interlake variance, summer chlorophyll responded positively to nutrients and negatively to herbivore size.

In the marine environment there have been only a few attempts to use cross-system comparisons to link higher tropic levels directly with biogeochemical processes. Armstrong (1982) found that shellfish harvest was directly related while finfish harvest was inversely related to nutrient loading and freshwater inflow in six Texas estuaries. Nixon (1982, 1992) found that total fisheries yield in a variety of marine and estuarine waters was correlated with primary production (figure 12-5). Nixon also pointed out that the fisheries yield per unit of primary production from marine waters was higher than the

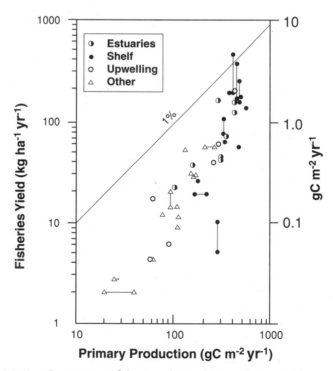

FIGURE 12-5   Cross-system fisheries relationships. Fisheries yield per unit area vs. primary production for various estuaries and marine systems (data sources and sites identified in Nixon 1992).

yield from lakes. Also, because in this figure the slope of the line was greater than 1, a greater percentage of the primary production was going into fisheries yield in more productive systems than in less productive systems.

The importance of the area of marsh or estuary to fisheries yield has also been investigated with the comparative approach. Turner (1977) related inter-tidal vegetation and latitude to the commercial harvest of penaeid shrimp. The variation of fish yield with the ratio of marsh area to open water suggested to Nixon (1980) that marsh food resources may not be the major factor driving the fishery and that habitat was important. Yet the positive relationship of total fish landings, not just estuarine-dependent fisheries, to total shoreline length led Hopkinson et al. (1991) to the opposite conclusion that implicated outwelling and food-web structural differences. Deegan et al. (1986) found a correlation between fish yield per unit of open water and river discharge, suggesting that the fisheries were responding to both habitat availability and nutrient inputs.

Strengths and Weaknesses

There are some obvious methodological difficulties in trying to examine the links between higher trophic levels and biogeochemical processes using a comparative approach. While we have a good knowledge of phytoplankton biomass and primary production in a large variety of aquatic ecosystems, good data on the biomass of higher trophic levels are not as readily available and information on secondary production is usually lacking. Even in the zoo-plankton, where production and biomass have been well studied in a number of systems, more recent data suggest that the omission of small size classes in older data may have led to considerable underestimates of production (Davis 1987). The problems become greater when attempting to measure fish or benthic production. First, the data available usually come from commercial fisheries statistics where catch effort, rather than biological productivity, may be a major factor that determines yield (Bayley 1988). Another problem is that fisheries data may include species from a number of different trophic levels; this obscures relationships between yield and biogeochemical processes. In addition, lags between biogeochemical processes and production need to be taken into account because fish may be several years old at harvest. One result of these lags is that coherence changes dramatically between primary production and piscivore populations depending on the time scale used. Carpenter (1988) demonstrated that the maximum coherence occurs for the time scale close to the piscivore life span. Finally, the fisheries data themselves may not be accurate, and may not include major consumers such as jellyfish. For these and other more substantive reasons, cross-ecosystem data of fisheries usually have a large amount of scatter, up to an order of magnitude or more. The data are often depicted in log-log plots, which limits their predictive utility. Further, trends of subsets may indicate different patterns than that suggested by all of the data, which introduces a philosophical dilemma in interpretation.

While the strong relationships that emerged between nutrient input and phytoplankton biomass or production did not prove causality, they did suggest that aquatic systems should show a predictable general response to changes in nutrient inputs. This response was confirmed through a variety of manipulative experiments where the additions of nutrients to whole lake systems, experimental ponds, marine embayments, and marine mesocosms consistently increased phytoplankton biomass and production. Further evidence came from managed systems where diversions of nutrient inputs reduced phytoplankton production in Lake Washington (Edmondson and Lehman 1981) and Kaneohe Bay (Smith 1981).

While the primary producers always respond strongly to biogeochemical manipulations, the responses of higher trophic levels have been less consistent (see later section titled "Experiments"). Often these higher trophic levels respond to top-down controls in fresh waters (Shapiro 1980; Carpenter et al. 1987) and in marine environments (Paine 1966; Menge 1976; Mann and Breen 1972; Sullivan et al. 1991). When it comes to fisheries, of course, it is likely that the decline in the world's fisheries is related to the top-down effect of overharvesting.

Recommendations

Does the importance of top-down control in aquatic ecosystems imply that an attempt to find linkages between biogeochemical processes and higher trophic levels using a cross-system approach will fail? It is now widely recognized that top-down and bottom-up control are not mutually exclusive (Menge 1992; Carpenter et al. 1992). The comparative approach has produced results that suggest clear trends, yet these relationships are sometimes contradictory among the comparisons themselves or in opposition to other experimental or observational data. However, the results of these comparisons have nevertheless usually been informative, sometimes provocative.

Cross-system comparisons, when properly done, can be used effectively to generate and perhaps reject hypotheses on the importance of bottom-up controls in estuaries. The fairly global nature of the relationship between nutrients and phytoplankton observed by Schindler (1978) and Nixon (1992), and the more local relationships observed with the MEI and fish yield (Schlesinger and Regier 1982), provide us with important information on the strengths of relationships at different levels in the food web.

Cross-system comparisons can be especially useful when combined with other approaches. When a number of long-term data sets are compared, the relative importance of controls within one system to the controls operating between systems can be better identified. For example, several long-term studies have shown relationships between fish growth or yield and discharge, but the slope of the relationship within systems differs and the systems show different relationships of discharge to yield (Hvidsten and Hansen 1988). Specific links between biogeochemical cycles and higher trophic levels can be tested by experiments; however, these links may not be the most important ones controlling higher trophic levels in nature. Carefully constructed cross-system comparisons can help determine the generality of experimental results. Some recent programs, such as the Land Margin Ecosystem Research (LMER 1993), specifically encourage groups studying different systems to work together on cross-system comparisons and synthesis (Hopkinson et al. 1998). While these projects do not all extend to the higher trophic level, useful experience may be gained in these collaborative efforts.

More accurate data on biomass and production of higher trophic levels would greatly strengthen our ability to do cross-system comparisons on linkages to biogeochemical processes. New acoustic (Demers, Brandt, Barry and Jech, chapter 15, this volume) and video (Davis et al. 1992) techniques may provide ecologists with more accurate and more economical ways of assessing the biomass of higher trophic levels in estuaries and coastal marine waters. New techniques examining cellular markers appear promising for examining the secondary production of small animals with rapid turnover (Moore 1996).

Utility for Management

In general, stronger relationships between biogeochemical processes and higher trophic levels have emerged from local rather than global comparisons. Managers have used the MEI and other empirical relationships to determine fish yield (Ryder et al. 1974), but this approach has been successful when the index is calibrated locally. Cross-system marine relationships so far are robust and significant, but variability of an order of magnitude or more may limit management applications to anticipating probable trends.

Computational Models

Status of the Approach

Various classes of computational models have been widely used to simulate and analyze ecological interactions at lower and upper trophic levels within estuarine food webs. In a few cases, these modeling approaches have been extended to reach across the food web from microbes to fish. Although there is great interest in generating models to simulate biogeochemical processes in

tandem with fish population dynamics, this objective is at the frontier of ecological modeling in estuaries. We are now in the early stages of this effort to
build bridging models; progress has been made, but there is much to be done.

For some time, numerical models have simulated photosynthesis, respiration, nutrient cycling, and other processes of plankton and benthic communities in relationship to physical exchanges of water (Kremer and Nixon 1978;
Baretta and Ruardij 1988). This approach assumes bottom-up controls on
estuarine ecosystem dynamics, and not surprisingly these models have been
successful in reproducing dynamics of processes at lower trophic levels. In a
few cases, the biomass of functional groups of organisms feeding high on the
food chain, such as fish, are also computed using simple relationships for
bioenergetics and animal movements (Bartell et al. 1988; Scavia et al. 1988;
Ross et al. 1994; Kemp et al. 1995). Rarely, populations of especially influential species have been coupled to plankton models in a mechanistic way, such
as seasonal invasions by menhaden (Kremer and Nixon 1978) or ctenophores
(Kremer and Kremer 1982). Detailed formulation of long-lived species present in the estuaries throughout the year is almost never attempted.

At the other extreme, computational models like those of Schaefer (1954)
and Beverton and Holt (1957) have been widely used for many years to compute changes in abundance of specific fish populations. These models may
focus on adults of a population or they may follow changes in number of individuals across the range of age groups, with primary emphasis on fishing mortality and physical controls on recruitment. Still another class of models
(Kitchell et al. 1977; Brandt and Hartman 1993) simulates the growth and
metabolism of animal populations, applying allometric relationships to compute detailed bioenergetics. A related approach involves individual-based
bioenergetic calculations, which can be coupled with behavioral models to
simulate movement, mortality, and life history. These individual calculations
track both biomass and numerical abundance, which can be extrapolated to
large populations, following them through recruitment, reproduction, and
mortality, and moving them across space (DeAngelis and Gross 1992). These
approaches have also been used successfully by Davis (1987) and Powell et al.
(1997) to analyze invertebrate populations at middle (herbivore-carnivore)
trophic levels.

Network models of flows of energy or materials among species and functional groups use an expanded version of the input-output matrix algebra
developed in econometrics to analyze structures of coastal marine food webs
(Ulanowicz 1983; Fasham 1985; Christensen and Pauly 1998). This method
(Hannon 1973), which employs a systematic steady-state computation to
integrate large empirical databases, addresses questions of trophic efficiencies
and pathway alternatives. These models calculate the partitioning of the flux
of biomass from lower to upper trophic levels at any level of aggregation (from

functional groups to species to age groups with species). They have been used to compute how trophic flows shift with alterations in community structure (Ulanowicz and Tuttle 1992) or with changes in external inputs (Deegan et al. 1994). However, the approach requires extensive data for all major species of an ecosystem. The calculations are static, with no predictive abilities.

Examples

For Chesapeake Bay, a long-term modeling project has been established to develop a simulation framework in which fish production and habitat quality are related to changes in nutrient and organic inputs to the estuary. Here, a suite of ecological simulations includes (1) process models that relate photosynthesis, biogeochemical cycles, and herbivorous production (Bartleson and Kemp 1991; Kemp and Bartleson 1991); (2) bioenergetic models relating habitat conditions and food availability to potential production (Lou and Brandt 1993); and (3) regression models to provide independent empirical validation of numerical constructs (Boynton et al. 1991).

These ecological models emphasize nutrient and organic matter sources and cycles, trophic interactions, and habitat structures. Together, the ecological models are meant to serve as a staging platform for the continuing evolution of a coupled hydrodynamic-water quality model for resource management. The ecological models incorporate ecological feedbacks (for example, fish predation on zooplankton, benthic suspension feeding on phytoplankton, seagrass-enhanced deposition of suspended particulates and associated water clarification) that may control the abundance and size structure of organisms at lower trophic levels. Published results of the hydrodynamically based water-quality model appear successful in relating nutrient loading to dissolved conditions in the estuary (see figure 14-11 in chapter 14 of this book; Di Toro et al. 1990; Cerco and Cole 1993; Johnson et al. 1993). However, the simulated rates of primary production are less consistent with field measurements (Di Toro, personal communication). Future efforts will focus more directly on how habitat alterations such as seagrass declines and deep-water hypoxia influence fish production.

For this same estuary, the structure of complex trophic interactions has been analyzed using network models. Analysis of a large matrix model of carbon flow in the mesohaline region of Chesapeake Bay indicated that (1) microbes were important sinks for organic matter; (2) a gelatinous zooplankter played a pivotal role; and (3) key fish species (striped bass, bluefish) were very sensitive to shifts in trophic flow between benthic and pelagic pathways (Baird and Ulanowicz 1989). Carbon and nitrogen had very different cycling rates and pathways (Baird et al. 1995). An aggregated version of this model subsequently demonstrated the impact of changes in the oyster culture and harvest industry on phytoplankton abundance via trophic feed-back effects (Ulanowicz and Tuttle 1992). The same model, calibrated for Chesapeake Bay and the Baltic Sea, revealed that the Baltic was more effi-cient in the efficiency of primary production transfer to fisheries (Ulanowicz and Wulff 1991).

One example of an ongoing management study using coupled models across trophic levels is the Across-Trophic-Level System Simulation (ATLSS) project, a system of models to investigate ecological effects of alternative hydraulic restoration scenarios in South Florida. The modeling goal is to relate nutrient and hydrological processes to dynamics of resident and migra-tory fish, bird, and other populations in the Everglades and Big Cypress Swamp. ATLSS employs different modeling approaches at each trophic level (figure 12-6). At lower trophic levels (primary producers, microbes, detritus, soil/sediment invertebrates) aggregated process models relate biogeochemistry to growth of primary and (simple) secondary producers. At intermediate trophic levels, age- and size-structured population models are used to simulate dynamics of functional groups of fish and invertebrates. At the level of top consumers individual-based models simulate charismatic megafauna—birds, alligators, deer, and panthers. Spatial patterns are simulated with a land-scape/vegetation model integrated with a plant succession model to predict changes in plant distributions with changing hydrology.

A final example is a management model under development for Waquoit Bay, a small (680 ha) shallow (~1 m) complex of tidal estuaries on Cape Cod (Massachusetts, USA). A computational model relates changes in land use in the Waquoit Bay watersheds to estuarine nitrogen loading. A dynamic simula-tion then estimates the ecological response of communities in the estuary. The nitrogen load resulting from the land-use mosaic is determined from a sequence of mechanistic calculations specifying the rates of delivery of nitro-gen in different land parcels and various removal processes (Valiela et al. 1997; Brawley et al. in press). Confidence limits for the loading estimate are based on propagation of uncertainty in the input parameters (Collins et al. in press). Inputs of nitrogen by groundwater and atmospheric deposition alter the growth rates and abundance of phytoplankton and benthic macroalgae. The next step is to improve the transferability of the model to different estuaries. This involves new approaches to the formulation for plankton and algal pro-duction based on robust empirical relationships rather than conventional detailed theoretical mechanisms. Using cross-system empirical relationships, simulated water-quality conditions are then related to critical environmental indicators such as water clarity, habitat suitable for eelgrass, the probability of anoxic events, and the diversity and abundance of resident fish populations (Kremer 1999).

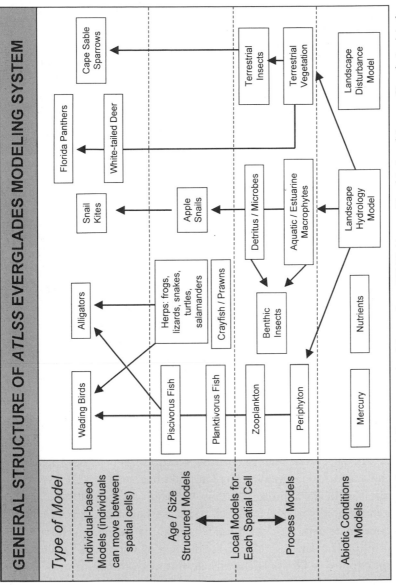

FIGURE 12-6 The modeling approach of the project ATLSS (Across Trophic Level System Simulation) is designed to integrate three approaches for different trophic levels of the Florida Everglades system. The overall objective of the modeling is to compare the future effects of alternative hydrologic scenarios on the biotic components of the systems. The figure is redrawn from Fleming et al. (1994). A similar figure may also be found in DeAngelis et al. (1998).

Strengths and Weaknesses

Many of the most pressing environmental problems and scientific questions currently facing coastal ecologists involve understanding relationships between processes at lower and upper trophic levels. How do changes in habit structure alter recruitment and production of specific fisheries? How do changes in mortality of keystone species such as native and exotic bivalves affect key biogeochemical processes such as denitrification? These and other related questions offer a major challenge for the next generation of estuarine scientists, a challenge the complexity of which will require multidisciplinary scientific teams for breadth of scope and numerical simulation models for integration of piecewise data collection. The simulation models must be capable of capturing dynamics at both ends of the trophic web. There are two alternative approaches for coupling analyses of dynamics at upper and lower trophic levels.

On one hand, different but compatible models that simulate processes at either end of the food chain could be numerically linked to create a coupled simulation. This would be similar in approach to existing coupled biophysical models that simulate water circulation and plankton processes together in real time (Hofmann, chapter 6 this book; Wroblewski and Hofmann 1989). Here there are significant numerical problems of matching time and space scales of linked simulations. These are, however, tractable problems that can be resolved.

The second approach involves creating a single model that captures the dynamics of key ecological processes across all levels of a trophic web. This can be done readily by simplifying and generalizing the nature of model variables, for example, using functional groups of organisms. This would be an acceptable approach for addressing broad questions, such as changes in potential fish production with ecosystem alteration addressed by Ross et al. (1994). However, to address questions at the population level, which includes many individuals within numerous age groups, would require a large, detailed, and data-hungry simulation. Although this represents a brute-force approach, it is now possible because of the dramatic improvements in computer hardware.

Each of these modeling approaches needs vast databases for calibration and validation and the ability to integrate such data. The data management effort needed to coordinate with large-scale modeling can be daunting (Hurtt and Armstrong 1996). The data assimilation methods of Lawson et al. (1996) are one promising approach for systematically dealing with the problem of data application for big models.

A related concern is the objective validation and documentation of large, complex models. The sheer size of these models is one problem. Large amounts of data are employed in formulating and tuning the many process routines of food web models. Increasingly, these process routines are coupled to fine-scaled, two-dimensional or three-dimensional hydrodynamic circulation regimes. Running simulations with the full model may be time consuming and costly. The sheer volume of mechanistic detail precludes objective validation. Another problem involves people. The structure of the integrated models grows in complexity over time, often involving a sequence of researchers who join and leave the project. By the time the models are deemed ready for management applications, adequate documentation of model assumptions, of actual structural details, and of attempts to validate the behavior of the many important processes may be unavailable and unrealistic to obtain. When the models are developed by government agencies or consulting companies, the model may never be published in the reviewed scientific literature.

Recommendations

There is an obvious and growing need for well-tested approaches for parallel simulation of ecological processes from bacteria to fish in estuarine ecosystems. Already there are numerous multidisciplinary projects attempting to address scientific and practical questions that involve interdependent changes in nutrient chemistry, biogeochemical cycles, community structures, and population dynamics of birds, fish, and other mobile carnivores. Few of these studies have the computational tools and approaches needed to integrate the relevant ecological information to analyze key interactions and understand broad relationships across time and space scales, from those relevant for bacteria to predatory birds and fish.

What is needed is for several long-term studies of coastal ecosystems to assign high priority to developing multiscale computational modeling approaches that either (1) couple an ensemble of models, each designed for different processes or organization levels; (2) construct a single hierarchically structured simulation system, which encompasses all relevant scales, processes, and organisms; or (3) explore the usefulness and generality of simple models as alternatives to increasingly complex process simulations. Modelers and empiricists need to work together toward this goal of an integrated understanding of relationships between biogeochemical processes and animal population dynamics. Different research programs working toward similar goals need to share experiences and compare and contrast results. The resulting models must be carefully documented and published in the scientific literature so that they are available to all. Such an effort could be sponsored by the emerging coastal initiative within the NSF-funded LTER program; however, even mission agencies such as EPA and NOAA could make this goal a priority for diverse studies in specific well-studied regions. It is germane to look critically at whether the models are generalizable and not specific to one estuary or one type of estuary.

Utility for Management

Ecosystem simulation models continue to be pressed into service for management applications. For example, closely related models have been employed for management in Chesapeake Bay (Cerco, chapter 14 this book; Cerco and Cole 1993), Long Island Sound (HydroQual 1996, 1997), and Boston Harbor (HydroQual 1995). The attractiveness is obvious, as these tools can be designed to make predictions corresponding to viable management scenarios offering a high degree of quantitative resolution (HydroQual 1997). There is little doubt, however, that the current uncertainty surrounding these predictions is substantial and not adequately evaluated; it is not even clear how to undertake this task in an objective and meaningful way. The shortcomings are easily identified (see Korfmacher 1998) but difficult to correct. One problem with this set of related models is that they have not been published in the scientific literature and openly evaluated. That is, modelers outside the consulting group and government agency that have use of these models cannot evaluate and use them. Because these related models dominate the market for large-scale coupled estuarine management models in the United States, this is not a trivial issue.

Continued application of ecosystem models to management questions requires diligent attention to the modeling basics: the adequacy of the conceptual model, the validity of the assumptions, the quality and appropriateness of the input data, and the careful objective evaluation of the model's predictions.

Nutrient and Energy Budgets and Mass-Balance Approaches

Status of the Approach

Mass-balance models based on the principles of conservation of mass and energy have been used extensively in estuarine synthesis. Ideally, in this approach, all input and output terms are quantified in a mass balance or budget. More often than not one or more terms are not known or inadequately quantified. However, the magnitude of a single unknown source or sink can be estimated when all other inputs and outputs are known. Even when there is more than one unquantified term, budgets may provide insight into the relative magnitude of the known compartments of a budget and may provide direction for future research.

Examples

Budgets of nutrients or energy have been used to indicate the relative importance of losses due to removal by fishing compared to nutrient inputs or primary production in estuaries. Annual mass balances of nitrogen and

phosphorus for a number of estuaries indicate that fisheries landings generally remove a small amount, 0.5–10%, of the nutrient inputs to estuaries
(Nixon et al. 1996; table 12-1). For example, the amount of nitrogen
removed from Narragansett Bay in fisheries landings is approximately 0.7%
of inputs (Nixon et al. 1996).

While fisheries landings generally are a minor sink for nitrogen and phosphorus in estuaries, seasonal migrations of anadromous fish transport large
masses of nutrients in their biomass into small streams (Durbin et al. 1979;
Kline et al. 1990; Richey et al. 1975). For example, in spawning streams the
transport of nitrogen from the ocean by Pacific salmon supplies 50 to 90% of
the nitrogen for periphyton growth (Kline et al. 1990).

Mass-balance energy models for estuaries indicate that a relatively small
percentage of the primary production becomes fish biomass (Weinstein et al.
1980; Bahr et al. 1982; Currin et al. 1984; Nixon et al. 1995). For example,
migratory menhaden export 5–10% of the carbon fixed in inshore primary
production offshore to the Gulf of Mexico (Deegan 1993). However, the total
amount of energy that is required to support fish production, given losses during trophic transfers, can be a major fraction of the primary production. Bahr
et al. (1982) found that approximately 60% of the primary production (12 $\times$
10^{13} Kcal/y) in all Louisiana coastal basins is required to support the inshore
fisheries (7 $\times$ 10^{13} Kcal/y).

While we might expect growth of planktivorous fish to be less than 1% of
primary production rates in terms of organic carbon, the percentage will be
higher in terms of nitrogen and phosphorus from the primary producers. This
occurs because the ratios of carbon to nitrogen and carbon to phosphorus
decrease in animals at higher levels of trophic chains. In addition, certain
planktivorous coastal fish, such as menhaden and mullet, obtain a substantial

TABLE 12-1

Nutrient inputs and the percentage of nitrogen and phosphorus inputs in various estuaries that are removed in fish landings (data from various sources,
Nixon et al. 1996).

System	Inputs (mmol m^{-2} yr^{-1}) N	P	Fish Landings (mmol m^{-2} yr^{-1}) N	P	Fish Landings (% of input) N	P
Baltic Sea	230	6.7	11	0.7	48	10.4
Chesapeake Bay	938	31	84	1.5	9.0	4.8
Delaware Bay	1900	158	small	small		
Narragansett Bay	1960	115	13	0.8	0.7	0.7
Guadalupe Estuary						
dry year	548	70	35	2.2	6.4	3.1
wet year	2058	171	85	5.3	4.1	3.1
Potomac Estuary	2095	27.6	10	1.8	0.5	6.5

proportion of their growth from herbivory on algae. In fact, recent analyses revealed that approximately 9% of the nitrogen and 5% of the phosphorus loading to Chesapeake Bay from watershed and atmosphere are removed in fisheries harvest, primarily menhaden (table 12-1; figure 12-7, Boynton et al. 1995). This calculation does not even consider the possible losses associated with net growth of emigrating fish such as the bay anchovy.

Unfortunately, there are insufficient examples of such mass balances that consider fish production, migration, and harvest to make initial cross-system comparisons. It would be of interest to understand what regulates the proportion of the total nitrogen and phosphorus loading that accumulates in fish production in different coastal environments. Is fish production actually limited by the amount of nitrogen or phosphorus available? Would variations in key biogeochemical processes, such as denitrification, significantly alter the proportion of estuarine nutrient inputs that lead directly to fish production? How does estuarine trophic condition affect this proportion?

Strengths and Weaknesses

The mass-balance approach can quantify a process or compartment that is very difficult to measure as long as all other inputs and outputs are known and there are steady-state conditions. Even when information on compartments is not precise, budgets indicate large differences between compartments or flows and which components are the most important (Hall 1972). The degree to which mass balance is attained in a budget can indicate whether or not there are large, still unknown terms. In this way, Nixon (1981) interpreted a large imbalance in a budget of nitrogen relative to phosphorus as the omission of a major flux (table 12-2). This flux turned out to be benthic denitrification. Constraints can be placed on the order of magnitude of an unknown compartment. For example, a physiological energy balance can provide a tool for estimating potential secondary production.

Though theoretically straightforward, the task of quantifying net material fluxes through inlets in tidal systems is daunting in a practical sense. As a result, it has been difficult to prove or disprove the importance of outwelling of organic matter from coastal to shelf waters (Nixon 1980). One reason for the difficulty is that generalizations drawn for any particular system may not be applicable for the great variety of coastal systems. Another major problem, pointed out by Kjerve et al. (1981), is that the complex circulation in coastal ponds and estuaries requires intensive measurements in space and time even to calculate fluxes of water and salt. Further, there is the chance that intermittent extraordinary events, such as storms, may overwhelm day-to-day flux balances. The conclusion is that it is nearly impossible to assess fluxes without instruments capable of continuous measurements of integrated transport

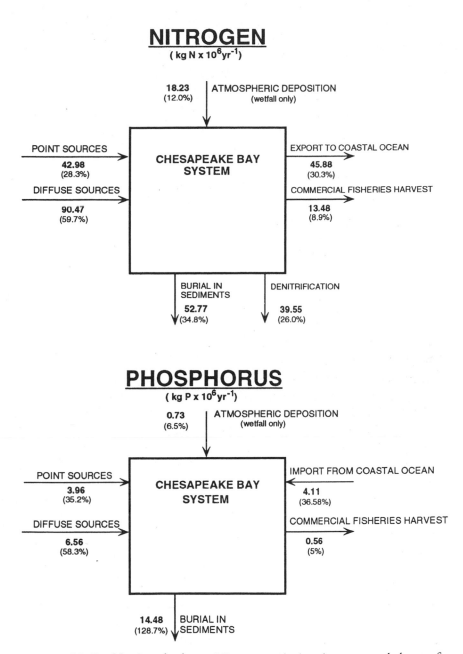

FIGURE 12-7 Nutrient budgets. Nitrogen and phosphorus mass balances for Chesapeake Bay (Boynton et al. 1995, used with permission of the Estuarine Research Federation).

TABLE 12-2

Budget of inputs and regeneration of nitrogen and phosphorus for Narragansett Bay (Nixon 1981). Note the conspicuous imbalance for N relative to P.

	Nitrogen in 10^6 moles yr^{-1}	Phosphorus in 10^6 moles yr^{-1}
Inputs		
Fixation	0.2	—
Rain	2.8	0.19
Runoff	16.2	0.8
Rivers	235	17.3
Sewage	278	21.7
Total Inputs	532	39.9
Regeneration		
Menhaden	0.8	0.1
Ctenophores	8.1	0.8
Net zooplankton	98.5	—
Benthos	264	41.1
Total Regeneration	371	42

through the entire cross section of inlets. The problems with dissolved constituents are aggravated for higher trophic levels. Although some successes have been reported for elemental budgets (Woodwell et al. 1977; Valiela and Teal 1979; Valiela 1983), taken together these hurdles will continue to limit the accuracy of material budgets based on integrated fluxes. A recent workshop (LMER 1993 pp. 6–15: Report of the workshop on estimation of fluxes through estuarine cross sections) was devoted specifically to attempts to surmount these difficulties.

Unfortunately, there are few examples of such mass balances that consider fish. The error associated with estimates of all terms may be sufficiently large that conclusions are unreliable. Especially if an unknown term is calculated based on mass balances, these cumulative errors are contained in the calculation of the unknown (error propagation). Three major assumptions of mass-balance calculations or budgets are that (1) all other terms are known; (2) all terms are accurately quantified; and (3) the system is in steady state. In reality, systems generally are not closed, have time lags, and are not in steady state. For example, estuarine fish faunas are dominated by species that move into the estuary as larvae, accumulate biomass, and then move offshore after growing to near adult size (Yanez-Arancibia 1985). There are often large year-to-year variations in the population size of fish, rates of nutrient input, and rates of primary production. Various components of a budget are often measured during different years or over different time and space scales (that is, daily primary production versus sedimentation over 50,000 years). Ideally all terms should be quantified on similar time and space scales but suitable spatial and temporal coverage is generally lacking. The use of in situ instrumentation for continuous sampling, incorporating optical fiber links, acoustic sampling, and satellite imagery, ultimately can provide the spatial and temporal coverage needed for many parameters.

Recommendations

Despite the potential shortcomings of the mass-balance approach, its use is highly recommended. Mass balances provide a strong reality check on the degree to which the system or process studied is understood. They may identify gaps or inadequacies in our knowledge. And they provide at least an initial quantitative evaluation of the relative magnitude of contributing factors. Technological advances capable of more intensive and integrated assessment of material fluxes should be actively developed and tested.

Utility for Management

Taken alone, mass balances provide little information directly applicable to most management issues. However, this approach does provide critical indication of the state of knowledge and the adequacy of available data for applied issues related to sources and fate of materials. Material budgets may thus provide a critical foundation from which to evaluate the implications of results arising from the same or related bodies of data.

Experiments

Status of the Approach

Manipulative experiments have played an important role in aquatic ecology by identifying causal links and mechanisms. Both these features are essential for synthesis about how an ecosystem may work. Applying a valid experimental approach to the issue of bottom-up control of higher trophic levels presents serious problems of scale. Yet the rigor and control provided with manipulative experiments is unique to this approach so that significant benefits can result.

Examples

Manipulative experiments in this field extend back nearly 60 years to the application of fertilizer to Scottish lochs to investigate the response of algae and animals to nutrient enrichment (Raymont 1950). This avenue of study eventually led to experiments that simultaneously evaluated the degree of control of plankton and benthic communities by nutrient supply and by fish (Hall et al. 1970).

Fresh waters, especially lakes, are more amenable to large-scale field experiments than marine systems. For this reason, the whole-lake manipulations by Schindler (1975) have made the Experimental Lake Area in Canada a center of synthesis for freshwater lake research. Experimental approaches have more recently also provided the underpinning of the synthetic ideas of trophic cascades, that is, top-down controls, and indirect effects of predation and nutrient supply in freshwater ecosystems (Carpenter et al. 1987). Stream experiments have also shown that nutrient additions can increase fish production in aquatic ecosystems (figure 12-8, Deegan and Peterson 1992).

In marine systems there has been less emphasis on field experiments, in part because manipulations are harder to do in energetic, turbulent waters and at the usually larger spatial scales of marine ecosystems. Nonetheless, manipulative approach led to major advances in understanding nutrient controls in coastal waters (Ryther and Dunstan 1971) and salt marshes (Valiela et al. 1985), and in understanding community structure on rocky shores (Paine 1966; Lubchenco 1978) and on soft sediments (Wiltse et al. 1984). All these experiments involved manipulating a given variable while assessing the effects on a variety of components or processes.

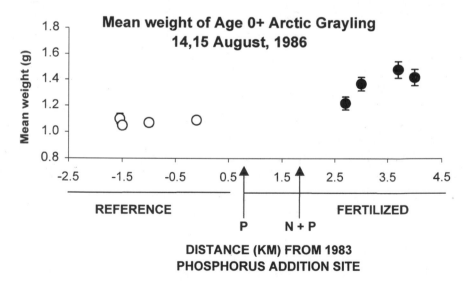

FIGURE 12-8   Field experiments. Mean weight (±SE) of young of the year Arctic grayling after growing for a summer in reaches of a river with (○) and without (●) nutrient fertilization (after Deegan and Peterson 1992).

Marine mesocosms have also been widely used (Grice and Reeve 1982). Two examples are notable. The Controlled Ecosystem Populations Experiment (CEPEX) used reinforced plastic bags enclosing up to 1,300 m³ (Grice et al. 1980). While most results related to plankton dynamics, these enclosures were large enough to support juvenile fish. Pollutant effects on salmon (Koeller and Wallace 1977) and feeding and growth of herring (Houde and Berkeley 1982) were studied experimentally.

The Marine Ecosystems Research Laboratory in Rhode Island (MERL) operates fourteen replicate land-based experimental tanks including 30 cm of sediment and 13 m³ of seawater from Narragansett Bay (Nixon et al. 1980). Experimental studies have examined petroleum hydrocarbon and trace metal impacts and various other manipulations. A 3-year study of eutrophication (Oviatt et al. 1986) explored the consequences of a range of nutrient-loading rates. The MERL experiments clearly demonstrated the links between nutrient supply and primary and total ecosystem production (see unnumbered triangles in figure 12-3B), identified couplings of plankton and benthos, and much more (Oviatt 1994).

Strengths and Weaknesses

There are at least three advantages of manipulative approaches. First and most important, manipulative experiments are the only unambiguous way to obtain information on causal relationships (Platt 1964). This ability to test hypotheses contrasts with comparative studies or natural experiments in which time or space effects are unavoidably confounded with the effects of the variable being studied. Second, in manipulative experiments the treatments can be specified, so that more explicit tests of specific questions are possible. For example, in a natural experiment where seasonal pulses of organic loads are used to test the effect of organic matter addition on a microbial process, there might be unintended effects of nutrients also added to the system. Good design of manipulative treatments may eliminate such problems. Third, manipulative treatments provide information on the response and controls of the processes and populations to a wide range of changed conditions. While the MERL mesocosms are not exact replicas of Narragansett Bay, the manipulations have provided an unequaled window on how processes that occur in the bay are controlled. As such, the MERL results have furnished the elements for synthesis of knowledge of the function and structure of Narragansett Bay.

There are at least two disadvantages to manipulative approaches. First, manipulations may impart artifacts. This is a hoary complaint, voiced by skeptics who from the very beginning of experimental science have challenged experiments as artificial. Experimenters aware of the problem of artifacts can overcome the problem, as is attested by the enormously accelerated progress of manipulative research (Platt 1964).

A second disadvantage is that manipulations often are not true whole-system experiments. Whole lakes do contain a complete ecosystem up to the top predators. But smaller experimental systems inevitably lack parts of the ecosystem, usually the large fish. This lack does not negate many of the insights gained about processes. Given these disadvantages, researchers should never claim that a mesocosm is a replica of a real system but should use mesocosms to gain understanding about ecological mechanisms and processes.

Recommendations

Experiments done at the whole-system scale, such as in the MERL meso-cosms or in the salt-marsh fertilization studies, can make use of integrative dependent variables, for example, total ecosystem production or denitrification, each of which synthesizes information for whole systems (Oviatt et al. 1986; Valiela and Teal 1979). In this sense, the experiment itself is a synthesis.

Combining experiments with other approaches may spur progress in ecological synthesis over the next decade. If we can model the effects of specific components or relationships based on experimental studies, we could then use the understanding in models to provide hypotheses to be tested using comparative approaches at scales larger than allowed by experiments. This would incorporate the advantages characteristic of the different approaches. For example, the best way to develop understanding of the microbial food web would be to model the events in a well-studied mesocosm. Yet this modeling is not possible with our current understanding despite millions of dollars spent in observing microbes in the sea. Clearly, to understand the ecology of microbes a great deal of research effort should be put into an experimental approach.

Utility for Management

Manipulative ecological experiments are uniquely able to test hypotheses bearing on cause and effect. The results include potential artifacts arising from differences in scale, complexity, and manipulation, but they have the notable benefit of utilizing living organisms within a replicated experimental framework. Well-designed experiments have the potential to provide unique and useful information on specific effects and impacts relevant to management.

Conclusions

1. Many approaches facilitate synthesis; to date, only some have been applied effectively to the question of how bottom-up controls affect higher trophic levels, the key topic of this working group.

2. Models are the skeletons on which all ecosystem syntheses are built. From conceptual models used to formulate thinking about empirical comparisons to formal numerical simulation, lessons from modeling are therefore especially relevant:

   a. Models are imperfect; iteration and revision is essential for meaningful results. Oft-proposed yet rarely realized, initiating real quantitative synthesis with various types of models as early as possible can shape development of working concepts and guide change in sampling strategy.

   b. Empirical analyses, such as statistical models and budgets, may arise from or suggest mechanisms but cannot confirm cause and effect. They frequently help pose testable questions.

   c. Mechanistic or process models help to explore presumed causal mechanisms, yet are incomplete and frequently ignore propagation of error resulting from uncertainty in parameters and formulations.

   d. Careful thought and parsimony should go into choosing what processes to include. Especially when approaching an issue as complex as bottom-up control of higher trophic levels, formulating all or even most interactions known to be relevant may not be a productive strategy. While increasing computing power makes very intricate models technically possible, modeling may be better directed at finding the 20% of the processes that explain 80% of the behavior.

3. Field data appropriate for large-scale synthesis require sample frequency and duration that challenge current limits. Attempts to extend these limits for a wide range of biotic and abiotic ecological variables are needed. Yet this raises a critical dilemma: the scales of rate processes differ dramatically from the scales of the long-term and regional responses of interest to meaningful synthesis. It is not clear that the best approach is more intensive sampling of rates and driving variables; assessment of meaningful averages and integrated responses may prove especially useful. More appropriate and accurate data on biomass and production of higher trophic levels is especially needed.

4. The five general approaches to synthesis discussed here each provide clues and aid understanding, yet are best seen as complementary. Effective synthesis at the ecosystem level will certainly result only from creative combination of diverse field and analytical approaches. Concurrent activities of collection of data and of modeling embedded in an adaptive, evolving program are likely to be especially productive.

5. Intelligent management demands scientific understanding of the forces controlling and changing estuarine biotic systems. Synthesis, especially

when it combines insights from different approaches, will provide information appropriate and essential to management.

References

Alpine, A.E., and J.E. Cloern. 1992. Trophic interactions and direct physical effects control phytoplankton biomass and production in an estuary. *Limnology and Oceanography* 37: 946–55.

Armstrong, N. E. 1982. Response of Texas estuaries to freshwater inflows. Pages 103–20 in V. S. Kennedy (editor), *Estuarine comparisons.* New York: Academic Press.

Bahr, L. M. Jr., J. W. Day Jr., and J. H. Stone. 1982. Energy costs-accounting of Louisiana fishery production. *Estuaries* 5: 209–15.

Baird, D., and R. Ulanowicz. 1989. The seasonal dynamics of the Chesapeake Bay ecosystem. *Ecological Monographs* 59: 329–64.

Baird, D., R. Ulanowicz, and W. Boynton. 1995. Seasonal nitrogen dynamics in Chesapeake Bay: A network approach. *Estuarine Coastal Shelf Science* 41: 137–62.

Baretta, J., and P. Ruardij (editors) 1988. *Tidal flat estuaries: Simulation and analysis of the Ems estuary.* New York: Springer-Verlag.

Bartell, S. A. Brenkert, R. O'Neill, and R. Gardner. 1988. Temporal variations in regulation of production in a pelagic food web model. Pages 101–18 in S. Carpenter (editor), *Complex interactions in lake communities.* New York: Springer-Verlag.

Bartleson, R. D., and W. M. Kemp. 1991. Preliminary ecosystem simulations of estuarine plankton-benthic interactions: The planktonic submodel. Pages 243–52 in J. A. Mihursky et al. (editors), *New perspectives in the Chesapeake Bay system.* Solomons, MD: Chesapeake Research Consortium Publications.

Bayley, P. B. 1988. Accounting for effort when comparing tropical fisheries in lakes, river-floodplains, and lagoons. *Limnology and Oceanography* 33: 963–72.

Beverton, R., and S. Holt. 1957. On the dynamics of exploited fish populations. *Fish. Invest. Min. Agr. Food* (Gr. Brit.) Ser. 2. 19.

Box, G. E. P., W. G. Hunter, and J. S. Hunter. 1978. *Statistics for experimenters.* New York: John Wiley & Sons.

Boynton, W. R., W. M. Kemp, J. M. Barnes, J. L. W. Cowan, S. E. Stammerhohn, L. L. Matteson, F. M. Rohland, and M. Marvin. 1991. Long-term characteristics and trends of benthic oxygen and nutrient fluxes in the Maryland portion of Chesapeake Bay. Pages 339–54 in J. A. Mihursky et al. (editors), *New perspectives in the Chesapeake Bay system.* Solomons, MD: Chesapeake Research Consortium Publications.

Boynton, W. R., W. M. Kemp, and C. W. Keefe. 1982. A comparative analysis of nutrients and other factors influencing estuarine phytoplankton production. Pages 209–30 in V. S. Kennedy (editor), *Estuarine comparisons.* New York: Academic Press.

Boynton, W. R., J. H. Garber, R. Summers, and W. M. Kemp. 1995. Inputs, transformations and transport of nitrogen and phosphorus in Chesapeake Bay and selected tributaries. *Estuaries* 18: 285–314.

Brandt, S. B., and K. J. Hartman. 1993. Innovative approaches with bioenergetics models: Future applications to fish ecology and management. *Transactions of the American Fisheries Society* 122: 731–35.

Brawley, J. W., G. Collins, J. N. Kremer, C. Sham, and I. Valiela. In press. A time-dependent model of nitrogen loading to estuaries from coastal watersheds. *Limnology and Oceanography.*

Carpenter, S. R., J. F. Kitchell, J. R. Hodgson, P. A. Cochran, J. J. Elser, M. M. Elser, D. M. Lodge, D. Kretchmer, X. He, and C. N. VonEnde 1987. Regulation of lake primary productivity by food web structure. *Ecology* 68: 1863–76.

Carpenter, S. R., T. M. Frost, J. F. Kitchell, T. K. Kratz, D. W. Schindler, J. Shearer, W. G. Sprules, M. J. Vanni, and A. P. Zimmerman. 1992. Patterns of primary production and herbivory in 25 North American lake ecosystems. Pages 67–96 in J. J. Cole, G. Lovett, S. Findlay (editors), *Comparative analyses of ecosystems: Patterns, mechanisms and theories.* New York: Springer-Verlag.

Carpenter, S. R. 1988. Transmission of variance through lake food webs. Pages 119–35 in S. R. Carpenter (editor), *Complex interactions in lake communities.* New York: Springer-Verlag,

Cerco, C., and T. Cole. 1993. Three-dimensional eutrophication model of Chesapeake Bay. *Journal of Environmental Engineering ASCE* 119: 1006–25.

Christensen, V., and D. Pauly. 1998. Changes in models of aquatic ecosystems approaching carrying capacity. *Ecological Applications* 8 (1) Suppl. S104–S109.

Cole, J. J., G. Lovett, and S. Findley 1992. *Comparative analyses of ecosystems: Patterns, mechanisms and theories.* New York: Springer-Verlag.

Colebrook, J. M. 1979. Continuous plankton records: Seasonal cycles of phytoplankton and copepods in the North Atlantic Ocean and North Sea. *Marine Biology* 51: 23–32.

Collins, G., J. N. Kremer, and I. Valiela. In press. Assessing uncertainty in estimates of nitrogen loading to estuaries for research, planning, and risk assessment. *Environmental Management.*

Cooper, S., and G. Brush. 1991. Long-term history of Chesapeake Bay anoxia. *Science* 254: 992–96.

Currin, B. J., J. P. Reed, and J. M. Miller. 1984. Growth, production, food consumption and mortality of juvenile spot and croaker: A comparison of tidal and nontidal nursery areas. *Estuaries* 7: 451–59.

Davis, C. 1987. Components of the zooplankton production cycle in the temperate ocean. *Journal of Marine Research* 45: 947–83.

Davis, C. S., S. M. Gallager, and A. R. Solow. 1992. Microaggregations of oceanic plankton by towed video microscopy. *Science* 257: 230–32.

DeAngelis, D. L., and L. J. Gross (editors) 1992. *Individual based models and approaches in ecology: Concepts and models.* New York: Routledge, Chapman and Hall.

DeAngelis, D. L., L. J. Gross, M. A. Huston, W. F. Wolff, D. M. Fleming, E. J. Comiskey, and S. M. Sylvester. 1998. Landscape modeling for Everglades ecosystem restoration. *Ecosystems* 1: 64–75.

Deegan, L. A. 1993. Nutrient and energy transport between estuaries and coastal marine ecosystems by fish migration. *Canadian Journal of Fisheries and Aquatic Sciences* 50: 74–79.

Deegan, L. A., and B. J. Peterson. 1992. Whole-river fertilization stimulates fish production in an Arctic tundra river. *Canadian Journal of Fisheries and Aquatic Sciences* 49: 1890-1901.

Deegan, L. A., J. W. Day, J. G. Gosselink, A. Yanez-Arancibia, G. S. Chavez, and P. Sanchez-Gil. 1986. Relationships among physical characteristics, vegetation distribution and fisheries yield in Gulf of Mexico estuaries. Pages 83–100 in V. Kennedy (editor) *Estuarine variability.* New York: Academic Press.

Deegan, L., J. Finn, C. Hopkinson, A. Giblin, B. Petersen, B. Fry, and J. Hobbie. 1994. Flow model analysis of the effects of organic matter-nutrient interactions on estuarine trophic dynamics. Pages 273–81 in K. Dyer and R. Orth (editors), *Change in fluxes in estuaries.* Fredensborg, Denmark: Olsen and Olsen.

Di Toro, D., P. Paquin, K. Suburamu, and D. Gruber. 1990. Sediment oxygen demand model: Methane and ammonia oxidation. *Journal of Environmental Engineering ASCE* 116: 945–86.

Dillon, P. J., and F. H. Rigler. 1975. A simple model for predicting the capacity of a lake for development based upon lake trophic status. *Journal of the Fisheries Research Board of Canada* 32: 1519–31.

Downing, J. A. 1992. Comparing apples with oranges: Methods of inter-ecosystem comparison. Pages 24–45 in J. J. Cole, G. Lovett, and S. Findlay (editors) *Comparative analyses of ecosystems: Patterns, mechanisms, and theories.* New York: Springer-Verlag.

Durbin, A. G., S. W. Nixon, and C. A. Oviatt. 1979. Effects of the spawning migration of the alewife, *Alosa pseudoharangus,* on freshwater ecosystems. *Ecology* 60: 8–17.

Eadie, B. B. McKee, M. Lansing, J. Robbins, S. Metz, and J. Trefry. 1994. Records of nutrient-enhanced coastal ocean productivity in sediments from Louisiana continental shelf. *Estuaries* 17: 754–65.

Edmondson, T., and J. Lehman. 1981. The effect of changes in the nutrient income on the condition of Lake Washington. *Limnology and Oceanography* 26: 1–29.

Fasham, M. 1985. Flow analysis of materials in the marine euphotic zone. Pages 139–62 in R. Ulanowicz and T. Platt (editors), *Ecosystem theory for biological oceanography.* Ottawa, Canada: Canadian Bulletin of Fisheries and Aquatic Sciences.

Fleming, D. M., D. L. DeAngelis, L. J. Gross, R. E. Ulanowicz, W. F. Wolff, W. F. Loftus, and M. A. Huston. 1994. ATLSS: *Across-Trophic-Level System Simulation for the freshwater wetlands of Everglades and Big Cypress Swamp.* Draft Workshop Report, Everglades National Park. Homestead, FL: National Biological Survey.

Fogarty, M. J., and S. A. Murawski. 1998. Large-scale disturbance and the structure of marine systems: Fishery impacts on Georges Bank. *Ecological Applications* 8: S6–S22.

Glover, R. S., G. Robinson, and J. Colebrook. 1972. Marine biological surveillance. *Environ. and Change* 2: 395.

Grice, G. D., and M. R. Reeve (editors) 1982. *Marine mesocosms: Biological and chemical research in experimental ecosystems.* New York: Springer-Verlag.

Grice, G. D., R. P. Harris, M. R. Reeve, J. F. Heinbokel, and C. O. Davis. 1980. Large scale enclosed water column ecosystem. An overview of Foodweb I, the final CEPEX experiment. *Journal of the Marine Biological Association of the United Kingdom* 60: 401–14.

Hall, C. A. S. 1972. Migration and metabolism in a temperate stream ecosystem. *Ecology* 53: 586–604.

Hall, D. J., W.E. Cooper, and E. E. Werner. 1970. An experimental approach to the production dynamics and structure of freshwater animal communities. *Limnology and Oceanography* 15: 839–928.

Hannon, B. 1973. The structure of ecosystems. *Journal of Theoretical Biology* 41:535–46.

Hanson, J. M., and W. C. Leggett. 1982. Empirical prediction of fish biomass and yield. *Canadian Journal of Fisheries and Aquatic Sciences* 39: 257–63.

Hanson, J. M., and R. H. Peters. 1984. Empirical predictions of crustacean zooplankton biomass and profundal macrobenthos biomass in lakes. *Canadian Journal of Fisheries and Aquatic Sciences* 41: 439–45.

Heal, O. W., and J. P. Grime. 1992. Comparative analysis of ecosystems: Past lessons and future directions. Pages 7–23 in J. J. Cole, G. Lovett, and S. Findlay (editors), *Comparative analyses of ecosystems: Patterns, mechanisms and theories.* New York: Springer-Verlag.

Hofmann, E. E., and T. M. Powell. 1998. Environmental variability effects on marine fisheries: Four case histories. *Ecological Applications* 8: S23–S32.

Hopkinson, C. S., R. D. Fallon, B. O. Jansson, and J. P. Schubauer. 1991. Community metabolism and nutrient cycling at Gray's reef, a hard bottom habitat in the Georgia Bight. *Marine Ecology Progress Series* 73: 105–20.

Hopkinson, C. S., I. Buffam, J. E. Hobbie, J. Vallino, M. Perdue, B. Eversmeyer, F. Prahl, J. Covert, R. Hodson, M. A. Moran, E. Smith, J. Baross, B. Crump, S. Findlay, and K. Foreman. 1998. Terrestrial inputs of organic matter to coastal ecosystems: An intercomparison of chemical characteristics and bioavailability. *Biogeochemistry* 43: 211–34.

Horne, J. K., and D. C. Schneider. 1997. Spatial variance of mobile aquatic organisms: Capelin and cod in Newfoundland coastal waters. *Philosophical Transactions of the Royal Society of London on Biological Sciences* 352: 633–42.

Horwood, J. W., and D. H. Cushing. 1978. Spatial distribution and ecology of pelagic fish. Pages 355–83 in J. H. Steele (editor), *Spatial pattern in plankton communities.* New York: Plenum Press.

Houde, E. D., and E. Rutherford. 1993. Recent trends in estuarine fisheries: Predictions of fish production and yield. *Estuaries.* 16: 161–76.

Houde, E. D., and S. A. Berkeley. 1982. Food and growth of juvenile herring, *Clupea harengus pallasi,* in CEPEX enclosures. Pages 239–49 in G. D. Grice and M. R. Reeve (editors), *Marine mesocosms: Biological and chemical research in experimental ecosystems.* New York: Springer-Verlag.

Hurtt, G., and R. Armstrong. 1996. A pelagic ecosystem model calibrated with BATS data. *Deep-Sea Research* 2 (43): 653–81.

Hvidsten, N. A., and L. P. Hansen. 1988. Increased recapture rate of adult salmon *Salmo salor* L. stocked as smolts at high water discharge. *Journal of Fish Biology* 32: 153–55.

HydroQual Inc. 1995. *A water quality model for Massachusetts and Cape Cod Bays: Calibration of the bays eutrophication model (BEM)*. Mahwah, NJ: HydroQual Inc.

———. *Water quality modeling analysis of hypoxia in Long Island Sound using LIS 3.0.* Mahwah, NJ: HydroQual Inc.

———. 1997. *Evaluation of nutrient management scenarios using LISS 3.0.* Mahwah, NJ: HydroQual Inc.

Johnson, B., K. Kim, R. Heath, B. Hsieh, and L. Butler. 1993. Validation of a three-dimensional hydrodynamic model of Chesapeake Bay. *Journal of Hydraulic Engineering ASCE* 199: 2–20.

Kemp, W. M., and R. D. Bartleson. 1991. Preliminary ecosystem simulations of estuarine plankton-benthic interactions: The benthic submodel. Pages 253–64 in J. A. Mihursky et al. (editors), *New perspectives in the Chesapeake Bay system.* Solomons, MD: Chesapeake Research Consortium Publications.

Kemp, W. M., and W. R. Boynton. 1992. Benthic-pelagic interactions: Nutrient and oxygen dynamics. Pages 149–221 in D. Smith, M. Leffler, and G. Mackiernan (editors), *Oxygen dynamics in Chesapeake Bay: A synthesis of research.* College Park, MD: Maryland Sea Grant College.

Kemp, W. M., W. R. Boynton, A. J. Hermann. 1995. Simulation models of an estuarine macrophyte ecosystem. Pages 262–78 in B. Patten and S. E. Jørgensen (editors), *Complex ecology.* Englewood Cliffs, NJ: Prentice Hall.

Kitchell, J., D. Stewart, and D. Weininger. 1977. Applications of a bioenergetics model to yellow perch (*Perca flavescens*) and walleye (*Stizostedion vitreum vitreum*). *Journal of the Fisheries Research Board of Canada* 34: 1922–35.

Kjerfve, D., L. H. Stevenson, J. A. Proehl, T. H. Chrzanowski, and W. M. Kitchesn. 1981. Estimation of material fluxes in an estuarine cross section: A critical analysis of spatial measurement density and errors. *Limnology and Oceanography* 26: 325–35.

Kline, T. C., J. J. Goering, O. A. Mathisen, P. H. Pie, and P. L. Parker. 1990. Recycling of elements transported upstream by runs of Pacific Salmon: 1. Del 15N and del 13C evidence in Sashin Creek, southeastern Alaska. *Canadian Journal of Fisheries and Aquatic Sciences* 47: 136–44.

Koeller, P., and G. T. Wallace. 1977. Controlled ecosystem pollution experiment: Effect of mercury on enclosed water columns. V. Growth of juvenile chum salmon (*Oncorhynchus keta*). *Mar. Sci. Comm.* 3: 395–406.

Korfmacher, K. S. 1998. Water quality modeling for environmental management: Lessons from the policy sciences. *Policy Sciences* 31: 35–54.

Kremer, J. N. 1999. *Quantifying critical responses to nutrient loading in the Waquoit Bay estuarine ecosystem.* Final report on an Ecological Risk Assessment of Waquoit Bay (Agreement X825675-01-0). Submitted to U.S. EPA, Office of Water. 25pp.

Kremer, J. N., and P. Kremer. 1982. A three trophic level estuarine model: Synergism of two mechanistic simulations. *Ecological Modelling* 15: 145–57.

Kremer, J. N., and S. W. Nixon. 1978. *A coastal marine ecosystem—simulation and analysis.* Ecological Studies. Vol. 24. Heidelberg, Germany: Springer-Verlag.

Lawson, L., E. Hofmann, and Y. Spitz. 1996. Time series sampling and data assimilation in a simple marine ecosystem model. *Deep-Sea Research 2* (43): 625–51.

Lehman, J. T. 1986. The goal of understanding in limnology. *Limnology and Oceanography* 31: 1160–66.

Likens, G. E. 1992. *The ecosystem approach: Its use and abuse.* Oldendorf, Germany: Ecology Institute Publications.

Livingston, R. J., and J. L. Duncan. 1979. Climatological control of a north Florida coastal system and impact due to upland forestry management. Pages 339–81 in R. J. Livingston (editor), *Ecological processes in coastal and marine systems.* New York: Plenum Press.

LMER. 1993. *Report of All Scientists' Meeting of Land Margin Ecosystems Research, Seaside, Oregon, 16–20 October.* Woods Hole, MA: Marine Biological Laboratory, LMER Coordinating Office, Ecosystems Center.

Lou, J., and S. B. Brandt. 1993. Bay anchovy, *Anchoa mitchilli,* production and consumption in mid-Chesapeake Bay based on a bioenergetics model and acoustic measures of fish abundance. *Marine Ecology Progress Series* 98: 223–36.

Lubchenco, J. 1978. Plant species diversity in a marine intertidal community: Importance of herbivore food preference and algal competitive ability. *American Naturalist* 112: 23–39.

Mackas, D., and C. Boyd. 1979. Spectral analysis of zooplankton spatial heterogeneity. *Science* 204: 62–64.

Mann, K. H., and P. A. Breen. 1972. The relation between lobster abundance, sea urchins and kelp beds. *Journal of the Fisheries Research Board of Canada* 29: 603–05.

Melack, J. M. 1976. Primary productivity and fish yields in tropical lakes. *Transactions of the American Fisheries Society* 105: 575–80.

Menge, B. A. 1976. Organization of the New England rocky intertidal community: The role of predation, competition, and environmental heterogenity. *Ecological Monographs* 42: 355–93.

———. 1992. Community regulation: Under what conditions are bottom-up factors important on rocky shores. *Ecology* 73: 755–65.

Moore, M. J., D. Shea, R. Hillman, and J. J. Stegeman. 1996 Trends in hepatic tumors and hydropic vacuolation, fin erosion, organic chemicals and stable isotope ratios in winter flounder from Massachusetts, USA. *Marine Pollution Bulletin* 32: 458–70.

Nixon, S. W. 1980. Between coastal marshes and coastal waters—A review of twenty years of speculation and research on the role of salt marshes in estuarine productivity and water chemistry. Pages 437–545 in P. Hamilton and K. B. McDonald (editors), *Estuarine and wetland processes.* New York: Plenum Press.

———. 1981. Remineralization and nutrient cycling in coastal marine ecosystems. Pages 111-38 in B. J. Neilson and L. E. Cronin (editors), *Estuaries and nutrients.* Clifton, NJ: Humana Press.

———. 1982. Nutrient dynamics, primary production, and fisheries yields of lagoons. *Oceanologica Acta* Special edition: Proceedings, International Symposium on Coastal Lagoons. Pages 357–71.

———. 1992. Quantifying the relationship between nitrogen input and the productivity of marine ecosystems. *Proc. Adv. Mar. Tech. Conf.* 5: 57–83.

Nixon, S. W., and M. E. Q. Pilson. 1983. Nitrogen in estuarine and coastal marine ecosystems. Pages 565–648 in E. J. Carpenter, and D. G. Capone (editors), *Nitrogen in the marine environment*. New York: Academic Press.

Nixon, S. W., D. Alonso, M. E. Q. Pilson and B. A. Buckley. 1980. Turbulent mixing in aquatic microcosms. Pages 881–49 in J. P. Giesy Jr. (editor), *Microcosms in ecological research*. DOE Symposium. Series 52, CONF–781101. Springfield, VA: NTIS.

Nixon, S. W., S. L. Granger, and B. L. Nowicki. 1995. An assessment of the annual mass balance of carbon, nitrogen, and phosphorus in Narragansett Bay. *Biogeochemistry.* 31: 15–61.

Nixon, S., J. W. Ammerman, L. P. Atkinson, V. M. Berounsky, G. Billen, W. C. Boicourt, W. R. Boynton, T. M. Church, D. M. DiToro, R. Elmgren, J. H. Garber, A. E. Giblin, R. A. Jahnke, N. J. P. Owens, M. E. Q. Pilson, and S. P. Seitzinger. 1996. The fate of nitrogen and phosphorus at the land-sea margin of the North Atlantic Ocean. *Biogeochemistry* 35: 141–80.

Oglesby, R. T. 1977. Relationships of fish yields to lake phytoplankton standing crop, production, and morphoedaphic factors. *Journal of the Fisheries Research Board of Canada* 34: 2271–79.

Oviatt, C. A. 1994. Biological considerations in marine enclosure experiments: Challenges and revelations. *Oceanography* 7: 45–51.

Oviatt, C. A., A. A. Keller, P. A. Sampou, and L. L. Beatty. 1986. Patterns of productivity during eutrophication: A mesocosm experiment. *Marine Ecology Progress Series* 28: 69–80.

Pace, M. L. 1986. An empirical analysis of zooplankton community size structure across lake trophic gradients. *Limnology and Oceanography* 31: 45–55.

———. 1992. Concluding remarks. Pages 24–45 in J. J. Cole, G. Lovett, and S. Findlay (editors), *Comparative analyses of ecosystems: Patterns, mechanisms, and theories*. New York: Springer-Verlag.

Paine, R. T. 1966. Food web complexity and species diversity. *American Naturalist* 100: 65–75.

Peters, R. H. 1991. *A critique for ecology.* Cambridge, England: Cambridge University Press.

———. 1988. The role of prediction in limnology. *Limnology and Oceanography* 31: 1143–59.

Peters, R. H., J. J. Armesto, B. Boeken, J. J. Cole, C. T. Driscoll, C. M. Duarte, T. M. Frost, J. P. Grime, J. Kolasa, E. Prepas, and W. G. Sprules. 1992. On the relevance of comparative ecology to the larger field of ecology. Pages 46–63 in J. J. Cole, G. Lovett, and S. Findlay (editors), *Comparative analyses of ecosystems: Patterns, mechanisms, and theories*. New York: Springer-Verlag.

Picket, W. 1989. Space-for-time substitution as an alternative to long-term studies. Pages 110–35 in G. E. Likens (editor), *Long-term studies in ecology: Approaches and alternatives.* New York: Springer-Verlag.

Platt, J. R. 1964. Strong inference. *Science* 146: 347–53.

Platt, T., and K. Denman. 1975. Spectral analysis in ecology. *Annual Review of Ecology and Systematics* 6: 189–210.

Powell, E. J. Klinck, E. Hofmann, and S. Ford. 1997. Varying the timing of oyster transplant: Implications for management from simulation studies. *Fish. Oceanogr.* 6: 213–37.

Raymont, J. E. G. 1950. A fish cultivation experiment in an arm of a sea loch. 4. The bottom fauna of Kyle Scotnich. *Proceedings of the Royal Society of Edinburgh Section B* 64 (B): 65–105.

Richey, J. E., M. A. Perkins, and C. R. Goldman. 1975. Effects of kokanee salmon (*Oncorhynchus nerka*) decomposition on the ecology of a subalpine stream. *Journal of the Fisheries Research Board of Canada* 32: 817–20.

Rose, G. A., and W. C. Leggett. 1990. The importance of scale to predator-prey spatial correlation: An example of Atlantic fishes. *Ecology* 71: 33–43.

Ross, A., W. Gurney, and M. Heath. 1994. A comparative study of the ecosystem dynamics of four fjords. *Limnology and Oceanography* 39: 318–43.

Ryder, R. A. 1965. A method for estimating the potential fish production of north-temperate lakes. *Transactions of the American Fisheries Society* 94: 214–18.

———. 1982. The morphoedaphic index—use, abuse, and fundamental concepts. *Transactions of the American Fisheries Society* 111: 154–64.

Ryder, R. A., S. R. Kerr, K. H. Loftus, and H. A. Regier. 1974. The morphoedaphic index, a fish yield estimator—review and evaluation. *Journal of the Fisheries Research Board of Canada* 31: 663–88.

Ryther, J. H., and W. M. Dunstan. 1971. Nitrogen, phosphorus and eutrophication in the coastal marine environment. *Science* 171: 1008–13.

Scarnecchia, D. L. 1981. Effects of streamflow and upwelling on yield of wild coho Salmon (*Oncorhynchus kisutch*) in Oregon. *Canadian Journal of Fisheries and Aquatic Sciences* 38: 471–75.

Scavia, D., G. Lang, and J. Kitchell. 1988. Dynamics of Lake Michigan plankton: A model evaluation of nutrient loading, competition and predation. *Canadian Journal of Fisheries and Aquatic Sciences* 45: 165–77.

Schaefer, M. B. 1954. Some aspects of the dynamics of populations important to the management of the commercial marine fisheries. *Bulletin of the Inter-American Tropical Tuna Commission* 1: 25–56.

Schindler, D. W. 1975. Whole lake eutrophication experiments with phosphorus, nitrogen and carbon. *Internationale Vereinigung fuer Theoretische und Angewandte Limnologie Verhandlungen* 19: 3221–31.

———. 1978. Factors regulating phytoplankton production and standing crop in the world's freshwaters. *Limnology and Oceanography* 23: 478–86.

Schlesinger, D. A, and H. A. Regier. 1982. Climatic and morphoedaphic indices of fish yields from natural lakes. *Transactions of the American Fisheries Society* 111: 141–50.

Shapiro, J. 1980. The importance of trophic-level interactions to the abundances and species composition of algae in lakes. Pages 105–15 in J. Barica and L. Mur (editors), *Hypertrophic ecosystems*. The Hague, The Netherlands: Dr. W. Junk Publishers.

Smith, S. V. 1981. Responses of Kaneohe Bay, Hawaii, to relaxation of sewage stress. Pages 391–410 in B. J. Neilson and L. E. Cronin (editors), *Estuaries and nutrients*. Clifton, NJ: Humana Press.

Steele, J. H., and E. W. Henderson. 1984. Modeling long-term fluctuations in fish stocks. *Science* 224: 985–87.

Sullivan, B. K., P. H. Doering, C. A. Oviatt, A. A. Keller, and J. B. Frithsen. 1991. Interactions with the benthos alter pelagic food web structure in coastal waters. *Canadian Journal of Fisheries and Aquatic Sciences* 48: 2276–84.

Sutcliffe, W. H. 1972. Some relations of land drainage, nutrients, particulate material and fish catch in two eastern Canadian bays. *Journal of the Fisheries Research Board of Canada* 29: 357–62.

Turner, R. E. 1977. Intertidal vegetation and commercial yields of Penaeid shrimp. *Transactions of the American Fisheries Society* 106: 411–16.

Ulanowicz, R. 1983. Identifying the structure of cycling in ecosystems. *Mathematical Biosciences* 65: 219–37.

Ulanowicz, R. E. and F. Wulff. 1991. Comparing ecosystem structures: The Chesapeake Bay and the Baltic Sea. Pages 140–66 in J. Cole, G. Lovett, and S. Findlay (editors), *Comparative analyses of ecosystems*. New York: Springer-Verlag.

Ulanowicz, R. E. and J. Tuttle. 1992. The trophic consequences of oyster stock rehabilitation in Chesapeake Bay. *Estuaries* 15: 298–306.

Valiela, I. 1983. Nitrogen in slat marsh ecosystems. Pages 649–78 in E. J. Carpenter and D. G. Capone (editors), *Nitrogen in the marine environment*. New York: Academic Press.

———. 1989. Conditions and motivations for long-term research: Studies on salt marshes and elsewhere. Pages 158–71 in G. E. Likens (editor), *Long-term studies in ecology: Approaches and alternatives*. New York: Springer-Verlag.

Valiela, I., and J. M. Teal. 1979. The nitrogen budget of a salt marsh ecosystem. *Nature* 280: 652–56.

Valiela, I., G. Collins, J. Kremer, K. Lajtha, M. Geist, B. Seely, J. Brawley, and C. Sham. 1997. Nitrogen loading from coastal watersheds to receiving estuaries: New method and application. *Ecological Applications* 7(2): 358–80.

Valiela, I., J. M. Teal, C. Cogswell, J. Hartman, S. Allen, R. Van Etten, and D. Goehringer. 1985. Some long-term consequences of sewage contamination in salt marsh ecosystems. Pages 301–16 in P. J. Godfrey, E. R. Kaynor, S. Pelczarski, and J. Benforada (editors), *Ecological considerations in wetlands treatment of municipal waste waters*. New York: Van Nostrand Reinhold Co.

Van Winkle, W., B. Kirk, and B. Rust. 1979. Periodicities in Atlantic coast striped bass (*Morone saxatilis*) commercial fisheries data. *Journal of the Fisheries Research Board of Canada* 36: 54–62.

Vollenweider, R. A. 1968. *The scientific basis of lake and stream eutrophication, with particular reference to phosphorous and nitrogen as eutrophication factors.* Technical Report OECD. Paris DAS/DSI/68. 27: 1–182.

Vollenweider, R. A., and P. J. Dillon. 1974. *The application of phosphorus loading concept to eutrophication research.* N.R.C. Technical Report 13690.

Weber, L. H., S. El-Syed, and I. Hampton. 1986. The variance spectra of phytoplankton, krill, and water temperature in the Antarctic Ocean south of Australia. *Deep-Sea Research* 33: 1327–43.

Weinstein, M. P., S. L. Weiss, and M. F. Walters. 1980. Multiple determinant of community structure in shallow marsh habitats, Cape Fear River Estuary, North Carolina, USA. *Marine Biology* 58: 227–43.

Williams, R. 1984. An overview of secondary production in pelagic ecosystems. Pages 361–405 in M. Fasham (editor), *Flows of energy and materials in marine ecosystems.* New York: Plenum Press.

Wiltse, W. I., K. Foreman, J. M. Teal, and I. Valiela. 1984. Importance of predators and food resources to the macrobenthos of salt marsh creeks. *Journal of Marine Research* 42: 923–42.

Woodwell, G. M., R. A. Houghton, C. A. S. Hall, D. E. Whitney, R. A. Moll, and D. W. Juers. 1977. The Flax Pond ecosystem study: The annual metabolism and nutrient budgets of a salt marsh. Pages 491–511 in R. L. Jeffries and A. J Davy (editors), *Ecological processes in coastal environments.* Oxford, England: Blackwell Scientific Publications Co.

Wroblewski, J., and E. Hofmann. 1989. U.S. interdisciplinary modeling studies of coastal-offshore exchange processes: Past and future. *Progress in Oceanography* 23: 65–99.

Yanez-Arancibia, A. (editor) 1985. *Fish community ecology in estuaries and coastal lagoons: Towards an ecosystem integration.* Mexico City, Mexico: DR (R) UNAM Press.

PART FOUR

Controls of Estuarine Habitats

PART IV DEALS WITH CONTROLS on the distribution and abundance of organisms in the estuary. This topic is key to many important resource management questions. Controls range from salinity and temperature to the abundance of food or conditions during a crucial stage of reproduction and early growth of organisms. Participants in the workshop reported in this part reviewed the state of current knowledge in this area and concluded that good management necessitates an extension of the concept of habitat to include the full range of environments that organisms encounter throughout their entire lives.

Compared with terrestrial environments, estuaries are extremely changeable and dynamic. For this reason, the habitat of an organism is not a single location or set of conditions. Although no one model incorporates all estuarine environments, it is possible to simulate some crucial features. For example, the dynamics of sediment deposition explain the distribution of rooted aquatic plants. The productivity and survival of algae and seagrasses in Chesapeake Bay, simulated through the use of coupled physical-biological models, are linked to the movement of nutrients throughout the entire bay. Remote sensing can be used to develop a picture of the distribution and abundance of small fish across transects in the estuary; this information allows use of a bioenergetics model to identify the habitats where predatory fish can survive and grow.

In chapter 13, Reed takes up the question of the factors that control the establishment of marsh vegetation and marshes. This is critical information for managers concerned with reestablishing marshes in disturbed areas or preserving marshes when sea level rises. Although there are fundamental processes affecting vegetation and marshes worldwide, local variations in the magnitude and scale of these processes are critical for successful management. For example, it is well known that marsh vegetation baffles flow and enhances sediment deposition. Recent work has shown that different species of plants cause different intensities of turbulence as the canopies flood.

Cerco describes a large model with physical and biological processes and illustrates one of its successes in chapter 14. It has been applied to environments as diverse as Massachusetts Bay, Chesapeake Bay, and Florida Bay. This approach is particularly useful for large systems where water slowly moves seaward and components are changed by estuarine processes acting over time. An example is the change in oxygen concentrations at various sites. The modeling simulation used calculations of circulation, availability of nutrients, temperature effects on respiration, and the formation and sinking of an algal bloom. This type of model is a valuable tool for synthesizing our knowledge of estuarine processes and applications. Such simulations have worked best so far for modeling physical factors and the processes and populations of the primary producers, the organisms at the lowest level of the food web.

The mechanistic or process-based modeling illustrated in chapter 14 is a good way to predict the impact of future events on estuarine ecology. Such a predictive effort would occur after the simulation of ecological processes is satisfactorily tested against field observations. Then model simulations can be made with different inputs, such as the changes in hydrology that would occur if atmospheric carbon dioxide concentrations and temperatures were to increase.

In chapter 15, Demers et al. present another approach to modeling the habitat of large fish in estuaries. In this approach, detailed data from field measurements of the distribution of fish and their prey are combined with a model of the growth of fish at various temperatures. Unfortunately this sort of detailed collection of data is hard to do routinely, despite its value for our understanding of the ecology of large fish. Not surprisingly, the investigators found that large fish can only exist where their prey is concentrated. The average abundance of prey found along a transect will not support large predaceous fish. These data show just how narrow a habitat window the large fish occupy.

Given the current state of habitat knowledge, it is not surprising that the participants in the workshop reported in chapter 16 called for more data collection and modeling. The data collection should take advantage of the rapidly evolving technology of instruments that make continuous measurements at anchored buoys. As for modeling, a conceptual model is needed that classifies estuarine habitats. Also needed are well-designed comparative studies that develop and make use of indicators of habitat integrity, dynamics, and variability.

CHAPTER 13

Coastal Biogeomorphology
An Integrated Approach to Understanding the Evolution, Morphology, and Sustainability of Temperate Coastal Marshes

Denise J. Reed

Abstract

Coastal biogeomorphology considers the role of plants and animals in geomorphic processes. Biogeomorphic interactions between vegetation and sediment depositional processes are important precursors to salt-marsh initiation. Where initial vegetation colonizers show vigorous growth, a more rapid transition from intertidal to emergent marsh can occur and the processes are common across latitudes and continents. Vegetative binding of creek-bank sediments plays an important role in the equilibrium balance between erosive flows and channel-bank resistance noted in most mature coastal marshes. This recognition is crucial to the planning of functional creek systems for restored coastal wetland systems. The role of vegetation in marsh elevation change can be both a direct contribution through the accumulation of organic material within the marsh soil and an indirect contribution via its role in baffling flows across the marsh surface, thus enhancing sediment deposition. Although several models of marsh vertical development acknowledge the importance of vegetative contributions, most do not explicitly incorporate the biogeomorphic aspect of system response to sea-level rise.

Introduction

The concept that interactions between biologic and geomorphic processes can be determinants of landscape form and process, especially at the micro or meso scale, is not a recent development. Viles (1988) draws attention to observations by several nineteenth-century naturalists, including Lyell, that both plants and animals can be responsible for geomorphic changes. The presence of biota in some form in all earthly environments, from algae and ice worms in ice sheet cryoconite holes (McIntyre 1984) to boring sea urchins on exposed limestone coasts (Trudgill et al. 1987), means that consideration of natural landscape form and change at least implicitly includes a biotic component. Indeed, it is frequently the difficulty of reproducing biological influences that confounds the application of laboratory experimentation to the understanding of real landscape processes.

Biogeomorphology calls for a more explicit consideration of the role of plants and animals in geomorphic processes. Viles (1988) identifies two foci for biogeomorphological studies: (1) the influence of landforms and geomorphology on the distribution and development of plants, animals, and microorganisms; and (2) the influence of plants, animals, and microorganisms on earth surface processes and landform development. Both of these interactions are present in most circumstances and distinguishing between the two to identify controlling factors is difficult in most circumstances. One of the most striking examples of biogeomorphology in action is the development of coral reefs, which requires a specific combination of physical and biological processes. Reefs are essentially physical, shallow-water constructions, whose growth is controlled primarily by light and exposure. For this reason, they can be threatened by sea-level rise and understanding their future fate or geologic development requires a biogeomorphic approach. The development of subaerial islands on atolls has been described by Roy and Connell (1991) as a result of vertical coral growth into a shallow zone where the substrate is susceptible to reworking by storms. Many low-lying atoll states, such as the Maldive Islands, are threatened by future sea-level rise scenarios. Assessing the nature and magnitude of the threat requires a biogeomorphological assessment of atoll and community response to sea-level rise. This call for a new emphasis is reiterated by IGBP-LOICZ, which has coastal biogeomorphology and sea-level rise as one of its foci (Pernetta and Milliman 1995).

One of the difficulties of developing the biogeomorphological approach to any landscape category is the diversity of the plant and animal life, which must be considered. On a site-by-site basis, significant understanding of interactions can be developed. However, whether the principles derived on such a basis can be more widely applied relies on a more synthetic consideration of landscapes across environmental gradients. This chapter will examine the potential role of biogeomorphology in understanding the sustainability of temperate coastal marshes in the face of sea-level rise. This will be achieved by synthesizing the findings of existing studies of marsh development and sea-level rise from a variety of physical settings with different vegetative communities. Three geomorphological aspects of coastal marshes will be examined: formation and evolution, channel network development, and marsh surface elevation change. Although the role of microorganisms and fauna in intertidal and marsh environments has been widely recognized (see, for example, Coles

1979; Kinler et al. 1987), this discussion will focus on interactions between geomorphic processes and vegetative communities. The objectives of the chapter are to develop a synthetic understanding of coastal marsh response to sea-level rise, to indicate where variability in system response can be explained by local biotic or physical factors, and to identify the value of the biogeomorphic approach to effective sustainable management of coastal marsh systems.

Evolution of Coastal Marshes

The transition from intertidal flat to coastal marsh is marked by colonization by emergent macrophytes, such as *Salicornia* spp. (Pethick 1980) or *Spartina* spp. (Pestrong 1965). Steers (1960) described the accumulation of fine sediments in sheltered coastal areas as being a necessary precursor for marsh initiation and observed that fine sediments were more likely to accumulate around the inner edge of the flat. Steers also specifically mentioned the likely presence of patches of algae, such as *Enteromorpha* spp., and their role as foci for sediment accumulation. However, these patches of algae can shift with storms, allowing redistribution of any accumulated sediments. Steers (1948) recognized that the critical phase of intertidal flat stabilization occurs with the spread of emergent vegetation, representing a more effective trap for suspended sediment than surficial microalgae or potentially mobile macroalgae. Where annual *Salicornia* spp. are the colonizers, as is frequently observed in northwestern European marshes (Steers 1948; Chapman 1960; Pethick 1980), enhanced trapping of sediments may be limited, both because of dieback during the winter (frequently a period of increased suspended sediment availability) and because of the susceptibility of the plant morphology to disruption by strong waves and tidal currents (Wiehe 1935). Consequently, the initial buildup of the emergent marsh surface is slow. French (1993) shows that in the early stages of infilling of the intertidal profile, a thin veneer of fine sediments accumulates over the preexisting intertidal surface (figure 13-1). The approximate time taken for marsh initiation in North Norfolk has been inferred by Pethick (1980) using historical maps. For several marshes on Scolt Head Island, the transition from intertidal flat with no creek or marsh boundary shown on the map, to a marsh shown as a vegetated surface with defined creeks, occurs in less than 100 years.

This model of biogeomorphic interactions between vegetation and sediment depositional processes leading to marsh initiation has been largely based on the marshes of north Norfolk, England. Where initial vegetation colonizers show more vigorous growth, a more rapid transition from intertidal to emergent marsh can occur. Pestrong (1965) observed *Spartina* spp. colonizing mudflats in San Francisco Bay and described luxuriant growth of dense stands. The efficacy of some *Spartina* species in trapping suspended sediments has

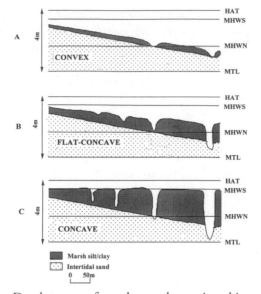

FIGURE 13-1   Development of marsh morphostratigraphic units in relation to tide levels with stages A, B, and C representing progressive infilling of intertidal unit and deepening of marsh creeks (after French 1993). (HAT = highest astronomical tide, MHWS = mean high-water springs, MHWN = mean high-water neaps, MTL = mean tide level).

been demonstrated in many areas, especially where invasive species have quickly covered large intertidal areas and increased elevations such as in Willapa Bay, Washington (Sayce 1991) and coastal China (Chung 1994). Marsh accretion measurements in colonies of *Spartina alterniflora* in Sawlog Slough in Willapa Bay are significantly higher (4.5 cm/yr) than in adjacent areas of *Triglochin maritimum*, the native low-marsh vegetation (1 cm/yr).

Reed (1989) has observed initial marsh formation in hollows between recently exposed moraine ridges close to a tidewater glacier in southern Chile. In this environment, a low rate of supply of suspended sediment slows vegetative growth and physical disruption of rafting icebergs limits vertical development. As a result of these physical factors, the transition rate from intertidal to emergent marsh would be low compared to estuarine marshes within the same coastal region (Reed 1989). These differences between colonizing species merely represent variations in the rate of transition from intertidal to emergent marsh. The essential role of emergent vegetation in the initial development of the marsh surface is common across latitudes and continents.

These examples are primarily instances of vegetative influence on landform change and physical processes (such as sediment accumulation). However, Orson and Howes (1992) have identified the other extreme of biogeomorphic interaction in their comparison of salt-marsh development in various coastal environments of Massachusetts, where geomorphic setting influences vegetation development. In salt marshes where connections to adjacent water are restricted by landforms such as tidal inlets, vegetation development is dominated by abiotic factors (for instance, storms, inlet dynamics). Systems like these, which are naturally isolated, will be strongly altered by an episodic event. However, marshes linked directly and efficiently to adjacent bodies of open water are exposed to hydrodynamic changes; plants are adjusted to changes. Thus, the same event that causes significant change in the isolated system represents less of a perturbation to the more dynamic system. The episodic nature of many coastal geomorphic changes, especially those related to storms, allows for periods of stability punctuated by periods of rapid change and, perhaps, recovery. Where colonization by annual species is the first stage of marsh initiation, recovery of the pioneer vegetation from storm impacts may be more rapid (that is, the next year given appropriate elevation and seed supply criteria) than where perennial species are disturbed. However, the contrast in physical robustness between, for example, *Salicornia* spp. (Wiehe 1935) and *Spartina* spp. (Chung 1994) suggests that the types of colonizing vegetation and geomorphic setting can lead to variations in the rate of marsh initiation but the essential interactions between geomorphology and vegetation are consistent.

Development of Tidal Creeks

As the marsh substrate accumulates through sediment deposition and vegetative growth, and the elevation of the surface above the preexisting intertidal flat increases (figure 13-1), marsh channel or creek development begins. Initial vegetation colonization is likely to occur in patches or clumps, controlled by surface irregularities in the original intertidal surface. When the elevation of vegetated areas increases, "tidal waters, instead of flowing as a thin film over the whole area, will be divided by these islands" (Steers 1948). Pestrong (1965), however, views the development of tidal channels in San Francisco Bay marshes to be analogous to rill development on fluvial slopes where sheetflow erodes the substrate at some critical distance away from the interfluve or drainage divide. Although Pestrong recognizes that creek banks, once established, become higher through sedimentation and accretion of the marsh surface, creeks are viewed as erosive forms. The strong vertical development observed in most systems as intertidal flats make the transition to emergent marshes suggests that creek systems evolve as both a result of erosion into the substrate and the increase in elevation of the adjacent vegetated surfaces (figure 13-1).

The pattern and form of tidal creeks has frequently been examined (Myrick and Leopold 1963; Pestrong 1965; Gardner and Bohn 1980; Collins et al.

1987; Shi et al. 1995) with some studies providing specific insights into the role of vegetation in creek morphology and planform (Garofalo 1980). Few studies, however, have attempted to address commonalities between sites despite the now widely recognized need for a more synthetic understanding of creek morphology for application to the construction and restoration of coastal wetlands (Coats et al. 1995). Some workers have suggested that tidal creeks within a marsh system can conform to a Davisian cycle of youth, maturity, and old age (Dame et al. 1992) with ephemeral creeks as the youngest end point and the ocean as the mature end point in a geohydrologic continuum. The concept assumes that young and mature creeks can develop their maturity as sea level rises and the younger components of the estuarine ecosystem move inland, transgressing upon more mature components of the upstream freshwater stream ecosystem. This transition at a particular creek section requires both ecological changes, as discussed with regard to oyster reefs by Dame et al. (1992), and some transformation of creek cross section to adapt to altered tidal prism.

Such adjustments might be assumed to occur in the same manner that a fluvial channel adjusts cross section to discharge variations (Leopold et al. 1964). However, the stability of planform in marsh creek systems observed in marshes at Barnstable, Massachusetts (Redfield 1972), suggests that in some systems antecedent drainage is a more important control on creek form than contemporary process-response. The stability of the Barnstable marsh creek systems compared with those in North Carolina and Georgia is attributed by Redfield (1972, p. 235) to the "dense turf" covering New England marshes. Although Pestrong (1965) finds no consistent relationship between changing channel cross sections and vegetation over a 2-year period in San Francisco Bay marshes, he did note that in small tributary channels, where flow is dominantly unidirectional, erosion and deposition were not restricted as much by vegetation as in higher-order streams with effective bidirectional flow. This increased role of vegetation in tidal headwaters is in contrast to Garofalo (1980), who observed differences in channel migration rates between New Jersey tidal marshes dominated by saline vegetation types and those dominated by freshwater communities. Migration rates were less in saline areas despite potentially higher erosive forces, suggesting a more significant role for salt-marsh vegetation (for example, *Spartina alterniflora, S. patens,* and *Distichlis spicata*) in binding channel banks, compared with areas dominated by *S. cynusuroides, Zizania aquatica,* or *Typha* spp. Van Eerdt (1985) has also noted that erosion rates on salt-marsh cliffs vary with vegetation type. In her study of marsh cliffs in the Oosterschelde, those cliffs bound by roots *of Spartina anglica* experienced erosion rates only 60% of those on similar cliffs dominated by *Limonium vulgare,* despite the greater length of *L. vulgare* roots. Pestrong (1965), however, also acknowledges the difficulty of defining cause and effect in relation to vegetation and creek stability and recognizes that sedimentation and hydraulic parameters may control channel geometry, which then influences the distribution and growth of vegetation. Collins et al. (1987), working in the same area, point specifically to the role of vegetation in influencing channel width in their study of the Petaluma marshes in San Francisco Bay. There, vigorous growth of *Salicornia virginica* (pickleweed) on the sides of first-order channels can effectively lead to open-channel closure and the development of subterranean channels (figure 13-2).

An approach developed to understanding adjustments in channel cross section in relation to velocity and discharge for fluvial streams, termed "hydraulic geometry" (Leopold and Maddock 1953), has been applied to tidal-creek systems in several studies. Myrick and Leopold (1963) applied a hydraulic-geometry approach to a tidal creek in Virginia using at-a-station measurements. They found that the primary difference between fluvial and tidal systems was that velocity changes more rapidly with discharge in tidal systems, and that this is compensated for by less rapid changes in depth and width. This implies greater stability of channel cross section under a range of discharges in tidal channels than in fluvial systems. Subsequent applications of these relationships to tidal creeks have been made by Leopold et al. (1993) in San Francisco Bay marshes, who generally support the findings of Myrick and Leopold (1963), and Zeff (1988) for marshes in New Jersey. Ashley and Zeff (1988) identified two types of tidal creeks within back barrier-lagoonal marshes in New Jersey: large through-flowing creeks, which act as conduits connecting water bodies such as the lagoon, ocean, or other tidal channels, and branching dead-end channels that flood and drain sections of marsh. Zeff (1988) found that larger channels (through-flowing and larger parts of the dead-end networks) showed hydraulic-geometry exponents similar to other tidal marshes, that is, that channels accommodate increased discharge by increasing velocity.

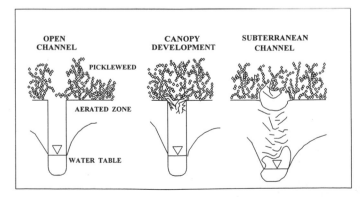

FIGURE 13-2   Subterranean channel development in Petaluma marshes as a result of pickleweed growth on channel banks (after Collins et al. 1987).

The smaller dead-end channels, however, are more similar to fluvial channels and accommodate increased discharge by adjustments in channel cross section, with channel depth increasing faster than channel width. Zeff (1988) noted that the dead-end channels experience maximum velocity at or close to bankfull (as in fluvial systems) while maximum velocities are reached in the larger through-flowing channels at mid-tide to low tide (a pattern similar to that in the Virginia tidal marshes originally studied by Myrick and Leopold).

It is interesting that in these geomorphic studies of hydraulic geometry in marsh channels, width is rarely responsive to changes in channel cross section. Pethick (1992) sees the shape and form of tidal creek networks in coastal marshes as a response to the need to dissipate tidal energy inputs. Pethick (1993) also notes the morphological changes in the tidal Blackwater Estuary in Essex that have occurred as a response to sea-level rise and increased water depths within the estuary. The channel has become wider, at the expense of intertidal flats and, to a certain extent, salt marshes at the estuary margins, but also shallower. The higher energy available in the open estuary and the fetch available to enhance wave attack on the salt-marsh margin (Pethick 1992) apparently allow adjustments in channel width in the Blackwater, which are limited in smaller tidal-marsh channels by lower velocities, a lack of erosive wave activity, and plant roots binding channel bank sediments. Adjustments to creek morphology in response to sea-level rise or alterations in local hydrology may be viewed as independent of marsh vegetation at the landscape level. However, at the scale of individual creeks the marsh vegetation, especially that colonizing the creek margins or levees, does play an important role in the response to sea-level rise and to altered hydrology. For example, vegetative binding of creek-bank sediments is important to the equilibrium balance between erosive flows and channel-bank resistance noted in most mature coastal marshes. This recognition is crucial to the planning of functional creek systems for restored coastal wetland systems. The approach adopted in many such projects is to construct several high-order creeks with dimensions approximating similar natural marshes, and to allow, even expect, progressive adjustments toward an equilibrium form (Coats et al. 1995). The importance of vegetative cover in then maintaining this equilibrium pattern must be incorporated into the planning process.

Marsh Surface Elevation

Future projections of sea-level rise associated with global climate change (Wigley and Raper 1992) indicate that unless there is an accompanying increase in vertical accretion rates in coastal marshes, coastal submergence and wetland loss will be widespread, particularly in deltas (Gornitz 1991). Presently, rates of elevation change in nondeltaic coastal marshes of the Atlantic and Gulf Coasts appear to be in balance with contemporary sea-level rise rates (Stevenson et al. 1986). However, these marshes are highly suscepti- ble to both submergence related to accelerated sea-level rise and decreases in the amount of sediment delivered to estuaries (and available for marsh accre- tion) because of dams and improved erosion control along rivers (Meade and Parker 1985; Kesel 1988). Of serious concern for coastal managers is the issue of whether the contributions to marsh elevation made by organic pro- duction, such as plant root production, can increase so as to compensate for any sediment deficiency in the face of rising sea level.

DeLaune et al. (1990) found primary production to be critical in a schematic model of the factors affecting marsh accretion (figure 13-3). Blum (1993) also notes that local differences in root production of *Spartina alterniflora* can influ- ence organic matter accumulation. Changes in soil conditions resulting from sea-level rise could alter the dynamics of belowground organic material and thus influence that component of accretion (figure 13-3). Although Cahoon et al. (1995) question whether examination of accretionary processes alone can pro- vide a valid assessment of the factors contributing to marsh surface-elevation change, the management of coastal marshes usually focuses on surficial processes. Any contributions to elevation change from either local shallow subsidence (Cahoon et al. 1995) or deeper regional subsidence (Penland and Ramsey 1990) are normally beyond the scope of human alteration and/or management.

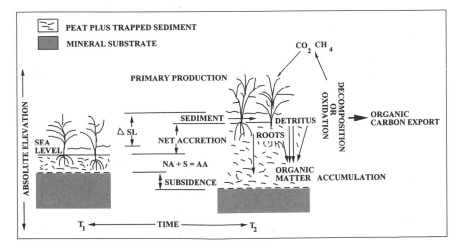

FIGURE 13-3   Schematic model of processes controlling marsh accretion (after DeLaune et al. 1990, ΔSL = sea level, NA = net accretion, S = subsidence, AA = ab- solute accretion)

The role of vegetation in accretionary processes can be both as a direct contribution through the accumulation of organic material within the marsh soil (figure 13-3) and indirectly via its role in baffling flows across the marsh surface, thus enhancing sediment deposition (Leonard and Luther 1995). Several authors have noted direct relationships between local marsh surface elevation and vegetative growth (McCaffrey and Thompson, 1980; Hatton et al. 1983) where clumps of *Spartina patens* appear to grow vertically in response in increased tidal flooding. Hartnall (1984) notes a similar development of "hummocky" topography in Gibraltar Point marshes in Lincolnshire, UK, but these are associated with seasonal cycles of growth and decay for *Puccinellia maritima*. The hummocks develop when *Puccinellia* clumps in low marsh areas collapse under autumn winds and tides and the aboveground biomass becomes "plastered" with trapped mud and incorporated into the soil. Hartnall's study also examined the effect of various plant types on marsh surface elevations, including the association of *Spartina anglica* with areas of elevation gain. Whether this is a result of organic matter accumulation or sediment trapping is not clear, but the study affirms measurements from Willapa Bay of enhanced accretion rates in *Spartina* compared to other salt-marsh plants.

Leonard and Luther (1995) examined the influence of several salt-marsh plants on flows across microtidal marshes in the Gulf of Mexico. They show a reduction in flow speeds associated with varying stem densities of both *Spartina alterniflora* and *Juncus roemerianus*. However, *S. alterniflora* shows a greater reduction in turbulence intensity with increasing stem density than *J. roemerianus*. This study quantifies the long-held understanding of the effect of marsh vegetation on baffling flow and providing effective traps for sediment deposition (Steers 1948). However, by demonstrating differences between plants with essentially similar gross morphologies (tall, erect structure) and identifying variations in turbulence intensity associated with various depths of canopy flooding, Leonard and Luther's (1995) study reiterates the need to understand biogeomorphic processes at the local scale. While our understanding of landscape development is not challenged by such detailed studies, the ability to apply that understanding to management and recreation of coastal marshes is both enhanced and questioned. Which vegetation should be planted? What is the necessary stem density to enhance deposition?

An optimal approach to resolving such questions in some systems may be modeling. So far, although several models of marsh vertical development acknowledge the importance of vegetative contributions (Allen 1990), most do not explicitly incorporate the biogeomorphic aspect of system response to sea-level rise (Krone 1987; French 1993). While the efforts of Goodwin et al.

(1992) and Woolnough et al. (1995) represent advances in the ability to predict flow dynamics and sediment deposition in coastal marshes, neither considers the biogeomorphic complexities of coastal marshes as demonstrated here. Predicting the future of marsh systems at the landscape scale requires the incorporation of detailed physical and sedimentological approaches into process-based ecosystem models of the type developed by Costanza et al. (1990). Our need to understand the future of marsh systems in the face of sea-level rise is apparent at a number of scales. At the landscape scale, protection of population centers from storm surges and coastal erosion is a critical issue. At the local scale, the long-term implications of dredging projects or water management schemes must be considered in effective coastal management policy.

Conclusions

The sustainability of coastal marshes in the face of relative sea-level rise is an issue of concern to coastal managers throughout the world (Lee 1993; Fankhauser 1995; Wolanski and Chappell 1996). The critical role of vegetation in determining the initiation, morphology, and vertical growth of coastal salt marshes has been demonstrated in this chapter. The commonalities that have been identified between marsh systems in distant parts of the world with different vegetative types illustrate that the differences between systems do not reflect differences in fundamental processes but variations in the magnitude and scale of process operation. Similar variations can exist within local marsh areas as hydrologic and substrate controls influence vegetative dynamics.

The strengths of the biogeomorphic approach to marsh processes are clear but the difficulty lies in implementing this approach. Comparative studies across marsh systems can assist in the refinement of synthetic understanding of biogeomorphic interactions, but the incorporation of this understanding into modeling tools is critical for effective management. The sensitivity of specific biogeomorphic processes to changes in sea level should be the critical determinant of whether additional synthetic field studies are required, or if efforts should focus on including existing knowledge into predictive models. The comparisons from systems around the world being carried out under IGBP-LOICZ present an opportunity to develop general synthetic models, identify needs for further research, and test model application to coastal management.

References

Allen, J. R. L. 1990. Salt-marsh growth and stratification: A numerical model with special reference to the Severn estuary. *Marine Geology* 95: 77–96.

Ashley, G. M., and M. L. Zeff. 1988. Tidal channel classification for a low-mesotidal salt marsh. *Marine Geology* 82: 17–32.

Blum, L. K. 1993. *Spartina alterniflora* root dynamics in a Virginia marsh. *Marine Ecology Progress Series* 102: 169–78.

Cahoon, D. R., D. J. Reed, and J. W. Day. 1995. Estimating shallow subsidence in microtidal salt marshes of the southeastern United States: Kaye and Barghoorn revisited. *Marine Geology* 128: 1–9.

Chapman, V. J. 1960. The plant ecology of Scolt Head Island. Pages 85–163 in J. A Steers (editor), *Scolt Head Island.* Cambridge, England: W. Heffer & Sons Ltd.

Chung, C.-H. 1994. Creation of *Spartina* plantations as an effective measure for reducing coastal erosion in China. Pages 443–52 in W. J. Mitsch (editor), *Global wetlands: Old world and new.* Amsterdam: Elsevier Science Publishing Co.

Coats, R. N., P. B. Williams, C. K. Cuffe, J. B. Zedler, D. J. Reed, S. M. Waltry, and J. S. Noller. 1995. *Design guidelines for tidal channels in coastal wetlands.* San Francisco, CA: Philip Williams & Associates Ltd.

Coles, S. M. 1979. Benthic microalgal populations on intertidal sediments and their role as precursors to salt marsh development. Pages 25–42 in R. L. Jefferies and A. J. Davy (editors), *Ecological processes in coastal environments: The first European ecological symposium and the 19th symposium of the British Ecological Society, Norwich, 12–16 September 1977.* New York: Blackwell Scientific Publications.

Collins, L. M., J. N. Collins, and L. B. Leopold. 1987. Geomorphic processes of an estuarine marsh: Preliminary results and hypotheses. Pages 1049–72 in V. Gardiner (editor), *International geomorphology 1986, Part 1.* New York: John Wiley & Sons Ltd.

Costanza, R., F. H. Sklar, and M. L. White. 1990. Modeling coastal landscape dynamics. *BioScience* 40: 91–107.

Dame, R., D. Childers, and E. Koepfler. 1992. A geohydrologic continuum theory for the spatial and temporal evolution of marsh-estuarine ecosystems. *Netherlands Journal of Sea Research* 30: 63–72.

DeLaune, R. D., W. H. Patrick Jr., and N. Van Breemen. 1990. Processes governing marsh formation in a rapidly subsiding coastal environment. *Catena* 17: 277–88.

Fankhauser, S. 1995. Protection versus retreat: The economic costs of sea-level rise. *Environment and Planning A* 27: 299–319.

French, J. R. 1993. Numerical simulation of vertical marsh growth and adjustment to accelerated sea-level rise, north Norfolk, U.K. *Earth Surface Processes & Landforms* 18: 63–81.

Gardner, L. R., and M. Bohn. 1980. Geomorphic and hydraulic evolution of tidal creeks on a subsiding beach ridge plain, North Inlet, S.C. *Marine Geology* 34: M91–M97.

Garofalo, D. 1980. The influence of wetland vegetation on tidal stream channel migration and morphology. *Estuaries* 3(4): 258–70.

Goodwin, P., J. Lewandowski and R. J. Sobey. 1992. Hydrodynamic simulation of small-scale tidal wetlands. Pages 149–61 in R. A. Falconer, K. Shiono, and R. G. S. Matthew (editors), *Hydraulic and environmental modelling: Estuarine and river waters*. Proceedings of the Second international Conference on Hydraulic and Environmental Modelling of Coastal, Estuarine and River Waters. Vol. 2. Aldershob, Hampshire, UK: Ashgate.

Gornitz, V. 1991. Global coastal hazards from future sea-level rise. *Palaeogeography, Palaeoclimatology and Palaeoecology* 89: 379–98.

Hartnall, T. J. 1984. Salt-marsh vegetation and micro-relief development on the New Marsh at Gibraltar Point, Lincolnshire. Pages 37–58 in M. W. Clark (editor), *Coastal research: UK perspectives* (based on a workshop of the Small Study Group on Nearshore Dynamics) Norwich, England: Geo Books.

Hatton, R. S., R. D. DeLaune, and W. H. Patrick. 1983. Sedimentation, accretion and subsidence in marshes of Barataria basin, Louisiana. *Limnology and Oceanography* 28: 494–502.

Kesel, R. H. 1988. The decline in the suspended load of the Lower Mississippi River and its influence on adjacent wetlands. *Environmental Geology and Water Science* 11: 271–81.

Kinler, N. W., G. Linscombe, and P. R. Ramsey. 1987. Nutria. Pages 328–43 in M. Novak et al. (editors), *Wild furbearer management and conservation in North America*. Ontario, Canada: Ministry of Natural Resources.

Krone, R. B. 1987. A method for simulating historic marsh elevations. Pages 316–23 in N. C. Kraus (editor), *Coastal sediments '87. Proceedings of a specialty conference on advances in understanding of coastal sediment processes*. New York: ASCE.

Lee, E. M. 1993. The political ecology of coastal planning and management in England and Wales: Policy responses to implications of sea-level rise. *The Geographical Journal* 159: 169–78.

Leonard, L. A., and M. E. Luther. 1995. Flow hydrodynamics in tidal marsh canopies. *Limnology and Oceanography* 40: 1474–84.

Leopold, L. B., J. N. Collins, and L. M. Collins. 1993. Hydrology of some tidal channels in estuarine marshland near San Francisco. *Catena* 20: 469–93.

Leopold, L. B., and T. Maddock Jr. 1953. *The hydraulic geometry of stream channels and some physiographic implications*. Geological Survey Professional Paper No. 252. Washington, DC: U. S. Government Printing Office.

Leopold, L. B., M. G. Wolman, and J. P. Miller. 1964. *Fluvial processes in geomorphology*. San Francisco, CA: W. H. Freeman and Company.

McCaffrey, R. J., and J. Thompson. 1980. A record of the accumulation of sediment and trace metals in a Connecticut salt marsh. *Advances in Geophysics* 22: 165–236.

McIntyre, N. F. 1984. Cryoconite hole thermodynamics. *Canadian Journal of Earth Science* 21: 152–56.

Meade, R. H., and R. S. Parker. 1985. Sediment in rivers of the United States. Pages 49–60 in *National water summary 1984*. U.S. Geological Survey Water–Supply Paper no. 2275. Washington, DC: U.S. Government Printing Office.

Myrick, R. M., and L. B. Leopold. 1963. *Hydraulic geometry of a small tidal estuary.* U.S. Geological Survey Professional Paper No. 422-B. Washington, DC: U.S. Government Printing Office.

Orson, R. A., and B. L. Howes. 1992. Salt marsh development studies at Waquoit Bay, Massachusetts: Influence of geomorphology on long-term plant community structure. *Estuarine, Coastal and Shelf Science* 35: 453–71.

Penland, S., and K. E. Ramsey. 1990. Relative sea-level rise in Louisiana and the Gulf of Mexico: 1908–1988. *Journal of Coastal Research* 6: 323–42.

Pernetta, J. C., and J. D. Milliman (editors). 1995. *Land-ocean interactions in the coastal zone, implementation plan.* IGBP Global Change Report No. 33. Stockholm, Sweden: The International Geosphere-Biosphere Programme: A Study of Global Change (IGBP) of the International Council of Scientific Unions (ICSU).

Pestrong, R. 1965. *The development of drainage patterns on tidal marshes.* Stanford University Publications, Geological Sciences 10(2). School of Earth Sciences. Stanford, CA: Stanford University.

Pethick, J. S. 1980. Salt-marsh initiation during the Holocene transgression: The example of the North Norfolk marshes, England. *Journal of Biogeography* 7: 1–9.

―――. 1992. Saltmarsh geomorphology. Pages 41–62 in J. R. L. Allen and K. Pye (editors), *Saltmarshes: Morphodynamics, conservation and engineering significance.* Cambridge, England: Cambridge University Press.

―――. 1993. Shoreline adjustments and coastal management: Physical and biological processes under accelerated sea-level rise. *The Geographical Journal* 159(2): 162–68.

Redfield, A. C. 1972. Development of a New England salt marsh. *Ecological Monographs* 42(2): 201–37.

Reed, D. J. 1989. Environments of tidal marsh deposition in Laguna San Rafael area, southern Chile. *Journal of Coastal Research* 5(4): 845–56.

Roy, P., and J. Connell. 1991. Climatic change and the future of atoll states. *Journal of Coastal Research* 7(4): 1057–75.

Sayce, J. R. 1991. *Spartina* in Willapa Bay: A case history. Pages 27–29 in T. F Mumford, et al. (editors), *Spartina workshop record, Seattle, Washington, November 14–15, 1990.* Washington Sea Grant Program. Seattle: University of Washington.

Shi, Z., H. F. Lamb, and R. L. Collin. 1995. Geomorphic change of saltmarsh tidal creek networks in the Dyfi Estuary, Wales. *Marine Geology* 128: 73–83.

Steers, J. A. 1948. *The Coastline of England & Wales.* Cambridge, England: Cambridge University Press.

————. 1960. Physiography and evolution. Pages 12–66 in J. A. Steers (editor), *Scolt Head Island.* Cambridge, England: W. Heffer & Sons Ltd.

Stevenson, J. C., L. G. Ward, and M. S. Kearney. 1986. *Vertical accretion in marshes with varying rates of sea-level rise.* Pages 241–59 in D. A. Wolfe (editor), *Estuarine variability.* Orlando, FL: Academic Press.

Trudgill, S. T., P. L. Smart, H. Friederich, and R. W. Crabtree. 1987. Bioerosion of intertidal limestone, Co. Clare, Eire. 1: *Paracentrotus lividus. Marine Geology* 74: 85–98.

van Eerdt, M. M. 1985. The influence of vegetation on erosion and accretion in salt marshes of the Oosterschelde, The Netherlands. *Vegetatio* 62: 367–73.

Viles, H. A., 1988. Introduction. Pages 1–8 in H. A. Viles (editor), *Biogeomorphology.* New York: Basil Blackwell Ltd.

Wiehe, P. O. 1935. A quantitative study of the influence of tide upon populations of *Salicornia. Journal of Ecology* 23: 323–33.

Wigley, T. M. L., and S. C. B. Raper. 1992. Implications for climate and sea level of revised IPCC emissions scenarios. *Nature* 357: 293–300.

Wolanski, E., and J. Chappell. 1996. The response of tropical Australian estuaries to sea level rise. *Journal of Marine Systems* 7: 267–79.

Woolnough, S. J., J. R. L. Allen, and W. L. Wood. 1995. An exploratory numerical model of sediment deposition over tidal salt marshes. *Estuarine, Coastal and Shelf Science* 41: 515–43.

Zeff, M. L. 1988. Sedimentation in a salt marsh–tidal channel system, southern New Jersey. *Marine Geology* 82: 33–48.

CHAPTER 14

Chesapeake Bay Eutrophication Model

Carl F. Cerco

Abstract

The CE-QUAL-ICM is a three-dimensional, time-variable eutrophication model. It incorporates twenty-two state variables that include physical properties; multiple forms of algae, carbon, nitrogen, phosphorus, and silica; and dissolved oxygen. The model is part of a larger package that includes a three-dimensional hydrodynamic model and a benthic sediment diagenesis model. Application to Chesapeake Bay over a three-year period, 1984–1986, indicates the model successfully simulates water-column and sediment processes that affect water quality. Phenomena simulated include formation of the spring algal bloom subsequent to the annual peak in nutrient runoff, onset and breakup of summer anoxia, and coupling of organic particle deposition with sediment-water nutrient and oxygen fluxes. The study demonstrates that complex eutrophication problems can be addressed with coupled three-dimensional hydrodynamic and water-quality models.

Introduction

The roots of the study described here extend back to 1983. In that year, a technical synthesis (Flemer et al. 1983) concluded that the volume of anoxic water in the bay had increased by an order of magnitude from 1950 to 1980. The conclusion was controversial; a previous synthesis (Heinle et al. 1980) reported that dissolved-oxygen concentration in the bay had not changed greatly. The existing database was insufficient to conclusively determine whether anoxia was increasing or not. Other indicators, however, including diminished fisheries harvest and shrinking distribution of submerged aquatic vegetation, supported the judgment that natural resources of the bay were deteriorating. The potential loss of resources spawned increased activity in bay monitoring, ecological research, and modeling for management purposes.

Chesapeake Bay

Physical Description

The Chesapeake Bay system (figure 14-1) consists of the mainstem bay, five major western-shore tributaries, and a host of lesser tributaries and embayments. Urban centers along the bay and tributaries include Norfolk and Richmond, Virginia, Washington, D.C., and Baltimore, Maryland. The mainstem is roughly 300 km long, 8 to 48 km wide, and 8 m in average depth. A deep trench with depths to 50 m runs up the center of the mainstem.

Total drainage area of the bay is 166,000 km². The primary source of fresh water to the system is the Susquehanna River (≈ 64% of total gauged freshwater flow), which empties into the northernmost extent of the bay. Other major freshwater sources are the Potomac (≈ 19%) and James Rivers (≈ 12%). The remaining western-shore tributaries, the York (≈ 3%), the Rappahannock (≈ 3%), and the Patuxent (< 1%), contribute only small fractions of the total freshwater flow to the bay.

The mainstem bay and major tributaries are classic examples of partially mixed estuaries (Pritchard 1967). When flows in these estuaries are averaged over lengthy time periods, generally more than 15 days, a net longitudinal circulation is evident. Longitudinal density gradients push bottom water upstream and enhance flow of surface water downstream. The volume of the density-induced flow vastly exceeds the volume of freshwater runoff.

Effects of the earth's rotation on long-term circulation induce flow predominately toward the right. The level of no net motion is tilted so that the depth of the downstream-flowing surface layer is less along the right-hand side, looking upstream, than along the left. In the mainstem, rotational effects are evident in surface salinity, which is higher along the eastern than along the western shore. Under some circumstances, the level of no net motion intersects the surface so that net surface flow is upstream along the eastern portion of the bay.

Eutrophication Processes

Bottom-water hypoxia. During the summer months, bottom waters of the mainstem bay are characterized by hypoxic (low dissolved oxygen) or anoxic (no dissolved oxygen) conditions (figure 14-2). Longitudinal and lateral extent of hypoxia are determined by the geometry of the trench that runs up the center of the bay. On occasion, hypoxic water extends from the bottom to

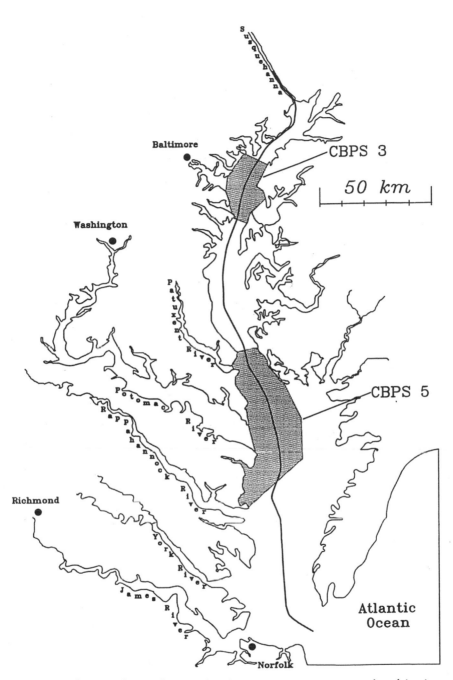

FIGURE 14-1 Chesapeake Bay showing program segments employed in time-series plots and transect employed in longitudinal plots.

Bottom Dissolved Oxygen, in Summer

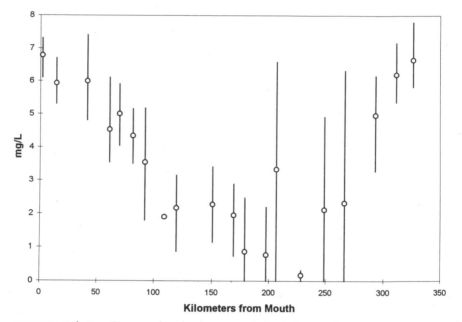

FIGURE 14-2   Bottom dissolved oxygen along mainstem bay transect, summer 1985. Data from Chesapeake Bay Program Office (CBPO) database.

within a few meters of the surface but summer-average oxygen concentration within the surface mixed layer is usually 6 g m^{-3} or greater.

Bottom-water hypoxia occurs at recurrent, predictable time intervals (figure 14-3). The onset is in late May when spring warming enhances respiration in benthic sediments. Decay of material deposited in spring and in previous years removes oxygen from bottom water. Density stratification prevents mixing of oxygenated surface water downwards. Low-oxygen conditions continue through the summer, maintained by respiration in bottom water. In mid-September, autumn winds end the hypoxic period by mixing surface water down to the bottom. Respiration in bottom water, diminished by cool temperatures, is insufficient to reestablish hypoxia following the mixing event.

Anoxic volume is strongly linked to runoff from the Susquehanna (Seliger and Boggs 1988). Two factors contribute to the linkage. First, large amounts of runoff induce strong density stratification, which inhibits oxygenation of bottom water. Second, nutrient loads contained in the runoff promote phytoplankton production. Excess production settles to the bottom of the bay, decays, and consumes oxygen. The relative influence of stratification versus loading cannot be determined since the two forcing functions covary

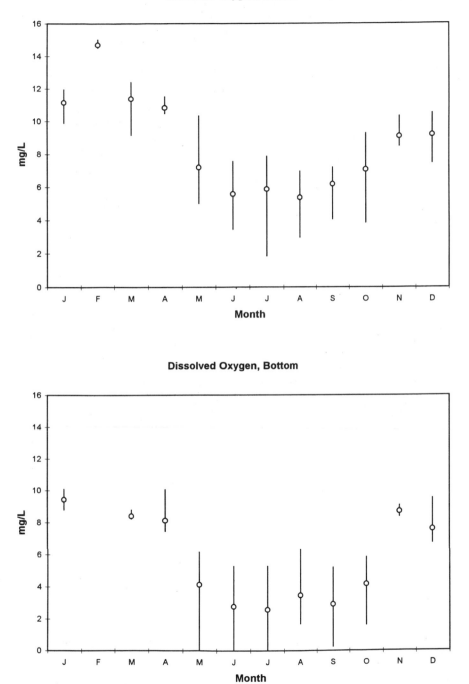

FIGURE 14-3 Dissolved oxygen in Chesapeake Bay Program Segment 3, 1985. Data from CBPO database.

(Boicourt 1992). The consensus of researchers is that year-to-year variations in anoxic volume, associated with variations in runoff, prevent detection of trends in the historic record (Smith et al. 1992).

Spring phytoplankton bloom. A second recurring phenomenon in the bay is the spring phytoplankton bloom. The spring bloom primarily consists of three genera of marine diatoms: *Cerataulina, Rhizosolina,* and *Thalassiosira.* The bloom usually commences in February and ends precipitously in late May (figure 14-4). The bloom is characterized by high chlorophyll concentrations throughout the water column. At times, a subsurface chlorophyll maximum occurs. During summer, chlorophyll concentrations and algal biomass are generally less than in spring and restricted to surface waters. Despite the disparity in biomass, however, primary production in summer exceeds production in spring (Malone et al. 1988).

Although the bloom occurs regularly, the magnitude and spatial extent of the bloom vary from year to year. Factors affecting this variability are not well known. The occurrence and apparent survival of viable algae at great depth, in the absence of light, also remain unexplained.

The occurrence and magnitude of the spring bloom are linked to subsequent bottom-water anoxia. A spring peak in carbon deposition to sediments results from the algal bloom (Boynton and Kemp 1985) and the decay of this fresh organic matter contributes to oxygen demand.

A subtle, and potentially more important link, is through a nutrient-trapping mechanism (Malone et al. 1988). Nutrients in spring runoff are taken up by algae during the bloom. Predation and algal mortality result in the transfer of nutrients, in particulate organic form, to benthic sediments. In summer, the nutrients are mineralized in the sediments and released to the water column. Nutrients released from the sediments support summer algal production. Carbon produced by algae settles to bottom waters, decays, and consumes oxygen. Diminished oxygen in bottom water enhances sediment nutrient release, especially of ammonium. The nutrient release continues the cycle of benthic release, algal production, and oxygen consumption.

Sediment-water interactions. Over time scales of years, benthic sediments are sinks of oxygen, nitrogen, phosphorus, and silica removed from the water column. Oxygen is consumed, directly or indirectly, in the oxidation of organic carbon and in the nitrification of ammonium. Substantial fractions of particulate nitrogen, phosphorus, and silica that settle into surficial sediments are buried to deep sediments from which recycling to the water column is impossible.

Over seasonal time scales, sediments can be significant sources of dissolved nutrients to the overlying water. The role of sediments in the systemwide nutrient budget is especially important in summer when seasonal low flows

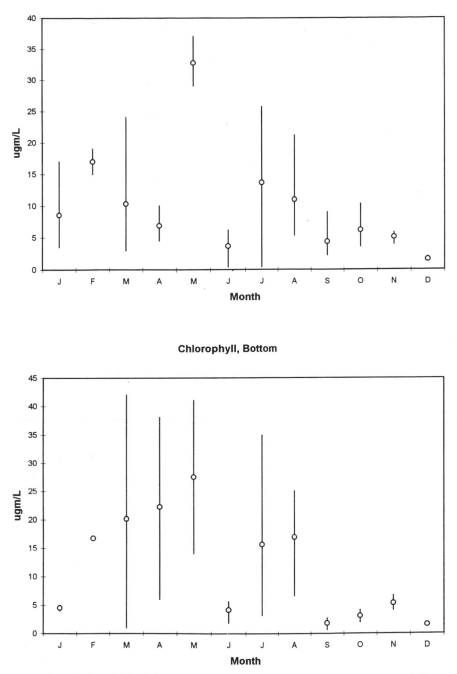

FIGURE 14-4 Chlorophyll in Chesapeake Bay Program Segment 3, 1985. Data from CBPO database.

diminish riverine nutrient input. During summer, warm temperature enhances biological processes in the sediments. Diagenesis (decay) of organic matter produces ammonium, phosphate, and silica that are released to the water column. Sediment ammonium release supplies an estimated 13 to 40% of algal nutrient requirements during August in Chesapeake Bay (Boynton and Kemp 1985). Silica nutrient limitation is not seen in summer because sediment release exceeds fluvial input by a factor of 5 or more (D'Elia et al. 1983).

Seasonal effects are also evident in sediment oxygen consumption. While temperature effects are significant, the seasonal pattern of sediment oxygen consumption is also affected by carbon supply to the sediments and by oxygen availability in the water column. Peak sediment oxygen consumption typically occurs in late spring (Kemp et al. 1992) when water warms, when fresh organic matter is available from the spring bloom, and when oxygen is freely available in the water column. During summer, sediment oxygen consumption in much of the bay is diminished from the spring peak. The depression occurs because organic matter is consumed in the water column before it reaches the sediments and because hypoxic conditions limit the supply of oxygen to the sediment-water interface. Under hypoxic conditions, oxygen demand created in the sediments is released to the water column, primarily in the form of sulfide.

Nitrogen and phosphorus budget. A variety of nutrient budgets have been constructed for the bay over the years (see USEPA 1982; Nixon 1987; Boynton et al. 1995). All budgets agree that loads from riverine runoff are the major nitrogen source. All budgets indicate that riverine loading is a significant phosphorus source but not as dominant as for nitrogen. All budgets agree that the system exports nitrogen to the ocean. Estimates of the fraction exported vary widely however, from a negligible quantity (USEPA 1982) to the majority of the load (Nixon 1987). Both direction and magnitude of net phosphorus flux at the mouth are controversial. The three budgets available indicate negligible import (USEPA 1982), significant export (Nixon 1987), and significant import (Boynton et al. 1995). The contradictory results are strongly affected by estimates of burial and the requirement to balance the budget.

Magnitude and direction of net loads at the bay mouth cannot be measured in the presence of tidal oscillations and short-term major events such as storm flows. Consequently, fluxes at the mouth are usually assigned to balance the budget once loads and burial are evaluated. Riverine loads, direct loads, and atmospheric loads can be quantified with reasonable accuracy. Estimation of the burial term has a large variance, however, which impacts the estimation of flux at the mouth. If large fractions of the loads are buried, little remains for export out the mouth. Calculation of net import may result if burial exceeds loading. If burial estimates are a small fraction of loading, then most load must be exported to the ocean.

The Chesapeake Bay Model Package

The first management model completed after the reported increase in anoxic volume was a two-dimensional, vertically averaged, hydrodynamic and water-quality model (Chen et al. 1984). The results of this rapidly completed study indicated greatly expanded efforts were required to produce a model sufficient for management of eutrophication in the bay. Required improvements included a fully three-dimensional model applied to an expanded, contemporary database.

The succeeding modeling effort (HydroQual 1987) applied three-dimensional hydrodynamic and eutrophication models to data collected in 1965, 1984, and 1985. Summer-average, steady-state conditions were modeled independently in each year. The model provided credible representations of historic and contemporary conditions in the bay and tributaries. Despite the success of the modeling effort, limitations were apparent. The chief limitation was absence of a predictive model of sediment-water interactions. The study indicated that sediment release was the dominant source of nitrogen and phosphorus during summer conditions. Sediment oxygen demand was a major dissolved-oxygen sink. No means existed to predict how these sediment processes would respond to nutrient load reductions, however. Neither was the time scale for completion of the responses predictable. A second limitation was the steady-state nature of the analysis. The steady-state model allowed no influence of conditions in previous years or seasons on summer-average water quality.

Based on the results of the preceding models, the EPA Chesapeake Bay Program Office formulated requirements for the present management model. Requirements included:

- ability to analyze the current origins and extent of anoxia;

- ability to predict the future of the bay as a result of management action or inaction; and

- ability to reconstruct the processes that have led the bay to its present condition.

A package of interactive models (figure 14-5) was assembled to comply with requirements and recommendations for the eutrophication model study. The CH3D-WES hydrodynamic model (Johnson et al. 1993) produced three-dimensional predictions of velocity, diffusion, surface elevation, salinity, and temperature on an intratidal (5-minute) time scale. A processor appended

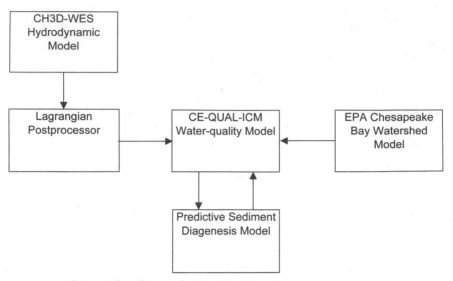

FIGURE 14-5   The Chesapeake Bay Model package.

to CH3D-WES filtered intratidal details from hydrodynamic output but maintained intertidal (12-hour) transport (Dortch et al. 1992). Intertidal flows, vertical diffusion, and surface elevation were written to disk for subsequent use by the water-quality model, dubbed CE-QUAL-ICM (Cerco and Cole 1993). Computation of eutrophication processes within the water-quality model was conducted on a 2-hour time scale. The water-quality model interacted dynamically with a predictive sediment diagenesis model (DiToro and Fitzpatrick 1993). Riverine loads and distributed loads were supplied to the eutrophication model by the EPA's Chesapeake Bay Watershed model (Linker at al. 1996).

The Conservation-of-Mass Equation

The hydrodynamic and eutrophication models operate by dividing the spatial continuum of the bay into a grid of discrete cells (figure 14-6). The grid for the present study contained 729 cells (roughly 5 × 10 km) in the surface plane and two to fifteen cells (1.5 to 2 m thick) in the vertical. Total number of cells in the grid was 4,073. CE-QUAL-ICM treats each cell as a control volume, which exchanges material with its adjacent cells. CE-QUAL-ICM solves, for each volume and for each state variable, the three-dimensional conservation-of-mass equation:

$$\frac{\delta V_i C_i}{\delta t} = \sum_{j=1}^{n} Q_j C_j^* + \sum_{j=1}^{n} A_j D_j \frac{\delta C}{\delta x_j} + \sum S_i$$

V_i = volume of ith control volume (m^3)
C_i = concentration in ith control volume (g m^{-3})

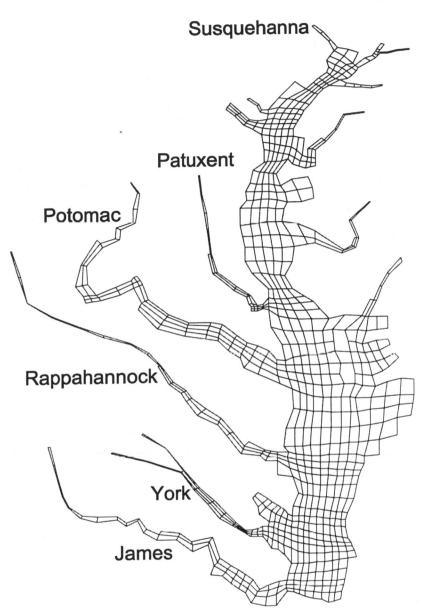

FIGURE 14-6   Plan view of 4,073-cell model grid.

Q_j = volumetric flow across flow face j of ith control volume (m³ sec⁻¹)
C^*_j = concentration in flow across flow face j (g m⁻³)
A_j = area of flow face j (m²)
D_j = diffusion coefficient at flow face j (m² sec⁻¹)
n = number of flow faces attached to ith control volume

S_i = external loads and kinetic sources and sinks in ith control volume (g sec^{-1})
t, x = temporal and spatial coordinates

Solution to the mass-conservation equation is via the finite-difference method using the QUICKEST algorithm (Leonard 1979) in the horizontal directions and a Crank-Nicolson scheme in the vertical direction.

The Eutrophication Model

The central issues in the water-quality model are primary production of carbon by algae and concentration of dissolved oxygen. Primary production provides the energy required by the ecosystem to function. Excessive primary production is detrimental, however, since its decomposition, in the water and sediments, consumes oxygen. Dissolved oxygen is necessary to support the life functions of higher organisms and is considered an indicator of the "health" of estuarine systems. In order to predict primary production and dissolved oxygen, a large suite of model-state variables is necessary (table 14-1).

Phytoplankton Kinetics

For phytoplankton, the sources and sinks in the conservation equation include production, metabolism, predation, and settling. These are expressed:

$$\frac{\sum S_i}{V} = \left(P_i - \mathrm{BM}_i - \mathrm{PR}_i - W_i \frac{\delta}{\delta z} \right) B_i$$

TABLE **14-1**
Water-quality model state variables.

Temperature	Salinity
Inorganic suspended solids	Cyanobacteria
Diatoms	Green algae
Dissolved organic carbon	Labile particulate organic carbon
Refractory particulate organic carbon	Ammonium
Nitrate	Dissolved organic nitrogen
Labile particulate organic nitrogen	Refractory particulate organic nitrogen
Total phosphate	Dissolved organic phosphorus
Labile particulate organic phosphorus	Refractory particulate organic phosphorus
Chemical oxygen demand	Dissolved oxygen
Particulate biogenic silica	Available silica

B_i = biomass of algal group i, expressed as carbon (g C m^{-3})
P_i = production of algal group i (d^{-1})
BM_i = basal metabolism of algal group i (d^{-1})
PR_i = predation on algal group i (d^{-1})
W_i = settling velocity of algal group i (m d^{-1})
z = vertical coordinate (m)

Production. Production by phytoplankton is determined by the availability of nutrients, the intensity of light, and the ambient temperature. The effects of each are considered to be multiplicative:

$$P = PM f(N) f(I) f(T)$$

PM = production under optimal conditions (d^{-1})
$f(N)$ = effect of suboptimal nutrient concentration ($0 \leq f \leq 1$)
$f(I)$ = effect of suboptimal illumination ($0 \leq f \leq 1$)
$f(T)$ = effect of suboptimal temperature ($0 \leq f \leq 1$).

Nutrients. Carbon, nitrogen, and phosphorus are the primary nutrients required for algal growth. Diatoms require silica as well. Inorganic carbon is usually available in excess and is not considered in the model. The effects of the remaining nutrients on growth are described by the formulation commonly referred to as "Monod kinetics" (Monod 1949). In the Monod formulation growth is dependent on nutrient availability at low nutrient concentrations but is independent of nutrients at high concentrations. A key parameter in the formulation is the "half-saturation concentration." Growth rate is half the maximum when available nutrient concentration equals the half-saturation concentration. Liebig's "law of the minimum" (Odum 1971) indicates growth is determined by the nutrient in least supply. For cyanobacteria and greens:

$$f(N) = \text{minimum} \left(\frac{NH_4 + NO_3}{KHn + NH_4 + NO_3}, \frac{PO_4 d}{KHp + PO_4 d} \right)$$

NH_4 = ammonium concentration (g N m^{-3})
NO_3 = nitrate concentration (g N m^{-3})
KHn = half-saturation constant for nitrogen uptake (g N m^{-3})
$PO_4 d$ = dissolved phosphate concentration (g P m^{-3})
KHp = half-saturation constant for phosphorus uptake (g P m^{-3})

Diatoms require silica, as well as nitrogen and phosphorus, for growth. For diatoms, the nutrient limitation is the minimum of the limitations expressed above or the following:

$$f(N) = \frac{SAd}{KHs + SAd}$$

SAd = dissolved silica concentration (g Si m^{-3})
KHs = half-saturation constant for silica uptake by diatoms (g Si m^{-3})

Light. Algal production increases as a function of light intensity until an optimal intensity is reached. Beyond the optimal intensity, production declines as intensity increases. Steele (1962) described this phenomenon:

$$f(I) = \frac{I}{Is} e^{\left(1 - \frac{I}{Is}\right)}$$

I = illumination rate (Langleys d^{-1})
Is = optimal illumination (Langleys d^{-1})

Within the model, an integrated form of Steele's relationship appropriate for daily-average illumination and discrete cell thickness is employed (DiToro et al. 1971).

Optimal illumination for photosynthesis depends on algal taxonomy, duration of exposure, temperature, nutritional status, and previous acclimation. Variations in optimal illumination are largely due to adaptations by algae intended to maximize production in a variable environment. Steele (1962) noted the result of adaptations is that optimal illumination is a consistent fraction ($\approx 50\%$) of daily illumination. Kremer and Nixon (1978) reported an analogous finding that maximum algal production occurs at a constant depth (≈ 1 m) in the water column. Their approach is adopted here so that optimal illumination is expressed:

$$Is = Io \; e^{-Ke \cdot Dopt}$$

Io   = surface illumination (Langleys d^{-1})
Dopt = depth of maximum algal production (m)
Ke   = light attenuation coefficient (m^{-1})

Temperature. Algal production increases as a function of temperature until an optimum temperature or temperature range is reached. Above the optimum, production declines until a temperature lethal to the organisms is attained. Numerous functional representations of temperature effects are available. Inspection of growth versus temperature curves indicates a function similar to a Gaussian probability curve provides a good fit to observations:

$$f(T) = e^{-KTg1(T-Tm)^2}, \quad \text{when } T \leq Tm$$

$$= e^{-KTg2(Tm-T)^2}, \quad \text{when } T > Tm$$

T = temperature (°C)
Tm = optimal temperature for algal growth (°C)

KTg1 = effect of temperature below Tm on growth ($°C^{-2}$)
KTg2 = effect of temperature above Tm on growth ($°C^{-2}$)

Basal metabolism. Basal metabolism is considered to be an exponentially increasing function of temperature:

$$BM = BMr \, e^{KTb\,(T-Tr)}$$

BMr = metabolic rate at Tr (d^{-1})
KTb = effect of temperature on metabolism ($°C^{-1}$)
Tr = reference temperature for metabolism ($°C$)

Predation. Detailed specification of predation rate requires predictive modeling of zooplankton biomass and activity. Zooplankton were not included in this version of the model. Instead, a constant predation rate was specified. Zooplankton activity is assumed to be influenced by temperature and is taken into account by incorporating an exponential temperature relationship into the predation term. The resulting predation formulation was identical to basal metabolism relationship. The difference in predation and basal metabolism was in the distribution of the end products of these processes.

The Model Carbon Cycle

A summary of the remainder of the model is presented here. Descriptions of the complete kinetics formulations and parameter evaluation can be found in Cerco and Cole (1994).

The model carbon cycle (figure 14-7) consists of the following elements:

- phytoplankton production;

- phytoplankton exudation;

- predation on phytoplankton;

- dissolution of particulate carbon;

- heterotrophic respiration;

- denitrification; and

- settling.

Algal production is the primary carbon source although carbon also enters the system through external loading. Predation on algae releases particulate and dissolved organic carbon to the water column. A fraction of the particulate organic carbon undergoes first-order dissolution to dissolved organic carbon. The remainder settles to the sediments. Dissolved organic carbon produced by

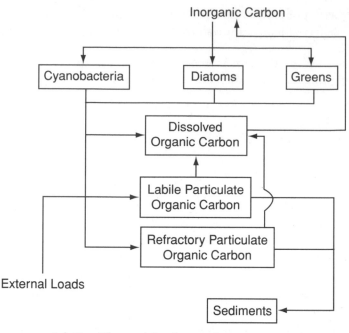

FIGURE 14-7   The model carbon cycle.

phytoplankton exudation, predation, and dissolution is respired or may be oxidized by denitrification to inorganic carbon. No carbon is recycled from the sediments to the water column although oxygen demand created by carbon diagenesis is included in the model.

The Model Nitrogen Cycle

The model nitrogen cycle (figure 14-8) includes the following processes:

- algal production and metabolism;
- predation;
- hydrolysis of particulate organic nitrogen;
- mineralization of dissolved organic nitrogen;
- settling;
- nitrification; and
- denitrification.

External loads provide the ultimate source of nitrogen to the system. Inorganic nitrogen is incorporated by algae during growth and released as ammonium and organic nitrogen through respiration and predation. A portion of the particulate organic nitrogen is hydrolyzed to dissolved organic

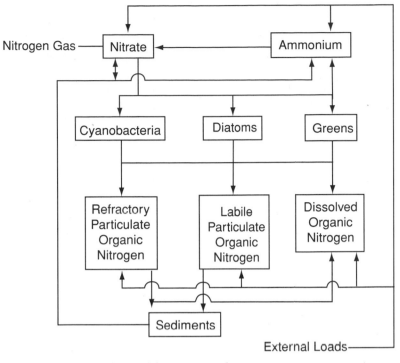

FIGURE 14-8 The model nitrogen cycle.

nitrogen. The balance settles to the sediments. Dissolved organic nitrogen is mineralized to ammonium. In an oxygenated water column, a fraction of the ammonium is subsequently oxidized to nitrate through the nitrification process. In anoxic water, nitrate is lost to nitrogen gas through denitrification. Particulate nitrogen that settles to the sediments is mineralized and recycled to the water column, primarily as ammonium. Nitrate moves in both directions across the sediment-water interface, depending on relative concentrations in the water-column and sediment interstices.

The Model Phosphorus Cycle

The model phosphorus cycle (figure 14-9) includes the following processes:

- algal production and metabolism;
- predation;
- hydrolysis of particulate organic phosphorus;
- mineralization of dissolved organic phosphorus; and
- settling.

External loads provide the ultimate source of phosphorus to the system. Dissolved phosphate is incorporated by algae during growth and released as phosphate and organic phosphorus through respiration and predation. A portion of the particulate organic phosphorus is hydrolyzed to dissolved organic phosphorus. The balance settles to the sediments. Within the sediments, particulate phosphorus is mineralized and recycled to the water column as dissolved phosphate. Dissolved organic phosphorus is mineralized to phosphate.

The Sediment Diagenesis Model

The need for a predictive benthic sediment model was made apparent by the results of the steady-state model study (HydroQual 1987) that preceded this one. The study indicated sediments were the dominant sources of phosphorus and ammonium during the summer period of minimum dissolved oxygen. An increase in sediment oxygen demand and nutrient release was implicated in a perceived decline in dissolved oxygen from 1965 to 1985. Simultaneously, basic scientific investigations were indicating the importance of sediment-water exchange processes in Chesapeake Bay and other estuarine systems (Boynton and Kemp 1985; Seitzinger et al. 1984; Fisher et al. 1982). For management purposes, a model was required with two fundamental capabilities: (1) to predict effects of management actions on sediment-water exchange

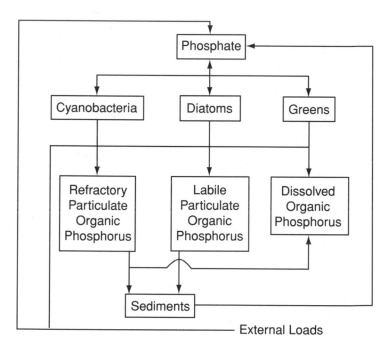

FIGURE 14-9   The model phosphorus cycle.

processes; and (2) to predict time scale for alterations in sediment-water exchange processes.

A sediment model to meet these requirements (DiToro and Fitzpatrick 1993) was developed based on an existing model of diagenetic processes in freshwater sediments (DiToro et al. 1990). The model (figure 14-10) was driven by net settling of organic matter from the water column to the sediments. In the sediments, the model simulated the diagenesis (decay) of the organic matter. Diagenesis produced oxygen demand and inorganic nutrients. Oxygen demand, as sulfide (in salt water) or methane (in fresh water), took three paths out of the sediments: export to the water column as chemical oxygen demand, oxidation at the sediment-water interface as sediment oxygen demand, and burial to deep, inactive sediments. Inorganic nutrients produced by diagenesis took two paths out of the sediments: release to the water column and burial to deep, inactive sediments. A listing of sediment model-state variables and of the predicted sediment-water fluxes is provided in table 14-2.

Coupling of the water-column and sediment models. Each column of cells in the water-quality model grid overlies a sediment model cell. Consequently, the number of sediment cells equals the number of surface cells in the water-quality model grid (729). The water-quality and sediment models interact on a time scale equal to the integration time step of the water-quality model.

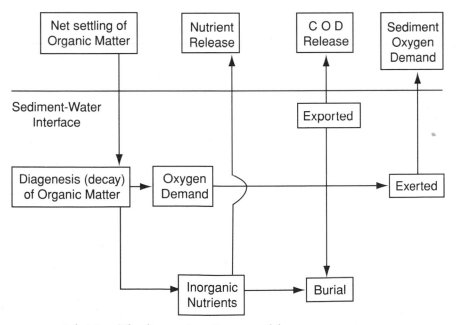

FIGURE 14-10 The diagenetic sediment model.

TABLE 14-2
Sediment model state variables and fluxes.

State Variable	Sediment-Water Flux
Temperature	
Particulate organic carbon	Sediment oxygen demand
Sulfide/methane	Release of chemical oxygen demand
Particulate organic nitrogen	
Ammonium	Ammonium flux
Nitrate	Nitrate flux
Particulate organic phosphorus	
Phosphate	Phosphate flux
Particulate biogenic silica	
Available silica	Silica flux

After each integration, predicted particle deposition, temperature, nutrient, and dissolved oxygen concentrations are passed from the water-quality model to the sediment model. The sediment model computes sediment-water fluxes of dissolved nutrients and oxygen based on predicted diagenesis and concentrations in the sediments and water. The computed sediment-water fluxes are incorporated by the water-quality model into appropriate mass balances and kinetic reactions.

Model Results

The model has been applied, in various configurations and in runs of varying duration, to years from 1959 to 1994 (Cerco and Cole 1994; Cerco 1995). Results are shown here for three primary calibration years, 1985–1987. Model computations and comparable data are presented in two formats: time series in surface and bottom waters at the center of the bay (Chesapeake Bay Program Segment 5 in figure 14-1) and a transect up the central bay axis (see figure 14-1).

The Mainstem Bay Water Column and Sediments

The model presented a realistic representation of conditions that prevailed in the mainstem bay from 1985 to 1987. Seasonal cycling and spatial distribution of concentrations in the water column were well represented as were major eutrophication processes.

Winter, December through February. Observations collected in winter were characterized by the onset of the annual phytoplankton bloom. In the lower bay chlorophyll usually attained its annual maximum during the months of January and February. In the mid-bay (figure 14-11) and upper bay, winter was usually the period of second-highest chlorophyll concentration. Inorganic nitrogen and silica were abundant in the upper 200 km of the bay because of runoff from the Susquehanna River. Inorganic phosphorus was also available for algal uptake although not in the excess concentrations apparent for nitrogen and silica. Dissolved oxygen was nearly at saturation concentrations throughout the water column.

The model reproduced the observed characteristics of the winter water column. In particular, the onset of the bloom and occurrence of diatoms at all depths were simulated (figure 14-11). Simulation of the bloom required several ad hoc modifications to algal kinetics. First, the base respiration and predation rates were reduced to 30% of the rates employed the rest of the year. Second, algae were permitted to settle through the water column but not into the sediments. These modifications presented insights into conditions necessary for occurrence of the spring bloom at all depths. First, predation pressure and other mortality factors during winter and spring must be very low. Second, diatoms must maintain their buoyancy in the water column above the sediments.

Model predictions (figure 14-12) indicated particle deposition was at an annual minimum during winter. Deposition was limited by algal buoyancy and by reduced predation, which diminished production of algal detritus. Sediment-water fluxes of nutrients were nearly zero, limited by temperature effects on sediment diagenesis. Sediment oxygen demand, <0.5 g m^{-2} day^{-1}, was apparent, however, caused by chemical oxidation of particulate sulfide accumulated in previous seasons.

Spring, March through May. The defining event in spring was the peak of the annual algal bloom. In the upper and middle bay, chlorophyll observations at all levels attained their annual peaks (figures 14-11, 14-13). Concentrations in the lower bay also remained high (figure 14-13) following the earlier peak in season 1. Inorganic nitrogen remained abundant because of continued spring runoff. Inorganic phosphorus approached depletion throughout the bay, however, as did silica in the lower 150 km. Seasonal-mean dissolved oxygen generally exceeded 5 g m^{-3} throughout the mainstem but at the head of the deep trench around km 250, the onset of bottom-water anoxia was apparent.

The model performed well in reproducing observations collected in spring, especially the occurrence of elevated chlorophyll concentrations at all depths (plate 2). Depletion of inorganic phosphorus was computed throughout the bay, in line with observations. The model also successfully computed the tremendous spatial gradients in dissolved inorganic nitrogen and dissolved silica between the riverine source and the oceanic sink, although computed dissolved nitrogen in the mid and lower bay was depleted earlier than observed (figure 14-11). Computed concentrations of inorganic phosphorus

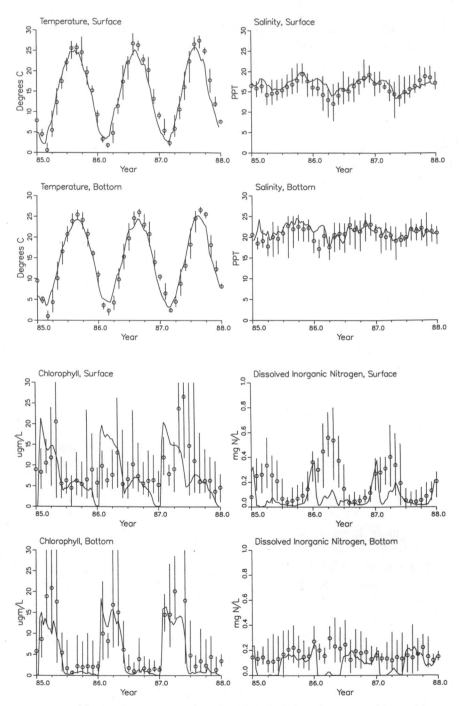

FIGURE 14-11 Time series of predicted and observed water quality in Chesapeake Bay Program Segment 5. Data from CBPO database.

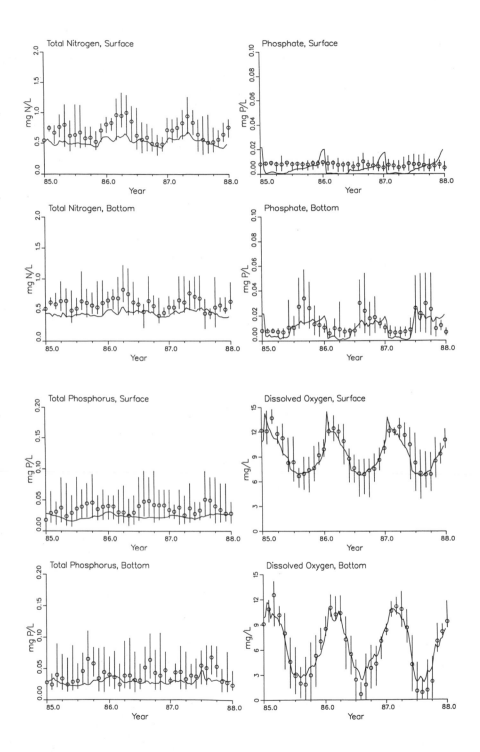

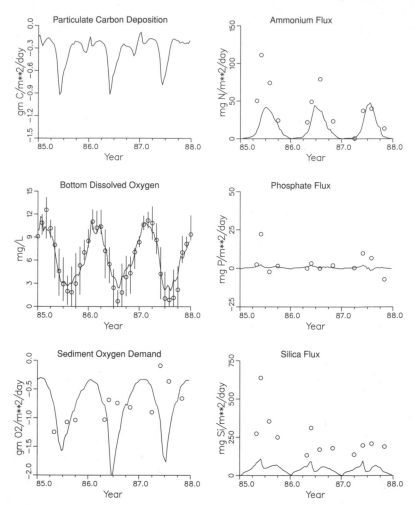

FIGURE 14-12   Time series of predicted and observed sediment-water fluxes in Chesapeake Bay Program Segment 5. Positive fluxes are from sediments to water; negative fluxes are from water to sediment. Data collected as part of the SONE Program (Boynton et al. 1986).

and of silica were limiting to algal growth, consistent with independent analyses (Fisher et al. 1992; Conley and Malone 1992). The model also successfully simulated the onset of bottom-water anoxia at the end of the season.

Predicted deposition to the sediments increased throughout spring and approached the annual maximum at the transition to summer, coincident with the collapse of the algal bloom (figure 14-12). Spatial trends in deposition were evident. Peak carbon, nitrogen, and phosphorus deposition occurred at the head of the bay, adjacent to the Susquehanna fall line. Secondary peaks in carbon, nitrogen, and phosphorus and the primary peak in silica deposition were at the head of the trench, beneath the maximum bloom. Observations revealed the onset of sediment nutrient releases. Observed sediment oxygen demand was substantial, ≈ 1 g m^{-2} day^{-1}. Consistent with trends in the observations, the model also indicated the onset of sediment nutrient releases and calculated annual maximum sediment oxygen demand during spring.

Summer, June through August. The transition from spring to summer was rapid and dramatic. Chlorophyll virtually disappeared from the bottom of the water column and was diminished in regions of the surface as well (figures 14-11, 14-14). Nitrate concentration attained an annual minimum because of diminished freshwater runoff. Ammonium and inorganic phosphorus were removed from surface water by algal uptake. These nutrients were abundant in subsurface water, however, because of sediment release (figure 14-12). Two factors contributed to silica availability. Diatoms comprised a lesser fraction of the algal community than in spring, thereby reducing demand, and temperature-enhanced sediment release occurred, thereby increasing supply. Most important for water quality, dissolved-oxygen minima in bottom waters were near zero for roughly 75 km of the mainstem (figure 14-14).

The model performed well in reproducing observed concentrations and processes during summer. As in the observations, the spring bloom died off rapidly and was replaced with an algal population concentrated at the water surface (color plate 2). Concentrations of ammonium, nitrate, and phosphate in surface waters were reasonably well predicted as were the differences between concentrations in surface and bottom waters (figure 14-14).

The interactions between system geometry and hypoxia were well represented in the model (color plate 3). Lowest predicted dissolved-oxygen concentrations occurred at the head of the deep trench and penetrated up the channel into Baltimore Harbor. Hypoxia followed the trench down the bay and penetrated into the connecting channel at the mouth of the Potomac. Substantial anoxia was also indicated in the secondary channel that runs along the Eastern Shore. Hypoxic water was virtually absent from shoal areas of the mainstem although hypoxia was indicated in deeper portions of some tributaries.

Observed sediment oxygen demand was erratic (figure 14-12). Although large demands were measured in summer, observations suggested that sediment oxygen demand was diminished for most of the season from a peak in spring or at the transition between seasons. Predicted sediment oxygen demand demonstrated a clear seasonal pattern. Maximum demand occurred at the transition between spring and summer, coincident with collapse of the spring algal

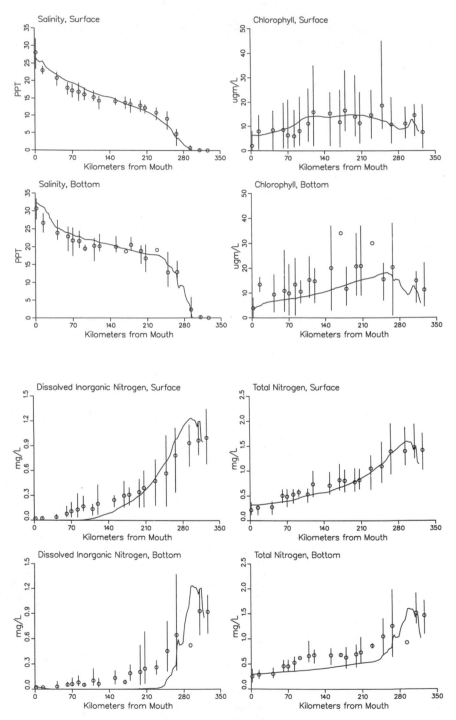

FIGURE 14-13 Predicted and observed water quality along mainstem bay transect, spring 1986. Data from CBPO database.

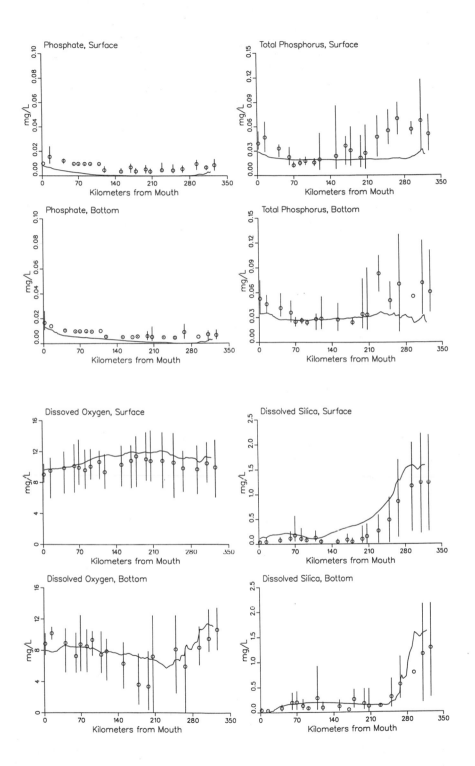

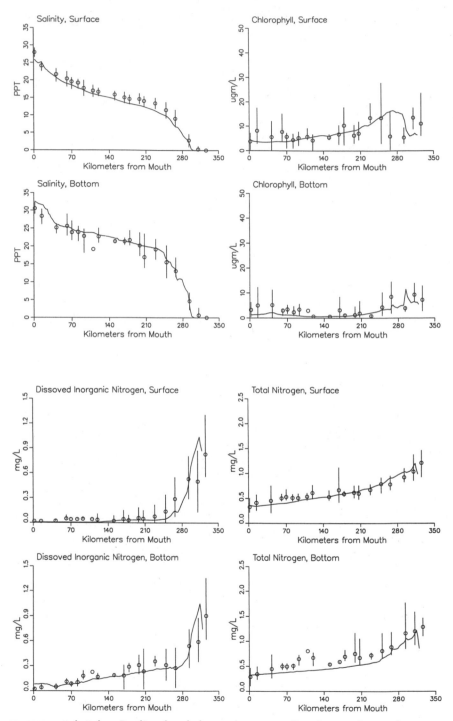

FIGURE 14-14 Predicted and observed water quality along mainstem bay transect, summer 1986. Data from CBPO database.

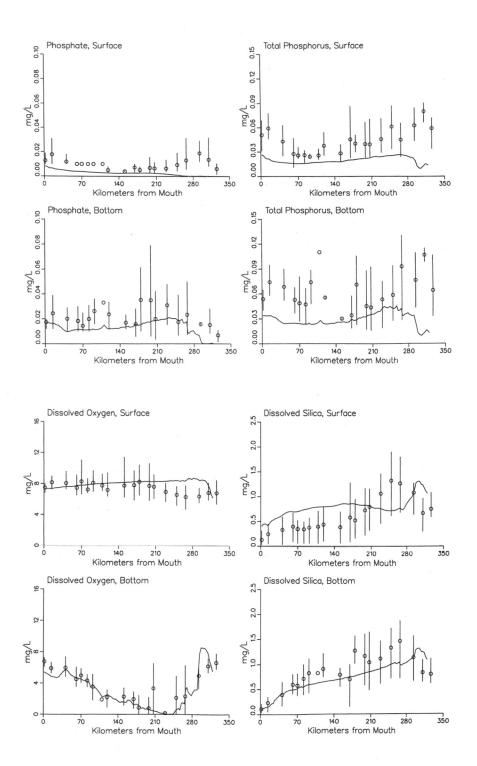

bloom. Following collapse of the bloom, computed sediment oxygen demand diminished, consistent with a computed decrease in carbon deposition.

Observed and computed sediment ammonium releases attained their maximum values during summer (figure 14-12). Three factors contributed to the maxima: the deposition of labile organic matter, temperature-enhanced diagenesis, and bottom-water hypoxia. Rapid diagenesis of fresh organic matter resulted in maximum production of ammonium while hypoxia blocked the nitrification/denitrification sequence so that a large fraction of ammonium was released.

In contrast to ammonium, no seasonal trend was evident in observed or computed sediment phosphorus release (figure 14-12). In this portion of the bay, the hypoxia is apparently not sufficient to generate the release that occurs when iron oxides in surface sediments are reduced to soluble forms, allowing free diffusion of sediment phosphate to overlying water. Consequently, bottom sediments act as a phosphorus trap and release little of the phosphorus deposited as a component of organic detritus.

The maximum observed silica release occurred at the transition from spring to summer, coincident with the collapse of the spring algal bloom (figure 14-12). The computed maximum release was in phase with observations but much less in magnitude. Model sensitivity runs indicated that the observed maximum could only be modeled by supplementing the silica deposited as algal biomass with an additional distributed load. The silica loads included in this version of the model were limited to dissolved silica in rivers and direct runoff. The sensitivity runs indicate the need to incorporate particulate silica contributed by the rivers and, potentially, bank erosion.

Autumn, September through November. Observations in autumn were characterized by annual minimum chlorophyll concentrations throughout the bay (figure 14-11). During early autumn in mid-bay, nutrients were scarce in surface waters. Later in the season, nitrate became more abundant as a result of fall storms.

The defining process in autumn was the fall overturn. In late September, autumn winds caused vertical mixing in the water column and ended the period of summer hypoxia (figure 14-11). Seasonal-mean dissolved oxygen was ≈ 5 g m^{-3} or greater everywhere although remnants of hypoxia were occasionally observed in bottom waters at the head of the trench.

Results of the model conformed to the pattern of the observations. Predicted autumn chlorophyll concentrations were minimal (figure 14-11). Computed inorganic nitrogen and phosphorus started the season at minimal values, then increased because of runoff events. The transition from hypoxic to oxygenated conditions was faithfully reproduced, indicating the accuracy of mixing processes computed in the hydrodynamic model.

Sediment nutrient releases in autumn were among the lowest observed (figure 14-12). Diminished release was caused by depletion of fresh organic matter in the sediments and by temperature effects on diagenesis. Substantial sediment oxygen demand was observed, 0.5 to 1 g m^{-2} day^{-1}, due largely to oxidation of sulfide previously stored in the sediments. Trends and magnitudes of fluxes in the model corresponded to the observations.

Nitrogen and Phosphorus Budgets

The initial model computational grid was cut off at the bay mouth (figure 14-6). We found that boundary conditions at the mouth were influenced by conditions inside the bay and in adjacent shelf waters. Consequently, a mass-balance algorithm was used to determine concentration boundary conditions (Cerco and Cole 1994). The algorithm employed a salt balance to distinguish between water leaving the bay, bay water recirculated from the shelf back into the bay, and new shelf water entering the bay. These concepts were retained when nutrient budgets for the system were computed.

The nitrogen budget for 1986 (figure 14-15), a year of roughly average hydrology, indicated that riverine loading was the largest nitrogen source to the system. The budget confirmed that the bay was a net exporter of nitrogen through the mouth. Two-thirds of the nitrogen entering the mouth was recycled nitrogen that originated in the bay. One-third of the nitrogen entering the mouth was newly imported from the shelf waters. The shelf load exceeded all loads to the system except riverine loads. Lesser loading sources were, respectively, wastewater (point sources), overland runoff (nonpoint sources), and atmospheric deposition to the water surface. The detailed budget revealed that half the atmospheric nitrogen load was returned to the atmosphere as a result of water-column denitrification.

Sediments were the greatest nitrogen sink, exceeding by far the net amounts lost to the ocean and through water-column denitrification. Within the sediments, denitrification exceeded burial and was the single largest pathway for nitrogen removal from the system.

The phosphorus budget for a year of average hydrology (figure 14-15) revealed that new phosphorus from the continental shelf was the largest load to the system, comprising nearly half the total load. Lesser loads came from rivers, wastewater, runoff, and the atmosphere. The greatest phosphorus sink was through burial. The only secondary sink was loss through the mouth. More than two-thirds of the phosphorus lost through the mouth was recycled and returned to the bay.

Nitrogen deposition to the sediments exceeded load to the system (figure 14-15). Excess of deposition over loading indicated nitrogen was rapidly cycled from the sediments to the water column and back to the sediments. On an annual basis, nearly half the deposited nitrogen was returned to the water column.

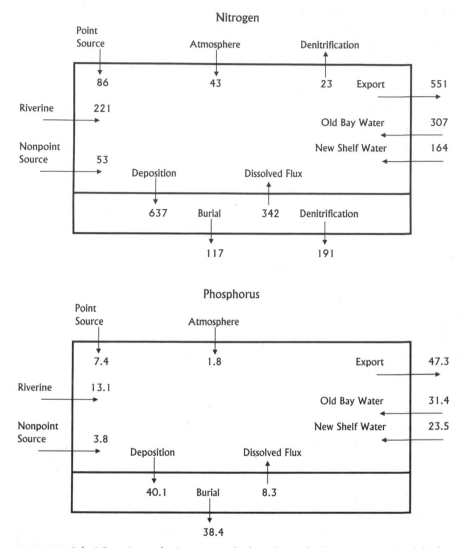

FIGURE 14-15   Annual nitrogen and phosphorus budgets (1,000 kg d⁻¹) for Chesapeake Bay system, 1986. Arrows pointing into boxes indicate sources. Arrows pointing out of boxes indicate sinks.

Phosphorus deposition was also substantial but represented a lesser fraction of total loads than for nitrogen (figure 14-15). Of more significance, only 20% of the deposited phosphorus was returned to the water. The retention was due to the strong tendency for phosphate to sorb onto sediment particles. By contrast, sorption provided negligible retention of ammonium.

Management Applications

Scenario analysis is a primary purpose of eutrophication modeling. Models are used to develop and test load-management strategies aimed at limiting eutrophication processes. More than thirty management scenarios (Thomann et al. 1994) were performed following calibration of the model. Each scenario was run a minimum of 10 years, to allow equilibration of sediments to reduced loading, under a variable hydrologic sequence that included years of high, average, and low riverine flows. Details of the scenario process were presented by Cerco (1994a).

One scenario of particular interest is the limit-of-technology (LOT) load-reduction scenario. The LOT scenario represented the optimal conditions obtained through management controls of nutrient loads throughout the system watershed. LOT point-source loads were computed based on effluent concentrations of 3 g m^{-3} total nitrogen and 0.075 g m^{-3} total phosphorus. LOT loads from riverine and direct nonpoint sources were obtained from the EPA Phase II watershed model (Donigian et al. 1991). The watershed model provided loads based on the application of best management practices to control nonpoint sources of nutrients. These practices included conservation tillage of cropland, removal of highly erodible land from tillage, implementation of structures such as vegetated filter strips, reduction of waste from animal feedlots, and control of urban loads. Existing atmospheric loads to the watershed and to the surface of the bay were employed. Investigation of benefits from controls on atmospheric deposition are planned following improvements to the watershed model and incorporation of an atmospheric deposition model (Appleton 1995) into the model package.

The largest nitrogen reduction under LOT controls is from point sources, followed by riverine loads (table 14-3). Absent the ocean, total reduction in nitrogen loading is 29%. If the ocean load is considered, total nitrogen reduction declines to 21%. Major, equivalent LOT phosphorus reductions come from riverine and point-source loads (table 14-3). The reduction in phosphorus loading is 53% if the ocean is not considered or 29% if it is.

Chlorophyll and Nutrient Limitations

The Monod relationship equation (4), employed in the computation of algal production, is used to quantify nutrient limitation. The potential range of $f(N)$ is from zero, indicating complete growth limitation, to unity, indicating no limitation. Half-saturation concentrations are 0.01 g m^{-3} for inorganic nitrogen, 0.001 g m^{-3} for dissolved phosphate, and 0.05 g m^{-3} for dissolved silica.

Under existing conditions, during the spring bloom (figure 14-16), the model indicates phosphorus is the most limiting nutrient from the fall line

TABLE 14-3

Nutrient Loads Averaged Over Ten-Year Scenario Cycle.

	Nitrogen		Phosphorus	
	Existing (kg d⁻¹)	LOT (kg d⁻¹)	Existing (kg d⁻¹)	LOT (kg d⁻¹)
Riverine	226,895	191,126	15,012	8,306
Nonpoint source	64,836	51,954	4,846	3,179
Point source	86,357	13,153	7,359	274
Atmospheric	43,025	43,025	1,823	1,823
Oceanic	164,000	164,000	23,500	23,500
Total, no ocean	421,113	299,259	29,040	13,583
Total with ocean	585,113	463,259	52,540	37,083

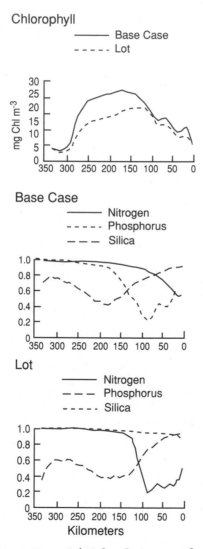

FIGURE 14-16 Spring surface chlorophyll concentration and nutrient limitations for existing conditions and LOT nutrient-reduction scenario.

down to km 125. Below km 125, silica is the most limiting nutrient. Under LOT controls, phosphorus remains the most limiting nutrient in the upper bay but nitrogen replaces silica as most limiting in the lower bay. The replacement occurs as silica, no longer utilized by algae in the upper bay, travels downstream to a previously silica-limited area. In response to the nutrient controls, maximum spring chlorophyll concentration decreases by roughly 40% from existing levels.

During summer (figure 14-17), under existing conditions, the model indicates phosphorus is most limiting from the fall line down to km 100. Below

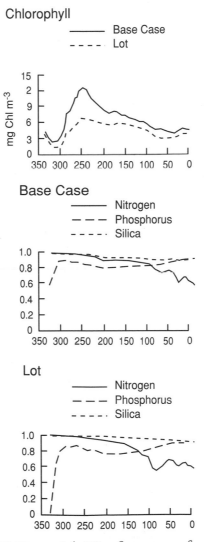

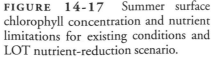

FIGURE 14-17 Summer surface chlorophyll concentration and nutrient limitations for existing conditions and LOT nutrient-reduction scenario.

km 100, nitrogen is most limiting. Careful examination of time-series plots for multiple years indicates the transition from phosphorus to nitrogen limitation is not clearly defined. The location depends on hydrology and time of the season. A more accurate description of summer is that phosphorus clearly limits above km 250, nitrogen clearly limits below km 100, and nitrogen and phosphorus limit from km 250 to 100. The regions of dominant nutrient limitations do not change under LOT controls although the phosphorus limit is stronger in the upper bay. The reduction of chlorophyll in response to LOT controls is most pronounced, roughly 50%, near km 250, in the vicinity of greatest sediment nutrient releases under base conditions.

Nitrogen, Phosphorus, and Dissolved Oxygen

Total nutrient concentrations were much more sensitive to load reductions near the fall line than near the mouth (figure 14-18). Although concentration at the mouth was free to fluctuate in response to load reductions, the concentration change was minimal. The limited response occurred, in part, because of the fixed loads (color plate 3; figure 14-15; table 14-3) from oceanic waters. Response was also limited because substantial fractions of nutrients discharged to the bay are lost through sedimentation rather than advected to the mouth. Transport analyses conducted with the model (Cerco and Cole 1994) indicated only 10% of nitrogen discharged above the Patuxent River leaves the upper bay under average hydrological conditions. Virtually all phosphorus discharged above the Patuxent River is trapped in the upper bay. Less than half of the nitrogen and phosphorus loads to the James River are conveyed to the mainstem, near the mouth. Loss of large load fractions in the upper bay and tributaries attenuates the influence of load controls on lower bay waters and on the mixture of waters entering the open mouth.

Under LOT nutrient controls (figure 14-18), average bottom-dissolved oxygen increased by ≈ 0.5 g m^{-3} throughout the mainstem from km 100 to 260. The improvement occurred primarily at depths greater than 12 m, in the trench that runs from the mouth to km 260. A substantial volume of water with dissolved-oxygen concentration below 1 g m^{-3} persisted, however. The dissolved-oxygen response was disappointing to managers who had hoped for complete elimination of anoxic (DO < 1 g m^{-3}) waters in response to nutrient controls. The response is reasonable, however, when it is understood that LOT controls eliminate only 21% of the total nitrogen loading and 29% of the total phosphorus loading to the system. Worth noting is that the model dissolved-oxygen predictions are consistent with a consensus formed by the scientific community (D'Elia et al. 1992) that a loading reduction of at least 40% is required to eliminate hypoxia. Additional improvements in dissolved oxygen require improved technology, controls on atmospheric deposition,

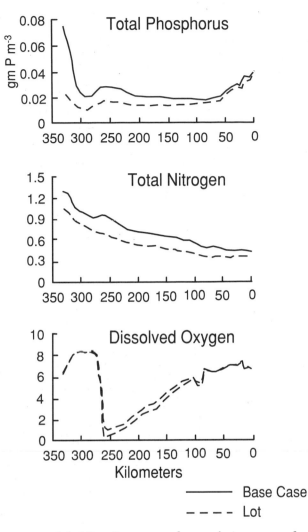

FIGURE 14-18   Summer surface total nitrogen, surface total phosphorus, and bottom dissolved-oxygen concentrations for existing conditions and LOT nutrient-reduction scenario.

and, perhaps, controls on sources that contribute to nutrient concentrations outside the bay mouth.

Present Syntheses

The present model accomplishes several levels of synthesis. First, the study demonstrates that the influences of three-dimensional transport processes and of complex geometry can be coupled with a eutrophication model. Salinity is an ideal indicator of integrated transport processes. The agreement between computed and observed salinity in time (figure 14-11), and space (figure 14-13) indicates the model provides an excellent representation of transport processes. The confinement of anoxia to deep waters in the central trench indicates geometrical effects are well represented (color plate 3).

The second synthesis is between eutrophication processes in the water column and in benthic sediments (figure 14-12). The study was one of the first to couple sediments and water column in a mechanistic, predictive fashion. Coupling of runoff events in one season to water quality in subsequent seasons and long-term residence of nutrients in sediments are now readily addressed in a modeling framework.

Future Prospects

At the time of writing, a second phase of Chesapeake Bay eutrophication modeling is nearing completion (Cerco and Meyers in press) and plans are under way for a third phase. The study under way expands the model treatment of biota and incorporates two size classes of zooplankton, deposit and filter-feeding benthos, and submerged aquatic vegetation. Spatially, the model domain was extended 30 km out onto the continental shelf.

In this second-phase model, computations of zooplankton, submerged aquatic vegetation, and benthos compare successfully with observations aggregated over annual time scales and at spatial scales on the order of 100 km². Mean total nitrogen and phosphorus at the bay mouth can be computed when constant concentrations of these substances are specified 30 to 75 km away from the bay mouth. A large portion of the phosphorus dynamics at the bay mouth can also be captured, apparently because total phosphorus at the mouth is influenced to a large extent by runoff to the bay. Little of the nitrogen dynamics is captured, however, indicating nitrogen at the bay mouth is strongly influenced by the environment outside the bay.

The expansion of the domain out onto the shelf was necessary because of problems encountered with specification of boundary conditions at the bay mouth. Expansion of the grid moved the problem of boundary condition specification but did not eliminate it. The nearly completed study indicates salinity on the northern study boundary is influenced by runoff in the Delaware River. By inference, nutrients at the extent of the present domain are also influenced by the Delaware and by loadings farther up the coast. As management of loadings within the bay reaches a maximum, nutrient sources outside the bay and its watershed must be considered. Consequently, an expansion of the grid to the entire East Coast, allowing examination of the most distant nutrient sources, is in the planning stages.

While the simulations of living resources currently under way are largely successful, they make apparent that many concepts of the model framework are stretched to their limits. Models such as the present one are founded on assumptions that are so ingrained in the conceptual process, they are rarely reconsidered: mass is conserved; substances are well mixed within computational cells; particles travel passively by means of advective and diffusive processes; all particles within a cell are identical. Of what use are these concepts in modeling fish that are discrete particles with age structure, behaviors, and self-propulsion? Population and bioenergetics models of the highest trophic levels are available or under development (Brandt et al. 1992). Coupling of these models with mass-balance models such as this one presents major hurdles. New concepts and paradigms must be developed. Nevertheless, the coupling must and will be accomplished. Synthesis of these diverse model approaches will allow computation of the impact of load management actions on the highest trophic levels and will represent a major advancement in the modeling field.

Acknowledgments

The mass-balance boundary condition algorithm was developed by Dominic DiToro of HydroQual Inc. Development of the Chesapeake Bay model package was sponsored by the U.S. Army Engineer District, Baltimore, Maryland, and the Chesapeake Bay Program Office, U.S. Environmental Protection Agency Annapolis, Maryland. Permission was granted by the Chief of Engineers to publish this information.

References

Appleton, E. 1995. Cross-media approach to saving Chesapeake Bay. *Environmental Science and Technology* 29(12): 550A–555A.

Boicourt, W. 1992. Influences of circulation processes on dissolved oxygen in the Chesapeake Bay. Pages 7–60 in D. Smith, M. Leffler, and G. Mackiernan (editors), *Oxygen dynamics in the Chesapeake Bay.* College Park, MD: Maryland Sea Grant College.

Boynton, W. R., and W. M. Kemp. 1985. Nutrient regeneration and oxygen consumption along an estuarine salinity gradient. *Marine Ecology Progress Series* 23: 45–55.

Boynton, W. R., W. M. Kemp, J. Garber, and J. Barnes. 1986. Ecosystem processes component (EPC) level I Data Report No. 3. [UMCEES]CBL Ref. No. 86–56. Solomons, MD: University of Maryland System Center for Environmental and Estuarine Studies.

Boynton, W. R., J. Garber, R. Summers, and W. M. Kemp. 1995. Inputs, transformations, and transport of nitrogen and phosphorus in Chesapeake Bay and selected tributaries. *Estuaries* 18: 285–314.

Brandt, S. B., D. Mason, and E. Patrick. 1992. Spatially-explicit models of fish growth rate. *Fisheries* 17(2): 23–31.

Cerco, C. F. 1994. Response of Chesapeake Bay to nutrient load reductions. *Journal of Environmental Engineering* 121(8): 549–57.

———. 1995. Simulation of long-term trends in Chesapeake Bay eutrophication. *Journal of Environmental Engineering* 121(4): 298–310.

Cerco, C. F., and T. Cole. 1993. Three-dimensional eutrophication model of Chesapeake Bay. *Journal of Environmental Engineering* 119(6): 1006–25.

———. 1994. Three-dimensional eutrophication model of Chesapeake Bay. Technical Report EL-94-4. Vicksburg, MS: U.S. Army Engineer Waterways Experiment Station.

Cerco, C. F., and M. Meyers. In press. Tributary refinements to the Chesapeake Bay model. *Journal of Environmental Engineering.*

Chen, H., P. Hyer, and Y. Unkulvasapaul. 1984. *Summary report on calibration of water quality models of the Chesapeake Bay system.* Gloucester Point, VA: Virginia Institute of Marine Science.

Conley, D., and T. Malone. 1992. Annual cycle of dissolved silicate in Chesapeake Bay: Implications for the production and fate of phytoplankton biomass. *Marine Ecology Progress Series* 81: 121–28.

D'Elia, C. F., D. Nelson, and W. R. Boynton. 1983. Chesapeake Bay nutrient and plankton dynamics: 3. The annual cycle of silicon. *Geochimica et Cosmochimica Acta* 47: 1945–55.

D'Elia, C. F., L. Harding, M. Leffler, and G. MacKiernan. 1992. The role and control of nutrients in Chesapeake Bay. *Water Science and Technology* 26(12): 2635–44.

DiToro, D., S. O'Connor, and R. Thomann. 1971. A dynamic model of the phytoplankton population in the Sacramento-San Joaquin delta. Pages 131–80 in J. D. Hem (editor), *Nonequilibrium systems in water chemistry.* Washington, DC: American Chemical Society.

DiToro, D., P. Paquin, K. Subburamu, and D. Gruber. 1990. Sediment oxygen demand model: Methane and ammonia oxidation. *Journal of Environmental Engineering* 116(5): 945–86.

DiToro, D., and J. Fitzpatrick. 1993. Chesapeake Bay sediment flux model. Contract Report EL-93-2. Vicksburg, MS: U.S. Army Engineer Waterways Experiment Station.

Donigian, A., B. Bicknell, A. Patwardhan, L. Linker, D. Alegre, C. Chang, and R. Reynolds. 1991. *Watershed model application to calculate bay nutrient loadings.* Annapolis, MD: U.S. Environmental Protection Agency, Chesapeake Bay Program Office.

Dortch, M., R. Chapman, and S. Abt. 1992. Application of three-dimensional Lagrangian residual transport. *Journal of Hydraulic Engineering* 118(6): 831–48.

Fisher, T. R., P. Carlson, and R. Barber. 1982. Sediment nutrient regeneration in three North Carolina estuaries. *Estuarine, Coastal and Shelf Science* 4: 101–16.

Fisher, T. R., E. Peele, J. Ammerman, and L. Harding. 1992. Nutrient limitation of phytoplankton in Chesapeake Bay. *Marine Ecology Progress Series* 82: 51–63.

Flemer, D., G. Mackiernan, W. Nehlsen, and V. Tippie. 1983. *Chesapeake Bay: A profile of environmental change.* Philadelphia, PA: U.S. Environmental Protection Agency, Region III.

Heinle, D., C. F. D'Elia, J. Taft, J. Wilson, M. Cole-Jones, A. Caplins, and E. Cronin. 1980. *A historical review of water quality and climatic data from Chesapeake Bay with emphasis on effects of enrichment.* CBP-TR-002E. Annapolis, MD: U.S. Environmental Protection Agency, Chesapeake Bay Program Office.

HydroQual. 1987. A steady-state coupled hydrodynamic/water quality model of the eutrophication and anoxia process in Chesapeake Bay. Final Report. Mahwah, NJ: HydroQual.

Johnson, B., K. Kim, R. Heath, B. Hsieh, and L. Butler. 1993. Validation of a three-dimensional hydrodynamic model of Chesapeake Bay. *Journal of Hydraulic Engineering* 199(1): 2–20.

Kemp, W. M., P. Sampou, J. Garber, J. Tuttle, and W. R. Boynton. 1992. Seasonal depletion of oxygen from bottom waters of Chesapeake Bay: Roles of benthic and planktonic respiration and physical exchange processes. *Marine Ecology Progress Series* 85: 137–52.

Kremer, J. N., and S. Nixon. 1978. *A coastal marine ecosystem simulation and analysis.* New York: Springer Verlag.

Leonard, B. 1979. A stable and accurate convection modelling procedure based on quadratic upstream interpolation. *Computer Methods in Applied Mechanics and Engineering* 19: 59–98.

Linker, L., C. Stigall, C. Chang, and A. Donigian. 1996. Aquatic accounting: Chesapeake Bay watershed model quantifies nutrient loads. *Water Environment and Technology* 8(1): 48–52.

Malone, T., L. Crocker, S. Pike, and B. Wendler. 1988. Influences of river flow on the dynamics of phytoplankton production in a partially stratified estuary. *Marine Ecology Progress Series* 48: 235–49.

Monod, J. 1949. The growth of bacterial cultures. *Annual Review of Microbiology* 3: 371–94.

Nixon, S. 1987. Chesapeake Bay nutrient budgets—a reassessment. *Biogeochemistry* 4: 77–90.

Odum, E. 1971. *Fundamentals of ecology,* 3d ed. Philadelphia, PA: W. B. Saunders.

Pritchard, D. 1967. Observations of circulation in coastal plain estuaries. Pages 37–44 in G. Lauff (editor), *Estuaries.* Washington, DC: American Association for the Advancement of Science.

Seitzinger, S. P., S. Nixon, and M. Pilson. 1984. Denitrification and nitrous oxide production in a coastal marine ecosystem. *Limnology and Oceanography* 29(1): 73–83.

Seliger, H., and J. Boggs. 1988. Long-term pattern of anoxia in the Chesapeake Bay. Pages 570–85 in M. Lynch and E. Krome (editors), *Understanding the estuary: Advances in Chesapeake Bay research.* Gloucester Point, VA: Chesapeake Research Consortium.

Smith, D., M. Leffler, and G. Mackiernan. 1992. *Oxygen dynamics in the Chesapeake Bay,* xiv–xv. College Park, MD: Maryland Sea Grant College.

Steele. J. H. 1962. Environmental control of photosynthesis in the sea. *Limnology and Oceanography* 7: 137–50.

Thomann, R., J. Collier, A. Butt, E. Casman, and L. Linker. 1994. *Response of the Chesapeake Bay water quality model to loading scenarios.* CBP/TRS 101/94. Annapolis, MD: U.S. Environmental Protection Agency, Chesapeake Bay Program Office.

U.S. Environmental Protection Agency. 1982. *Chesapeake Bay program technical studies, a synthesis.* Washington, DC: U.S. Environmental Protection Agency.

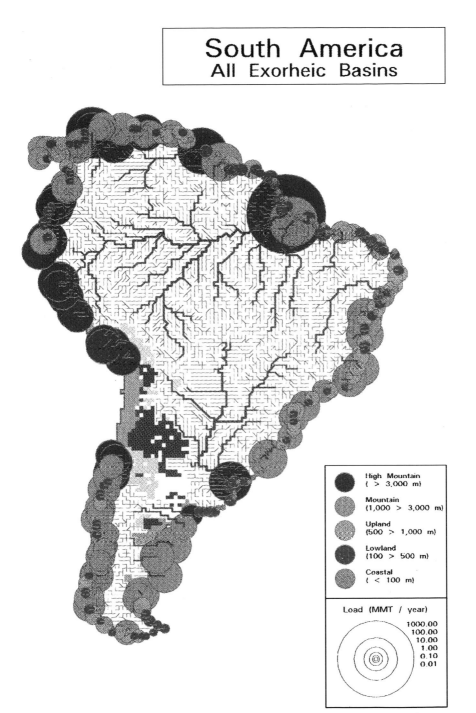

PLATE 1 A geographically specific transformation of the relationships shown in figure 3-10 for South America to river basins discharging to the ocean (exorheic basins, to predict suspended sediment load.) (Accompanies chapter 3.)

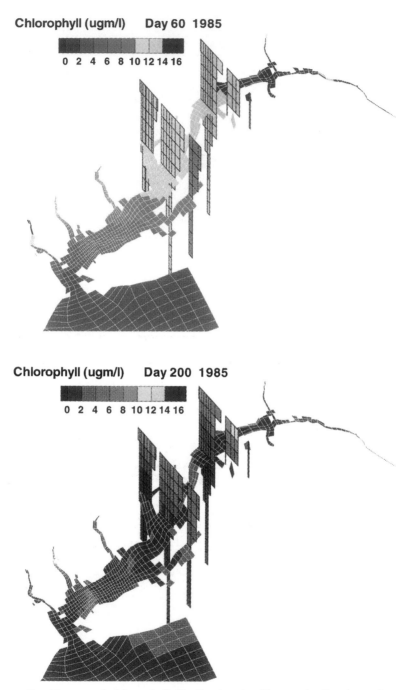

PLATE 2 Computed chlorophyll distribution in Chesapeake Bay in spring and summer 1985. The longitudinal-lateral distribution is shown by a cutting plane at depth = 7.5 m. The vertical-lateral distribution is shown by cutting planes located 121 km and 197 km from the bay mouth. (Accompanies chapter 14.)

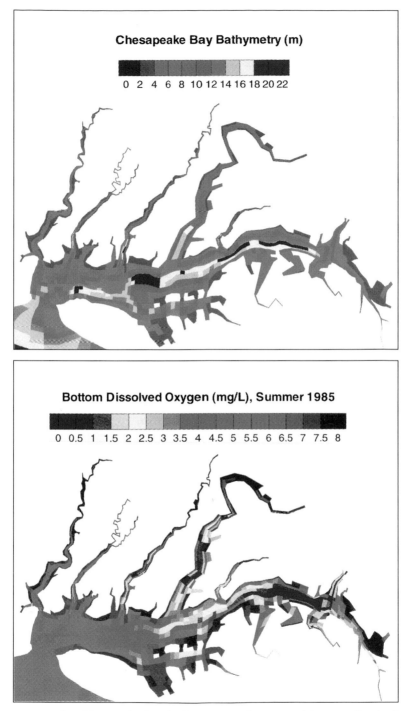

PLATE 3 Chesapeake Bay bathymetry and computed bottom dissolved oxygen, summer 1986. (Accompanies chapter 14.)

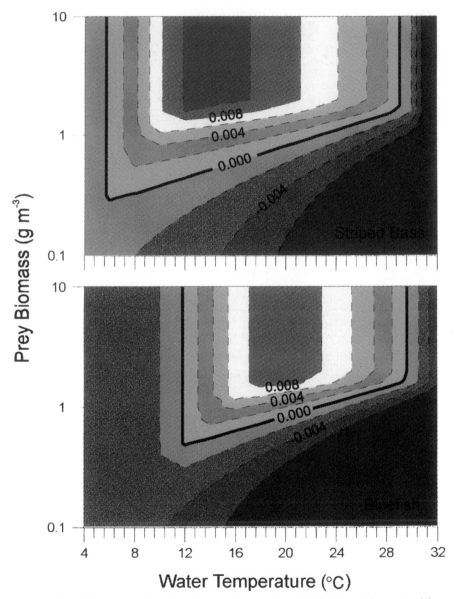

PLATE 4 Contours of predicted growth rates (g g^{-1} d^{-1}) for a 2 kg striped bass (top panel) and a 2 kg bluefish (bottom panel) at different water temperatures and prey biomass (g m^{-3}). The fish biomass scale is logarithmically spaced (base 10). (Accompanies chapter 15.)

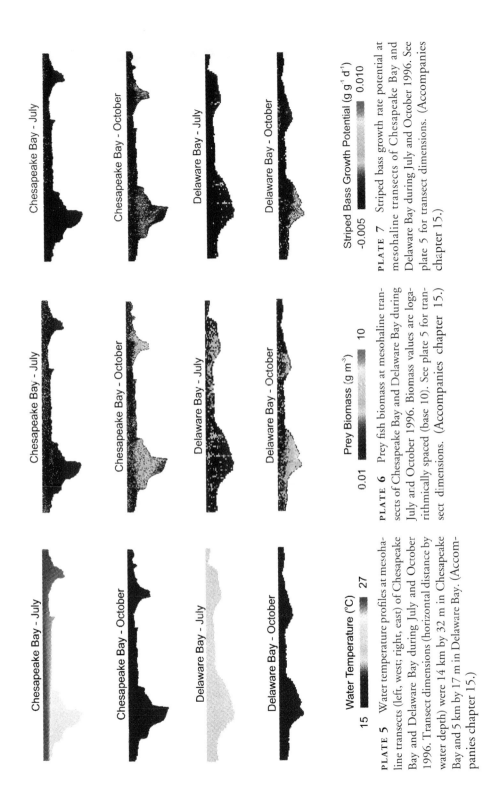

PLATE 5 Water temperature profiles at mesohaline transects (left, west; right, east) of Chesapeake Bay and Delaware Bay during July and October 1996. Transect dimensions (horizontal distance by water depth) were 14 km by 32 m in Chesapeake Bay and 5 km by 17 m in Delaware Bay. (Accompanies chapter 15.)

Water Temperature (°C)

15   27

PLATE 6 Prey fish biomass at mesohaline transects of Chesapeake Bay and Delaware Bay during July and October 1996. Biomass values are logarithmically spaced (base 10). See plate 5 for transect dimensions. (Accompanies chapter 15.)

Prey Biomass (g m⁻³)

0.01   10

PLATE 7 Striped bass growth rate potential at mesohaline transects of Chesapeake Bay and Delaware Bay during July and October 1996. See plate 5 for transect dimensions. (Accompanies chapter 15.)

Striped Bass Growth Potential (g g⁻¹ d⁻¹)

-0.005   0.010

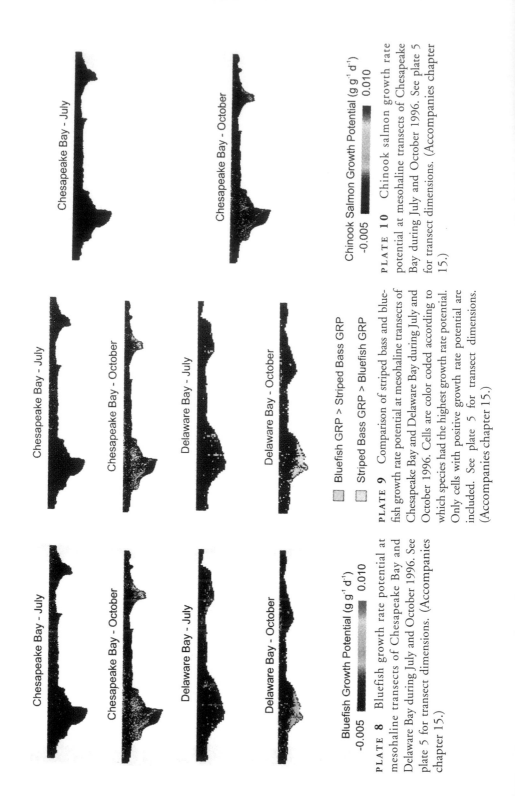

Chesapeake Bay - July

Chesapeake Bay - October

Chesapeake Bay - July

Chesapeake Bay - October

Delaware Bay - July

Delaware Bay - October

Chesapeake Bay - July

Chesapeake Bay - October

Delaware Bay - July

Delaware Bay - October

Chinook Salmon Growth Potential (g g^{-1} d^{-1})

-0.005   0.010

PLATE 10   Chinook salmon growth rate potential at mesohaline transects of Chesapeake Bay during July and October 1996. See plate 5 for transect dimensions. (Accompanies chapter 15.)

Bluefish GRP > Striped Bass GRP

Striped Bass GRP > Bluefish GRP

PLATE 9   Comparison of striped bass and bluefish growth rate potential at mesohaline transects of Chesapeake Bay and Delaware Bay during July and October 1996. Cells are color coded according to which species had the highest growth rate potential. Only cells with positive growth rate potential are included. See plate 5 for transect dimensions. (Accompanies chapter 15.)

Bluefish Growth Potential (g g^{-1} d^{-1})

-0.005   0.010

PLATE 8   Bluefish growth rate potential at mesohaline transects of Chesapeake Bay and Delaware Bay during July and October 1996. See plate 5 for transect dimensions. (Accompanies chapter 15.)

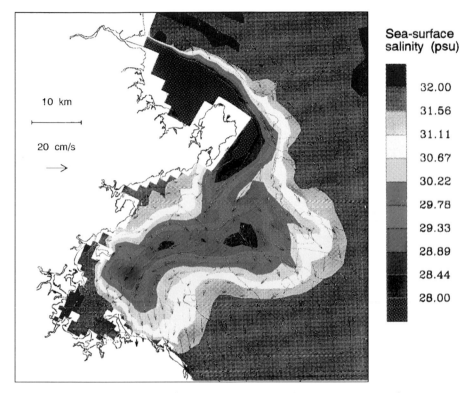

PLATE 11   Modeled surface current and salinity response of western Massachusetts to a runoff event from the Merrimack River. From Signell et al. 1993, with permission. (Accompanies chapter 16.)

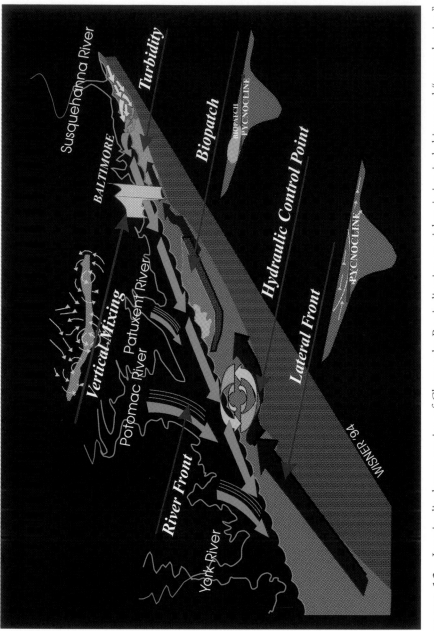

PLATE 12 Longitudinal cross section of Chesapeake Bay indicating spatial variation in habitats and "control points" that can be focal points of intense habitat-biota interactions. Figure courtesy of Walt Boynton, UM-CEES. (Accompanies chapter 16.)

CHAPTER 15

Spatially Explicit Models of Growth Rate Potential

Linking Estuarine Fish Production to the Biological and Physical Environment

Eric Demers, Stephen B. Brandt, Karen L. Barry, and J. Michael Jech

Abstract

Estuaries are high-yield fishing areas that are characterized by spatial heterogeneity in physical and biological conditions. Models of fish production have traditionally been based on systemwide averages of environmental conditions, but habitat heterogeneity can substantially influence fish growth. Growth rate potential (GRP) provides a spatially explicit approach that integrates the heterogeneous nature of estuaries into a simple modeling framework. In this chapter, we describe and illustrate the application of GRP to compare potential growth of two piscivores and to determine the potential growth of a non-native species introduced into an estuary. Acoustically derived prey distributions and temperature profiles were merged in a spatially explicit analysis to estimate and compare GRP of striped bass and bluefish in Chesapeake and Delaware Bays. In this analysis, bluefish grew better in the thermal regimes and prey biomass available during midsummer while striped bass had higher potential growth during fall. This suggests that, although striped bass and bluefish use similar prey resources, they may be thermally and temporally segregated, thereby reducing competitive overlap. In our second example, GRP results indicate that the suitability of Chesapeake Bay for the growth of chinook salmon (a non-native, hypothetical invader) was very low during summer, whereas in October, water temperature and prey availability could possibly support chinook salmon growth. This spatially explicit approach proved to be a valuable tool to study fish production in estuarine systems where heterogeneous conditions can affect populations at systems levels.

Introduction

Estuaries are very productive ecosystems widely recognized as high-yield fishing areas (McHugh 1967). Indeed, more than 50% of the total U.S. fishery harvest includes species that are estuarine or estuarine dependent at some life stage (Houde and Rutherford 1993). The large nutrient- and sediment-rich water plumes of estuaries also influence continental shelf and offshore fisheries (Day et al. 1989). The ecological importance of estuarine habitats to fish includes providing productive feeding areas (Goshorn and Epifanio 1991), spawning sites for adults (Shepherd and Grimes 1983), and nursery grounds for larvae and juveniles (Grecay and Targett 1996). Many estuaries are impacted by anthropogenic activities such as changes in land-use patterns, nutrient enrichment, contaminant loading, recreational and commercial fisheries, shipping, general recreation, and aquaculture. Such activities can lead to biological and physical alterations to the environment that can adversely affect fish production.

Fish production depends on changes in the number of individuals in populations over time (caused by mortality, migration, and recruitment), and in growth rates of individuals within the population. The growth rate of an individual fish is a highly pliant, species- and size-specific response to environmental conditions and food availability. Growth is an important parameter, since high growth has been linked to improved survivorship and larger size at age (Houde 1987). Growth rate is also directly related to reproductive potential because larger females often produce more and larger eggs, which can enhance larval survival (Monteleone and Houde 1990).

Models of population production are typically based on mean conditions (water temperature, prey density) over large areas, but recent work suggests that local biological and physical processes that occur at relatively small spatial scales can significantly affect population processes including trophic interactions, mortality, and system-level production (see Kotliar and Wiens 1990). Traditional models of fish production have also assumed homogeneity and constancy of the environment, but natural systems are neither homogeneous nor constant in time and space. Systemwide averages of predator and prey abundances that do not incorporate spatial heterogeneity of the environment may be insufficient for understanding production dynamics or predator-prey interactions. Moreover, heterogeneity in the environment results in a mosaic of rate-determining habitats that often have nonlinear effects on the ecological and physiological processes that regulate fish growth and survival. For example, systemwide averages of water temperature and prey density may be meaningless to fish growth rate if the overlap in the distribution of predator and prey populations across a rate-determining heterothermal environment is not considered. Rose and Leggett (1990) showed that correlations in the spatial distribution of Atlantic cod (*Gadus morhua*) and their prey, capelin (*Mallotus villosus*), were positive, negative, or insignificant depending on the scale of observation. Since predator-prey and foraging models are sensitive to scale effects (Wiens 1989), scale-dependent linkages of biological function to the biological and physical structure must be evaluated to understand the mechanisms regulating production processes and dynamics in aquatic systems (Legendre and Demers 1984; Carpenter 1988).

Because of inherent difficulties in measuring underwater spatial and temporal heterogeneity, there have been relatively few studies of spatial processes in aquatic systems. However, recent advances in underwater remote sensing techniques (such as Doppler current profilers, optical plankton counters, underwater acoustics) now provide sufficient spatial information about aquatic environments to formulate and test hypotheses. The development of techniques for continuously measuring phytoplankton and zooplankton provides a framework to define the spatial characteristics and dynamic linkages between biology and physics at lower trophic levels (Bennett and Denman 1985). For fish, underwater acoustics provide one of the few means to continuously measure fish abundances across large bodies of water. Acoustic techniques have been applied routinely for fish stock assessment (see Brandt et al. 1991) and have also been used to directly measure spatial patchiness of fish distributions and its correlation with physical structure (Nero et al. 1990). The spatial information inherent in acoustic data can also be integrated in spatially explicit ecological models of fish production.

Spatially explicit modeling of fish growth rate potential has recently been proposed as a quantitative tool for linking species- and size-specific production to the physiological and behavioral requirements of fish, and to the prevailing biological and physical conditions of the environment (Brandt et al. 1992; Mason et al. 1995). Growth rate potential is defined as the expected growth rate of a predator if placed in a particular volume of water with known physical (temperature) and biological (food resources) characteristics. Thus, growth rate potential reflects the individual's response to environmental conditions, metabolic requirements, and overall activity level. This measure has advantages over simply measuring fish growth in the field in that it (1) provides a measurement of the environment itself; (2) is a mechanistic and process-based approach that allows interpretation of the causes of changes in growth rate; (3) is independent of the actual predator distribution and actual growth rates; and (4) allows predictions of the potential responses of a native or non-native species introduced into a new habitat.

In this chapter, we (1) review the application of spatially explicit models of growth rate potential to functionally define and quantify fish growth with respect to the physiological needs of targeted fish species; and (2) show with examples how this approach might be applied to estuarine fisheries issues.

Our first example compares the pelagic environment of two economically and ecologically important species, striped bass (*Morone saxatilis*) and bluefish (*Pomatomus saltatrix*) in Chesapeake Bay and Delaware Bay. In this example, we compare the pelagic environment for these two predatory species across estuaries and seasons and evaluate the degree of growth rate potential overlap between species. Our second example includes a determination of the suitability of Chesapeake Bay for chinook salmon (*Oncorhynchus tshawytscha*) as an example of a hypothetical species introduction and to illustrate how growth rate potential does not require the predator to be present in the modeled environment.

General Approach

Spatial Modeling

Spatial complexity is an inherent and scale-dependent attribute of pelagic ecosystems (Legendre and Demers 1984; Wiens 1989), yet it is often ignored in studies of biological processes and production, particularly at higher trophic levels. Biological processes that occur at relatively small spatial scales ($1–10$ m^3) may significantly affect production at the whole-system level (Kotliar and Wiens 1990). For example, Lasker (1978) showed that survival and growth of larval northern anchovies (*Engraulis mordax*) depended on the existence of food patches, whereas average values of prey density would lead to starvation. Therefore, the distribution of a predator and its prey across heterogeneous environments may strongly influence the predator's consumption, growth, and, by extension, overall production at the ecosystem level.

By coupling spatially explicit modeling with remote-sensing techniques (underwater acoustics) that provide a high-resolution view of the environment (see Brandt et al. 1992), the bias associated with standard spatially averaged models can be overcome. Incorporating the spatial environment into a modeling framework has improved our understanding of how production processes and predator-prey interactions respond to the underlying biological and physical structure of the environment (Carpenter 1988).

Spatially explicit models of fish growth rate potential have been used for a wide variety of applications and environments. These models have been used to (1) evaluate growth and habitat quality of adult salmonines in the North American Great Lakes (Goyke and Brandt 1993; Mason et al. 1995; Hondorp and Brandt 1996); (2) map seasonal patterns of striped bass growth rate (Brandt and Kirsch 1993) and bay anchovy (*Anchoa mitchilli*) planktivory (Luo and Brandt 1993; Luo et al. 1996) in Chesapeake Bay; (3) evaluate the effects of thermal fronts on fish growth (Brandt 1993); and (4) assess the effects of spatial scale of observation on fish growth by borrowing approaches used in landscape ecology (Brandt and Kirsch 1993; Brandt and Mason 1994; Mason and Brandt 1996). The spatially explicit modeling approach has been extended to include individual-based models and ideal free distributions to predict habitat selection of foragers in spatially complex environments (Tyler and Hargrove 1997; Tyler and Rose 1997; Tyler and Brandt in review). Most of these studies have shown that fish growth is sensitive to the spatial heterogeneity of physical and biological characteristics of environments, and that scales of sampling and modeling critically affect interpretations of biological process.

Model Structure

Spatially explicit models subdivide the habitat into small homogeneous volumes or cells. Each cell is treated individually and is characterized by its own set of measured or simulated attributes including water temperature, prey density, and prey size. The same process-oriented simulation models are run in each cell to produce a growth rate potential, but each cell is parameterized with its own attributes. Fish growth rate potential in each cell is defined by the relationship between the supply of prey afforded by the habitat (prey density and sizes) and the amount of prey that the predator could consume (predator demand) based on its physiological capabilities under the environmental characteristics (water temperature) of each cell. Thus, predator growth rate potential within each cell depends on the innate growth potential of the predator and on the constraints imposed by the habitat in the cell. These constraints include inadequate prey availability, suboptimal water temperatures, and foraging limitations.

The grid model has a particular volume of water, V_i, as the basic cell unit. In our case, the number and size of the cells are determined by water depth, horizontal extent of sampling, and the technological limitations of underwater acoustic sampling. Field measures of prey density (D_i), prey size (S_i), and interpolated water temperature (T_i) for each cell i are used as inputs to foraging and growth (bioenergetics) models that are run in each cell. The high-resolution prey data are acquired with acoustic techniques that measure fish densities and sizes throughout the water column on a near-continuous basis (MacLennan and Simmonds 1992; Brandt 1996). The foraging submodel computes predator consumption rate potential (C_i) from measured prey densities and sizes. The bioenergetics model estimates predator growth rate potential (G_i) from the consumption rate and the physiology of the predator. Results are displayed as cross-sectional maps of predator growth rate potential in each cell as if the predator were to occupy that particular volume of water. Recent developments in interactive data visualization (see Platt and Sathyendranath 1988) are used to illustrate the spatial relationships of fish growth rates. Brandt et al. (1992) provide a general introduction and discussion of the overall approach.

Bioenergetics Model

Fish growth rates are highly sensitive to water temperature and food supply (Bartell et al. 1986) and thus, bioenergetics models are useful for evaluating the effects of changes in temperature and prey abundances on consumption, growth, and ultimately trophic interactions (Brandt and Hartman 1993). Bioenergetics models are species-specific, energy-balance approaches that describe the flow of energy through an individual fish and how that energy is partitioned among consumption (foraging), growth (somatic and reproductive), and losses (respiration, egestion, excretion, and specific dynamic action). Bioenergetics models can be used to estimate food consumption based on observed growth rates, thermal history, and diet, or alternately to simulate growth of fish under different conditions of diet, prey availability, or water temperature. We use the "Wisconsin" bioenergetics model (Kitchell et al. 1977; Hanson et al. 1997) to compute a fish's growth rate potential under given environmental conditions. This model has been used for a wide variety of applications to fisheries management and ecology (Ney 1993; Brandt and Hartman 1993). Sensitivity analyses (Bartell et al. 1986) and model validation studies (Rice and Cochran 1984; Beauchamp et al. 1989) have shown that bioenergetics models generally provide accurate and robust estimates of fish consumption and growth.

A species-specific bioenergetics model is run in each cell to determine the growth rate potential of an individual fish that might occupy that cell. Growth rate (G_i) of an individual fish in cell volume V_i depends on consumption (C_i), metabolic costs (R_i), and the energy losses due to excretion (U_i), egestion (F_i), and specific dynamic action (H_i):

$$G_i = C_i - (R_i + F_i + U_i + H_i).$$

Consumption and respiration are influenced nonlinearly by the water temperature (T_i) in the cell volume and the weight (W) of the predator, and energy losses $(U_i, F_i, \text{ and } H_i)$ are typically expressed as proportions of food consumed. We assume equal energy densities $(J g^{-1})$ for prey and predator since sensitivity analyses have shown bioenergetics models to be insensitive to energy density (Bartell et al. 1986).

Foraging Model

Several foraging models can be used to define the relationship of prey density to predator consumption (Stephens and Krebs 1986). Generally, foraging is considered a random process whereby the predator randomly encounters prey, and prey are assumed to be randomly distributed. Normally, the latter occurs only in a small local volume or cell. To simplify the problem of spatial patchiness of the prey, we reduce the scale of observation to a volume of water sufficiently small so as to assume randomly distributed prey. Our approach does not assume discrete food patches, but rather uses a nearly continuous spatial measure of food density determined with underwater acoustics. The predator consumption rate (C_i) is a general function of encounter rate (E_i, the number of prey encountered per unit time) and the combined probabilities (k) of prey detection (recognition), attack, capture, and ingestion (Fuiman and Gamble 1989):

$$C_i = E_i \cdot k$$

We define the latter combined probabilities (k) as foraging efficiency.

Predator encounter rate with prey is assumed to depend directly on predator reaction distance (RD, the distance at which an individual prey is recognized), swimming speeds of the predator (v) and prey (u, $u < v$), and prey density (D_i) (Gerritsen and Strickler 1977). We generally apply the encounter rate model developed by Gerritsen and Strickler (1977):

$$E_i = \frac{\pi(\text{RD})^2}{3} \cdot \frac{3v^2 + u^2}{v} \cdot D_i$$

Predator consumption rate (C_i) is constrained by water temperature and weight-dependent maximum consumption ($C_{\max}$); that is, an individual fish cannot consume more than it can physiologically contain, assimilate, and evacuate in a single day. Feeding rate is thus bounded by a maximum consumption function.

Measurements of Prey Density

Underwater acoustics provide one of the few means to obtain high-resolution and near-continuous measures of the spatial distribution of fish density over large areas. This technique can effectively sample pelagic ecosystems that are often characterized by short-term dynamics and extreme spatial patchiness through the water column (Brandt 1996). General reviews of acoustics can be found in MacLennan and Simmonds (1992), Brandt (1996), and Medwin and Clay (1998). Acoustic systems sample the water column by sending repetitive pulses (for example, 3 pulses s^{-1}) of sound in a directed beam downward through the water column as the survey vessel moves across the surface. Acoustic pulses are generated by short (0.2–1.0 ms) bursts of high-frequency (for example, 12–420 kHz) voltage from a beam-forming array of pressure transducers. The resulting pressure wave propagates from the transducer at the speed of sound in water. When the sound wave encounters a fish, or any other acoustic scatterer, an echo is reflected from the target and received at the surface. Echoes contain information on the number of targets, target location, and target size. The strength of an echo can be related to the biomass of the target. Since sound travels in sea water at ca. 1,500 m s^{-1}, the entire water column can be sampled quickly, and a continuous map of fish densities can be obtained. Acoustic data are combined into cells by defining the depth intervals and horizontal distances over which data are pooled.

Application Examples

The above model produces spatial arrays of fish growth rate potential that define the environment from the point of view of the predator. We provide two examples of applications from estuarine systems: (1) a comparison of growth rate potential for striped bass and bluefish in two Atlantic Coast estuaries, the Chesapeake Bay and Delaware Bay; and (2) a determination of growth rate potential for chinook salmon in Chesapeake Bay as an example of a hypothetical species introduction.

Striped Bass and Bluefish Growth Potential in Chesapeake and Delaware Bays

Chesapeake Bay and Delaware Bay are large, productive estuaries that support extensive commercial and recreational fisheries. These estuaries are highly dynamic ecosystems (as exhibited, for instance, in changes in commercial catches, increasing hypoxia in bottom waters of Chesapeake Bay), and they are constantly influenced by management activities (nutrient-loading reductions, enhancement or restoration of fisheries production). Any perturbations to these ecosystems that affect lower trophic levels could eventually be manifested in the higher trophic levels occupied by economically important fishes (bottom-up control; Carpenter 1988). Similarly, changes in overall predator consumption rates caused by changes in the environment (such as the seasonal cycle of water temperatures or increased hypoxia) or in the abundance of predators (such as stocking, migration, mortality) could cause subsequent changes in lower trophic levels (top-down control; Carpenter et al. 1985). Many of the ecologically and economically valuable fishes of these two estuaries are piscivores that depend on the pelagic environment for survival. Although some progress has been made toward understanding the upper trophic levels in these ecosystems (see Baird and Ulanowicz 1989), the functional linkages between prey abundances, environmental characteristics, and predator production remain to be quantitatively defined.

For this example, we evaluate the growth rate potential of a 2 kg striped bass and a 2 kg bluefish in Chesapeake Bay and Delaware Bay. Striped bass and bluefish support valuable recreational and commercial fisheries in these two estuaries, and their abundance in Atlantic Coast estuaries has fluctuated in the last few decades (NMFS 1997). Both predators commonly feed on the same prey fish species, namely bay anchovy, Atlantic menhaden (*Brevoortia tyrannus*), Atlantic silversides (*Menidia menidia*), Atlantic croaker (*Micropogonias undulatus*), and spot (*Leiostomus xanthurus*) during their estuarine residency (Hartman and Brandt 1995a), and thus competitive interactions likely occur between these two species. Our modeling approach can provide an evaluation of the estuarine environment and potential competitive overlap for these two estuarine predators.

We use the species-specific bioenergetics models developed by Hartman and Brandt (1995b) for these two species based on laboratory experiments. Striped bass and bluefish differ in their basic energetics (figure 15-1): under unlimited food availability, striped bass is capable of positive growth from 5 to 31°C, whereas positive growth for bluefish occurs at water temperatures of 10–33°C. Also, optimal temperature ranges for maximum growth differ between species (13–19° and 19–24°C for striped bass and bluefish, respectively). Although bluefish are capable of higher individual consumption than striped bass, this is somewhat offset by higher metabolic costs for bluefish (Hartman and Brandt 1995b).

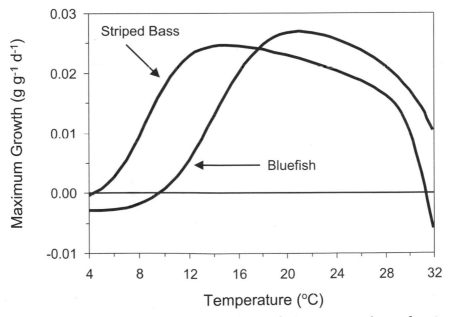

FIGURE 15-1   Bioenergetics model predictions of maximum growth rates for a 2 kg striped bass and a 2 kg bluefish at various water temperatures and under conditions of unlimited food availability.

Field measurements of prey density and water temperature were made at night during July and October 1996. Sampling was conducted along lateral transects taken in the mesohaline, middle portion of each estuary (Chesapeake Bay: 37°46'N, 76°10'W; Delaware Bay: 39°08'N, 75°13'W). Transect lengths were 14 km in Chesapeake Bay and 5 km in Delaware Bay. Water temperatures were interpolated from profiles taken with conductivity-temperature-depth sensors (CTD) at the beginning and end of each transect, and at two additional equidistant points along the transects in Chesapeake Bay. Prey densities were measured with underwater acoustics. The acoustic methods are detailed in Brandt (1990) and Barry et al. (1997) and are summarized here. Acoustic data were collected continuously along each transect using a split-beam 120 kHz Simrad echosounder. The transducer was mounted on a 2.5-m-long aluminum vessel towed near-surface (1.5 m) along the research vessel at speeds of 2.0–2.5 m s^{-1}. Digital information was stored directly in a personal computer and processed using the processing software Digital Echo Visualization and Information System (DEVIS 3.0) developed by Jech and Luo (in review). This system processes digital acoustic data and translates integrated echo and individual target information into two-dimensional matrices of fish density and fish size. Fish densities (fish m^{-3}) were measured throughout the water column using narrow depth intervals (0.5 m) and averaged over constant time intervals (15 sec, ~30–40 m). Fish size (total length) was computed from measured target strengths using Love's (1977) equation for clupeid fishes. Fish densities were converted to fish biomass (g m^{-3}) using species-specific length-weight relationships (Hartman and Brandt 1995c) applied to the mean fish size and fish density within each cell. The acoustic data were screened for prey too small or too large to be consumed by predators. Prey size ranges were determined from diet data for Chesapeake Bay (Hartman 1993); they were set to 40–250 mm for striped bass and 40–300 mm for bluefish. Species composition was assessed with 2–4 midwater trawls taken near the acoustic transects. Bay anchovy comprised 80% and 95% of total fish caught in midwater trawls in Chesapeake Bay during July and October, respectively. Atlantic croaker and spotted seatrout (*Cynoscion nebulosus;* July only) comprised most of the remaining fish caught. In Delaware Bay, more than 99% of all fish caught in midwater trawls during both July and October were bay anchovy.

We estimated growth rate potential for each species, estuary, and time of year using the GRP Map Maker software developed by Tyler (1998). This software integrates field measures of prey density and water temperature with bioenergetics and foraging model parameters to produce spatially explicit estimates of growth rate potential within each cell along our acoustic transects. The results are displayed as maps of growth rate potential as a function of horizontal position and water-column depth along the transect.

Given the different foraging behaviors of bluefish and striped bass, some of the foraging model parameters differed between species. Predator weights were converted to total length to facilitate use in the foraging model. A 2 kg predator corresponded to a total length of 567 mm for striped bass and 584 mm for bluefish (K. J. Hartman, unpublished data). Swimming speeds (body length s^{-1}; BL s^{-1}) of striped bass and bluefish were set at 1.0 and 1.25 BL s^{-1}, respectively. The higher swimming speed for bluefish was based on their perceived higher relative mobility in comparison with striped bass (Freadman 1979). Data on the reactive distances (RD) of striped bass and bluefish are unavailable, so we assumed that RD was one body length (BL) for all encounter rate calculations. Foraging efficiencies (k) for these two species under field conditions are unknown. We assumed that the probability of prey detection, attack, and capture were each equal to 0.10, and that the probability of prey ingestion once captured was equal to 1.0. Thus, the product of these probabilities (that is, foraging efficiency) was set to 0.001.

The overall relationship of fish growth rate potential to prey density and water temperature (color plate 4) shows that growth rate potential can be determined if any combination of temperature and prey density are known. Positive growth can occur only if water temperature is within the range for positive growth, and prey density is high enough for food consumption to exceed metabolic and waste losses. If prey density exceeds the level necessary for maximum consumption at any given water temperature, surplus food does not contribute to additional growth.

Temperature and prey fish distribution. Water temperatures varied seasonally and between estuaries but were relatively uniform (range <2.5 °C) within transects (color plate 5). During July, water temperatures were higher in Chesapeake Bay (24.5–26.7 °C) than in Delaware Bay (21.4–23.0 °C). By contrast, water temperatures were very similar between Chesapeake Bay and Delaware Bay during October (15.2–16.2°C) and varied by <1°C within each estuary. Water temperatures during July corresponded to the range where bluefish are capable of higher growth rates than striped bass, whereas temperatures during October were more favorable to striped bass growth (figure 15-1).

Prey fish biomass was higher in October than in July for both estuaries (color plate 6). During July in Chesapeake Bay, prey fish were scattered relatively evenly throughout the upper half of the transect (0–15 m), but there was virtually no prey in the deeper water. The absence of prey in deeper water was likely due to hypoxic conditions at those depths (dissolved oxygen <1 mg l^{-1}). Hypoxia did not occur in Delaware Bay, and prey were distributed throughout the transect during July. Transect-wide mean prey fish biomass during July (table 15-1) was lower in Chesapeake Bay (0.11 g m^{-3}) than in Delaware Bay (0.17 g m^{-3}). Similarly, only 10% of the environment sampled in Chesapeake Bay during July contained prey fish (that is, 685 of 6,758 cells with biomass > 0 g m^{-3}), compared to 24% for Delaware Bay (table 15-1). During October, patterns of prey biomass were more similar between estuaries, and the highest levels occurred in the deeper channel areas. Although mean prey fish biomass differed between estuaries during October (0.33 and 0.20 g m^{-3} for Chesapeake Bay and Delaware Bay, respectively), the proportion of the environment with prey fish was similar for both estuaries (46% for Chesapeake Bay and 43% for Delaware Bay; table 15-1).

Growth rate potential. For both estuaries, striped bass growth rate potential was highest during October (color plate 7), due to better temperatures for growth and higher prey fish biomass. During July, striped bass growth rate potential was low across most of the transects in both estuaries, and a few higher growth rate potential levels were limited to areas with elevated levels of prey biomass. Only 1–3% of the environments could support positive growth during July (table 15-1). Although prey biomass was similar between estuaries in July, mean striped bass growth rate potential was higher in Delaware Bay because lower temperatures were better for striped bass growth. During October, striped bass growth rate potential was high in the deeper channel areas of both estuaries where prey fish biomass was also highest. Approximately 11–14% of all cells supported positive growth (table 15-1),

TABLE 15-1

Mean and percentage of cells with positive prey biomass and growth rate potential for Chesapeake Bay and Delaware Bay during July and October 1996.

Measurement	Chesapeake Bay		Delaware Bay	
	July	October	July	October
Biomass				
Mean	0.107	0.330	0.171	0.195
% cells greater than 0	10.1	45.6	23.8	42.9
Striped Bass Growth Rate Potential				
Mean	−0.011	−0.003	−0.008	−0.003
% cells greater than 0	1.5	13.8	2.5	11.2
Bluefish Growth Rate Potential				
Mean	−0.015	−0.005	−0.011	−0.005
% cells greater than 0	1.7	13.4	2.7	10.8

and 5% of the transects in both estuaries contained prey biomass levels that were high enough for striped bass to achieve maximum food consumption and hence maximum growth rate potential.

Bluefish growth rate potential did not follow the same pattern as for striped bass (color plate 8). The highest levels of growth rate potential for bluefish in both estuaries occurred during July, but they were limited to a few areas with elevated levels of prey biomass. Mean bluefish growth potential in July was slightly higher in Delaware Bay since water temperatures were closer to the optimal level for bluefish growth (table 15-1). Only 2–3% of all cells could support positive growth during July, but most of these cells had prey biomass high enough for bluefish to achieve maximum growth. During October, bluefish growth rate potential was high in the deeper channel areas where prey fish biomass was also highest. Highest growth rate potential during October was lower than during July because of cooler water temperatures, which were less favorable for bluefish growth. Nonetheless, about 11–13% of the environment supported positive growth (table 15-1). In general, bluefish reached higher growth rate potential levels during July because of favorable water temperatures, but higher prey biomass during October resulted in a larger proportion of the environment being suitable for positive growth.

For both species, transect-wide averages of growth rate potential were always negative (table 15-1), even though spatial maps of growth rate potential displayed numerous areas with positive growth. This exemplifies the importance of having spatially explicit information on distributions of water temperature and prey biomass. Simply using average values of water temperature and prey density across the environment would lead to the erroneous conclusion that the habitat could not support positive growth for striped bass or bluefish. Positive growth by these two species in Chesapeake Bay and Delaware Bay is likely linked to the existence of areas with appropriate food resources and environmental conditions.

Growth rate potential overlap. Because striped bass and bluefish feed on similar prey fish species during their estuarine residency (Hartman and Brandt 1995a), they may be competing for food resources. Indeed, distributions of growth rate potential for striped bass and bluefish were often similar across environments; that is, cells with elevated growth rate potential for one species usually had elevated growth rate potential for the other species as well. To quantify the extent of growth rate potential overlap between striped bass and bluefish, we compared growth rate potential levels between species and for each environment. Cells were color coded (light blue for bluefish and yellow for striped bass) according to which species had the highest growth rate potential. Only cells with positive growth rate potential were included in this analysis.

In both estuaries, growth rate potential was higher for bluefish than for striped bass across transects during July (color plate 9). The warm water temperatures during July in both estuaries were more favorable for bluefish growth. During October, striped bass had higher growth rate potential than did bluefish in about 56% and 52% of cells with positive growth in Chesapeake Bay and Delaware Bay, respectively. Cooler water temperatures during October were more favorable for striped bass growth than for bluefish growth. At water temperatures of 15–16°C, bluefish growth is largely limited by their reduced maximum possible consumption rates. Although prey fish biomass was high during October, bluefish growth was restricted by the temperature-related physiological limits imposed on maximum consumption. However, at prey fish biomass concentrations that were suboptimal for both species (that is, $0–1 \text{ g m}^{-3}$), bluefish had higher growth rate potential. Higher growth rate potential for bluefish in those cells was largely due to their more active foraging behavior in the model, which enabled them to search a larger volume and encounter more prey in a given amount of time. In turn, higher encounter rates resulted in higher consumption rates and a higher growth rate potential.

Measures of growth rate potential support the concept that fish use of estuaries may be driven by both thermal physiology and prey availability. Striped bass were better suited to the thermal regimes and prey biomass available during October, while their growth potential during July was limited by high water temperatures. In contrast, thermal conditions were suitable for bluefish during July but were limiting maximum growth rate potential during October even though prey biomass was relatively high. Such patterns of growth rate potential for these two estuarine species are consistent with their reported use of Atlantic Coast estuaries. Bluefish typically inhabit estuaries when water temperatures are above 15°C (usually May to October) and they overwinter in the coastal waters off the southeastern United States from November to April, where temperatures are more favorable during that time (Kendall and Walford 1979). Bluefish grow rapidly during their summertime estuarine residency, and age-1 and age-2 fish can double their weight between June and October (Hartman and Brandt 1995a). Striped bass are annual residents of Chesapeake Bay and Delaware Bay, and their highest growth rates typically occur during spring and fall with little or no growth during summertime (Hartman and Brandt 1995a). Although striped bass and bluefish likely compete for similar prey in estuaries, competitive overlap may be reduced by their different thermal physiologies.

Growth Rate Potential for Chinook Salmon in Chesapeake Bay: Example of Species Introduction

Because growth rate potential is based on environmental conditions and predator physiology and because it is independent of actual predator distribution and growth rates, this approach can be particularly useful to evaluate the potential response (for instance, growth) of a species that has been purposefully or accidentally introduced into an habitat. One can evaluate the suitability of a particular environment for a species based simply on measurements of the environment's thermal structure and prey distribution. Estimates of growth rate potential can reveal whether conditions are adequate to allow introduced individuals to achieve positive growth and for the population to be sustained or to expand.

To illustrate the application of growth rate potential models for species introductions, we estimated growth rate potential for a 2 kg chinook salmon in Chesapeake Bay as a hypothesized case of the introduction of a non-native species into a new ecosystem. We estimated chinook salmon growth rate potential during July and October at the same mesohaline transect used above for Chesapeake Bay. Growth rates were calculated using the bioenergetics model for chinook salmon developed by Stewart and Ibarra (1991). We assumed that chinook salmon in Chesapeake Bay would be piscivorous predators of pelagic fish, and hence that our acoustic estimates of prey fish biomass would be adequate representations of prey availability. We also assumed that the foraging behavior of chinook salmon would be similar to that of bluefish and used similar foraging parameters.

Chinook salmon growth rate potential. Maps of growth rate potential indicate that chinook salmon growth in the Chesapeake Bay would differ greatly between seasons (color plate 10). During July, chinook salmon would not be capable of achieving any positive growth since water temperatures at that time are well in excess of the maximum temperature (20°C) for chinook salmon positive growth. However, favorable water temperatures and sufficient prey fish biomass during October resulted in positive chinook salmon growth rate potential in 6% of the environment. Water temperatures in Chesapeake Bay during October were near the optimal growth range (9–15°C) for chinook salmon growth.

Our results indicate that chinook salmon could perhaps survive and grow in Chesapeake Bay when water temperatures are below 20°C (November to May) and if prey biomass is adequate. However, no positive growth could occur from June to October because of excessive water temperature and the absence of cool-water refuges. Although we used an unlikely species for stocking in Chesapeake Bay, this analysis provides an example of how growth rate potential models can be used to evaluate the potential response of an introduced species to its new environment. This approach can be particularly useful for a species targeted for reintroduction into a rehabilitated environment.

Summary and Future Applications

Spatially explicit models of fish growth rate potential provide a mechanism for linking biological and physical environments, and for defining the distributional patterns of the aquatic environment at different spatial scales. By coupling high-resolution data on prey density, prey sizes, and thermal structure with bioenergetics and foraging models, the spatially explicit modeling approach goes beyond the simple (but essential) correlation of biological and physical structures to model the expected functional responses of fishes to their physical and biological habitats. This type of approach may be particularly useful in heterogeneous habitats where density-dependent (biological) and density-independent (physical) processes occurring at relatively small spatial scales can substantially affect population processes and production at the system level (Possingham and Roughgarden 1990).

Growth rate potential modeling has a wide range of possible applications. As we have shown with our examples, this approach provides a robust means for comparing growth potential for targeted species among habitat and time of year. Similarly, comparisons can be made between species that occupy a similar habitat, making it possible to evaluate the seasonal and spatial overlap in growth potential of competing species. Growth rate potential also provides a relative measure of ecological efficiency for fish under a specific set of habitat characteristics. That is, by comparing actual growth rates and habitat use with predictions from growth rate potential models, it is possible to assess the ecological efficiency of the species within that particular environment. A highly efficient predator in areas of the environment that contain optimal water temperatures and prey density will maximize its growth rate. By contrast, a less efficient predator located in areas with suboptimal conditions of temperature, prey, or both may have lower growth rates.

The transformation of prey fish density and size into estimates of consumption rate for calculation of growth rate potential remains the least well-known component to quantify growth rate potential (Bartell et al. 1986; Mason et al. 1995). In our examples, we assumed a constant reactive distance (RD = 1 body length) and foraging efficiency (k = 0.001) in the foraging model. Certainly, the reactive distance of a predator will change in response to diet and seasonal light levels and turbidity (Benfield and Minello 1996), and foraging efficiency will vary as a function of prey size, prey species, predator experience, hunger level, and predator preference (Stephens and Krebs 1986). Light level at depth, prey size structure, and prey species composition often differ greatly within and among estuaries and likely affect growth rate potential estimates. Future investigations should address behavioral interactions between predators and their prey, and how changes in the physical environment (such as water temperature or turbidity) influence the intensity of predator-prey interactions.

Despite existing model limitations, the spatially explicit approach provides a solid foundation for evaluating species-specific habitat needs based on the physiological requirements of targeted species. Our approach provides information on the physiological response of fishes that cannot be determined from prey abundance, prey size, or water temperature alone. Complex spatial details of the biological and physical environment are explicitly considered for the evaluation of growth rate potential. Extensive high-resolution and remote-sensing data are now routinely collected by agencies throughout the world. Data from such monitoring programs can provide the information necessary for quantifying and assessing changes in species- and size-specific fish growth rate potential in estuaries.

We contend that growth rate potential models are useful tools to help us understand the relationships between prey supply and predator demand in aquatic systems and to forecast the effects of proposed management actions. Possible effects of stocking decisions, population fluctuations, invasion by exotics, nutrient reduction strategies, and weather can be modeled with the approach outlined in this chapter. If planktivorous fish abundances decline in Chesapeake Bay (as they have in the last decades), the pelagic community may change greatly. Spatial models can help assess how changes in the fish communities and the environment will affect piscivore production and the health of the ecosystem. As such, growth rate potential can also serve as an indicator of ecosystem health or habitat quality in which maximum growth rate reflects the highest habitat quality.

Acknowledgments

This research was funded by grants from the Electrical Power Research Institute (WO914001), the U.S. Environmental Protection Agency, NOAA's Coastal Ocean Program (COASTES), and NSF's Land Margin Ecosystem Research Program (DEB-9412113). We thank J. Hagy and S. Jung for providing water temperature and mid-water-trawl data for Chesapeake Bay, J. A. Tyler for help with growth rate potential modeling, and the multitude of ship crews and scientific personnel for help at sea. This chapter is contribution No. 1098 of the NOAA–Great Lakes Environmental Research Laboratory, Ann Arbor, Michigan.

References

Baird, D., and R. E. Ulanowicz. 1989. The seasonal dynamics of the Chesapeake Bay ecosystem. *Ecological Monographs* 59: 329–64.

Barry, K. L., E. Demers, J. M. Jech, and S. B. Brandt. 1997. *Acoustic measurements of fish abundance in the Delaware Bay*. Final Report to the Electrical Power Research Institute, Palo Alto, CA.

Bartell, S. M., J. E. Breck, R. H. Gardner, and A. L. Brenkert. 1986. Individual parameter perturbation and error analysis of fish bioenergetics models. *Canadian Journal of Fisheries and Aquatic Sciences* 43: 160–68.

Beauchamp, D. A., D. J. Stewart, and G. L. Thomas. 1989. Corroboration of a bioenergetics model for sockeye salmon. *Transactions of the American Fisheries Society* 118: 597–607.

Benfield, M. C., and T. J. Minello. 1996. Relative effects of turbidity and light intensity on reactive distance and feeding of an estuarine fish. *Environmental Biology of Fishes* 46: 211–16.

Bennett, A. F., and K. L. Denman. 1985. Phytoplankton patchiness: Inferences from particles statistics. *Journal of Marine Research* 43: 307–35.

Brandt, S. B. 1990. *Acoustic quantification of fish abundance in the Chesapeake Bay*. Final Report to Maryland Department of Natural Resources, Power Plant Research Program, Annapolis, MD.

———. 1993. The effect of thermal fronts on fish growth: A bioenergetics evaluation of food and temperature. *Estuaries* 16: 142–59.

———. 1996. Acoustic assessment of fish abundance and distribution. Pages 385–432 in B. R. Murphy and D. W. Willis (editors), *Fisheries' techniques*. Bethesda, MD: American Fisheries Society.

Brandt, S. B., and K. J. Hartman. 1993. Innovative approaches with bioenergetics models: Future applications to fish ecology and management. *Transactions of the American Fisheries Society* 122: 731–35.

Brandt, S. B., and J. Kirsch. 1993. Spatially explicit models of striped bass growth rate potential in Chesapeake Bay. *Transactions of the American Fisheries Society* 122: 845–69.

Brandt, S. B., and D. M. Mason. 1994. Landscape approaches for assessing spatial patterns in fish foraging and growth. Pages 211–38 in D. J. Stouder, K. L. Fresh, and R. J. Feller (editors), *Theory and application in fish feeding ecology*. Columbia: University of South Carolina Press.

Brandt, S. B., D. M. Mason, E. V. Patrick, R. L. Argyle, L. Wells, P. Unger, and D. J. Stewart. 1991. Acoustic measures of the abundance and size of pelagic planktivores in Lake Michigan. *Canadian Journal of Fisheries and Aquatic Sciences* 48: 894–908.

Brandt, S. B., D. M. Mason, and E. V. Patrick. 1992. Spatially explicit models of fish growth rate. *Fisheries* 17: 23–33.

Carpenter, S. R. 1988. *Complex interactions in lake communities*. New York: Springer-Verlag.

Carpenter, S. R., J. F. Kitchell, and J. R. Hodgson. 1985. Cascading trophic interactions and lake productivity. *BioScience* 35: 634–38.

Day, J. W., C. A. S. Hall, W. M. Kemp, and A. Yanez-Arancibia. 1989. *Estuarine ecology.* New York: John Wiley & Sons.

Freadman, M. A. 1979. Swimming energetics of striped bass (*Morone saxatilis*) and bluefish (*Pomatomus saltatrix*): Gill ventilation and swimming metabolism. *Journal of Experimental Biology* 83: 217–30.

Fuiman, L. A., and J. C. Gamble. 1989. Influence of experimental manipulations on predation of herring larvae by juvenile herring in large enclosures. Rapport et Procès-Verbaux des Réunions. *Conseil International pour l'Exploration de la Mer* 191: 359–65.

Gerritsen, J., and J. R. Strickler. 1977. Encounter probabilities and community structure in zooplankton: A mathematical model. *Journal of the Fisheries Research Board of Canada* 34: 73–82.

Goshorn, D. M., and C. E. Epifanio. 1991. Diet of larval weakfish and prey abundance in Delaware Bay. *Transactions of the American Fisheries Society* 120: 684–92.

Goyke, A. P., and S. B. Brandt. 1993. Spatial models of salmonine growth rates in Lake Ontario. *Transactions of the American Fisheries Society* 122: 870–83.

Grecay, P. A., and T. E. Targett. 1996. Spatial patterns in condition and feeding of juvenile weakfish in Delaware Bay. *Transactions of the American Fisheries Society* 125: 803–8.

Hanson, P. C., T. B. Johnson, D. E. Schindler, and J. F. Kitchell. 1997. *Fish bioenergetics 3.0.* WISCU-T-97-001. Madison, WI: University of Wisconsin, Sea Grant Institute.

Hartman, K. J., 1993. Striped bass, bluefish, and weakfish in the Chesapeake Bay: Energetics, trophic linkages, and bioenergetics model applications. Ph.D. thesis, University of Maryland.

Hartman, K. J., and S. B. Brandt. 1995a. Trophic resource partitioning, diets, and growth of sympatric estuarine predators. *Transactions of the American Fisheries Society* 124: 520–37.

———. 1995b. Comparative energetics and the development of bioenergetics models for sympatric estuarine piscivores. *Canadian Journal of Fisheries and Aquatic Sciences* 52: 1647–66.

———. 1995c. Estimating energy density of fish. *Transactions of the American Fisheries Society* 124: 347–55.

Hondorp, D. W., and S. B. Brandt. 1996. Spatially explicit models of fish growth rate: Tools for assessing habitat quality. *Great Lakes Research Review* 2: 11–19.

Houde, E. D. 1987. Fish early life dynamics and recruitment variability. *American Fisheries Society Symposium* 2: 17–29.

Houde, E. D., and E. S. Rutherford. 1993. Recent trends in estuarine fisheries: Predictions of fish production and yield. *Estuaries* 16: 161–76.

Jech, J. M., and J. Luo. (in review) Digital echo visualization and information system (DEVIS) for processing spatially explicit fisheries acoutic data. *Journal of Fisheries Research.*

Kendall, A. W., and L. W. Walford. 1979. Sources and distribution of bluefish (*Pomatomus saltatrix*) larvae and juveniles off the East Coast of the United States. *Fisheries Bulletin U.S.* 77: 213–27.

Kitchell, J. F., D. J. Stewart, and D. Weininger. 1977. Applications of a bioenergetics model to yellow perch (*Perca flavescens*) and walleye (*Stizostedion vitreum vitreum*). *Journal of the Fisheries Research Board of Canada* 34: 1922–35.

Kotliar, N. B., and J. A. Wiens. 1990. Multiple scales of patchiness and patch structure: A hierarchical framework for the study of heterogeneity. *Oikos* 59: 253–60.

Lasker, R. 1978. The relation between oceanographic conditions and larval anchovy in the California current: Identification of factors contributing to recruitment failure. Rapport et Procès-Verbaux des Réunions. *Conseil International pour l'Exploration de la Mer* 173: 212–30.

Legendre, L., and S. Demers. 1984. Towards dynamic biological oceanography and limnology. *Canadian Journal of Fisheries and Aquatic Sciences* 41: 2–19.

Love, R. H. 1977. Target strength of an individual fish at any aspect. *Journal of the Acoustical Society of America* 62: 397–403.

Luo, J., and S. B. Brandt. 1993. Bay anchovy *Anchoa mitchilli* production and consumption in mid-Chesapeake Bay based on a bioenergetics model and acoustic measures of fish abundance. *Marine Ecology Progress Series* 98: 223–36.

Luo, J., S. B. Brandt, and M. J. Klebasko. 1996. Virtual reality of planktivores: A fish's perspective of prey selection. *Marine Ecology Progress Series* 140: 271–83.

MacLennan, D. N., and E. J. Simmonds. 1992. *Fisheries Acoustics.* London: Chapman and Hall.

Mason, D. M., and S. B. Brandt. 1996. Effects of spatial scale and foraging efficiency on the predictions made by spatially explicit models of fish growth rate potential. *Environmental Biology of Fishes* 45: 283–98.

Mason, D. M., A. P. Goyke, and S. B. Brandt. 1995. A spatially explicit bioenergetics measure of habitat quality for adult salmonines: Comparison between Lakes Michigan and Ontario. *Canadian Journal of Fisheries and Aquatic Sciences* 52: 1572–83.

McHugh, J. L. 1967. Estuarine nekton. Pages 581–620 in G. H. Lauff (editor), *Estuaries.* American Association for the Advancement of Science Publication 83. Washington, DC: AAAS.

Medwin, H., and C. S. Clay. 1998. *Fundamentals of acoustical oceanography.* New York: Academic Press.

Monteleone, D. M., and E. D. Houde. 1990. Influence of maternal size on survival and growth of striped bass (*Morone saxatilis* Walbaum) eggs and larvae. *Journal of Experimental Marine Biology and Ecology* 140: 1–11.

Nero, R. W., J. J. Magnuson, S. B. Brandt, T. K. Stanton, and J. M. Jech. 1990. Finescale biological patchiness of 70 kHz acoustic scattering at the edge of the Gulf Stream-Echofront 85. *Deep-Sea Research* 37: 999–1016.

Ney, J. J. 1993. Bioenergetics modeling today: Growing pains on the cutting edge. *Transactions of the American Fisheries Society* 122: 736–48.

NMFS (National Marine Fisheries Service). 1997. *Fisheries of the United States, 1996.* Silver Spring, MD: National Marine Fisheries Service.

Platt, T., and S. Sathyendranath. 1988. Oceanic primary production: Estimation by remote sensing at local and regional scales. *Science* 241: 1613–20.

Possingham, H. P., and J. Roughgarden. 1990. Spatial population dynamics of a marine organism with a complex life cycle. *Ecology* 71: 973–85.

Rice, J. A., and P. A. Cochran. 1984. Independent evaluation of a bioenergetics model for largemouth bass. *Ecology* 65: 732–39.

Rose, G. A., and W. C. Leggett. 1990. The importance of scale to predator-prey spatial correlations: An example of Atlantic fishes. *Ecology* 71: 33–43.

Shepherd, G. R., and C. B. Grimes. 1983. Reproduction of weakfish, *Cynoscion regalis*, in the New York bight and evidence for geographically specific life history characteristics. *Fishery Bulletin U.S.* 82: 501–11.

Stephens, D. W., and J. R. Krebs. 1986. *Foraging theory.* Princeton, NJ: Princeton University Press.

Stewart, D. J., and M. Ibarra. 1991. Predation and production by salmonine fishes in Lake Michigan, 1978–88. *Canadian Journal of Fisheries and Aquatic Sciences* 48: 909–22.

Tyler, J. A. 1998. *GRP Map Maker: A user's guide to spatial models of fish habitat combining acoustic data and bioenergetics models. (GRP Map Maker v. 2.8).* NOAA Technical Memorandum ERL GLERL-110. Great Lakes Environmental Research Laboratory, Ann Arbor, MI.

Tyler, J. A., and W. W. Hargrove 1997. Predicting forager distributions over large resource landscapes: A modeling analysis of the ideal-free distribution. *Oikos* 79: 376–86.

Tyler, J. A., and K. A. Rose. 1997. Individual-based model of fish cohort growth, movement, and survival in a spatially explicit environment. *Journal of Animal Ecology* 66: 122–36.

Tyler, J. A., and S. B. Brandt (in review). Spatial models of fish habitat quality: Does growth rate potential relate to fish growth? *Ecological Modeling.*

Wiens, J. A. 1989. Spatial scaling in ecology. *Functional Ecology* 3: 385–97.

CHAPTER 16

Habitat-Biotic Interactions

Charles A. Simenstad, Stephen B. Brandt, Alice Chalmers, Richard Dame,
Linda A. Deegan, Robert Hodson, and Edward D. Houde

Abstract

Conventional concepts of estuarine and near-shore coastal habitat are generally inadequate descriptions of processes and organisms that respond to habitat variability and change or integrate far larger and more complex "habitat landscapes." In particular, the role of structure within and among habitats, networks through which organisms and critical processes that influence secondary production operate, and the role of estuarine circulation "control point" features in shaping food-web structure and variability are poorly known. In response to these gaps, the Habitat-Biotic Interactions Working Group established as their goal to *identify approaches needed to synthesize a mechanistic understanding of how habitat structure influences estuarine secondary production and food webs.* We recommend eight major steps to enhance synthesis of natural and anthropogenic changes in estuarine production related to habitat-biotic interactions: (1) develop an estuarine habitat classification scheme that relates habitat structure to estuarine production and food-web processes; (2) examine existing long-term data sets to identify the scope and frequency of variability in habitat structure; (3) implement comparative studies of habitat function; (4) develop and link habitat and landscape models that capture the dynamics of biota and process interactions over large estuarine scales; (5) develop indicators of ecosystem habitat integrity, dynamics, and variability; (6) link site-specific field experiments and modeling approaches to scale processes and process understanding across ecosystems and landscapes; (7) apply advanced measurement technologies to give details of distributions and abundances of secondary consumers not now achievable; and (8) link habitat structure to assessment and prediction of resource management scenarios. Addressing ecosystem change in response to habitat structure, as well as the impacts of coastal zone management impacts upon ecosystems, will require innovative syntheses at much more expanded time and space scales than heretofore considered.

Introduction

To a large degree, the nature and structure of estuarine and coastal habitats control ecosystem productivity and community structure. Resources such as nutrients and light that limit primary production can be extensively influenced by inputs from land and rivers (chapter 4, Fisher et al.; chapter 11, Boynton and Kemp; chapter 2, Howarth et al.; and chapter 3, Vörösmarty and Peterson) or, to a lesser extent, the shelf and open ocean. On the other hand, many of the processes within estuaries and coastal zones that affect nutrient cycling and the conversion of the responding primary production into secondary producer biomass are likely related more to habitat structure. Thus, the source and amount of organic matter at the base of estuarine food webs are determined by diverse allochthonous and autochthonous processes, and prone to considerable spatial and temporal variability, but habitat structure influences the pathways and efficiencies of transfer to consumer organisms at higher levels in the food web. This dichotomy may explain some of the noise and uncertainty inherent in estimating high-trophic-level production based only on nutrient budgets and stoichiometry.

Overall, the scientific view of habitats has until recently been relatively restricted and narrow in terms of taxonomic groupings, ecosystems, and scales (McCoy and Bell 1991). Although we have not conducted a comprehensive review of the literature similar to that of McCoy and Bell, our familiarity with the contemporary literature suggests this myopic view is particularly true for estuarine and coastal ecosystems. For example, the areal extent of specific estuarine and coastal habitats such as marshes and submerged aquatic vegetation has long been used as an index of fish and wildlife production, especially when addressing rearing (nursery) habitats of economically important fish and shellfish, but examination of interactions of these biota *among* habitats over time is infrequent and ecosystem-scale analyses are rare. Although within-habitat linkages to production of higher trophic levels are gradually emerging (such as, fish growth and survival relative to salt-marsh structure), estimates of processes across the estuarine landscape, or comparisons among estuaries with different habitat structures, are wanting. Yet the strength of interactions between biota and their habitat, whether physical (larval transport, aggregation of food particles) or ecological (feeding efficiency, refugia from predators), may be the ultimate determinant of productive capacity *and* composition of consumers at the apex of estuarine and coastal food webs.

Understanding change in the structure of habitat, its influence on fundamental ecosystem processes, responses of habitat-associated biota, and effects of the anthropogenic components of change is essential to both managing our exploitation of coastal resources and minimizing the impacts on these resources of our occupation of the coastal margin. However, variability and change in biotic interactions associated with natural variation in habitat structure can be difficult to separate from anthropogenic degradation or shifts in habitats. Habitat structure naturally varies and changes over a hierarchy of spatial and temporal scales among different estuaries or within any one estuary. Different types of estuaries reflect dramatically different occurrences and distributions of habitats because of geologic origins and histories and contemporary climatic settings that embrace thousands of square kilometers and years. Within short time and space scales (for example, annual-decadal, meters), variability and change in habitat can also introduce variability in the quantity and quality of productivity at all trophic levels. In what are perceived as the most stable of habitats, such as salt marshes, habitat structure changes over different space and time scales, some of which are relatively predictable (for example, seasonal and successional changes in biota) and others highly stochastic (for example, disturbance events). While scientists have long recognized that variability and change in habitat structure is reflected in variable levels of production by estuarine biota, understanding of the underlying processes is meager. As a result, the diagnostic and, preferably, predictive capability that is necessary to separate natural dynamics in estuarine habitat change and production from changes imposed by anthropogenic effects is insufficient for coastal management decisions. As has been implied already in chapter 12 (Kremer et al.), ecologically complete syntheses require both mechanistic and predictive understanding of the underlying processes that control the composition and production of higher trophic levels. Management of human influences on important ecosystem processes will not improve until interactions between estuarine habitats and dependent biotic resources, such as fish, shellfish, and wildlife, are understood and at least somewhat predictable at larger spatial and temporal scales.

The Habitat-Biotic Interactions Working Group's objective was to *assess how more synthetic approaches should be brought to bear on fundamental questions about the role of habitat structure and variability in estuarine ecosystem processes.* While there is a breadth of valid questions that readily emerge around this issue (table 16-1), the Working Group focused on processes that affected biotic production, and specifically higher trophic level production, which to a large degree defines our socioeconomic and cultural dependence on estuarine ecosystems. We immediately concluded that the ecosystem context of these questions precluded the traditional single-species, single-habitat approach to understanding the role of habitat structure. Understanding the association between estuarine biota and habitats demands life history, physiological, and ecological knowledge of estuarine communities as they interact

TABLE 16-1

Predominance of pelagic vs. benthic food webs?
   What estuarine characteristics determines which predominates?
   When variable, what controls switching between pathways?
   What is the role of benthic-pelagic couplings?

Is biotic diversity related to estuarine productivity?
   Do species introductions change food webs?
   Does food web efficiency change with biodiversity?

Are there estuarine indicators of estuarine productivity and food web structure and efficiency?
   What intrahabitat structural attributes (e.g., edge) influence secondary production?
   What landscale-scale features (e.g., habitat linkages, configurations, location) affect secondary production?

How prevalent and variable are regional, extraestuarine influences on estuarine food web structure and production?

What is the importance and variability of higher trophic level (top-down) controls on different descriptors of estuarine biotic communities?
   Composition?
   Food web structure?
   Food web efficiency?
   Productivity?
   Direct vs. indirect effects?

How have anthropogenic changes in disturbance regimes changed biotic interactions, community structure, and production dynamics?

across the mosaic of habitats that comprise estuarine landscapes. New perspectives and analytical tools are required to gain such an understanding. Consequently, the Working Group formulated the question, What are the effects of habitat structure and change on interactions at ecosystem space-time scales? Implicit in this question are the concepts of (1) strength, (2) variability, and (3) time/space scales of interactions between either individual habitats or overall estuarine habitat structure and estuarine production.

   In the Working Group deliberations and the summary below, our goal was to *identify approaches needed to synthesize a mechanistic understanding of how habitat structure influences estuarine secondary production and food webs*. This goal complements the goal of the Linking Biogeochemical Processes to Estuarine Food Webs Working Group (Kremer et al., chapter 12) by examining the relative importance of non-biogeochemical controls on the transfer of organic matter to higher trophic levels. In this respect, the overall "carry-home" message of this volume should ultimately involve integration of recommendations by both of these working groups.

The Concept of Habitat and Habitat Structure

Safriel and Ben-Eliahu (1991) provide a comprehensive definition of habitat that is particularly appropriate to estuarine and coastal ecosystems: "the environment of a community confined to a portion of the landscape," including three components: (1) physicochemical features such as salinity and temperature; (2) resources such as food and space; and (3) interacting organisms other than those functioning as resources, such as predators, competitors, and mutualists. While the concept of habitat is commonly recognized, definitions of habitat structure are more numerous and diverse depending on different taxa and scale viewpoints but may be defined as the arrangement of objects or features in the environment (McCoy and Bell 1991).

Habitat Landscapes

In conventional usage, *habitat* has been defined by the occupation of a particular organism, that is, the locality, site, and particular type of local environment in which an organism is found. However, a broader ecosystem perspective would suggest that habitat embraces any part of the environment on which consumer organisms, directly or indirectly, depend during the estuarine portion of their life histories. Motile consumers (nekton) integrate different habitats in moving among spawning grounds, nursery areas, refugia from predation, feeding zones, and migration routes (Ayvazian et al. 1992). This use of "habitat landscapes" involves multiple scales of space (meters to hundreds of meters) and time (hours to seasons), as aptly illustrated (figure 16-1) for tidal marsh nekton by Kneib (1997). Over a hierarchy of spatial scales within and among habitats, the movement of nekton across the landscape is theorized to account for a "trophic relay" of intertidal primary production that changes over the habitat range and life stage of the nekton (for example, juvenile and adult resident, transient predatory) and habitat (figure 16-2, Kneib 1997). At the extremes of the estuary scale, many species and particularly top-level carnivores account for major fluxes of organic matter and nutrients into or out of the estuarine ecosystem as larvae, juveniles, and adults, with accompanying variability due to both biotic dynamics (for example, population biology, behavior) and physical dynamics (such as river flow and coastal currents; Deegan 1993).

It may be a useful convention to visualize habitats as discrete segments of the environment that fit together to make a place for communities of organisms to persist. These parts include the physical environment (such as substrate provided by plants, sediments, and even organisms, such as oysters), and the chemical environment (such as temperature, salinity, and dissolved oxygen) and the many organisms (such as plants and invertebrates) that comprise a food web. If portions are missing, the habitat is incomplete or damaged and unable to function at full capacity or be sustainable. Similarly, adopting a life history and ecosystem perspective compels ecologists to examine habitat-biotic interactions among habitats across habitat landscapes. Given the multidimensionality of factors defining habitats, we can conceptualize habitats as having

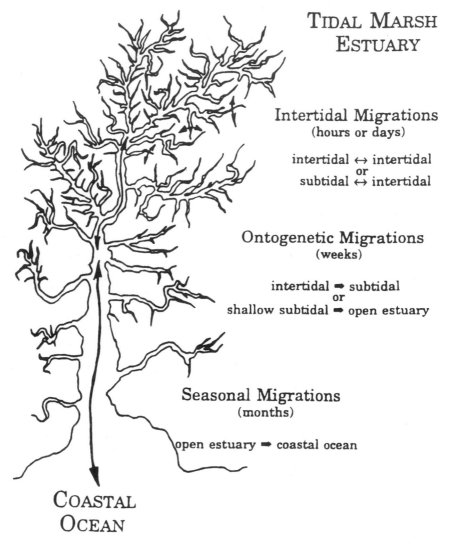

FIGURE 16-1   Nekton movement among estuarine tidal-marsh habitats demonstrating range of short-term (behavioral movement between intertidal and subtidal) to longer-term (ontogentic [life-history] and growth-related) space and time scales (from Kneib 1997, used with permission).

three primary components: (1) complexity; (2) heterogeneity; and (3) scale, where complexity defines the individual habitat elements, heterogeneity the complex of different habitat elements (and habitats), and scale the scale at which they are integrated (figure 16-3). An important contribution to habitat structure is often the "architectural" influence of the organisms themselves, exemplified by the communities that develop around large, hard-shelled

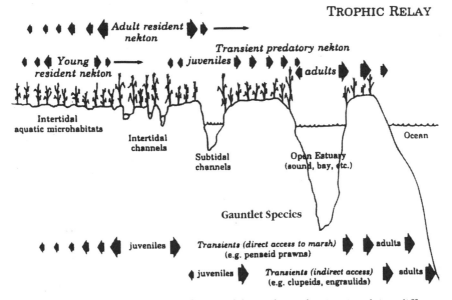

FIGURE 16-2 Trophic relay of intertidal marsh production involving different life stages of resident and transient nekton across habitat landscapes (from Kneib 1997, used with permission); gauntlet species are nekton, particularly forage fishes, whose juveniles encounter a gauntlet of predators during their large-scale migrations from the spawning ground to the estuary and back.

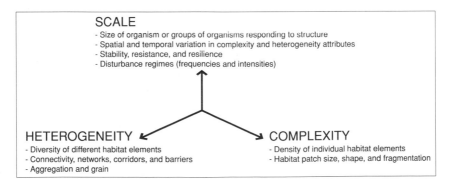

FIGURE 16-3 Components of habitat structure. Modified from McCoy and Bell 1991.

bivalves such as oysters and mussels that form reefs, and tubes, burrows, and castings of soft-substratum macrofauna such as polychaetes and bivalves (Sebens 1991).

The relationship between an organism and its habitat is also highly dependent on the organism's taxonomy, size, and trophic status. The habitat of a protozoan, which might entirely encompass the sphere of an estuarine aggregate <50 μm in diameter, is vastly different from striped bass habitat, which includes several different locations within estuaries as well as near-coastal waters outside the estuary. From an ecological perspective, however, there can be strong linkages between the protozoan and striped bass because the striped bass may depend on a food-web pathway initiated by the protozoan.

In defining habitat, the primary descriptors have been physical location, geomorphology, and biological features because these provide a convenient organizational typology. Thus, a habitat is most directly described by its position along a salinity gradient, tidal flooding regime (or tidal elevation as a indicator) substrate type, energy (wave, current) exposure, and dominant biological components, as demonstrated by the U.S. Fish and Wildlife Service's National Wetland Inventory (NWI; Cowardin et al. 1979). However, the biological components often become the primary descriptors, such as marshes, tidal forested swamps, oyster reefs, kelp beds, eelgrass meadows, mangroves, and intertidal and subtidal mudflats or sandflats. Less obvious, but equally important, examples are structural features inherent in different types of estuarine circulation, such as tidal fronts separating different water masses or plumes of low-salinity water produced by large rivers (color plate 11).

Physicochemical characteristics that determine physiological and behavioral responses of estuarine biota also delimit organisms' habitats. Environmental properties such as temperature, salinity, and dissolved oxygen greatly influence the scope of use of habitats, especially where extreme variation occurs, as in most intertidal habitats. Evolutionary and real-time responses limit organisms' recruitment, foraging, and general movement. This is illustrated in the estimate of growth potential of striped bass in Chesapeake Bay, as estimated by bioenergetic models using temperature, prey density, and prey size distributions (Brandt and Kirsch 1993; see also figure 15-5 in this volume).

Connectivity and Control Points

Landscape features that link habitats influence the flux of dissolved and suspended constituents and the management of organisms. Networks and corridors, such as dendritic tidal rivulet and channel systems, play particularly important roles because they link a geohydrologic continuum of habitats with varying geochemical functions (for example, constitute sinks or sources of nutrients and dissolved and particulate materials; Dame et al. 1992) and provide corridors of access for nekton foraging along the marsh edge and seeking refuge within marsh vegetation (Rozas and Odum 1987; Rozas et al. 1988; Minello et al. 1994; Kneib 1997). The continuum of habitats and linking corridors become particularly important to anadromous and catadromous nekton species, such as salmon, which must traverse the entire estuarine gradient at several stages in their life history (Simenstad and Cordell, in revision). Therefore, habitat structure at the ecosystem scale encompasses the configuration, and arrangement and connectivity, of habitat elements that typify most estuaries (figure 16-4).

Unique water-column features of estuarine hydrodynamics also contribute to habitat structure. In addition to tidal fronts, where concentration of plankton and other food particles provide a distinct foraging interface for fish and seabirds, circulation "control points" along the estuarine gradient can be sites of intense geochemical and biological activity (color plate 12). Control points may be described as circulation discontinuities that are caused by focusing of advective and diffusive processes. These locally intensify chemical and biological interactions either directly because of circulation (for instance, upwelling) or indirectly because of organisms' volitional responses to the feature (such as fish and bird foraging). One of the better examples of such control points is

FIGURE 16-4   Complex estuarine continuum of forested watershed, freshwater tidal marsh, salt marsh, creeks and pannes, Plum Island Sound (Massachusetts), barrier beach, maritime forest, and the Atlantic Ocean.

estuarine turbidity maxima (ETM), where residual circulation and other effects produce a region of intense trapping and aggregation of inorganic and detritus particles, zooplankton, and larval fish, and high bacterial activity (Uncles and Stephens 1993; Baross et al. 1994; Bernat et al. 1994; Jay 1994; Simenstad et al. 1994a, 1994b; Morgan et al. 1997; Crump and Baross 1996). ETM have been shown to account for a disproportionally high contribution to the food web in the Columbia River estuary (Simenstad et al. 1990), and to be associated with recruitment processes of anadromous fish and blue crab growth in Chesapeake Bay (Boynton et al. 1997). Because of the intense concentration of material and activity at control points, they likely form an important lateral extension of shoreline habitats, such that consumers in peripheral shallow and wetlands benefit from the proximity to ETM through increased food resources.

Habitat Variability, Natural and Anthropogenic Change

Habitat variability is a consequence of changes in geomorphology (water, circulation, basin morphology), watershed, and water quality (light attenuation, chemistry), and biological characteristics (the structural effects of organisms such as submerged aquatic vegetation, oysters, and the like) influencing estuarine and near-coastal environments. A knowledge of these three factors and their influence on food-web pathways is required to assess levels, efficiency, and variability in estuarine productivity. Estuaries vary in their fundamental characteristics that relate forcing (for example, tidal, river flow, wind) to physical processes (circulation and sediment transport; see Jay et al., chapter 7). Some of the variables from estuary to estuary that can affect food webs and productivity are water inputs, tidal influences, water and particle residence times, stratification characteristics, and basin morphologies. For example, the Columbia and Chesapeake estuaries, while sharing some common estuarine features, are notable for their distinct differences in habitat structure, which result in contrasting levels and quality of productivity.

Numerous documented and anecdotal observations of changes in estuarine productivity are associated with changes in habitat. However, while we have considerable knowledge of productivity and food-web processes in some estuarine systems, we lack synthetic ability to extend that knowledge to other estuaries with similar or different habitats. Many obvious effects of major habitat changes introduced by humans, such as deforestation, wetland loss, damming of rivers, increased nutrient loading, and introductions of exotic species, are known for many estuaries. Comparably dramatic natural changes, such as from hurricanes (Ruzecki et al. 1976), have also been described. More subtle, indirect effects of habitat change that may occur progressively or incrementally over a variety of time and space scales are less well known and quite often are indistinguishable from changes due to natural sources. Furthermore, there have been very few comparisons of habitat change across diverse estuarine ecosystems that lead toward a general understanding of the consequences to estuarine food webs.

Acute Habitat Change

Acute habitat change tends to result from direct impacts, usually intense natural disturbances (for example, hurricanes, freshwater flooding) or activities by humans (such as dredging or filling). These can cover extensive areas, as in hurricane effects (Chabreck and Palmisano 1973), but often do not result in a permanent change in estuarine processes. The persistence of natural estuarine processes facilitates natural recovery of the pre-disturbance habitat structure. However, both continual removal of estuarine habitat and alteration of natural physical and geomorphic processes results in a comparatively permanent shift in production. Maryland and Connecticut have lost 74% of their estuarine wetlands, California, 91% (Dahl and Johnson 1991). Dredging, diking, and filling of emergent wetlands alone is principally responsible for a loss of greater than 94% of San Francisco Bay's historic marsh habitat (Nichols et al. 1986), and an average of 42.5% of these important, highly productive habitats have been removed from Pacific Northwest estuaries, with many having lost over 60% (Bortleson et al. 1980; Boulé 1981; Boulé et al. 1983; Boulé and Bierly 1987). Dredging of canals and navigation channels alone has caused a direct loss of ~16% (12,000 ha) of Louisiana's coastal marshes (Turner and Cahoon 1988; Boesch et al. 1994). The most recent analysis of *Status and Trends of Wetlands in the Conterminous United States* (Dahl and Johnson 1991; Dahl and Allord 1997) suggests that these dramatic rates of intertidal habitat loss are declining, but between 1985 and 1995 an alarming 3,760 ha of total estuarine wetland habitat, and 734 ha of estuarine subtidal habitat, disappeared.

Chesapeake Bay has suffered significant habitat losses and modifications that impact fisheries productivity and have diminished quality of bay habitats for overall biological productivity. Early declines in anadromous fish (shads, river herrings, striped bass) were a consequence of blockages and dams on spawning tributaries (Hildebrand and Schroeder 1928), but later declines that became precipitous in the 1970s probably were attributable to multiple causes, including fishing, habitat loss, and declines in habitat quality. For example, 2.5% of the wetlands in the bay watershed were lost during the period 1982 to 1989 (Tiner 1994). Although not easily assignable to any specific cause, declines in fisheries productivity followed Hurricane Agnes in 1972 and were concurrent with increased nutrient loadings, losses of seagrass meadows, and concomitant effects on habitat quality. American shad, once the object of the most valuable fishery in the bay, dwindled to levels that required a moratorium on fishing (Anonymous 1989). Striped bass stocks recovered under greatly reduced fishing mortality, demonstrating that habitat deterioration alone was not the cause of the species' collapse (Richkus et al. 1992). Oysters have declined to disastrously low levels in the bay. While the causes are debatable, heavy fishing, destruction of habitat by the fishing process, siltation of oyster beds, and the prevalence of disease in the stressed, remnant stock (Rothschild et al. 1994) indicated that habitat destruction was one significant contributor to the disastrous decline. The role of seagreasses in fisheries productivity has not been clarified for Chesapeake Bay, but recent signs of modest recoveries in bay seagrasses (Anonymous 1995) are viewed with optimism by ecologists and fishery managers concerned about the bay's fisheries resources.

Subtle Habitat Change

Subtle, long-term habitat changes can involve more chronic or indirect mechanisms, including (1) degradation of water quality from eutrophication (for example, increased areas and persistence of hypoxia/anoxia); (2) loss/reduction of sediment sources (such as shifting from accretion to erosion regimes in marshes); (3) shifts in estuarine circulation (for example, salinity distribution) due to river diversion and regulation; (4) coastal subsidence and/or sea-level rise; (5) disturbance or removal of keystone species or other critical habitat components by fisheries or other mechanisms (such as disease), and (6) introductions and expansions of exotic species. Compared to acute habitat changes, these more subtle, chronic changes typically produce nonlinear and often highly complex responses by estuarine biota that are often difficult to distinguish from natural fluctuations.

These more subtle anthropogenic influences can significantly alter the processes that control higher trophic level production independent of obvious structural changes such as measurable loss of habitat. A well-documented example of this is the effect of eutrophication on fish production in submerged aquatic vegetation (SAV) habitats (Deegan and Buchsbaum in press). Eutrophication increases algal growth (plankton, epiphytes, and macroalgae) preferentially over vascular plants. This change in plant community affects food acquisition and survivorship of juvenile fish that use the SAV habitat as a critical nursery area. Food webs are altered as the invertebrate community responds to changing plant assemblages, in part as invertebrate susceptibility to predation by small fishes declines as increased algal cover provides more refuge. The declines in vascular plant density, biomass, height, and health increase the susceptibility of small fish to predation. The integrated result of these shifts in habitat primary production and architecture is a decline in the abundance and growth of small fish (Deegan and Buchsbaum in press). These changes occur long before the SAV habitat disappears.

More subtle differences in the character or quality of a habitat may also be reflected in varying production. The growth of juvenile fish, for example, is demonstrably different in different estuarine habitats (see Sogard 1992, 1994; Sogard and Able 1992). Recruitment growth and survival that varies as a function of river flow may relate to control-point dynamics and position, or relationship to other habitats (such as nurseries; Stevens 1977).

Examples of Documented Habitat Changes and Estuarine Responses

Examples of some of the better-documented changes in estuarine function associated with habitat alteration include (1) impacts of major water withdrawals from river flow into San Francisco Bay, the Gulf of Mexico, and the Sea of Azov; (2) construction and operation of the High Aswan Dam; (3) eutrophication of Chesapeake Bay; (4) extensive colonization of exotic species in San Francisco Bay; and (5) a complex combination of natural and anthropogenic changes in the Mississippi River delta.

River inflow to estuaries fluctuates naturally and both drought and flood extremes alter habitat-associated production and food-web relationships. For example, drought effects on fish production, caused primarily by changing optimum habitat utilization, has been documented to alter secondary productivity and trophic structure (Austin and Bonzek 1996; Livingston et al. 1997). However, anthropogenic diversion and regulation of river flow can impose far-reaching impacts, over diverse time scales, to fundamental habitat-structuring processes in estuaries that, unlike natural fluctuations, may be difficult to reverse (Simenstad et al. 1992). Alterations of the natural hydroperiod can affect estuarine circulation on different time and magnitude scales, ranging from short-term (diel, peak power-generation cycles) to longer-term (seasonal or interannual) changes. As a result, longitudinal and vertical estuarine habitat structure is altered by changes in mixing and salinity regimes, and the position and dynamics of control-point features altered. Truncated flooding can reduce sediment and nutrient inputs to estuarine wetlands. Reduction in flooding reduces disturbance frequency and intensity in modifying estuarine geomorphology, allowing wetlands and other intertidal habitats to evolve toward higher successional stages. Loss of riparian habitat flooding may also reduce the flux of organic matter flux to the estuary.

Even partial loss of freshwater flow imposes stresses on natural estuarine habitat functions. Losses of 40–60% of the freshwater flow (and higher diversions during high flow periods), and 60–75% of the sediment input, into north San Francisco Bay have impacted the production of estuarine biota, particularly those associated with the critical mesohaline mixing zone and estuarine turbidity maxima (Cloern et al. 1983; Nichols et al. 1986; Rozengurt et al. 1987; Kimmerer et al. 1998). In some cases, information about the naturally fluctuating relationships between freshwater inflow and estuarine secondary production may be sufficient to minimize impacts or even optimize production of desirable species. Harvest of fisheries resources associated with Texas estuaries vary naturally with estuarine inflow (see Hackney 1978; Schroeder 1978) in relationships that are sufficiently predictable to be usable for maximizing annual commercial harvest performance curves by inflow management (Powell and Matsumoto 1994).

The effects of large-scale diversions of fresh water on estuarine function are more definitive. Diversion of Nile River water and riverborne sediments by the High Aswan Dam of Egypt, which was completed in 1964, has almost eliminated fresh water reaching the Mediterranean Sea (Stanley and Warne 1993). As a result, not only have erosion, salinization, and pollution wreaked havoc on the northern Nile delta, but extensive changes in estuarine/coastal habitats due to erosion, subsidence, shifts in nutrient availability, and salinity intrusion have coincided with extensive declines in lagoon and coastal fish production (Aleem 1972) and likely contribute to the invasion of the water hyacinth (Waterbury 1979). Diversion of the Donets and Kuban Rivers from the Sea of Azov is associated with comparably dramatic declines in freshwater inflow, sediments, and nutrient inputs, sponsoring radical changes in habitats and biota, reduced productivity of planktonic and benthic communities, and reductions in anadromous fish catches by 90–95% and in spawning by 50–80% (Rozengurt and Haydock 1981; Bronfman 1977; Volovik 1986; Remisova 1984; Rosengurt et al. 1987; Micklin 1988).

Much of the habitat change in Chesapeake Bay, especially among SAV and oyster reef habitats, has been driven by a combination of overfishing, including the disturbance effect of harvesting methods (Rothschild et al. 1994), and eutrophication. This eutrophication and associated anoxia (Kemp et al. 1983, 1992; Kemp and Boynton 1984; Officer et al. 1984) have been attributed to nutrient loading from wastewater discharge, fertilizer leaching into freshwater sources, and atmospheric deposition (Nixon et al. 1986; Nixon 1995; Boynton et al. 1995). Far-reaching and persistent shifts in habitat structure have been associated with seasonal hypoxia events in Chesapeake Bay, resulting not only in dramatic shifts in benthic community structure but also changes in the availability and function of habitats available to nekton in the bay (Breitburg 1992, 1994; Breitburg et al. 1994) or those that can provide optimum growth (Brandt and Kirsch 1993). By reducing reef height, fishery harvesting of oyster reefs further compounds the effects of hypoxia on the oyster reef community (Lenihan and Peterson 1998). Shifts in the extent and distribution of submerged aquatic vegetation and bivalve (oyster, mussel) reefs imply significant alterations in ecosystem processes because of their manifold roles in production and nutrient cycling processes (Dame and Patten 1981; Dame et al. 1989, 1992) as well as harboring prey resources and providing refugia from predation for fish and shellfish (Dame 1979; Breitburg et al., 1994). Oyster overharvest, sedimentation, and eutrophication have resulted in an ecosystem "phase shift" from dominance by benthic filterfeeders to plankton communities in Chesapeake Bay (Newell 1988; Dame 1996; Ulanowicz and Tuttle 1992). A similar shift has been postulated for San Francisco Bay (Dame 1996), although the introduction of exotic, suspension-feeding bivalves (*Potamocorbula amurensis*) has likely produced a dramatic reverse-phase shift to benthic control on phytoplankton production (Cloern 1982; Nichols 1985; Nichols et al. 1990; Alpine and Cloern 1992).

In addition to direct habitat loss and the indirect effects of dramatic reduction in water inflow, San Francisco Bay and many other West Coast estuaries are also suffering extensive habitat changes because of the introduction of exotic species. By 1979, at least 255 foreign invertebrate species had become established in San Francisco Bay (Carlton 1979) and, more recently, phenomenally large populations of the Asian clam *Potamocorbula amurensis* have almost completely dominated the bay's benthic habitats and appear to be altering the pelagic communities through their extensive filter feeding (Carleton et al; 1990, Nichols et al. 1990). Intertidal habitats are equally under pressure from invasive plants, particularly the exotic smooth cordgrass, *Spartina alterniflora*, which is rapidly converting unvegetated mudflats to cordgrass marshes (Callaway and Josselyn 1992); such marsh expansion is not necessarily an unnatural process per se, but it is occurring at an unnaturally accelerated rate (Simenstad and Thom 1995) and altering the natural ratio and pattern of unvegetated to vegetated intertidal habitats.

The combination of naturally subsiding remnant deltas and massive engineering of water and sediment transport across the Mississippi delta has combined with construction of canals and navigational channels and oil and gas extraction to accelerate coastal wetland loss in Louisiana to levels as high as ~114 km^2 yr^{-1} from 1956 to 1978 and, while this loss rate is declining, coastal wetland loss rate continues to be ~65–90 km^2 yr^{-1} (Barras et al. 1994; Boesch et al. 1994). Fragmentation of coastal marshes and their conversion to subtidal habitats are the most symptomatic characteristics of this wetland loss. The strong correlative association documented between the extent of intertidal wetlands and commercial catches of penaeid shrimps (Turner 1977), likely related to increased abundance, and perhaps production, of shrimp, other crustaceans, and some fishes as a function of high vegetation-water interface (Rozas 1992), predicts eventual declines in production with continued marsh disintegration (Browder et al. 1988, 1989).

Action Items and Recommendations

Need for Ecosystem Approach

Among the nine issues that the National Research Council (1994) identified as posing significant threats to the integrity of coastal ecosystems, five (eutrophication, habitat modification, hydrologic and hydrodynamic disruption, introduction of nonindigenous species, global climate change and variability) directly involve the issue of habitat changes on estuarine secondary production. Recent emphasis placed on "essential fish habitat" under reauthorization of the Magnuson-Stevens Fisheries Act further illustrates acknowledgment of a direct link between fisheries production and the integrity of aquatic habitats. However, estuarine science has rarely dealt in a diagnostic or predictive manner with assessing change, either within-habitat variability over a landscape, or much less, the consequences to sustainable resources. Continued investment in habitat-specific applications of models, comparisons, and experiments will not alone provide the scientific and management community with the information needed to scale the effects of habitat change on production to the level of estuarine ecosystems. This "ecosystem approach" requires understanding of the fundamental linkages among ecosystem components, biological responses to physical and geochemical processes, rates and variability of these underlying processes, and the effect of disturbance and other modes of ecosystem change. But even more holistic approaches are required, especially those that approach habitats as landscapes and enable the discrimination of natural from anthropogenic change. Inquiry into the importance of estuarine habitat structure to production processes requires a highly integrated understanding of fundamental physical, chemical, and biological processes derived from the iteration between data and synthesis (figure 4-2, Fisher et al., chapter 4).

Recommendations

The Working Group recommends eight major initiatives that would advance syntheses of natural and anthropogenic changes in habitat-biotic interactions on estuarine production:

1. develop an estuarine habitat classification scheme that relates habitat structure to estuarine production and food-web processes;

2. examine existing long-term data sets to better define variability in secondary productivity as a function of change in habitat structure and integrity;

3. implement comparative studies of habitat function and change among and within estuaries;

4. develop and link habitat and habitat landscape models;

5. develop indicators of ecosystem habitat integrity, dynamics, and variability;

6. link site-specific field experiments and modeling approaches to scale processes across ecosystems and landscapes;

7. apply advanced measurement technologies to give details of distributions and abundances of secondary consumers not now achievable; and

8. link habitat structure to assessment and prediction of resource management scenarios.

The focus of these initiatives is to develop and synthesize information on estuarine habitat-biotic interactions that can be employed to compare habitat differences and production processes in divergent estuarine ecosystems.

1. Develop an estuarine habitat classification scheme that relates habitat structure to estuarine production and food-web processes. Any synthetic approach to assessing habitat-biotic interactions requires an estuarine classification scheme that incorporates physical, geomorphic, and geochemical attributes of habitats that have ecological significance (Jay et al., chapter 7). Systematic habitat classification is particularly important for ecosystem comparisons. With the exception of the recent emergence of the hydrogeomorphic (HGM) system (Brinson 1993), estuarine habitat classification has not progressed significantly since development of the estuarine system structure under the Cowaradin et al. (1979) National Wetland Inventory system. While HGM incorporates more geomorphic attributes, it does not classify variability in within- or among-habitat structure.

An ecosystem- or habitat landscape-based classification scheme should utilize diagnostics of both within-habitat and interhabitat processes, such that we can progress beyond the persistent assumption that biotic production is linearly related to habitat area. Little attention has been paid to landscape characteristics, such as position along the estuarine gradient, connections between similar habitats, and habitat patch size and shape. Furthermore, such an estuarine habitat classification scheme needs to be placed in a hierarchical context within a higher-order classification of the type of estuary and watershed (Jay et al., chapter 7).

2. Examine existing long-term data sets. Within-estuary fish and macroinvertebrate catches need to be examined and compared in order to understand the variability in secondary production across different types of estuaries with different habitat composition and extent. Most of the "controls" on various estuarine processes are actually the products of interactive processes. For example, high river discharge has a different effect in estuaries with vast expanses of salt marsh as compared to estuaries with large, shallow, unvegetated areas. We need to develop broad-scale relationships between multiple and interconnected habitat configurations and measure secondary production over sufficient periods to capture the effects of interannual variability.

Similarly, covariability relationships between growth and survival of consumer life-history stages and their residence in control-point and other habitat features should be explored relative to the temporal and spatial dynamics of these features. For instance, the position and intensity of tidal fronts and estuarine turbidity maxima vary in part as a function of river-flow forcing. The recruitment and year-class strength of fish and macroinvertebrates that are specifically associated with these habitat features, or occupy adjacent habitats, could relate directly or with some predictable lags to variation in intra- and interannual river flow.

3. Implement comparative studies. Cross-system comparisons of estuaries with different habitat structure, which may be related to inherent differences in land-margin characteristics (see first recommendation, Jay et al., chapter 7) as well as anthropogenic changes, provide one of the most powerful mechanisms to explain differences or changes in estuarine production processes. In particular, variability in fish and shellfish production (for instance, commercial catches) that is evident in comparisons of nutrient loading (Nixon et al. 1986; Boynton et al. 1995) may be elucidated further by relating both the quantity and quality of habitats that are known or suspected to affect the survival and production of fisheries species. Other system contrasts that would be relevant to habitat-biotic interactions include freshwater inflow regimes; quantity and quality of organic matter fluxes; estuarine circulation features (such as estuarine turbidity maxima); and invasions and changes in prominence of exotic species. One example of an insightful comparison is the inherent difference in processes dominating estuarine food webs, or in phase shifts among dramatically different processes, relative to the role of physical and geomorphic characteristics of different estuaries. Dame (1996) illustrated this relationship between estimated estuarine residence time and bivalve clearance rates that discriminates where suspension feeding does or does not regulate system trophic structure (figure 16-5).

4. Develop and link habitat and habitat landscape models. Recent technical and conceptual developments in spatially explicit modeling (see Demers et al., chapter 15) provide the means to relate habitat structure to estuarine processes. It is now feasible to incorporate habitat-specific submodels into

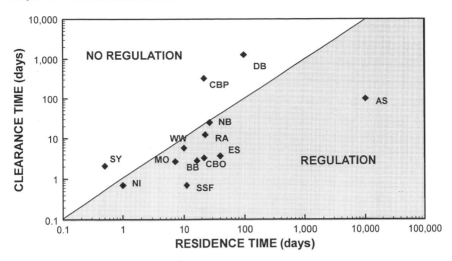

FIGURE 16-5   Relationship between bivalve clearance time and residence time, and potential for suspension feeding to dominate trophic structure, in a number of estuaries (from Dame 1996, used with permission from CRC Press); AS = Asko, Baltic Sea; BB = Bay of Brest; CBO = Chesapeake Bay, past; CBP = Chesapeake Bay, present; DB = Delaware Bay; ES = Eastern Scheldt; NB = Narragansett Bay; NI = North Inlet; MO = Marennes-Oleron Bay; RA = Ria de Arosa; SSF = South San Francisco Bay; SY = Sylt; WW = Western Wadden Sea.

two- and three-dimensional hydrodynamic models and other large ecosystem models to explicitly incorporate relevant space and time scales. Although still complex and requiring extensive code and input data, such ecosystem models can be designed to address fundamental questions about long-term, subtle and nonlinear responses of estuarine biota to habitat change. Examples of synthesis approaches that are amenable to spatially explicit ecosystem models include (1) examining scenarios or comparing differences in estuarine structure at the estuary or watershed scale; (2) testing hypotheses concerning whole-ecosystem processes such as benthic-pelagic couplings; (3) modeling natural experiments such as the impact of different nutrient-loading regimes on bottom-up (verses top-down) effects; (4) evaluating the significance of taxonomic or functional group changes by testing the sensitivity of model predictions to different species, size, or functional groups; and (5) predicting habitat and associated biotic responses to changes in circulation (such as residence time and stratification/mixing) due to river inflow manipulation.

5. Develop indicators of ecosystem habitat integrity, dynamics, and variability. Despite the plethora of paradigms about the importance of estuarine habitats to secondary production, particularly fisheries and aquaculture resources,

there are few cases in which the habitat attributes related to secondary production have been effectively quantified. Most indices incorporate measures
of habitat quantity rather than quality, are site- rather than ecosystem- or
landscape-specific, and do not assess the scope and pattern of natural variability in the indices. Indices of estuarine ecosystem integrity designed to
detect effects of habitat differences on fish assemblage structure (Deegan et
al. 1997) need be applied to more habitats and systems.

Potential short-term rate measurements, such as DNA/RNA, should be
explored as well as long-term, more integrative measurements such as stable
isotopes and other biomarkers. Particular emphasis should be given to evaluating indices that can be generated from paleochronologies such as estuarine
marsh sediment cores that may encompass stratigraphic patterns over several
hundred years, including periods before the influence of European man.

Other approaches, such as functional group analyses and indices of biological integrity (Karr 1987, 1991), also offer potentially powerful indicators of
current conditions and would be particularly useful in comparing estuaries
with fundamentally different habitat structures or quality. Both manipulative
experimental and natural contrasts can also be incorporated into indices of
short-term responses to habitat quality. For instance, daily fish growth using
otolith microstructure can be determined from fish exposed to different habitats or habitat quality once the response rate and bioenergetics of otolith deposition is known (Sogard and Able 1992; Secor et al. 1995). Indices of processes
at lower trophic level also need to be considered because the processes that
generate variability that is propagated to higher trophic levels often originate
at the autotrophic steps. For example, indices of microbe species and functional group diversity, and types and numbers of trophic interactions, may
provide insight into the chemical form and overall net flux of carbon, nitrogen, and energy from DOC and POC to higher trophic levels. Newly available
molecular techniques can be now be used to compare similarity between
whole bacterial populations and quantitative and qualitative differences in
species-specific (based on DNA homology) or biogeochemical group-specific
differences (based on molecular probing and in situ amplification/hybridiza-
tion) at the individual organism and microhabitat level. For the first time,
techniques exist for direct observation (for example, by intracellular PCR
amplification and labeling of indicator nucleic acid sequences coupled to
existing methods of epifluorescence microscopy) of phylogenetic and functional diversity of estuarine bacterial communities at the microhabitat level.
Comparisons may not be possible of such indicators as free-living planktonic
bacteria versus cells associated with sediments and aggregates, other higher
organisms, or POC. In addition, we can now examine in situ the level of
expression (or lack thereof?) of specific genetic capabilities under changing
estuarine conditions.

In all cases, such indicators of habitat and ecosystem integrity need to be sensitive to changes in within-habitat processes due to landscape position, biogeographic community, or human alteration and disturbance.

6. Link site-specific field experiments and modeling approaches to scale processes across ecosystems and landscapes. Habitat-biotic interactions are seldom scaled to larger ecosystems or landscapes. However, assessment of the effect of habitat change on such critical estuarine processes as secondary production demands tests of the assumption that habitat-specific relationships can be extrapolated over the broader scale of the estuarine landscape. In conjunction with a habitat-landscape classification scheme and spatially explicit modeling, the applicability and procedure for scaling habitat-specific processes must be assessed relative to environmental gradients, landscape and estuarine circulation features, and other characteristics of estuaries that could be influencing habitat organization.

7. Apply advanced measurement and assessment technologies. We need to develop measurement and assessment technologies that are capable of providing details of distributions and abundances of secondary producers and consumers not now achievable. A most critical need is the means to better assess populations of higher trophic level organisms in coastal ecosystems. Historically, relationships between ecosystem function and higher-trophic-level production have been based on commercial catches of fish and shellfish. While commercial fisheries catch records have provided important clues to these relationships, catch records are at best imperfect representations of the commercial populations and highly prone to biases of fishing effort and change in fishing strategy. In addition, relationships between environmental conditions and fish production based on commercial species may not be typical or representative of all the species in an estuary. Small fishes (such as herrings, anchovies, silversides, and sand lances), which are important forage fish for commercial fishes (as well as many other higher-trophic-level consumers such as seabirds and marine mammals), and which may dominate estuarine fish populations, probably have different habitat requirements. New applications of bioacoustics and related technology need to be developed to assess the abundance of these populations, especially in open-water, pelagic habitats.

8. Link habitat structure to assessment and prediction of resource management scenarios. Coastal-zone and resource management policies ultimately determine habitat structure and change. Both conceptual and numerical methods need to be developed to aid managers and the public in understanding the consequences of alternative management scenarios on estuarine habitats and the resulting ecosystem responses. The use of "ecological performance indices" such as posed by Done and Reichelt (1998), relating the temporal and spatial scale and intensity of fishery disturbance to habitat distributions and successional stages, presents one such approach to evaluating the alternatives between degraded and desirable ecosystem states.

References

Aleem, A. A. 1972. Effect of river and flow management on marine life. *Marine Biology* 15: 200–208.

Alpine, A. E., and J. E. Cloern. 1992. Trophic interactions and direct physical effects control phytoplankton biomass and production in an estuary. *Limnology and Oceanography* 37: 946–55.

Anonymous. 1989. *Chesapeake Bay alosid management plan.* Annapolis, MD: Chesapeake Bay Program.

Anonymous. 1995. *Trends in the distribution, abundance, and habitat quality of submerged aquatic vegetation in Chesapeake Bay and its tidal tributaries: 1971–1991.* Annapolis, MD: Chesapeake Bay Program. CBP/TRS 137/95.

Austin, H. M., and C. F. Bonzek. 1996. Effects of the June 1995 freshet on the main Virginia tributaries to the Chesapeake Bay. *Virginia Journal of Science* 47: 251–80.

Ayvazian, S. G., L. A. Deegan, and J. T. Finn. 1992. Comparison of habitat use by estuarine fish assemblages in the Acadian and Virginian zoogeographic provinces. *Estuaries* 15: 368–83.

Barras, J. A., P. E. Bourgeois, and L. R. Handley. 1994. *Land loss in coastal Louisiana 1956–90.* Open File Report 94-01. National Biological Survey, National Wetlands Research Center.

Baross, J. A., B. Crump, and C. A. Simenstad. 1994. Elevated microbial loop activities in the Columbia River estuarine turbidity maxima. Pages 459–64 in K. Dyer and B. Orth (editors), *Changing particle flux in estuaries: Implications from science to management.* ECSA22/ERF Symposium, Plymouth, U. K., September 1992. Fredensborg, Denmark: Olsen & Olsen Press.

Bernat, N., B. Kopcke, S. Yasseri, R. Thiel, and K. Wolfstein. 1994. Tidal variation in bacteria, phytoplankton, zooplankton, mysids, fish, and suspended particulate matter in the turbidity zone of the Elbe estuary: Interrelationships and causes. *Netherlands Journal of Aquatic Ecology* 28: 467–76.

Boesch, D. F., M. N. Josselyn, A. J. Mehta, J. T. Morris, W. K. Nuttle, C. A. Simenstad, and D. J. P. Swift. 1994. Scientific assessment of coastal wetland loss, restoration and management in Louisiana. *Journal of Coastal Research* (spec. issue 20).

Bortleson, G. C., M. J. Chrzastowske, and A. K. Helgerson. 1980. *Historical changes of shoreline and wetland at eleven major deltas in the Puget Sound region, Washington.* Atlas No. HA-617. Washington, DC: U. S. Geological Survey.

Boulé, M. E. 1981. Tidal wetlands of the Puget Sound region, Washington. *Journal of Society of Wetland Scientists* 2: 47–59.

Boulé, M. E., T. Miller, and N. Olmstead. 1983. *Inventory of wetland resources and evaluation of wetland management in western Washington*. Olympia, WA: Washington Department of Ecology.

Boulé, M. E., and K. F. Bierly. 1987. History of estuarine wetland development and alteration: What have we wrought? *Northwest Environ. J.* 3: 43–61.

Boynton, W. R., J. H. Garber, R. Summers, and W. M. Kemp. 1995. Inputs, transformations, and transport of nitrogen and phosphorus in Chesapeake Bay and selected tributaries. *Estuaries* 18: 285–314.

Boynton, W. R., W. Boicourt, S. Brandt, L. Harding, E. Houde, D. V. Holliday, M. Jech, W. M. Kemp, C. Lascara, S. D. Leach, A. P. Madden, M. Roman, L. Sanford, and E. M. Smith. 1997. Interactions between physics and biology in the estuarine turbidity maximum (ETM) of Chesapeake Bay, USA. *Int. Council Explor. Sea,* ICES CM 1997/S:11.

Boynton, W. R., and W. M. Kemp. "Influence of River Flow and Nutrient Loads on Selected Ecosystem Processes: A Synthesis of Chesapeake Bay Data." Chap. 11., this volume.

Brandt, S. B., and J. Kirsch. 1993. Spatially explicit models of striped bass growth in the mid-Chesapeake Bay. *Transactions of the American Fisheries Society* 122: 845–69.

Breitburg, D. L. 1992. Episodic hypoxia in the Chesapeake Bay: Interacting effects of recruitment, behavior and physical disturbance. *Ecological Monographs* 62: 525–46.

Breitburg, D. L. 1994. Behavioral responses of fish larvae to low oxygen risk in a stratified water column. *Marine Biology* 120: 615–25.

Breitburg, D. L., N. Steinberg, S. DuBeau, C. Cooksey, and E. D. Houde. 1994. Effects of low oxygen on predation on estuarine fish larvae. *Marine Ecology Progress Series* 104: 235–46.

Brinsom, M. M. 1993. *A hydrogeomorphic classification for wetlands*. Wetlands Research Program Technical Report WRP-DE-4. Washington, DC: U. S. Army Corps of Engineers.

Bronfman, A. M. 1977. The Azov Sea water economy and ecological problems: investigation and possible solutions. Pages 39–58 in G. F. White (editor), *Environmental effects of complex river development*. Boulder, CO: Westview Press.

Browder, J. A., H. A. Bartley, and K. S. Davis. 1984. A probabilistic model of the relationship between marsh-land-water interface and marsh disintegration. *Ecological Modelling* 29: 245–60.

Browder, J. A., L. N. May Jr., A. Rosenthal, J. G. Gosselink, and R. H. Baumann. 1989. Modeling future trends in wetland loss and brown shrimp production in Louisiana using thematic mapper imagery. *Remote Sensing of Environment* 28: 45–59.

Callaway, J. C., and M. N. Josselyn. 1992. The introduction and spread of smooth cordgrass (*Spartina alterniflora*) in south San Francisco Bay. *Estuaries* 15: 218–26.

Carlton, J. T. 1979. Introduced invertebrates of San Francisco Bay. In T. J. Conomos (editor), *San Francisco Bay: The urbanized estuary*. San Francisco, CA: Pacific Division American Association Advanced Science.

Carlton, J. T., J. K. Thompson, L. E. Shemel, and F. H. Nichols. 1990. Remarkable invasion of San Francisco Bay (California, USA) by the Asian clam *Potamocorbula amurensis*. 1. Introduction and dispersal. *Marine Ecology Progress Series* 66: 81–94.

Chabreck, R. H., and A. W. Palmisano. 1973. The effects of Hurricane Camille on the marshes of the Mississippi River delta. *Ecology* 54: 1118–23.

Cloern, J. E. 1982. Does the benthos control phytoplankton biomass in South San Francisco Bay? *Marine Biology Progress Series* 9: 191–202.

Cloern, J. E., A. Alpine, B. Cole, R. Wong, J. Arthur, and M. Ball. 1983. River discharge controls on phytoplankton dynamics in the northern San Francisco Bay estuary. *Estuarine, Coastal and Shelf Science* 16: 415–29.

Cowardin, L. M., V. Carter, F. C. Golet, and E. T. LaRoe. 1979. *Classification of wetlands and deepwater habitats of the United States*. FWS/OBS-79/31. Washington, DC: U. S. Fish and Wildlife Service, Office of Biological Services.

Crump, B. C., and J. A. Baross. 1996. Particle-attached bacteria and heterotrophic plankton associated with the Columbia River estuarine turbidity maxima. *Marine Ecology Progress Series* 138: 265–73.

Dahl, T. E., and C. E. Johnson. 1991. *Status of trends of wetlands in the conterminous United States, mid-1970's to mid-1980's*. Washington, DC: U. S. Fish and Wildlife Service.

Dahl, T. E., and G. J. Allord. 1997. *Technical aspects of wetlands: History of wetlands in the conterminous United States*. Paper No. 2425. National Wat. Sum. Wetland Res., U. S. Geol. Surv. Wat. Suppl.

Dame, R. F. 1979. The abundance, diversity, and biomass of macrobenthos on North Inlet, South Carolina, intertidal oyster reefs. *Proceedings of the National Shellfisheries Association* 69: 6–10.

———. 1996. *The ecology of marine bivalves: An ecosystem approach*. Boca Raton, FL: CRC Press.

Dame, R. F., and B. C. Patten. 1981. Analysis of energy flows in an intertidal oyster reef. *Marine Ecology Progress Series* 5: 115–24.

Dame, R. F., J. D. Spurrier, and T. G. Wolaver. 1989. Carbon, nitrogen, and phosphorus processing by an oyster reef. *Marine Ecology Progress Series* 54: 249–56.

Dame, R. F., J. D. Spurrier, and R. G. Zingmark. 1992. *In situ* metabolism of an oyster reef. *Journal of Experimental Marine Biology and Ecology* 164: 147–59.

Dame, R., D. Childers, and E. Koepfler. 1992. A geohydrologic continuum theory for the spatial and temporal evolution of marsh-estuarine ecosystems. *Netherlands Journal of Sea Research* 30: 63–72.

Deegan, L. A. 1993. Nutrient and energy transport between estuaries and coastal marine ecosystems by fish migration. *Canadian Journal of Fisheries and Aquatic Sciences* 50: 74–79.

Deegan, L. A., and R. N. Buchsbaum. In press. The effect of habitat loss and degradation of fisheries in R. N. Buchsbaum, W. E. Robinson, and J. Pederson (editors), *The decline of fisheries resources in New England: Evaluating the impact of overfishing, contamination, and habitat degradation*. Massachusetts Bays Program. Amherst: University of Massachusetts Press.

Deegan, L. A., J. T. Finn, and J. Buonaccorsi. 1997. Development and validation of an estuarine biotic integrity index. *Estuaries* 20: 601–17.

Demers, E., S. B. Brandt, K. L. Barry, and J. M. Jech. Spatial models of fish productivity throughout Chesapeake Bay. Chap. 15, this volume.

Done, T. J., and R. E. Reichelt. 1998. Integrated coastal zone and fisheries ecosystem management: Generic goals and performance indices. *Ecological Applications* 8: S110–S118.

Fisher, T. R., D. Correll, R. Costanza, J. T. Hollibaugh, C. S. Hopkinson, R. W. Howarth, N. N. Rabalais, J. E. Richey, C. J. Vörösmarty, and R. G. Wiegert. Synthesizing drainage basin inputs to coastal systems. Chap. 4, this volume.

Hackney, C. T. 1978. *Summary of information: Relationship of freshwater inflow to estuarine productivity along the Texas coast.* Washington, DC: U. S. Fish and Wildlife Service, Biological Services Program FWS/OBS-78/73.

Hildebrand, S. F., and W. C. Schroeder. 1928. *Fishes of Chesapeake Bay.* (Reprint by T. F. H. Publications, Neptune, NJ, for Smithsonian Insitution, 1972.)

Howarth, R. W., N. Jaworski, D. Swaney, A. Townsend, and G. Billen. Natural and anthropogenic influences on fluxes of carbon, nitrogen, and phosphorus from the landscape to the estuaries: An evaluation for the northeastern United States. Chap. 2, this volume.

Jay, D. A. 1994. Particle trapping in estuarine turbidity maxima. *Journal of Geophysical Research* 99: 446–61.

Jay, D. A., W. R. Geyer, and D. R. Montgomery. An ecological perspective on estuarine classification. Chap. 7, this volume.

Karr, J. A. 1987. Biological monitoring and environmental assessment: A conceptual framework. *Environmental Management* 11: 249–56.

———. 1991. Biological integrity: A long neglected aspect of water resource management. *Ecological Applications* 1: 66–84.

Kemp, W. M., and W. R. Boynton. 1984. Spatial and temporal coupling of nutrient inputs to estaurine primary production: The role of particulate transport and decomposition. *Bulletin of Marine Science* 35: 242–47.

Kemp, W. M., W. R. Boynton, R. R. Twilley, J. C. Stevens, and J. C. Means. 1983. The decline of submerged vascular plants in upper Chesapeake Bay: Summary of results concerning possible causes. *Marine Technology Society Journal* 17: 78–89.

Kemp, W. M., P. A. Sampou, J. Garber, J. Tuttle, and W. R. Boynton. 1992. Seasonal depletion of oxygen from bottom waters of Chesapeake Bay: Roles of benthic and planktonic respiration and physical exchange processes. *Marine Ecology Progress Series* 85: 137–52.

Kimmerer, W. J., J. R. Burau, W. A. Bennett. 1998. Tidally oriented vertical migration and position maintenance of zooplankton in a temperate estuary. *Limnology and Oceanography* 43: 1697–1709.

Kneib, R. T. 1997. The role of tidal marshes in the ecology of estuarine nekton. *Oceanography and Marine Biology Annual Review* 35: 163–220.

Kremer, J. N., W. M. Kemp, A. E. Giblin, I. Valiela, S. P. Seitzinger, E. Hofmann. Linking biogeochemical processes to estuarine food webs. Chap. 12, this volume.

Lenihan, H. S., and C. H. Peterson. 1998. How habitat degradation through fishery disturbance enhances impacts of hypoxia on oyster reefs. *Ecological Applications* 8: 128–40.

Livington, R. J., X. Niu, F. G. Lewis III, and G. C. Woodsum. 1997. Freshwater input to a gulf estuary: Long-term control of trophic organization. *Ecological Applications* 7(1): 277–299.

Micklin, P. P. 1988. Desiccation of the Aral Sea: A water management disaster in the Soviet Union. *Science* 241: 1170–76.

Minello, T. J., R. J. Zimmerman, and R. Medina. 1994. The importance of marsh edge for natant macrofauna in a created salt marsh. *Wetlands* 14: 184–98.

Morgan, C. A., J. R. Cordell, and C. A. Simenstad. 1997. Sink or swim? Copepod population maintenance in the Columbia River estuarine turbidity maxima region. *Marine Biology* 129: 309–17.

McCoy, E. D., and S. S. Bell. 1991. Habitat structure: The evolution and diversification of a complex topic. Pages 3–27 in S. S. Bell, E. D. McCoy, and H. R. Mushinsky (editors), *Habitat structure: The physical arrangement of objects in space*. London: Chapman and Hall.

National Research Council. 1994. *Priorities for coastal ecosystem science*. Committee to Identify High-Priority Science to Meet National Coastal Needs, Ocean Studies Board, Commission on Geosciences, Environment and Resources. Washington, DC: National Academy Press.

Newell, R. I. E. 1988. Ecological changes in Chesapeake Bay: Are they the result of over-harvesting the American oyster, *Crassostrea virginica*? Pages 536–46 in M. P. Lynch and E. C. Krome (editors), *Understanding the estuary: Advances in Chesapeake Bay research*. Solomons, MD: Chesapeake Research Consortium.

Nichols, F. H. 1985. Increased benthic grazing: An alternative explanation for low phytoplankton biomass in northern San Francisco Bay during the 1976–1977 drought. *Estuarine Coastal Shelf Science* 21: 379–88.

Nichols, F. H., J. E. Cloern, S. N. Luoma, and D. H. Peterson. 1986. The modification of an estuary. *Science* 231: 567–73.

Nichols, F. H., J. K. Thompson, and L. E. Schemel. 1990. Remarkable invasion of San Francisco Bay (California, USA) by the Asian clam *Potamocorbula amurensis*. 2. Displacement of a former community. *Marine Ecology Progress Series* 66: 95–101.

Nixon, S. W. 1995. Coastal marine eutrophication: A definition, social causes, and future concerns. *Ophelia* 41: 199–219.

Nixon, S. W., C. A. Oviatt, J. Frithsen, and B. Sullivan. 1986. Nutrients and the productivity of estuarine and coastal marine ecosystems. *Journal of the Limnological Society of Southern Africa* 12: 43–71.

Officer, C. B., R. B. Biggs, J. L. Taft, L. E. Cronin, M. A. Tyler, and W. R. Boynton. 1984. Chesapeake Bay anoxia: Origin, development and significance. *Science* 23: 22–27.

Powell, G. L., and J. Matsumoto. 1994. Texas Estuarine Mathematical Programming Model: A tool for freshwater inflow management. Pages 401–6 in K. R. Dyer and R. J. Orth (editor), *Changes in fluxes in estuaries: Implications from science to management*. ESCA22/ERF Symposium. Fredensborg, Denmark: Olsen & Olsen.

Remisova, S. S. 1984. Water balance of the Sea of Azov. *Journal of Water Res.* 1: 109–21.

Richkus, W. A., H. M. Austin, and S. J. Nelson. 1992. Fisheries assessment and management synthesis: Lessons for Chesapeake Bay. Pages 75–114 in S. Nelson, C. McManus, P. Elliott, and B. Farquhar (editors), *Perspectives on Chesapeake Bay, 1992: Advances in estuarine sciences*. Chesapeake Research Consortium Publication 143. Annapolis, MD: Chesapeake Bay Program.

Rothschild, B. J., J. S. Ault, P. Goulletquer, and M. Hèral. 1994. Decline of the Chesapeake Bay oyster population: A century of habitat destruction and overfishing. *Marine Ecology Progress Series* 111: 29–39.

Rozas, L. P. 1992. Comparison of nekton habitats associated with pipeline canals and natural canals in Louisiana salt marshes. *Wetlands* 12: 136–46.

Rozas, L. P., and W. E. Odum. 1987. Use of tidal freshwater marshes by fishes and macrofaunal crustaceans along a marsh stream-order gradient. *Estuaries* 10: 36–43.

Rozas, L. P., C. C. McIvor, and W. E. Odum. 1988. Intertidal rivulets and creekbanks: Corridors between tidal creeks and marshes. *Marine Ecology Progress Series* 47: 303–7.

Rozengurt, M., and I. Haydock. 1981. Method of computation of ecological regulation of the salinity regime in estuaries and shallow water seas in connection with water regulation for human requirements. Pages 474–506 in R. D. Cross and D. L. Williams (editors), *Proceedings of national symposium on freshwater flow to estuaries*. FWS/OBS-81-04. Vol. 2. U. S. Fish and Wildlife Service.

Rozengurt, M., M. J. Hertz, and M. Josselyn. 1987. The impact of water diversions on the river-delta-estuary-sea ecosystems of San Francisco Bay and the Sea of Azov. In D. M. Goodrich (editor), *San Francisco Bay: Issues, resources, status and management*. NOAA Estuary of the Month Series 6. Washington, DC.

Ruzecki, E. P., J. R. Schubel, R. J. Huggett, A. M. Anderson, M. L. Wass, R. J. Marasco, and M. P. Lynch. 1976. *The effects of tropical storm Agnes on the Chesapeake Bay estuarine system*. The Chesapeake Bay Consortium, Inc. Publication No. 54. Baltimore, MD: Johns Hopkins University Press.

Safriel, U. N., and M. N. Ben-Eliahu. 1991. The influence of habitat structure and environmental stability on the species diversity of polychaetes in vermetid reefs. Pages 349–69 in S. S. Bell, E. D. McCoy, and H. R. Mushinsky (editors), *Habitat structure: The physical arrangement of objects in space*. London: Chapman and Hall.

Schroder, W. W. 1978. Riverine influence on estuaries: A case study. Pages 347–64 in M. Wiley (editor), *Estuarine interactions*. New York: Academic Press.

Sebens, K. P. 1991. Habitat structure and community dynamics in marine benthic systems. Pages 211–34 in S. S. Bell, E. D. McCoy, and H. R. Mushinsky (editors), *Habitat structure: The physical arrangement of objects in space*. London: Chapman and Hall.

Secor, D. J., A. Henderson-Arzapalo and P. M. Piccoli. 1995. Can otolioth microchemistry chart patterns of migration and habitat utilization in anadromous fishes? *J. Exp. Mar. Biol. Ecol.* 192: 15–33.

Signell, R. P., A. F. Blumberg, and H. L. Jenter. 1993. Modeling transport processes in the coastal ocean. *Journal of Marine and Environmental Engineering* 1: 31–52.

Simenstad, C. A., and K. L. Fresh. 1995. Influence of intertidal aquaculture on benthic communities in Pacific Northwest estuaries: Scales of disturbance. *Estuaries* 18: 43–70.

Simenstad, C. A., and R. M. Thom. 1995. *Spartina alterniflora* (smooth cordgrass) as an invasive halophyte in Pacific Northwest estuaries. *Hortus Northwest* 6: 9–12, 38–40.

Simenstad, C. A., and J. R. Cordell (in revision). Ecological assessment criteria for restoring anadromous salmonids' habitat in Pacific Northwest estuaries. *Ecological Engineering.*

Simenstad, C. A., C. D. McIntire, and L. F. Small. 1990. Consumption processes and food web structure in the Columbia River estuary. *Progress in Oceanography* 25: 271–98.

Simenstad, C. A., D. A. Jay, and C. R. Sherwood. 1992. Impacts of watershed management on land-margin ecosystems: The Columbia River estuary as a case study. Pages 266–306 in R. J. Naiman (editor), *Watershed management: Balancing sustainability and environmental change.* New York: Springer-Verlag.

Simenstad, C. A., D. J. Reed, D. A. Jay, J. A. Baross, F. G. Prahl, and L. F. Small. 1994a. Land-margin ecosystem research in the Columbia River estuary: Investigations of the couplings between physical and ecological processes within estuarine turbidity maxima. Pages 437–44 in K. Dyer and B. Orth (editors), *Changing particle flux in estuaries: Implications from science to management.* ECSA22/ERF Symposium, Plymouth, U. K., September 1992. Fredensborg, Denmark: Olsen & Olsen Press.

Simenstad, C. A., C. A. Morgan, J. R. Cordell, and J. A. Baross. 1994b. Flux, passive retention, and active residence of zooplankton in Columbia River estuarine turbidity maxima. Pages 473–82 in K. Dyer and B. Orth (editors), *Changing particle flux in estuaries: Implications from science to management.* ECSA22/ERF Symposium, Plymouth, U.K., September 1992. Fredensborg, Denmark: Olsen & Olsen Press.

Sogard, S. M. 1992. Variability in growth rates of juvenile fishes in different estuarine habitats. *Marine Ecology Progress Series* 85: 163–72.

———. 1994. *Use of suboptimal foraging habitats by fishes: Consequences to growth and survival.* Pages 103–31 in D. L. Stouder, K. L. Fresh, and R. J. Feller (editors), *Theory and application in fish feeding ecology.* Belle W. Baruch Lib. Mar. Sci. 18. Columbia: University of South Carolina Press.

Sogard, S. M., and K. W. Able. 1992. Growth variation of newly settled winter flounder (*Pseudopleuronectes americanus*) in New Jersey estuaries as determined by otolith microstructure. *Netherlands J. Sea Res.* 29: 163–72.

Stanley, D. J., and A. G. Warne. 1993. Nile delta: Recent geological evolution and human impact. *Science* 260: 628–34.

Stevens, D. E. 1977. Striped bass (*Morone saxatilis*) year class strength in relation to river flow in the Sacramento-San Joaquin estuary, California. *Transactions of the American Fisheries Society* 106: 34–42.

Tiner, R. 1994. *Recent wetland status and trends in the Chesapeake watershed: 1982–1989.* Annapolis, MD: Chesapeake Bay Program.

Turner, R. E. 1977. Intertidal vegetation and commercial yields of penaeid shrimp. *Transactions of the American Fisheries Society* 106: 411–16.

Turner, R. E., and D. R. Cahoon (editors). 1988. Causes of wetland loss in the coastal central Gulf of Mexico. Vol. 1. Executive summary. OCS Study MMS 87-0120. Final Report to U. S. Minerals Management Service, New Orleans, LA.

Ulanowicz, R. E., and J. H. Tuttle. 1992. The trophic consequences of oyster stock rehabilitation in Chesapeake Bay. *Estuaries* 15: 257–65.

Uncles, R. J., and J. A. Stephens. 1993. Nature of the turbidity maximum in the Tamar Estuary, U. K. *Estuarine Coastal Shelf Science* 36: 413–31.

Volovik, S. P. 1986. Changes in the ecosystem of the Azov Sea in relation to economic development in the basin. *J. Ichthyology* 26: 1–15.

Vörösmarty, C. J., and B. J. Peterson. Macro-scale models of water and nutrient flux to the coastal zone. Chap. 3, this volume.

Waterbury, J. 1979. *Hydropolitics of the Nile Valley*. Syracuse, New York: Syracuse University Press.

PART FIVE

Synthesis for Estuarine Management

THE CHAPTERS IN THIS PART SHARE THE PREMISE that the most effective environmental regulations and policies will be based on adequate scientific understanding of the interactions of organisms and their environments. This understanding must be predictive; that is, it must allow the development of scenarios of the effects of various choices for management. For estuaries in the twenty-first century, this predictive science must include the role of human activities as factors of change in the natural world.

In chapter 17, Costanza and Voinov show how human activities can be incorporated into a model of coastal ecosystems. They describe a large, spatially explicit model that integrates economics and ecology in an investigation of the Patuxent River watershed in Maryland. They argue that adaptive management requires consideration of social forces in addition to ecological and economic factors. Adaptive management, they note, views "development policy and management as experiments where interventions at several scales are made to achieve understanding and to identify and test policy options."

This model is still under development a decade after it was designed. One lesson is that a complex model takes a long time to build and test. Another is how the tremendous amount of detailed environmental information available in geographic information systems can be used to develop spatially explicit models. This chapter also shows how modelers simplify their creations whenever possible.

The workshop presented as chapter 18 in this part reports a number of successful efforts in which synthesis of scientific knowledge is being applied to the management of a large aquatic system. Sites include the Great Lakes, Chesapeake Bay, and San Francisco Bay. Although participants in a large project on environmental problems in Florida Bay are just beginning to synthesize environmental data, they are making good use of all available information to plan their research and modeling.

Successful management does not automatically happen because good science is a part of field studies, modeling, and planning. Workshop participants pointed out additional factors that have contributed to successful management. These factors include the presence of key individuals, long-term databases, environmental calamities that capture public attention, a whole-ecosystem perspective, and the presence of a good institutional structure for management.

Workshop participants concluded that the scientific management of estuarine systems would be greatly improved by the formal application of a series of steps. They include:

- designating a lead agency within each estuarine management program to coordinate the activities of other agencies;

- communicating among stakeholders in government, the public, and the scientific community;

- implementing a scientific advisory process that incorporates synthesis of data and comparisons across a variety of systems;

- providing sustained support at the national level for basic research on estuarine processes; and

- encouraging comparison and synthesis within national networks of projects and programs to produce generalized models for managers.

CHAPTER 17

Integrated Ecological Economic Regional Modeling

Linking Consensus Building and Analysis for Synthesis and Adaptive Management

Robert Costanza and Alexey Voinov

Abstract

Both understanding and adaptive management of regional systems require a true integration of ecological, economic, and social factors. This chapter describes a process of adaptive regional management that utilizes (1) integrated, dynamic, spatially explicit ecological economic modeling to test understanding, explore future scenarios, and evaluate policy options; (2) workshops involving major stakeholder groups to design, build, and evaluate the models (and in the process build consensus about the nature of the problems); and (3) adaptive policy implementations that are treated as experiments, with the results monitored and fed back into the consensus building, modeling, and policy-formulation process. A case study in the Patuxent River watershed in Maryland is described.

Integrated Ecological Economic Systems Modeling

New understanding about system dynamics and predictability that has emerged from the study of "complex systems" is creating new tools for modeling interactions between human and natural systems. A range of techniques has become available through advances in computer speed and accessibility, and by implementing a broad, interdisciplinary systems view.

Systems are groups of interacting, interdependent parts linked together by exchanges of energy, matter, and information. *Complex* systems are characterized by (1) strong (usually nonlinear) interactions between the parts; (2) complex feedback loops that make it difficult to distinguish cause from effect; and (3) significant time and space lags, discontinuities, thresholds, and limits; all resulting in (4) the inability to simply "add up" or aggregate small-scale behavior to arrive at large-scale results (von Bertalanffy 1968; Rastetter et al. 1992). Ecological and economic systems both independently exhibit these characteristics of complex systems. Taken together, linked ecological economic systems are devilishly complex.

While almost any subdivision of the universe can be thought of as a "system," modelers of systems usually look for boundaries that minimize the interaction between the system under study and the rest of the universe in order to make their job easier. The interactions between ecological and economic systems are many and strong. So, while splitting the world into separate economic and ecological systems is possible, it does not minimize interactions and is a poor choice of boundary.

Classical (or reductionist) scientific disciplines tend to dissect their subject into smaller and smaller isolated parts in an effort to reduce the problem to its essential elements. In order to allow the dissection of system components, it must be assumed that interactions and feedbacks between system elements are negligible or that the links are essentially linear so they can be added up to give the behavior of the whole (von Bertalanffy 1968). Complex systems violate the assumptions of reductionist techniques and therefore are not well understood using the perspective of classical science. In contrast, *systems analysis* is the scientific method applied across many disciplines, scales, resolutions, and system types in an integrative manner.

In economics, for example, a typical distinction is made between partial-equilibrium and general-equilibrium analysis. In partial-equilibrium analysis, a subsystem (a single market) is studied with the underlying assumption that there are no important feedback loops from other markets. In general-equilibrium analysis, on the other hand, the totality of markets is studied in order to bring out the general interdependence in the economy. The large-scale, whole-economy, general-equilibrium effects are usually quite different from the sum of the constituent small-scale, partial-equilibrium effects. Add to this the further complication that in reality "equilibrium" is never achieved, and one can begin to see the limitations of classical, reductionist science in understanding complex systems.

Economic and ecological analysis needs to shift away from implicit assumptions that eliminate links within and between economic and natural systems, because, due to the strength of the real-world interactions between these components, failing to link them can cause severe misperceptions and indeed policy failures (Costanza 1987). Since reductionist thinking fails in the quest to understand complex systems, new concepts and methods must be devised.

To achieve a comprehensive understanding that is useful for modeling and prediction of linked ecological economic systems requires the synthesis and integration of several different conceptual frames. As Levins (1966) has described this search for robustness: "we attempt to treat the same problem with several alternative models each with different simplifications. . . . Then, if these models, despite their different assumptions, lead to similar results we have what we call a robust theorem which is relatively free of the details of the model. Hence our truth is the intersection of independent lies."

Existing modeling approaches can be classified according to a number of criteria, including scale, resolution, generality, realism, and precision. The most useful approach within this spectrum of characteristics depends on the specific goals of the modeling exercise. We describe a few examples of how one might match model characteristics with several of the possible modeling goals relevant for ecological economic systems, and claim that a better appreciation of the range of possible model characteristics and goals can help to more optimally match characteristics and goals.

Complex systems analysis offers great potential for generating insights into the behavior of linked ecological economic systems. These insights will be needed to change the behavior of the human population toward a sustainable pattern, a pattern that works in synergy with the life-supporting ecosystems on which it depends. The next step in the evolution of ecological economic models is to fully integrate the two fields and not just transfer methods between them. Clark's (1976, 1981, 1985) bioeconomics work was the beginning of this recognition of the importance of linking the mutually interacting sub-parts. But much work remains to be done to bring the two fields and the technology that supports them to the point where their models can adequately interact. Transdisciplinary collaboration and cooperative synthesis among natural and social scientists and others will be essential (Norgaard 1989).

Computers and Modeling

Until computers became available, the equations that described the dynamics of systems had to be solved analytically. This severely limited the level of complexity (as well as the resolution) of the systems that could be studied and the complexity of the dynamics that could be studied for any particular system. Table 17–1 shows the limits of analytical methods in solving various classes of mathematical problems in general.

As table 17-1 shows, only relatively simple linear systems of algebraic or differential equations can, in general, be solved analytically. The problem is that most complex, living systems (like economies and ecosystems) are decidedly *nonlinear,* and efforts to approximate their *dynamics* with linear equations have been of only limited usefulness. In addition, complex systems often exhibit discontinuous and chaotic behavior (Rosser 1991).

We differentiate here between the use of linear systems of equations to model complex system *dynamics* (which does not work well) versus the use of linear systems to understand system *structure* (which may work reasonably

TABLE 17-1

The limits of analytical methods in solving mathematical problems (after von Bertalanffy 1968). The thick solid line divides the range of problems that are solvable with analytical methods from those that are very difficult or impossible using analytical methods and require numerical methods and computers to solve. "Systems" problems are typically nonlinear and fall in the range that requires numerical methods. It should be noted that while some special problems that fall in areas labeled "impossible" in the table are actually possible to solve using analytical methods (frequently requiring special tricks); in general one cannot depend on a solution being available. Computers have *guaranteed* that a solution will not be general and will represent system dynamics only for the specified set of parameters.

Equations	Linear			Nonlinear		
	One Equation	Several Equations	Many Equations	One Equation	Several Equations	Many Equations
Algebraic	Trivial	Easy	Difficult	Very difficult	Very difficult	Impossible
Ordinary differential	Easy	Difficult	Essentially impossible	Very difficult	Impossible	Impossible
Partial differential	Difficult	Essentially impossible	Impossible	Impossible	Impossible	Impossible

well). Integrating these views of structure and dynamics is a key item for research on complex ecological economic systems.

In recent years, computers have become not only faster but also much more accessible. This has allowed researchers to develop methods to allow adaptive, evolutionary, dynamic solutions. For example, Holland and Miller (1991) describe how recent advances in computers and machine learning (a form of artificial intelligence) have spawned "artificial adaptive agents," computer programs that can simulate evolution and acquire sophisticated behavioral patterns. In these programs, individual agents (processes, elements, pieces of computer code) in networks of interacting agents reproduce themselves in the next time period based on some measure of their performance in the current time period. The system exhibits changing group behavior over time and mimics evolution. To exhibit this adaptive behavior, the actions of the agents must be valued, and the agents must act to increase this value over time. Algorithms like these can provide a realistic representation of ecological and economic processes.

Another useful technique is "metamodeling," in which more general models are developed from more detailed ones. Richard Cabe, Jason Shogren, and colleagues (1991) have developed this technique to link models of agricultural production and economic behavior that could not normally be used together because, for one, they run at different time and space scales. Their models, which cover the entire midwestern farm belt of the United States, provide a method for a quick and cost-efficient evaluation of ecological economic policies.

Computers have been also used to further the analytical tools for model analysis (Voinov and Tonkikh 1987; Kuznetsov and Rinaldi 1996). When a straightforward analytical solution becomes impossible, computers may be used to generate phase portraits of model behavior and make qualitative conclusions about system dynamics similar to those an analytical solution would give. In this case the numerical methods are employed to analyze certain behavioral patterns of the system (stability of steady points, bifurcation, chaos) rather than generate just the trajectories of the system for particular sets of parameters.

Computer hardware advances such as CRAY supercomputers and Connection Machines (massively parallel supercomputers) facilitate the modeling of complex systems using advanced numerical computation algorithms (such as finite difference and finite element routines, cellular automata algorithms, and emerging methods that employ at least a modicum of artificial "intelligence"). For example, parallel computers make high-spatial-resolution, regional and global ecological economic models computationally feasible (Costanza et al. 1990; Costanza and Maxwell 1993) and allow the types and resolution of evolutionary and metamodeling approaches to expand dramatically. These new capabilities, linked with a more realistic and pluralistic view of the various roles and limitations of models in understanding and decision making, can dramatically increase the effectiveness of modeling.

Purposes of Models

Models are analogous to maps. Like maps, they have many possible purposes and uses, and no one map or model is right for the entire range of uses (Levins 1966; Robinson 1991). It is inappropriate to think of models or maps as anything but crude (but in many cases absolutely essential) abstract representations of complex territory, whose usefulness can best be judged by their ability to help solve the navigational problems faced. Models are essential for policy evaluation but are often also misused since there is ". . . the tendency to use such models as a means of legitimizing rather than informing policy decisions. By cloaking a policy decision in the ostensibly neutral aura of scientific forecasting, policy-makers can deflect attention from the normative nature of that decision . . ." (Robinson 1991).

In the case of modeling ecological economic systems, purposes can range from developing simple conceptual models, in order to provide a general understanding of system behavior, to detailed realistic applications aimed at evaluating specific policy proposals. It is inappropriate to judge this whole range of models by the same criteria. At minimum, the three criteria of *realism* (simulating system behavior in a qualitatively realistic way), *precision* (simulating behavior in a quantitatively precise way), and *generality* (representing a broad range of systems' behaviors with the same model) are necessary. Holling (1964) first described the fundamental trade-offs in modeling between these three criteria. Later Holling (1966) and Levins (1966) expanded and further applied this classification. No single model can maximize all three of these goals, and the choice of which objectives to pursue depends on the fundamental purposes of the model. Several examples in the literature of ecological and economic models demonstrate the various ways in which trade-offs are made between realism, precision, and generality.

High-generality Conceptual Models

In striving for generality, models must give up some realism and/or precision. They can do this by simplifying relationships and/or reducing resolution. Simple linear and nonlinear economic and ecological models tend to have high generality but low realism and low precision (Clark and Munro 1975; Brown and Swierzbinski 1985; Lines 1989, 1990; Kaitala and Pohjola 1988). Examples include the Lotka-Volterra predator-prey model (Volterra 1931) and its further generalizations (Svirezhev and Logofet 1983), "minimal" models of eutrophication (Voinov and Svirezhev 1984; Voinov and Tonkikh 1987) and of global biosphere processes (Svirezhev and von Bloh 1998),

Holling's "4-box" model (Holling 1987), the "ecological economy" model of Brown and Roughgarden (1992), most conceptual macroeconomic models (Keynes 1936; Lucas 1975), economic growth models (Solow 1956), and the "evolutionary games" approach. For example, the "ecological economy" model of Brown and Roughgarden (1992) contains only three state variables (labor, capital, and "natural resources"), and the relationships between these variables are highly idealized. But the purpose of the model was not to create high realism or precision, but rather to address some basic, general questions about the limits of economic systems in the context of their dependence on an ecological life-support base.

High-precision Analytical Models

Often high precision is wanted (quantitative correspondence between data and model) and realism and generality can be sacrificed. One strategy here is to keep resolution high but to simplify relationships and deal with short time frames. Some models strive to strike a balance between mechanistic small-scale models, which trace small fluctuations in a system, and more general whole-system approaches that remove some of the noise from the signal but do not allow the modeler to trace the source of system changes. The alternative some ecologists have devised is to identify one or a few properties that characterize the system as a whole (Wulff and Ulanowicz 1989). For example, Hannon and Joiris (1987) used an economic input-output model to examine relationships between biotic and abiotic stocks in a marine ecosystem and found that this method allowed them to show the direct and indirect connection of any species to any other and to the external environment in this system at high precision (but low generality and realism). Also using input-output techniques, Duchin's (1988, 1992) aim was to direct development of industrial production systems to efficiently reduce and recycle waste in the manner of ecological systems. Large econometric models (Klein 1971) used for predicting short-run behavior of the economy belong to this class of models since they are constructed to fit existing data as closely as possible (at the sacrifice of generality and realism).

High-realism Impact Analysis Models

When the goal is to develop realistic assessments of the behavior of specific complex systems, generality and precision must be relaxed. High-realism models are concerned with accurately representing the underlying processes in a specific system, rather than precisely matching quantitative behavior or being generally applicable. Dynamic, nonlinear, evolutionary systems models at moderate to high resolution generally fall into this category. Coastal physical-biological-chemical models (Wroblewski and Hofmann 1989) that are used to investigate nutrient fluxes and contain large amounts of site-specific data fall into this category, as do micromodels of behavior of particular business activities. Another example is Costanza et al.'s (1990) model of coastal landscape dynamics that included high spatial and temporal resolution and complex nonlinear process dynamics. This model divided a coastal landscape into 1 km² cells, each of which contained a process-based dynamic ecological simulation model. Flows of water, sediments, nutrients, and biomass from cell to cell across the landscape were linked with internal ecosystem dynamics to simulate long-term successional processes and responses to various human impacts in a very realistic way. But the model was very site-specific and of only moderate numerical precision.

Moderate-generality and Moderate-precision Indicator Models

In many types of systems modeling, the desired outcome is to accurately determine the overall magnitude and direction of change, trading off realism for some moderate amount of generality and precision. For example, aggregate measures of system performance such as standard GNP, environmentally adjusted net national product (or "green NNP"), which includes environmental costs (Mäler 1991), and indicators of ecosystem health (Costanza et al. 1992) fit into this category. The microcosm systems employed by Taub (1989) allow some standardization for testing ecosystem responses and developing ecosystem performance indices. Taub (1987) notes, however, that many existing indicators of change in ecosystems are based on implicit ecological assumptions that have not been critically tested, either for their generality, realism, or precision.

Scale and Hierarchy

In modeling complex systems, the issues of scale and hierarchy are central (O'Neill et al. 1989). Some claim that the natural world, the human species included, contains a convenient hierarchy of scales based on interaction-minimizing boundaries; scales ranging from atoms to molecules to cells to organs to organisms to populations to communities to ecosystems (including economic and human-dominated ecosystems) to bioregions to the global system and beyond (Allen and Starr 1982; O'Neill et al. 1986). By studying the similarities and differences between different kinds of systems at different scales and resolutions, one might develop hypotheses and test them against other systems to explore their degree of generality and predictability.

The term "scale" in this context refers to both the resolution (spatial grain size, time step, or degree of complication of the model) and extent (in time, space, and number of components modeled) of the analysis. The process of "scaling" refers to the application of information or models developed at one scale to problems at other scales. In both ecology and economics, primary information and measurements are generally collected at relatively small scales (that is, small plots in ecology, individuals or single firms in economics) and that information is then used to build models at radically different scales (in other words, regional, national, or global). The process of scaling is directly tied to the problem of aggregation (the process of adding or otherwise combining components), which, in complex, nonlinear, discontinuous systems (like ecological and economic systems), is far from a trivial problem (O'Neill and Rust 1979; Rastetter et al. 1992).

For example, in applied economics basic data sets are usually derived from the national accounts, which contain data that are linearly aggregated over individuals, companies, or organizations. Sonnenschein (1974) and Debreu (1974) have shown that, unless very strong and unrealistic assumptions are made about the individual units, the aggregate (large-scale) relations between variables have no resemblance to the corresponding relations on the smaller scale.

Rastetter et al. (1992) describe and compare three basic methods for scaling that are applicable to complex systems. All of their methods are attempts to utilize information about the nonlinear small-scale variability in the large-scale models. They list (1) partial transformations of the fine-scale mathematical relationships to coarse scale using a statistical expectations operator that incorporates the fine-scale variability; (2) partitioning or subdividing the system into smaller, more homogeneous parts (that is, spatially explicit modeling); and (3) calibration of the fine-scale relationships to coarse-scale data, when these data are available. They go on to suggest a combination of these methods as the most effective overall method of scaling in complex systems.

A primary reason for aggregation error in scaling complex systems is the nonlinear variability in the fine-scale phenomenon. For example, Rastetter et al. (1992) give a detailed example of scaling a relationship for individual leaf photosynthesis as a function of radiation and leaf efficiency to estimate the productivity of the entire forest canopy. Because of nonlinear variability in the way individual leaves process light energy, one cannot simply use the fine-scale relationships among photosynthesis, radiation, and efficiency along with the average values for the entire forest to get total forest productivity without introducing significant aggregation error. One must somehow understand and incorporate this nonlinear fine-scale variability into the coarse-scale equations using some combination of the three methods mentioned above. Method 1 (statistical expectations) implies deriving new coarse-scale equations that incorporate the fine-scale variability. The problem is that incorporation of this variability often leads to equations that are extremely complex and cumbersome (Rastetter et al. 1992). Method 2 (partitioning) implies subdividing the forest into many relatively more homogenous levels or zones and applying the basic fine-scale equations for each partition. This requires a method for adjusting the parameters for each partition, a choice of the number of partitions (the resolution), and an understanding of the effects of the choice of resolution and parameters on the results. Method 3 (recalibration) implies simply recalibrating the fine-scale equations to coarse-scale data. It presupposes that coarse-scale data are available (as more than simply the aggregation of fine-scale data). In many important cases, however, this coarse-scale data is either extremely limited or not available. Thus, while a judicious application of all three aggregation methods is necessary, from the perspective of complex systems modeling, the partitioning approach (Method 2) seems to hold particular promise, because it can take fullest advantage of emerging computer technologies and databases.

From the scaling perspective, hierarchy theory is a potentially useful tool for partitioning systems in ways that minimize aggregation error. According to hierarchy theory, nature can be partitioned into "naturally occurring" levels that share similar time and space scales and that interact with higher and lower levels in systematic ways. Each level in the hierarchy sees the higher levels as constraints and the lower levels as noise. For example, individual organisms see the ecosystem they inhabit as a slowly changing set of constraints and the operation of their component cells and organs is what matters most to them. However, Norton and Ulanowicz (1992) suggest that what appears to be "noise" at a lower level could be turned into significant perturbations on the higher level. This can happen when a critical mass of components participate in a "trend," a behavioral pattern, which affects the slower processes at the higher level. The rapid and extensive human uses of fossil fuels could be seen as such a trend, causing perturbations at the global atmospheric level, which might feed back and radically alter the framework of action at the lower level.

Shugart (1989) explains the relationship between scales: "Clearly, natural patterns in environmental constraints contribute substantially to the spatial pattern and temporal dynamics of particular ecosystems . . . these patterns, especially temporal ones, may resonate with natural frequencies of plant growth forms (i.e., phenology and longevity) to amplify environmental patterns." The simplifying assumptions of hierarchy theory may ease the problem of scaling by providing a common (but somewhat generalized) set of rules that could be applied at any scale in the hierarchy.

Fractals and Chaos

The concept of fractals (Mandelbrot 1977) can be seen as another related approach to the problem of scaling, based on the fundamental principle of "self-similarity" between scales. This concept implies a regular and predictable relationship between the scale of measurement (here meaning the resolution of measurement) and the measured phenomenon. For example, the measured length of a coastline is an increasing function of the resolution at which it is measured. At higher resolutions, one can "see" and measure more of the small-scale bays and indentations in the coast and the total length measured increases.

The relationship between length and resolution usually follows a regular pattern that can be summarized in the following equation:

$$L = k\, s^{(1 - D)} \tag{1}$$

where:

L = the length of the coastline or other "fractal" boundary
s = the size of the fundamental unit of measure or the resolution of the measurement
k = a scaling constant
D = the fractal dimension.

Phenomenon that fit equation (1) are said to be "self-similar" because as resolution is increased one perceives patterns at the smaller scale similar to those at the larger scale. This convenient "scaling rule" [equation (1)] has proven very useful in describing many kinds of complex boundaries and behaviors (Mandelbrot 1983; Turner et al. 1989; Olsen and Schaffer 1990; Sugihara and May 1990; Milne 1991). One test of the principle of self-similarity is that it can applied to production of computer-generated shapes that have a decidedly "natural" and organic look to them (Mandelbrot 1977).

Certain nonlinear dynamic systems models exhibit behaviors whose phase plots (x(t) vs. x(t-dt)) are fractals. These "chaotic" attractors, as they are called, are among four possible pure types of attractors that can be used to classify system dynamics. The other three are (1) point attractors (indicating stable, non-time-varying behavior); (2) periodic attractors (indicating periodic time behavior); and (3) noisy attractors (indicating stochastic time behavior). Real system behavior can be thought of as representing some combination of these four basic types.

The primary questions about the range of applicability of fractals and chaotic systems dynamics to the practical problems of modeling ecological economic systems are the influence of scale, resolution, and hierarchy on the mix of behaviors observed in systems. This is a key problem for extrapolating from small-scale experiments or simple theoretical models to practical applied models of ecological economic systems.

Resolution and Predictability

The significant effects of nonlinearities raise some interesting questions about the influence of resolution (including spatial, temporal, and component) on the performance of models, in particular their predictability. For example, the relationship between the degree of complication (the number of components included) and the predictability of models is an important input to model design. Hofmann (1991) discusses this concern in the context of scaling coastal models to the global scale. The difficulty of using aggregate models that integrate over many details of finer resolution models is that the aggregated models may not be able to represent biological processes on the space and time scales necessary. Hofmann suggests that coupled detailed models (where the output of one model becomes the input for another) may be a more practical method for scaling models to larger systems.

Costanza and Maxwell (1993) analyzed the relationship between spatial resolution and predictability and found that while increasing resolution provides more descriptive information about the patterns in data, it also increases the difficulty of accurately modeling those patterns. There may be limits to the predictability of natural phenomenon at particular resolutions, and "fractal-like" rules that determine how both "data" and "model" predictability change with resolution.

Predictability (Colwell 1974) measures the reduction in uncertainty about one variable, given knowledge of others using categorical data. One can define spatial *auto-predictability* (P_a) as the reduction in uncertainty about the state of a pixel in a scene, given knowledge of the state of adjacent pixels in that scene, and spatial *cross-predictability* (P_c) as the reduction in uncertainty about the state of a pixel in a scene, given knowledge of the state of corresponding pixels in other scenes. P_a is a measure of the internal pattern in the data, while P_c is a measure of the ability of a model to represent that pattern.

Some limited testing of the relationship between resolution and predictability (by resampling land-use map data at different spatial resolutions) showed a strong linear relationship between the log of P_a and the log of resolution (measured as the number of pixels per square kilometer). This fractal-like characteristic of "self-similarity" with decreasing resolution implies that predictability, like the length of a coastline, may be best described using a unitless dimension that summarizes how it changes with resolution. One can define a "fractal predictability dimension" (D_P) in a manner analogous to the normal fractal dimension, that summarizes this relationship. D_P allows convenient scaling of predictability measurements taken at one resolution to others.

Cross-predictability (P_c) can be used for pattern matching and testing the fit between map scenes. In this sense it relates to the predictability of models versus the internal predictability in the data revealed by P_a. While P_a generally increases with increasing resolution (because more information is being included), P_c generally falls or remains stable (because it is easier to model aggregate results than fine-grained ones). Thus we can define an optimal resolution for a particular modeling problem that balances the benefit in terms of increasing data predictability (P_a) as resolution increases, with the cost of decreasing model predictability (P_c). Figure 17-1 shows this relationship in generalized form.

These results may be generalized to all forms of resolution (spatial, temporal, and number of components) and may shed some interesting light on "chaotic" behavior in systems. When looking across resolutions, chaos may be the low level of model predictability that occurs as a natural consequence of high resolution. Lowering model resolution can increase model predictability by averaging out some of the chaotic behavior, at the expense of losing detail about the phenomenon. For example, Sugihara and May (1990) found chaotic dynamics for measles epidemics at the level of individual cities, but more predictable periodic dynamics for whole nations.

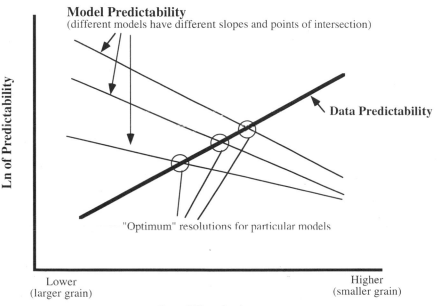

FIGURE 17-1   Hypothetical relationship between resolution and predictability of data and models, plotted on log-log axes (from Costanza and Maxwell 1994). Data predictability is a measure of the internal pattern in the data (for example, the degree to which the uncertainty about the state of landscape pixels is reduced by knowledge of the state of adjacent pixels in the same map). Model predictability is a measure of the correspondence between data and models (for example, the degree to which the uncertainty about the state of pixels is reduced by knowledge of the corresponding state of pixels in a model of the system). In general data predictability rises with increasing resolution, because more internal patterns are perceived, while model predictability falls, because it becomes more difficult to match the high-resolution patterns. Particular types of models and data sets would fall on different lines, and certain types of models would require certain types of data. An "optimal" resolution occurs where the data and model predictability lines intersect.

Evolutionary Approaches

In modeling the dynamics of complex systems it is impossible to ignore the discontinuities and surprises that often characterize these systems and the fact that they operate far from equilibrium in a state of constant adaptation to changing conditions (Rosser 1991; Holland and Miller 1991; Lines 1990; Kay 1991). The paradigm of evolution has been broadly applied to both ecological and economic systems (Boulding 1981; Arthur 1988; Lindgren 1991; Maxwell and Costanza 1993) as a way of formalizing understanding of adaptation and learning behaviors in nonequilibrium dynamic systems. The general evolutionary paradigm posits a mechanism for adaptation and learning in complex systems at any scale using three basic interacting processes: (1) information storage and transmission; (2) generation of new alternatives; and (3) selection of superior alternatives according to some performance criteria.

The evolutionary paradigm is different from the conventional optimization paradigm popular in economics in at least four important respects (Arthur 1988): (1) evolution is path dependent, meaning that the detailed history and dynamics of the system are important; (2) evolution can achieve multiple equilibria; (3) there is no guarantee that optimal efficiency or any other optimal performance will be achieved, due in part to path dependence and sensitivity to perturbations; and (4) "lock-in" (survival of the first rather than survival of the fittest) is possible under conditions of increasing returns. While, as Arthur (1988) notes, "conventional economic theory is built largely on the assumption of diminishing returns on the margin (local negative feedbacks)," life itself can be characterized as a positive-feedback, self-reinforcing, autocatalytic process (Kay 1991; Guenther and Folke 1993), and we should expect increasing returns, lock-in, path dependence, multiple equilibria, and suboptimal efficiency to be the rule rather than the exception in economic and ecological systems.

Cultural versus Genetic Evolution

In biological evolution, the information storage medium is the genes, the generation of new alternatives is by sexual recombination or genetic mutation, and selection is performed by nature according to a criteria of "fitness" based on reproductive success. The same *process* of change occurs in ecological, economic, and cultural systems, but the elements on which the process works are different. For example, in cultural evolution the storage medium is the culture (the oral tradition, books, film, or other storage medium for passing on behavioral norms), the generation of new alternatives is through innovation by individual members or groups in the culture, and selection is again based on the reproductive success of the alternatives generated, but reproduction is carried out by the spread and copying of the behavior through the culture rather than through biological reproduction. One may also talk of "economic" evolution, a subset of cultural evolution dealing with the generation, storage, and selection of alternative ways of producing things and allocating that which is produced. The field of evolutionary economics has grown up in the last decade or so based on these ideas (cf. Day and Groves 1975; Day 1989). Evolutionary theories in economics have already been successfully applied to problems of technical change, to the development of new institutions, and to the evolution of means of payment.

For large, slow-growing animals like humans, genetic evolution has a built-in bias toward the long run. Changing the genetic structure of a species requires that characteristics (phenotypes) be selected and accumulated by differential reproductive success. Behaviors learned or acquired during the lifetime of an individual cannot be passed on genetically. Genetic evolution is therefore usually a relatively slow process requiring many generations to significantly alter a species' physical and biological characteristics.

Cultural evolution is potentially much faster. Technical change is perhaps the most important and fastest evolving cultural process. Learned behaviors that are successful, at least in the short term, can be almost immediately spread to other members of the culture and passed on in the oral, written, or video record. The increased speed of adaptation that this process allows has been largely responsible for *Homo sapiens'* amazing success in appropriating the resources of the planet. Vitousek et al. (1986) estimate that humans now directly control from 25 to 40% of the total primary production of the planet's biosphere, and this is beginning to have significant effects on the biosphere, including changes in global climate and in the planet's protective ozone shield.

Thus, the costs of this rapid cultural evolution are potentially significant. Like a car that has increased speed, humans are in more danger of running off the road or over a cliff. Cultural evolution lacks the built-in long-run bias of genetic evolution and is susceptible to being led by its hyperefficient short-run adaptability over a cliff into the abyss.

Another major difference between cultural and genetic evolution may serve as a countervailing bias, however. As Arrow (1962) has pointed out, cultural and economic evolution, unlike genetic evolution, can at least to some extent employ foresight. If society can see the cliff, perhaps it can be avoided.

While market forces drive adaptive mechanisms (Kaitala and Pohjola 1988), the systems that evolve are not necessarily optimal, so the question remains: What external influences are needed and when should they be applied in order to achieve an optimum economic system via evolutionary adaptation? The challenge faced by ecological economic systems modelers is to first apply the models to gain foresight, and to respond to and manage the system feedbacks in a way that helps avoid any foreseen cliffs (Berkes and

Folke 1994). Devising policy instruments and identifying incentives that can translate this foresight into effective modifications of the short-run evolutionary dynamics is the challenge (Costanza 1987).

Evolutionary Criteria

A critical problem in applying the evolutionary paradigm in dynamic models is defining the selection criteria a priori. In its basic form, the theory of evolution is circular and descriptive (Holling 1987). Those species or cultural institutions or economic activities survive which are the most successful at reproducing themselves. But we only know which ones were more successful *after the fact.* To use the evolutionary paradigm in modeling, we require a quantitative measure of fitness (or more generally *performance*) in order to drive the selection process.

Several candidates have been proposed for this function in various systems, ranging from expected economic utility to thermodynamic potential. Thermodynamic potential is interesting as a performance criteria in complex systems because even very simple chemical systems can be seen to evolve complex nonequilibrium structures using this criteria (Prigogine 1972; Nicolis and Prigogine 1977, 1989), and all systems are (at minimum) thermodynamic systems (in addition to their other characteristics) so that thermodynamic constraints and principles are applicable across both ecological and economic systems (Eriksson 1991).

This application of the evolutionary paradigm to thermodynamic systems has led to the development of far-from-equilibrium thermodynamics and the concept of dissipative structures (Prigogine 1972). An important research question is to determine the range of applicability of these principles and their appropriate use in modeling ecological economic systems.

Many dissipative structures follow complicated transient motions. Schneider and Kay (1993) propose a way to analyze these chaotic behaviors and note that, "Away from equilibrium, highly ordered stable complex systems can emerge, develop and grow at the expense of more disorder at higher levels in the system's hierarchy." It has been suggested that the integrity of far-from-equilibrium systems has to do with the ability of the system to attain and maintain its (set of) optimum operating point(s) (Kay 1991). The optimum operating point(s) reflect a state where self-organizing thermodynamic forces and disorganizing forces of environmental change are balanced. This idea has been elaborated and described as "evolution at the edge of chaos" by Bak and Chen (1991) and Kauffman and Johnson (1991).

The concept that a system may evolve through a sequence of stable and unstable stages leading to the formation of new structures seems well suited to ecological economic systems. For example, Gallopin (1989) stresses that to understand the processes of economic impoverishment ". . . The focus must necessarily shift from the static concept of poverty to the dynamic processes of impoverishment and sustainable development within a context of permanent change. The dimensions of poverty cannot any longer be reduced to only the economic or material conditions of living; the capacity to respond to changes, and the vulnerability of the social groups and ecological systems to change become central." In a similar fashion Robinson (1991) argues that sustainability calls for maintenance of the dynamic capacity to respond adaptively, which implies that we should focus more on basic natural and social processes than on the particular forms these processes take at any time. Berkes and Folke (1994) have discussed the capacity to respond to changes in ecological economic systems, in terms of institution building, collective actions, cooperation, and social learning. These might be some of the ways to enhance the capacity for resilience (increase the capacity to recover from disturbance) in interconnected ecological economic systems.

Modeling as a Consensus-building Tool in Adaptive Management

Integrated modeling of large regional systems, from watersheds to continental-scale systems and ultimately to the global scale, requires input from a very broad range of people. We need to see the modeling process as one that involves not only the technical aspects, but also the sociological aspects involved with using the process to help build consensus about the way the system works and which management options are most effective. This consensus needs to extend both across the gulf separating the relevant academic disciplines and across the even broader gulf separating the science and policy communities, and the public. Appropriately designed and used integrated ecological economic modeling exercises can help to bridge these gulfs.

The process of modeling can (and must) also serve this consensus-building function. It can help to build mutual understanding, solicit input from a broad range of stakeholder groups, and maintain a substantive dialogue between members of these groups. In the process of adaptive management, integrated modeling and consensus building are essential components (Gunderson et al. 1995).

The potential to use modeling as a way to build consensus has been greatly expanded in recent years by the advent of new, much easier-to-use computers and modeling software. As just one example, it is now possible, with graphic, icon-based modeling software packages (such as STELLA [HPS 1995] from High Performance Systems Inc.), to involve a group of relative modeling novices in the construction of relatively complex models, with a few people competent in modeling acting as facilitators. STELLA uses the simple basic model components of stocks, flows, auxiliary variables, and functional connections. The graphic representations of these units are connected and manipulated on the screen to build the basic structure of the model. This process can be transparent to a group when the computer screen is projected onto the wall. Participants can then both follow the model construction process and contribute their knowledge to the process. After the basic model structure is developed, the program requires more detailed decisions about the functional connections between variables. This process is also transparent to the group, using well-designed dialogue boxes, and the potential for both graphic and algebraic input. Once preliminary versions of the model have been constructed, it can be run to develop understanding of its dynamics and sensitivity, to compare its behavior to data for the system, and to help decide where best to put additional effort in improving the model. This can be thought of as an initial "scoping" step that facilitates broad-based input and consensus.

Based on this initial "consensus-building" model development stage, which focuses on generalism as described above, it may be appropriate and desirable to move to a more realistic or precise modeling stage. This stage could involve more traditional "experts" and is more concerned with analyzing the details of specific scenarios or policy options. The case study of the Patuxent River watershed described below has moved into this "policy analysis" stage of model development and use. It is still critical to maintain stakeholder involvement and interaction in this stage with regular workshops and meetings to discuss model progress and results.

While integrated models aimed at realism and precision are large, complex, and loaded with uncertainties of various kinds (Costanza et al. 1990; Groffman and Likens 1994), our abilities to understand, communicate, and deal with these uncertainties are rapidly improving. It is also important to remember that while increasing the resolution and complexity of models expands what we can say about a system, it also limits how accurately we can say it. Model predictability tends to fall with increasing resolution due to compounding uncertainties as described above (Costanza and Maxwell 1993). What we are after are models that optimize their "effectiveness" (Costanza and Sklar 1985) by choosing an intermediate resolution where the product of predictability and resolution (effectiveness) is maximized

It is also necessary to place the modeling process within the larger framework of adaptive management (Holling 1978) if it is to be effective. We need to view the implementation of policy prescriptions in a different, more adaptive way that acknowledges the uncertainty embedded in our models and allows participation by all the various stakeholder groups. "Adaptive management" views regional development policy and management as "experiments," where interventions at several scales are made to achieve understanding and to identify and test policy options (Holling 1978; Walters 1986; Lee 1993; Gunderson et al. 1995). This means that models and policies based on them are not taken as the ultimate answers, but rather as guiding an adaptive experimentation process with the regional system. More emphasis is placed on monitoring and feedback to check and improve the model, rather than using the model to obfuscate and defend a policy that is not corresponding to reality. Continuing stakeholder involvement is essential in adaptive management.

A Case-study Integrated Model: The Patuxent Watershed

The Patuxent Landscape Model (PLM) was designed to serve as a tool in a systematic analysis of the interactions among physical and biological dynamics of the Patuxent watershed, conditioned on socioeconomic behavior in the region. A companion socioeconomic model of the region's land-use dynamics was developed to link with the PLM and provide a means of capturing the feedbacks between ecological and economic systems (figure 17-2). By coupling the two models and exchanging information and data between them, the socioeconomic and ecological dynamics can be incorporated. Whereas in most ecosystem models the socioeconomic development is fed into the model in the form of scenarios or forcing functions, a coupled model can explore dynamic feedbacks, adjusting the socioeconomic change in response to the ecological perturbations.

To run the ecological and economic modules in concert, one needs to account for specifics of both modules in their design and make assumptions about how the information will be exchanged. In particular, the spatial representation of both modules should be matched such that land-use or land-cover transformations in one module can be communicated to the other one directly inside the model. In this case it would be difficult to employ the approach based on spatial aggregation to larger units, called elementary landscapes, elementary watersheds, elementary areas of pollution or hillslopes (Beven and Kirkby 1979; Krysanova et al. 1989; Band et al. 1991; Sasowsky and Gardner 1991), which are considered homogeneous and form the basis for the hydrologic flow network. In these models the boundaries between spatial units are fixed and cannot be modified during the course of the simulation. A more mechanistic approach seems to be better suited when the landscape is aggregated as a grid of relatively small homogeneous cells and process-based simulations are run for each cell with relatively simple rules for material fluxing among nearest neighbors (Sklar et al. 1985; Burke et al. 1990; Costanza et al. 1990; Engel et al. 1993). This fairly straightforward approach requires extensive spatial data sets and high computational capabilities in

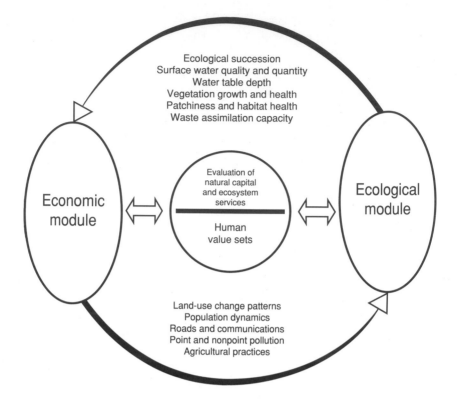

FIGURE 17-2   Relationships and linkages between the economic and ecological subsystems. The ecological and economic modules provide essential feedbacks that are instrumental to create a realistic system of values and to learn to measure these values.

terms of both storage and speed. However, it provides for quasi-continuous modifications of the landscape, where habitat boundaries may change in response to socioeconomic transformations.

The economic module of the PLM is presented elsewhere (Bockstael 1996; Bockstael and Bell 1997; Geoghegan et al. 1997). Here we focus on describing the ecological module, paying special attention to those aspects of the model that were made necessary by the integrated nature of the entire research effort. We first outline the overall model design in terms of its spatial, temporal, and structural organization. Then we look at the single-cell (unit) ecological model and focus on some of the recent modifications of the General Ecosystem Model (GEM) necessary for the Patuxent application. Next we consider the spatial implementation of the model and discuss some aspects relating to scale and resolution. We conclude with a review of the results and potential applications of the model. A fairly complete and up-to-date status report on the model and its results can be found at our Web site: http://iee.umces.edu/PLM/

Model Structure

The PLM may be considered as an outgrowth of the approach first developed in the Coastal Ecosystem Landscape Spatial Simulation (CELSS) model (Sklar et al. 1985; Constanza et al. 1990) and later applied to a series of wetland areas, including the Florida Everglades (Fitz et al. in press; Fitz and Sklar in press). The modeled landscape is partitioned into a spatial grid of square unit cells. The model is hierarchical in structure, incorporating the ecosystem-level unit model that is replicated in each of the unit cells representing the landscape (figure 17-3). With this approach, the model builds on the format of a raster-based geographic information system (GIS), which is used to store all the spatially referenced data included in the model. Thus, the model can be considered an extension of the analytical function of a GIS, adding dynamics and knowledge of ecological processes to the static snapshots stored in a GIS.

Although the same unit model runs in each cell, individual models are parameterized according to habitat type and georeferenced information for a particular cell. The habitat-dependent information is stored in a parameter database that includes initial conditions, rate parameters, stoichiometric ratios, and the like. The habitat type and other location-dependent characteristics are referenced through links to GIS files. In this sense, the PLM is one of several site-specific ecological models that are process-based and designed to apply to a range of habitats. Some other models within this category are CENTURY (Parton et al. 1988), TEM (Vörösmarty et al. 1989), and BIOME-BGC (Running and Coughlan 1988). All these models can be adapted to a particular site through parameterization of initial stocks and flux rates among various ecosystem components. These models vary in complexity and capabilities, which makes one model more suitable for certain applications than others. As a rule of thumb, more complex models will resolve issues in more detail but are more difficult and time-consuming to calibrate and run (Costanza and Maxwell 1994). The unit model in the PLM aims for an intermediate level of complexity so that it is flexible enough to be applied to a range of ecosystems but is not so cumbersome that it requires a super-computer.

The unit models in each cell exchange matter and information across space. The horizontal fluxes that join the unit models together are defined by surface and subsurface hydrology. Alternative horizontal fluxes could be movement of air, animals, and energy, such as fire and tidal waves. At this stage, the PLM fluxes only water and entrained material. The spatial hydrology module calculates the amount of water fluxed over the surface and in the saturated sediment. The fluxes are driven by cell-to-cell hydraulic head differences of surface water and saturated sediment water, respectively. The fluxes of water between cells carry dissolved and suspended material. At each time step, first the unit

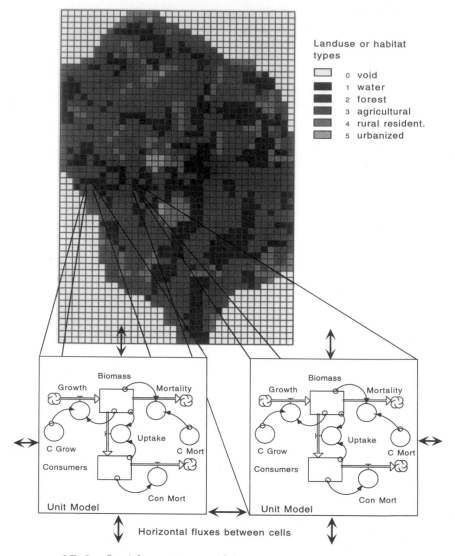

FIGURE 17-3 Spatial organization of the Patuxent watershed model. The unit model is replicated in each of the cells on the study area. Different habitat types are characterized by different parameters in the unit model. Hydrological fluxes connect the unit models horizontally.

model updates the stocks within each cell due to vertical fluxing and then cells communicate to flux matter horizontally, simulating flows and determining ecological condition across the landscape.

Figure 17-4 presents how the various modeled events are distributed in time when simulated in the PLM. The model employs a time step of 1 day, so that most of the ecological variables are updated daily. However, certain processes can be run at longer or shorter time steps. For example, some spatial hydrologic functions may need an hourly time step, whereas certain external forcing functions are updated on a monthly or yearly basis.

This explicit spatial and flexible temporal design of the PLM ecological module is instrumental for linkage with a companion economic model that predicts the probability of land-use change within the seven counties of the Patuxent watershed (Bockstael 1996). The economic model allows human decisions to be modeled as a function of both economic and ecological spatial variables. Based on empirically estimated parameters, spatially heterogeneous probabilities of land conversion are modeled as functions of predicted land values in residential and alternative uses, and costs of conversion. Land-value predictions are modeled as functions of local and regional characteristics. The predictive model of land-use conversion generates the relative likelihood of conversion of cells, and thus the spatial pattern of greatest development pressure. To predict the absolute amount of new residential development, the probabilistic land-use conversion model is further combined with models of regional growth pressure. As a result, a new land-use map is generated and fed into the ecological model on a yearly basis.

Unit Model

The General Ecosystem Model or GEM (Fitz et al. 1996) was used as the initial unit model in developing the PLM. However, since GEM was developed and calibrated mostly for wetland ecosystems, certain modifications were made to provide for a smooth transfer to the predominantly terrestrial habitats of the Patuxent watershed.

The GEM unit model is structured according to a modular concept, which is enforced by the semantics of the development tools used (Maxwell and Costanza 1997). Different modules (or sectors) can be designed independently and linked together. This process facilitates the reuse of modules and a cleaner subdivision of the development effort. Different sectors in the GEM represent hydrology, nutrient movement and cycling, terrestrial and estuarine primary productivity, and aggregated consumer dynamics (figure 17-5). The hydrology sector of the unit model is fundamental to modeled processes since it links the climatic forcing functions to chemical and biotic processes, and allows feedbacks between sectors. Phosphorus and nitrogen are cycled through plant uptake and organic matter decomposition, with the latter simulated in the sector that describes the sediment and soil dynamics. The sector for macrophytes includes growth response to various environmental constraints (including water and nutrient availability), changes in leaf canopy structure (influencing water transpiration), mortality, and other basic plant dynamics. Feedbacks among the biological, chemical, and physical model components structure habitat and influence ecosystem response to changing conditions.

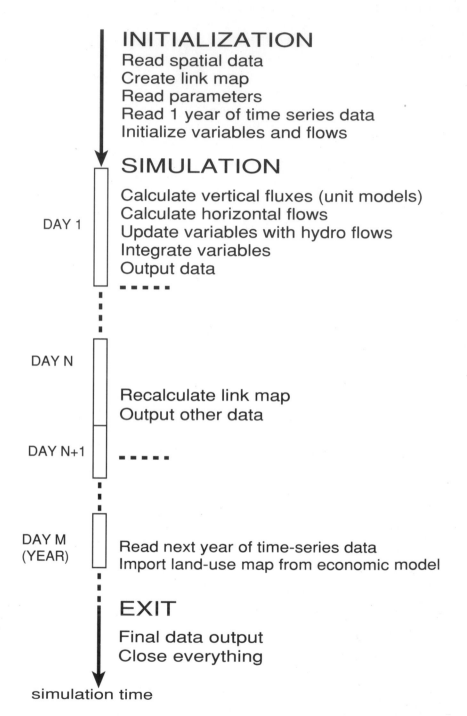

FIGURE 17-4 Temporal course of events in PLM. The Spatial Modeling Environment (SME, see p. 487) offers certain flexibility in scheduling simulation events. Individual time steps can be assigned to different modules.

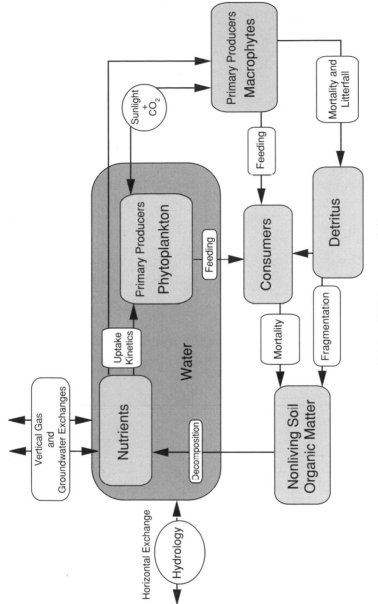

FIGURE 17-5 GEM conceptual diagram.

Spatial Implementation

Once the local ecological processes were described, we needed to decide on the algorithms that put the local dynamics within a spatial context. For watersheds in general and for the Patuxent in particular, hydrologic fluxes seem to be the most important mechanism linking the cells together and delivering the suspended and dissolved matter across the landscape.

The importance of hydrologic transport has been long recognized and considerable effort has been put into creating adequate models for various landscapes (Beven and Kirkby 1979; Beasley and Huggins 1980; Grayson et al. 1992). Nevertheless there are no off-the-shelf universal models that can be easily adapted for a wide range of applications. As a part of a more complicated modeling structure, the hydrologic module is required to be simple enough to run within the framework of the integrated physical-ecological model yet sufficiently detailed to incorporate locally important processes. As a result, some hydrologic details need to be sacrificed to make the whole task more feasible, and these details may differ from one application to another, depending on the size of the study area, the physical characteristics of the slope and surface, and the goals and priorities of the modeling effort.

To simplify hydrologic calculations, we merged process-based and quasi-empirical algorithms (Voinov et al. 1998). First, given the cell size within the model (200 m or 1 km), every cell is assumed to have a stream or depression where surface water can accumulate. Therefore, the whole area becomes a linked network of channels, where each cell contains a channel reach that discharges into a single adjacent channel reach along the elevation gradient. An algorithm generates the channel network from a link map that connects each cell with its one downstream neighbor chosen from the eight possible nearest neighbors.

Second, since most of the landscape is characterized by an elevation gradient, the flow is assumed to be unidirectional, fluxing water downstream. In the simplified algorithm, a portion of water is taken out of a cell and added to the next one linked to it downstream (figure 17-6A). To comply with the Courant condition (Chow et al. 1988), this operation is reiterated many (10–20) times a day, effectively generating a smaller time step to allow faster river flow. The number of iterations needed for the hydrologic module is calibrated so that the water flow rates match gauge data.

This procedure was further simplified by allowing the water to flow through more than one cell over one iteration (figure 17-6B) and then generalized by assuming a variable number of cells in the downstream link (figure 17-6C), as a function of the amount of water in the donor cell. This was adopted to allow for a faster flow when more water is available on the surface (Voinov et al. 1999). It increased flexibility in describing individual hydro-

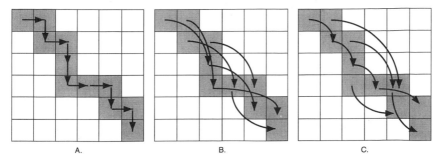

FIGURE 17-6   Algorithms of spatial hydrologic fluxing in the model. (A) Fixed linking to the next cell downstream. Hydrologic fluxing is reiterated n times over the same link map. (B) Fixed linking over n cells downstream. Instead of n iterations of fluxing, the water is moved directly to the n-th cell downstream. (C) Dynamic linking. The length of the link path is determined as a function of the stage in a cell.

graphs and in generalizing them over longer time periods and larger watershed areas.

For groundwater movement we used a linear Darcy approximation, which moves water among adjacent cells in proportion to a conductivity coefficient and the head difference. The groundwater movement provides the slow water flow that generates the river baseflow. Surface-water runoff is the major determinant of the peak flow observed.

Software Development

The sophisticated structure of PLM is supported by several general-purpose software packages developed and refined to meet the needs of PLM. Figure 17-7 presents the interaction between the software modules and the data involved. The modules shown in gray are the ones with which the user needs to interface.

The unit model is developed using the off-the-shelf application STELLA (HPS 1995), which has been widely used for dynamic simulation modeling and allows simple icon-based model building and preliminary analysis. The Spatial Modeling Environment (SME; Maxwell and Costanza 1994; Maxwell 1995; Maxwell and Costanza 1995) links icon-based modeling environments with distributed computing resources.

The model equations generated from STELLA are exported to the SME through a program that translates the STELLA file into the Modular Modeling Language (MML; Maxwell and Costanza 1997). This language provides the semantic structures needed to facilitate archiving and reuse of modules by other researchers. MML-specified model components can be combined hierarchically and are converted by the Code Generator into a C++

object hierarchy within the SME. The C++ objects are then compiled and linked with SME libraries to generate a stand-alone simulation driver.

The SME driver is a simulation environment that runs the spatial simulation on a number of possible parallel or serial computers. It is implemented as a set of distributed C++ objects that exchange data among themselves using network-based multiprocessing. A spatial simulation running within the SME driver is structured as a set of independent modules, each with a potentially unique spatiotemporal representation, defined by a data structure called a Frame. The Frame specifies the topology of the module and is implemented as a set of Points (cells) with (intercellular) links, as well as algorithms for transferring and translating data to and from other Frames. Examples of Frames used in PLM include two-dimensional grids (spatial coverages such as soil maps, land-use maps, and the like), graphs and networks (such as for rivers and streams), and Point sets (such as for running unit models for a single aggregated set of conditions rather than across a heterogeneous space). With

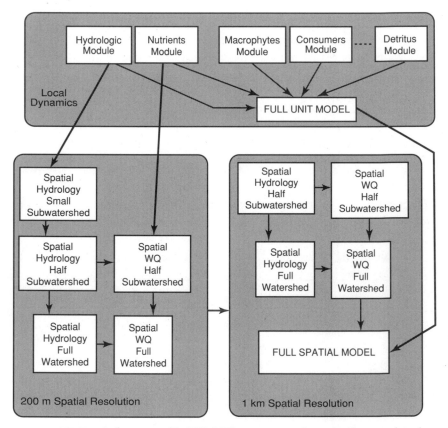

FIGURE 17-7   Software used in PLM. The gray areas show the front-end packages with which the user needs to interact while developing and running the model.

this method, simulations involving modules with disparate spatiotemporal scales can be executed transparently, since the implementation of each module's Frame allows incoming data to be remapped to fit its topology and remaps outgoing data into a universal format.

Unsophisticated users do not need to understand all the details of SME as long as they are running the front-end package, which is the SME user interface, implemented as a module in the Collaborative Modelling Environment (CME; Villa 1997a). CME allows users to define projects, simulation models, input/output configurations, and simulation runs with different calibration data. All of these objects are stored in the database for sharing by collaborating individuals or groups.

Calibration and Testing

Much of the time involved in developing spatial process-based models is devoted to calibration and testing of the model behavior against known historical or other data (Costanza et al. 1990). Calibrating and running a model of this level of complexity and resolution requires a multi-stage approach. We performed the calibration and testing at several time and space scales (figure 17-8). Taking advantage of the model modularity, these tests were carried out for various parts of the model as well as for the whole model. Initial unit model calibrations were handled in STELLA, then the fine-tuning was performed using the Model Performance Index (MPI; Villa 1997b; Villa et al. submitted). At the same time certain modules were put into the spatial context and calibration of the spatial model was carried out. The data against which we test and calibrate the model have been summarized by Costanza et al. (submitted) and Voinov et al. (1999) and can be also found on our Web

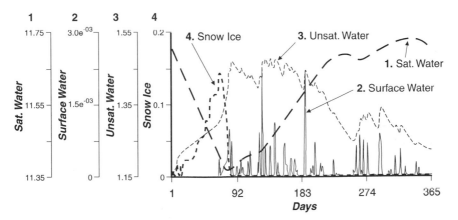

FIGURE 17-8 Output of the unit hydrologic model, calibrated to qualitative local data.

page at http://iee.umces.edu/PLM. In what follows we mostly illustrate this multi-tier calibration procedure with some of the results obtained for the hydrologic module of the model.

First a ballpark calibration was performed for the unit model hydrology. The unit model simulates the hydraulic head of Surface Water, Snow/Ice, water in the unsaturated layer (Unsat. Water) and water in the saturated sediment (Sat. Water; figure 17-9). The latter two variables represent the actual amount of water in the saturated and unsaturated layers, as if it were completely "squeezed" out of the sediment and then its hydraulic head measured. To calculate the real depth of the water table, the amount of saturated water is multiplied by the soil porosity. At this stage it was important to reproduce the qualitative picture of water dynamics in a cell, making sure that there are no long-term trends and that the stages remain within certain limits over several years of model runs.

The results were quite sensitive to the horizontal flow rates of surface water and groundwater (figure 17-10), which could be only parameterized rather than calculated in the unit model. Fairly small changes in values of these parameters (<5%) produced visible variations in the state variables, hiding the variability due to changes in the other parameters responsible for local vertical dynamics such as infiltration and evapotranspiration. Therefore, more detailed calibration of the hydrologic model in the local scale did not make much sense. Besides, there were no reliable local data with which to compare the unit model output.

For a spatial implementation, we chose two scales at which to run the model—a 200 m and a 1 km cell resolution. The 200 m resolution is more appropriate for capturing some of the ecological processes associated with land-use change but is too detailed and requires too much computer processor

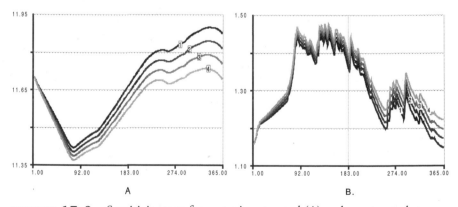

FIGURE 17-9   Sensitivity tests for water in saturated (A) and unsaturated storage (B). High sensitivity to horizontal groundwater flow rate demonstrates the importance of spatial hydrologic processes for the adequate model performance in the local scale. (1, flow rate = 0.0027 m/day; 2, 0.0028; 3, 0.0029; 4, 0.003.)

time to perform the numerous model runs required for calibration and scenario evaluation. The 1 km resolution reduced the total number of model cells in the watershed from 58,905 to 2,352 cells.

We also identified a hierarchy of subwatersheds for calibration at different spatial extents. The Patuxent watershed has been divided into a set of nested

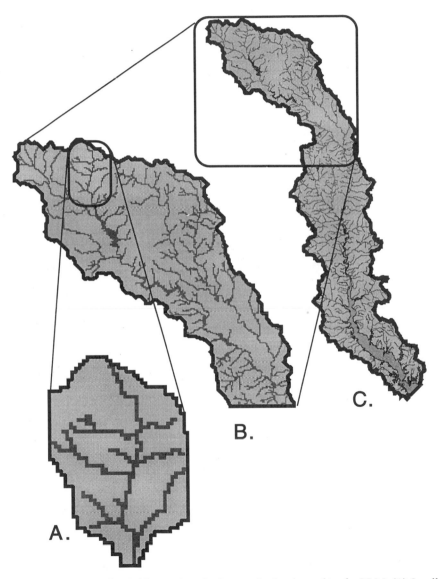

FIGURE 17-10   Spatial hierarchy of subwatersheds adopted in the PLM: (A) Small Cattail Creek subwatershed; (B) upper part of the watershed draining near Bowie USGS gauging station; (C) the full Patuxent River watershed.

subwatersheds to perform analysis at three scales (figure 17-11). A small
(23 km²) subwatershed of Cattail Creek in the northern part of the Patuxent
Basin was used as a starting point. The next-larger watershed was the upper
nontidal half of the Patuxent watershed that drained to the USGS gauge at
Bowie (940 km²). Ultimately we examined the whole Patuxent watershed
(2,352 km²). The number of total model cells grew from 566 cells initially, to
23,484 cells for the half watershed, and then to 58,905 cells for the entire
study area at the 200 m resolution.

A set of experiments was staged with the small Cattail Creek subwatershed
to test the sensitivity of the surface-water flux. Three crucial parameters controlled surface flow in the model: infiltration rate, horizontal conductivity,

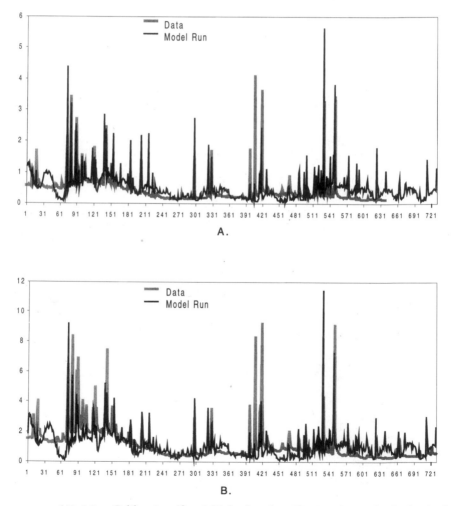

FIGURE 17-11 Calibration (first 365 days) and verification (second 365 days) of
the spatial hydrologic module based on the 1980–1981 data for two gauging stations
on Cattail Creek: (A) near Cooksville station; (B) near Glenwood station.

and number of iterations per time step in the hydrologic model. River-flow peak height was strongly controlled by the infiltration rate. The conductivity determined river levels between storms and the number of iterations or the linkage length modified the width of the storm peaks.

Surface-water flow was calibrated against the thirteen USGS gauging stations in the area that have data concurrent with the climatic data series (1980–1990). The model results for the Cattail Creek subwatershed were in fairly good agreement with the gauge data (figure 17-12). The model parameters were adjusted over a time period of 1 year, and then they were fixed for the second-year run. Since the initial conditions were roughly approximated it took several months for the model to adjust. In comparing the results to the data it should be noted that we were not greatly concerned about simulation of individual hydrographs or rainfall events. Rather we were trying to reproduce the overall water-flow patterns and total volumes fluxed in the area over the time period of a year and more. Some of the flow statistics for the model calibration over a time period of 5 years are presented in table 17-2.

The calibration for the small subwatershed did not hold as well for the half-watershed area, when the same model was applied. For the gauging station used to calibrate Cattail Creek, the error was quite small; however, for downstream gauging stations on the Patuxent River the error was considerably higher. We failed to capture the peak flows in the model.

The number of iterations (effective time step) was adjusted to increase the accuracy of the fit for the larger watershed. Sensitivity analysis performed for

TABLE 17-2

Model verification statistics for the Cattail subwatershed and the Half subwatershed draining at Bowie. The max 10% of flow is the sum of those daily flows making up the upper 10% or the maximum values. It gives an estimate of the peak flows. The min 50% of flow is the sum of the daily flows making up the lower 50% or minimum values. This characterizes the baseflow of the stream. Some models do well on baseflow but err on peaks. Others simulate peaks well but misrepresent the total flow. Here we are mostly concerned with the total flow and try to keep the max 10% errors within reasonable limits.

	Cattail			Bowie		
	Data	Model	% Error	Data	Model	% Error
Total flow	2,510.41	2,527.58	0.68	36,617.43	37,978.78	3.6
max 10%	930.2	925.79	−0.48	12,497.58	16,546.70	27.9
min 50%	587.3	596.25	1.50	7,917.98	6,582.62	−18.4
Total 1986	326.16	282.24	−15.56	4,752.94	4,352.84	−8.8
Total 1987	472.83	469.25	−0.76	6,446.08	7,041.22	8.8
Total 1988	482.01	414.22	−16.37	6,751.99	5,841.62	−14.5
Total 1989	660.62	748.29	11.72	10,507.98	11,881.88	12.3
Total 1990	568.78	611.31	6.96	8,158.45	8,861.23	8.3

the Cattail Creek subwatershed indicated that this parameter could be opti-
mized at fifteen iterations. In the larger watershed a better fit was obtained by
increasing the number of iterations. At this scale we were moving water farther
and therefore needed to increase water movement by increasing the number of
iterations to better simulate the short-term high peaks. In this case, the variable
linkage length approach became crucial and significantly improved the model
performance. The simulation results after these adjustments are presented in
figure 17-12. The results still were not as good as for the Cattail Creek subwatershed (table 17-2); however, the total volumes and average flow patterns
matched well enough. This model behavior illustrates that different scales present new emergent behavior of the system, and that rescaling is always a delicate
process that cannot be done mechanically until there is a greater understanding
of the processes involved. Unless adaptation to changing scale is embedded in
the model structure (for example, the self-adjusting linkage length), running
the model at varying scales will require recalibration to account for additional
data and function that potentially appear in the larger scales.

Of the many possible sources for changes in performance, it is likely that the
spatial or temporal representation of climatic data is an important factor. In
the PLM the spatial rainfall and other data were interpolated from daily
records of seven stations distributed over the study area. The smaller Cattail
Creek hydrology was driven by one climatic station whereas the half-
watershed model incorporated data from three stations. The lack of data on
the true variability of the meteorological data in space and time hinders the
model's ability to accurately represent short-term or localized response in river
flow. However, the general hydrologic trends seemed to be well captured by the
model.

The other model component significantly altered by spatial dynamics was
the nutrients module, especially the part that represented the fate of nitrogen
available for plant uptake. Phosphorus is more easily absorbed than nitrogen
and is less available for horizontal transport in terrestrial ecosystems, whereas
nitrogen is more easily dissolved and the phosphorus is closely related to
hydrologic fluxes. Therefore, another submodel was considered that in addition to hydrology contained the nitrogen module. This was the model that we
referred to as the Water Quality (WQ) submodel (figure 17-8), and which
was calibrated at the same scales and resolutions as the hydrologic model
alone. Unfortunately, the data available for dissolved nitrogen were limited to
observations of NO_3 content in the estuarine part of the Patuxent River. The
northernmost point in that data set was close enough to be extrapolated to the
outlet point of the study area in the half-watershed scale. The calibration was
performed for this station and then for the full watershed.

Spatial dynamic output is best represented as color animation; therefore,
we refer the reader to our Web page at http://iee.umces.edu/PLM, which further describes the model and gives a better idea of its performance.

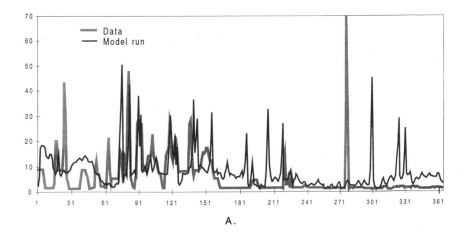

A.

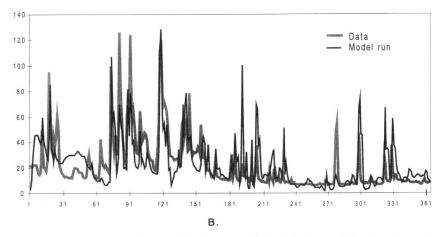

B.

FIGURE 17-12 Verification for the spatial hydrologic module based on the 1980–1981 data for two gauging stations in the upper subwatershed: (A) near Laurel station. This station is located immediately after a reservoir, whose operation schedule is not accounted for in the model. This explains the flat baseflow rate measured at summer, as well as the high flow on day 275 caused by opening the tainter gate in the dam; (B) near Bowie station.

Full Ecological Model

The multi-tier calibration approach assumes that modules can be calibrated independently, at least to a certain extent. The hydrologic module was much more dependent on the spatial implementation than the full ecological unit model that presented local dynamics within particular habitat types. Therefore, while most of the hydrologic calibrations had to be carried out spatially, the ecological unit model could be rigorously studied and calibrated for the local conditions within the spatially homogeneous cell. The calibration presented here simulates a 10-year time period using a constant weather regime from 1986 for each year. Field monitoring at twelve forested sites located within the eastern United States (Johnson and Lindberg 1992) provided mean flux rates and organic matter nutrient contents for input and calibration. Biomass and species composition for the Patuxent area were derived through the Forest Inventory and Analysis database (FIA; Hansen et al. 1997). The forest association was oak-hickory with 0.6% coniferous trees and the rest of the parameters were queried from the database for this association. The consumer sector was made inactive in anticipation of stronger supporting data currently being developed.

The calibration was run for three different stages in forest development. At the first or young stage the forest biomass was set at 10% of the maximum attainable biomass, which is based on the 75th percentile value for oak-hickory in the FIA. The second stage (intermediate) was set at 50% of the maximum biomass, while the third stage (old) was set at 90% of the maximum biomass. Ten-year averages of inorganic phosphate concentrations (PO_4^{3-}-P), dissolved inorganic nitrogen concentrations (DIN), net primary production (NPP; table 17-3, detrital matter, and nonliving soil organic matter (NLOM) (figure 17-13) are compared with similar values available through the FIA database for the Patuxent watershed, or literature on temperate forests.

Similar unit calibration procedures were performed for the other habitat types on the watershed. The PLM currently distinguishes between five land-use types: (1) open water and wetlands; (2) agricultural land; (3) forests; (4) urbanized land (which includes commercial, high-density residential and

TABLE 17-3

Ten-year averages for three forest model variables compared to literature values (NPP = net primary production; DIN = dissolved inorganic nitrogen).

Model output	NPP (kg·m⁻²·y⁻¹) Mean	SD	PO_4^{3-}-P(µg/l) Mean	SD	DIN (µg/l) Mean	SD
Young	0.039	0.006	0.017	0.004	4.1	5.5
Intermediate	0.29	0.014	0.025	0.019	2.7	2.6
Old	0.497	0.014	0.031	0.027	4.2	3.5
All forest ages	0.27	0.190	0.024	0.02	3.7	4.1
Reference data						
All forest ages	0.14[a]	0.67	0.185[b]	0.165	5[c]	5

[a]Derived through the FIA Database for the Patuxent watershed.
[b]Midpoint and maximum deviation reported by Stevenson for sandy soils.
[c]Midpoint and maximum deviation reported by Aber (1992) for deciduous forests.

industrial land); and (5) low-density and medium-density residential land. After calibrating the unit model for the five "habitat" types, we could start running and recalibrating the model spatially for the whole area and for all the processes included.

The full spatially explicit ecological model, including the full unit model and the spatial hydrologic model as described above, was run for several years using historical climate inputs for calibration purposes. Two methods were used to compare the model performance with the available data. On the one hand, certain modeled variables, or indices that aggregate model variables, were compared with point time-series data. In this case spatial dynamics were integrated into time-series data. On the other hand, we generated raw spatial data (map coverages), which could be compared to data when available.

Several time-series data sets such as stream-flow, nutrient concentration in the streams, and historical tree-ring data for the region, were available to calibrate longer-term runs of the model with these data sets. Model output was compared with field data by visually inspecting superimposed graphs and comparing annual mean and total values.

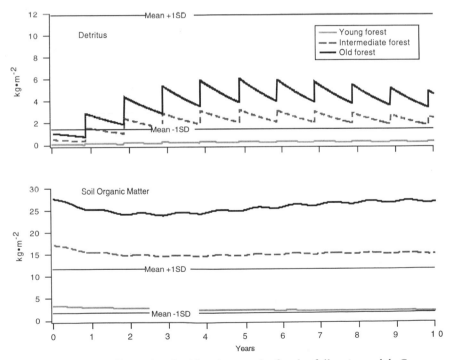

FIGURE 17-13 Example of calibration results for the full unit model. Comparison of detritus and soil organic matter dynamics with 10-year mean values found in the literature.

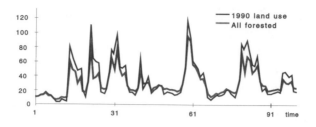

FIGURE 17-14   An example of a scenario run with the PLM. Simulation of water flow from Cattail subwatershed under 1990 land-use patterns and for the all-forested landscape.

Comparison of raw spatial data is a much more difficult and less-studied procedure. Data are scarce and rarely match the spatial extent and resolution required by the model. One of the few spatial data sets available for comparison with model output are the data derived from Advanced Very High Resolution Radiometer (AVHRR) satellites, the Normalized Difference Vegetation Index (NDVI) or "greenness" index. It was used to calibrate the full model's predictions of primary production for intra-annual effects. We created an index from the NDVI data to compare the magnitude of NDVI change with the magnitude of NPP change between cells in time and space. This was useful for qualitative spatial calibrations, even though the correlation between NDVI and NPP is not yet agreed upon. Visual comparison showed fairly good agreement between the model output and the data currently available. The NDVI data for at least a 5-year period, from 1990 to 1995, will be used for further model verification. Among others, example output for plant primary production and total photosynthetic biomass from the model can be seen on our Web page (http://iee.umces.edu/PLM). It shows the typical pattern of seasonal growth in the region.

The model development has now reached the stage when we can start running and analyzing various scenarios. An example of one such experiment is presented in figure 17-14, where we have compared the simulated flow from Cattail Creek area under the land-use patterns that existed in 1990 with the flow that the same landscape would generate if it were completely forested (the precolonization conditions). As was anticipated, the forested watershed produced more baseflow and lower flow peaks.

Conclusions

Linked ecological economic modeling is a potentially important tool for addressing issues of land-use change. Because of the high complexity and large uncertainties in parameters and processes, any numerical estimates are intended to be used with caution, and trade-offs between model generality, realism, and precision must be acknowledged. Nevertheless, this approach to modeling can offer useful information to those currently addressing degradation of ecological systems at the landscape scale. Most important, the model integrates our current understanding of ecological and economic processes to give the best available estimates of the effects of land-use or land-management change. The model also highlights areas where knowledge is lacking and where further research could be targeted for the most impact.

The high data requirements and computational complexities of applications like the PLM can slow model development and implementation. The PLM tries to find a balance between realism, precision, and generality by keeping unit model complexity moderate while providing enough process-oriented, spatially and temporally explicit information to be useful for management purposes. Spatial data is becoming increasingly available for these types of analyses and our modeling framework is able to take advantage of spatial and dynamic data in its relatively raw form without being forced to use complex spatial or temporal aggregation schemes. System dynamics strongly influences ecosystem processes, with processes changing in dominance over time. Ecological analyses will be inadequate unless we incorporate these dynamics.

Our experience with GEM implementation indicates that general models should be applied with caution. While they may be extremely useful for cross-ecosystem comparisons and intercalibrations (Fitz et al. 1996), general models may become either redundant or inadequate in particular applications. The goal of a given study ultimately justifies the application of a certain modeling approach. In the case of large watersheds with complex and diverse ecosystem dynamics and extensive data requirements, the model inevitably needs fine-tuning to the peculiarities of local ecological processes and the specifics of available information. Therefore, an approach based on flexible modeling systems that offer the possibility to build models from existing functional blocks, libraries of modules, functions, and processes (Voinov and Akhremenkov 1990; Maxwell and Costanza 1997) seems to be a good one for watershed modeling.

The Modular Modeling Language we use offers the promise that models of varying degrees of detail can be archived and made available for interchange during new model development. Then, for implementing a model for a particular area, modules can be selected based on the relative importance of local processes; fine detail can be used where needed and otherwise avoided. The flexibility of rescaling the model spatially, temporally, and structurally allows construction of a hierarchical array of models varying in their resolution and complexity to suit the needs of particular studies and challenges, from local to global ones. With each aggregation level and scheme chosen, we can view the output within the framework of other hierarchical levels and keep track of what we gain and lose.

Acknowledgments

Initial funding for this research came from the U.S. EPA Office of Policy, Planning and Evaluation (Coop. Agreement No. CR821925010, Michael Brody and Mary Jo Keely, project officers). Funding has been provided by the U.S. EPA/NSF Water and Watersheds Program through the U.S. EPA Office of Research and Development (Grant No. R82-4766-010). Our sincere thanks go to those researchers who shared their data and insights with us, including economic modelers Nancy Bockstael, Jacqueline Geoghegan, Elena Irwin, Kathleen Bell, and Ivar Strand, and other researchers including Walter Boynton, Jim Hagy, Robert Gardner, Rich Hall, Debbie Weller, Joe Tassone, Joe Bachman, Randolph McFarland, and many others at the Maryland Department of Natural Resources, Maryland Department of the Environment, and University of Maryland Center for Environmental Studies. Our thanks are due to Roelef Boumans, Helena Voinov, Ferdinando Villa, and Lisa Wainger, our collaborators on the Patuxent Model, whose insights, data, and comments are very much appreciated. We specially thank Carl Fitz and Thomas Maxwell for their valuable input and feedback on this chapter.

References

Aber, J. D. 1992. Nitrogen cycling and nitrogen saturation in temperate forest ecosystems. *Trends in Ecology and Evolution* 7(7): 220–24.

Allen, T. F. H., and T. B. Starr. 1982. *Hierarchy*. Chicago, IL: University of Chicago Press.

Arrow, K. 1962. The economic implications of learning by doing. *Review of Economic Studies*. 29: 155–73.

Arthur, W. B. 1988. *Self-reinforcing mechanisms in economics*. Pages 9–31. in P. W. Anderson, K. J. Arrow, and D. Pines (editors), *The economy as an evolving complex system*. Redwood City, CA: Addison Wesley.

Bak, P., and K. Chen. 1991. Self-organized criticality. *Scientific American* 264: 46.

Band, L. E., D. L. Peterson, S. W. Running, J. Coughlan, R. Lammers, J. Dungan, and R. Nemani. 1991. Forest ecosystem processes at the watershed scale: Basis for distributed simulation. *Ecological Modelling* 56: 171–96.

Berkes, F., and C. Folke. 1994. Investing in cultural capital for a sustainable use of natural capital. In A. M. Jansson, M. Hammer, C. Folke, and R. Costanza (editors) *Investing in natural capital: An ecological economics approach to sustainability*. Washington DC: Island Press.

Beasley, D. B., and L. F. Huggins. 1980. *ANSWERS (Areal Nonpoint Source Watershed Environment Response Simulation), user's manual*. West Lafayette, IN: Purdue University.

Beven, K. J., and M. J. Kirkby. 1979. A physically-based, variable contributing area model of basin hydrology. *Hydrological Sciences Bulletin* 24(1): 43–69.

Bockstael, N. E. 1996. Economics and ecological modeling: The importance of a spatial perspective. *American Journal of Agricultural Economics* 78(5): 1168–80.

Bockstael, N., and K. Bell. 1997. Land use patterns and water quality: The effect of differential land management controls. In: S. N. R. Just (editor), *International water and resource economics consortium: Conflict and cooperation on trans-boundary water resources.*

Boulding, K. E. 1981. *Evolutionary economics.* Beverly Hills, CA: Sage.

Brown, G. M., and J. Roughgarden. 1992. An ecological economy: Notes on harvest and growth. Beijer Discussion Paper Series No. 12. Stockholm, Sweden: Beijer International Institute of Ecological Economics.

Brown, G. M., and J. Swierzbinski. 1985. Endangered species, genetic capital and cost-reducing R&D. Pages 111–27 in D. O. Hall, N. Myers, and N. S. Margaris (editors), *Economics of ecosystems management.* Dordrecht, The Netherlands: Dr. W. Junk Publishers.

Burke, I. C., D. S. Schimel, C. M. Yonker, W. J. Parton, L. A. Joyce, and W. K. Lauenroth. 1990. Regional modeling of grassland biogeochemistry using GIS. *Landscape Ecology* 4(1): 45–54.

Cabe, R., J. Shogren, A. Bouzaher, and A. Carriquiry. 1991. Metamodels, response functions, and research efficiency in ecological economics. Working paper No. 91-WP 79. Center for Agricultural and Rural Development. Ames: Iowa State University.

Chow, V. T., D. R. Maidment, and L. W. Mays. 1988. *Applied hydrology.* New York: McGraw-Hill.

Clark, C. W. 1976. *Mathematical bioeconomics.* New York: John Wiley & Sons.

———. 1981. Bioeconomics of the ocean. *BioScience* 31: 231–37.

———. 1985. *Bioeconomic modelling and fisheries management.* New York: Wiley and Sons.

Clark, C. W., and G. R. Munro. 1975. The economics of fishing and modern capital theory: A simplified approach. *Journal of Environmental Economics and Management* 2: 92–106.

Colwell, R. K. 1974. Predictability, constancy, and contingency of periodic phenomena. *Ecology* 55: 1148–53.

Costanza, R. 1987. Social traps and environmental policy. *BioScience* 37: 407–12.

Costanza, R., and T. Maxwell. 1994. Resolution and predictability: An approach to the scaling problem. *Landscape Ecology* 9: 47–57.

Costanza, R., and F. H. Sklar. 1985. Articulation, accuracy, and effectiveness of mathematical models: A review of freshwater wetland applications. *Ecological Modeling* 27: 45–68.

Costanza, R., F. H. Sklar, and M. L. White. 1990. Modeling coastal landscape dynamics. *BioScience* 40: 91–107.

Costanza, R., B. Norton, and B. J. Haskell (editors). 1992. *Ecosystem health: New goals for environmental management.* Washington, DC: Island Press.

Costanza, R., A. Voinov, R. Boumans, T. Maxwell, F. Villa, L. Wainger, H. Voinov, and N. Bockstael. Submitted. Integrated ecological economic modeling of the Patuxent River watershed, Maryland. *Ecological Monographs.*

Day, R. H. 1989. Dynamical systems, adaptation and economic evolution. MRG Working Paper No. M8908. Los Angeles: University of Southern California.

Day, R. H., and T. Groves (editors). 1975. *Adaptive economic models.* New York: Academic Press.

Debreu, G. 1974. Excess demand functions. *Journal of Mathematical Economics* 1: 15–23.

Duchin, F. 1988. Analyzing structural change in the economy. Pages 113–128 in M. Ciaschini (editor), *Input-output analysis: Current developments.* London: Chapman and Hall.

Duchin, F. 1992. Industrial input-output analysis: Implications for industrial ecology. *Proceedings of the National Academy of Sciences USA* 89: 851–855.

Engel, B. A., R. Srinivasan, and C. Rewerts. 1993. A Spatial decision support system for modeling and managing agricultural non-point source pollution. Pages 230–237 in M. F. Goodchild, B. O. Parks and L. T. Steyaert (editors), *Enviromental modeling with GIS.* New York: Oxford University Press.

Eriksson, K-E. 1991. Physical foundations of ecological economics. Pages 186–96 in L. O. Hansson and B. Jungen (editors) *Human responsibility and global change.* Göteborg, Sweden: University of Göteborg Press.

Fitz, H. C., E. DeBellevue, R. Costanza, R. Boumans, T. Maxwell, L. Wainger, and F. Sklar. 1996. Development of a general ecosystem model for a range of scales and ecosystems. *Ecological Modelling* 88(1/3): 263–95.

Fitz, H. C., R. Costanza, and A. Voinov. A dynamic spatial model as a tool for integrated assessment of the Everglades, USA. In R. Costanza (editor), *Ecological economics for integrated modeling and assessment.* In press.

Fitz, H. C., and F. H. Sklar. Ecosystem analysis of phosphorus impacts in the Everglades: A landscape modeling approach. In R. Reddy (editor), *Phosphorus biogeochemistry in Florida ecosystems.* In press.

Gallopin, G. C. 1989. Global impoverishment, sustainable development and the environment: A conceptual approach. *International Social Science Journal* 121: 375–97.

Geoghegan, J., L. Wainger, and N. Bockstael. 1997. Spatial landscape indices in a hedonic framework: An ecological economics analysis using GIS. *Ecological Economics* 23(3): 251–64.

Grayson, R. B., I. D. Moore, and T. A. McMahon. 1992. Physically based hydrologic modeling: 1. A Terrain-based model for investigative purposes. *Water Resources Research* 28(10): 2639–58.

Groffman, P. M., and G. E. Likens (editors). 1994. *Integrated regional models: Interactions between humans and their environment.* New York: Chapman and Hall.

Gunderson, L., C. S. Holling, and S. Light (editors). 1995. *Barriers and bridges to the renewal of ecosystems and institutions.* New York: Columbia University Press.

Guenther, F., and C. Folke. 1993. Characteristics of nested living systems. *Journal of Biological Systems* 1 (3): 257–74.

Hannon, B., and C. Joiris. 1987. A seasonal analysis of the southern North Sea ecosystem. *Ecology* 70: 1916–34.

Hansen, M. H., T. Frieswyk, J. F. Glover, and J. F. Kelly. 1997. The Eastwide Forest

Inventory and Analysis Data Base: Users Manual. *http://www.srsfia.usfs.msstate.edu/ewman.htm.*

Hofmann, E. E. 1991. How do we generalize coastal models to global scale? Pages 401–417 in R. F. C. Mantoura, J. M. Martin, and R. Wollast (editors), *Ocean margin processes in global change.* New York: John Wiley & Sons.

Holland, J. H., and J. H. Miller. 1991. Artificial adaptive agents in economic theory. *American Economic Review* 81: 365–70.

Holling, C. S. 1964. The analysis of complex population processes. *The Canadian Entomologist* 96: 335–47.

———. 1966. The functional response of invertebrate predators to prey density. *Memoirs of the Entomological Society of Canada* No. 48.

———. (editor). 1978. *Adaptive environmental assessment and management.* London: John Wiley & Sons.

———. 1987. Simplifying the complex: The paradigms of ecological function and structure. *European Journal of Operational Research* 30: 139–46.

HPS. 1995. STELLA: High performance systems.

Johnson, D. W., and S. E. Lindberg. 1992. *Atmospheric deposition and forest nutrient cycling: A synthesis of the integrated forest study.* New York: Springer-Verlag.

Kaitala, V., and M. Pohjola. 1988. Optimal recovery of a shared resource stock: A differential game model with efficient memory equilibria. *Natural Resource Modeling* 3: 91–119.

Kauffman, S. A., and S. Johnson. 1991. Coevolution to the edge of chaos: Coupled fitness landscapes, poised states, and coevolutionary avalanches. *Journal of Theoretical Biology* 149: 467–505.

Kay, J. J. 1991. A nonequilibrium thermodynamic framework for discussing ecosystem integrity. *Environmental Management* 15: 483–95.

Keynes, J. M. 1936. *General theory of employment, interest and money.* London: Harcourt Brace.

Klein, L. R. 1971. Forecasting and policy evaluation using large-scale econometric models: The state of the art. In M. D. Intriligator (editor), *Frontiers of quantitative economics.* Amsterdam, The Netherlands: North-Holland Publishing Co.

Krysanova, V., A. Meiner, J. Roosaare, and A. Vasilyev. 1989. Simulation modelling of the coastal waters pollution from agricultural watershed. *Ecological. Modelling* 49: 7–29.

Kuznetsov, Y. A., and S. Rinaldi. 1996. Remarks on food chain dynamics. *Mathematical Biosciences* 134(1): 1–33.

Lee, K. 1993. *Compass and gyroscope.* Washington, DC: Island Press.

Levins, R. 1966. The strategy of model building in population biology. *American Scientist* 54: 421–31.

Lindgren, K. 1991. Evolutionary phenomena in simple dynamics. Pages 295–312 in C. G. Langton, C. Taylor, J. D. Farmer and S. Rasmussen (editors), *Artificial life, SFI studies in the sciences of complexity.* Vol. 10. New York: Addison-Wesley.

Lines, M. 1989. Environmental noise and nonlinear models: A simple macroeconomic example. *Economic Notes* 19: 376–94.

———. 1990. Stochastic stability considerations: A nonlinear example. *International Review of Economics and Business* 3: 219–33.

Lucas, R. E. 1975. An equilibrium model of the business cycle. *Journal of Political Economy* 83: 1113–45.

Mäler, K. G. 1991. National accounts and environmental resources. *Environmental and Resource Economics* 1: 1–15.

Mandelbrot, B. B. 1977. *Fractals. Form, chance and dimension.* San Francisco, CA: W. H. Freeman and Co.

———. 1983. *The fractal geometry of nature.* San Francisco, CA: W. H. Freeman and Co.

Maxwell, T. 1995. Distributed modular spatio-temporal simulation. *http://kabir.cbl.cees.edu /SME3.*

Maxwell, T., and R. Costanza. 1994. Spatial ecosystem modeling in a distributed computational environment. In J. v. d. Bergh and J. v. d. Straaten (editor), *Toward sustainable development: Concepts, methods, and policy.* Washington, DC: Island Press.

———. 1995. Distributed modular spatial ecosystem modelling. *International Journal of Computer Simulation: Special Issue on Advanced Simulation Methodologies* 5(3): 247–62.

———. 1997. A language for modular spatio-temporal simulation. *Ecological Modelling* 103(2,3): 105–14.

Milne, B. T. 1991. Lessons from applying fractal models to landscape patterns. Pages 199–235 in M. G. Turner and R. Gardner (editors), *Quantitative methods in landscape ecology.* Ecological Studies 82. New York: Springer-Verlag.

Nicolis, G., and I. Prigogine. 1977. *Self-organization in non-equilibrium systems.* New York: John Wiley & Sons.

———. 1989. *Exploring Complexity.* New York: W. H. Freeman.

Norgaard, R. B. 1989. The case for methodological pluralism. *Ecological Economics* 1: 37–57.

Norton, B. G., and R. E. Ulanowicz. 1992. Scale and biodiversity policy: A hierarchical approach. *Ambio* 21: 244–249.

O'Neill, R. V., and B. Rust. 1979. Aggregation error in ecological models. *Ecological Modelling* 7: 91–105.

O'Neill, R. V., D. L. DeAngelis, J. B. Waide, and T. F. H. Allen. 1986. *A hierarchical concept of ecosystems.* Princeton, NJ: Princeton University Press.

O'Neill, R. V., A. R. Johnson, and A. W. King. 1989. A hierarchical framework for the analysis of scale. *Landscape Ecology* 3: 193–205.

Olsen, L. F., and W. M. Schaffer. 1990. Chaos versus noisy periodicity: Alternative hypotheses for childhood epidemics. *Science* 249: 499–504.

Parton, W. J., J. W. B. Stewart, and C. V. Cole. 1988. Dynamics of C, N, P, and S in grassland soils: A model. *Biogeochemistry* 5: 109–131.

Prigogine, I. 1972. Thermodynamics of evolution. *Physics Today* 23: 23–28.

Rastetter, E. B., A. W. King, B. J. Cosby, G. M. Hornberger, R. V. O'Neill, and J. E. Hobbie. 1992. Aggregating fine-scale ecological knowledge to model coarser-scale attributes of ecosystems. *Ecological Applications* 2: 55–70.

Robinson, J. B. 1991. Modelling the interactions between human and natural systems. *International Social Science Journal* 130: 629–47.

———. 1992. Of maps and territories: The use and abuse of socio-economic modelling in support of decision-making. *Technological Forecast and Social Change* 42(2): 147–64.

Rosser, J. B. 1991. *From catastrophe to chaos: A general theory of economic discontinuities.* Amsterdam, The Netherlands: Kluwer.

Running, S. W., and J. C. Coughlan. 1988. General model of forest ecosystem processes for regional applications. 1. Hydrologic balance, canopy gas exchange and primary production processes. *Ecological Modelling* 42: 125–54.

Sasowsky, C. K., and T. W. Gardner. 1991. Watershed configuration and geographic information system parameterization for SPUR model hydrologic simulations. *Water Resources Bulletin* 27(1): 7–18.

Schneider, E. D., and J. J. Kay. 1993. Life as a manifestation of the second law of thermodynamics. *Mathematical and Computer Modelling* 19(6–8): 25–48.

Shugart, H. H. 1989. The role of ecological models in long-term ecological studies. Pages 90–109 in G. E. Likens (editor), *Long-term studies in ecology: Approaches and alternatives.* New York: Springer-Verlag.

Sklar, F. H., R. Costanza, and J. W. Day Jr. 1985. Dynamic spatial simulation modeling of coastal wetland habitat succession. *Ecological Modelling* 29: 261–81.

Solow, R. M. 1956. A contribution to the theory of economic growth. *Quarterly Journal of Economics* 70: 65–94.

Sonnenschein, H. 1974. Market excess demand functions. *Econometrica* 40: 549–563.

Stevenson, F. J. 1986. *Cycles of soil: Carbon nitrogen, phosphorus, sulfur, micronutriets.* New York: John Wiley and Sons.

Sugihara, G., and R. M. May. 1990. Nonlinear forecasting as a way of distinguishing chaos from measurement error in time series. *Nature* 344: 734–741.

Svirezhev, Y. M., and D. O. Logofet. 1983. *Stability of biological communities.* Moscow: Mir.

Svirezhev, Y. M., and W. von Bloh. 1998. A zero-dimensional climate-vegetation model containing global carbon and hydrological cycle. *Ecological Modelling* 106(2/3): 109–29.

Taub, F. B. 1987. Indicators of change in natural and human impacted ecosystems: Status. Pages 115–44 in S. Draggan, J. J. Cohrssen, and R. E. Morrison (editors), *Preserving ecological systems: The agenda for long-term research and development.* New York: Praeger.

———. 1989. Standardized aquatic microcosm—development and testing. Pages 47–92 in A. Boudou and F. Ribeyre (editors), *Aquatic ecotoxicology: Fundamental concepts and methodologies.* Vol 2. Boca Raton, FL: CRC Press.

Turner, M. G., R. Costanza, and F. H. Sklar. 1989. Methods to compare spatial patterns for landscape modeling and analysis. *Ecological Modelling* 48: 1–18.

Villa, F. 1997a. Guide to the Spatial Modeling Environment graphical user interface. *http://kabir.cbl.umces.edu/~villa/sme_int/sme_int.html.*

———. 1997b. Usage of the Model Performance Evaluation software. Internal report, Institute for Ecological Economics, University of Maryland.

Villa, F., R. M. J. Boumans, and R. Costanza. Submitted. Design and use of a Model Performance Index (MPI) for the calibration of ecological simulation models. *Journal of Environmental Modeling and Software.*

Vitousek, P., P. R. Ehrlich, A. H. Ehrlich, and P. A. Matson. 1986. Human appropriation of the products of photosynthesis. *BioScience* 36: 368–73.

Voinov, A. A., and Y. M. Svirezhev. 1984. A minimal model of eutrophication in freshwater ecosystems. *Ecological Modelling* 23: 277–92.

Voinov, A. A. and A. P. Tonkikh. 1987. Qualitative model of eutrophication in macrophyte lakes. *Ecological Modelling* 35: 211–26.

Voinov, A., and A. Akhremenkov. 1990. Simulation modeling system for aquatic bodies. *Ecological Modelling* 52: 181-205.

Voinov, A., C. Fitz, and R. Costanza. 1998. Surface water flow in landscape models: 1. Everglades case study. *Ecological Modelling* 108(1–3): 131–44.

Voinov, A., H. Voinov, and R. Costanza. 1999. Landscape modeling of surface water flow: 2. Patuxent case study. *Ecological Modelling* 119: 211–30.

Volterra, V. 1931. *Leçons sur la Théorie Mathématique de la Lutte pour la Vie.* Paris: Gauthier-Villars.

von Bertalanffy, L. 1968. *General system theory: Foundations, development, applications.* New York: George Braziller.

Vörösmarty, C. J., B. Moore III, A. L. Grace, M. P. Gildea, J. M. Melillo, and B. J. Peterson. 1989. Continental scale model of water balance and fluvial transport: An application to South America. *Global Biogeochemical Cycles* 3: 241–65.

Walters, C. J. 1986. *Adaptive management of renewable resources.* New York: Macmillan.

Wroblewski, J. S., and E. E. Hofmann. 1989. U.S. interdisciplinary modeling studies of coastal-offshore exchange processes: Past and future. *Progress in Oceanography* 23: 65–99.

Wulff, F., and R. E. Ulanowicz. 1989. A comparative anatomy of the Baltic Sea and Chesapeake Bay ecosystems. Pages 232–56 in F. Wulff, J. G. Field, and K. H. Mann (editors), *Network analysis of marine ecology: Methods and applications.* Coastal and Estuarine Studies Series. Heidelberg, Germany: Springer-Verlag.

CHAPTER 18

Scientific Synthesis in Estuarine Management

Donald F. Boesch, Joanna Burger, Christopher F. D'Elia, Denise J. Reed,
and Donald Scavia

Science and environmental management are, in a sense, relatively new acquaintances. To be sure, ancient Romans developed rudimentary understanding of the science and engineering necessary to manage the flow and disposal of sewage wastes. Eighteenth-century Londoners were acutely aware that something was rotten in the malodorous Thames—and that somebody had to do something about it. But, in general, the "solution to pollution was dilution," and serious scientific and technical solutions to environmental problems the world over only began to ensue well into the Industrial Age with the advent of secondary-treatment of sewage, and with a concomitant understanding that cause-and-effect relationships exist between human activities and environmental impacts. Even then, problems found solutions only on a case-by-case basis.

It was not until the period of heightened awareness of environmental problems, from 1969 to 1972, that the U.S. Congress passed a series of sweeping and ambitious environmental laws, such as the National Environmental Policy Act (NEPA) of 1970 and the Federal Water Pollution Control Act (FWPCA) of 1972, which committed the nation to approach environmental problems systematically and in doing so set a challenging example for the rest of the world. In some cases, these laws placed new demands on environmental sciences. For example, NEPA required lengthy Environmental Impact Statements that must summarize knowledge of the affected environment and assess the impact of the proposed activity and alternatives. But in other cases, such as the FWPCA, goals were set to eliminate discharges or control them to the limits of technology without regard to knowledge of actual impacts on the environment. Despite the many obvious successes in environmental management and pollution control in the United States, the fact that much of this seminal environmental legislation is based on uncertain science is now purported by some detractors to show that existing laws and regulations should be abolished.

For a great variety of reasons related to scientific uncertainty, the complexity of nature, the mismatch between the pace of science and the urgency of management decisions, differences in the cultures of practitioners, and political forces, there exists what has been referred to as a gap or "impedance mismatch" between science and environmental management (National Research Council 1995). This difficult interface is a source of frustration and complaint by scientists and managers alike. Nonetheless, modern management of coastal and estuarine environments is increasingly stressing the need for scientific knowledge. This is manifest in such programs as the National Estuary Program (NEP) and the Chesapeake Bay, Great Lakes, and Gulf of Mexico programs, and in the embracing of ecosystem management or integrated coastal management approaches (Healey and Hennessey 1994).

Toward the goal of improved development and use of synthetic scientific knowledge in estuarine management, in this chapter we first consider several cases studies of the use of science in estuarine or coastal management. We attempt to identify factors that have led to success and factors that have impeded it. We explore how synthetic scientific knowledge about estuaries can be developed and how it can be applied in site-specific management. Finally, we discuss the roles that federal and state agencies (both those that manage estuaries and those that support science) and the scientific community must play if we are to advance on this goal. Our premise is that environmental policies and regulations based on better science will be more effective and, accordingly, better regarded by all.

Accomplishments and Challenges

The examples we provide below are purely anecdotal: they have been chosen arbitrarily and consequently do not represent either a systematic or an exhaustive inventory of possible candidates for consideration. Nonetheless, we feel that they adequately represent a wide scope of situations in terms of both jurisdictional diversity and environmental complexity that valid general conclusions can be drawn from them.

Great Lakes

Probably the first national recognition of aquatic and coastal management problems occurred when the Cayahoga River at Lake Erie caught fire. While that event caught the public's eye, it was the extent of historically rich research results and the development and refinement of conceptual and mathematical models that refocused attention on the more pervasive overenrichment of the Great Lakes. This original focus on eutrophication gave way, after significant science-based management actions were enacted, to toxic contaminant, habitat, and fishery issues that were also identified primarily by the scientific community. Continued emphasis on mass-balance and modeling perspectives have provided effective interfaces between scientific discovery and management and policy action. Nongovernmental organizations (NGOs) have been very active and intergovernmental institutions have been developed to coordinate the jurisdictions involved. The International Joint Commission serves to coordinate science and policy between Canada and the United States. The EPA's Great Lakes Program coordinates state and federal activities within the United States. There is even a scientific society, the International Association of Great Lakes Research, which focuses heavily on the Laurentian Great Lakes and interacts closely with the IJC and other intergovernmental agencies (Francis and Regier 1995).

Today, the Great Lakes are hardly to be viewed as beyond environmental concern. However, there is clear recognition and acknowledgment that scientifically based actions initiated two decades ago have proved successful in confronting the problems identified at that time. For example, Lake Erie has benefited greatly from nutrient-control strategies that have been developed and is no longer in danger of severe eutrophication (Francis and Regier 1995).

Chesapeake Bay

As early as 1972 there were emerging concerns among the public and political leaders, boosted by some scientific observations and considerable speculation, about the effects of human activities on Chesapeake Bay. These concerns were heightened by the rapid population growth the region was witnessing. A Bi-State Conference on the Chesapeake Bay (Chesapeake Research Consortium 1977) identified the "continuing input of large quantities of chemicals which can be overenriching . . . baywide failure of oyster reproduction . . . and decline of rooted aquatic vegetation" as serious concerns and discussed population growth and point and nonpoint inputs of nutrients as possible causes. There was little hard scientific evidence of these linkages at the time, although later analysis of historical data did corroborate such an interpretation (Malone et al. 1993). Since then, scientists and managers have been engaged in a dynamic, occasionally contentious—but nonetheless productive—interaction that has resulted in major commitments of the multiple political jurisdictions in the Chesapeake Bay watershed for the treatment of point-source discharges, the control of nonpoint sources, and the use of increasingly sophisticated management tools, such as ecosystem models of the bay and its

Box 18.1

"Mature" Estuarine Management: Chesapeake Bay

Malone et al. (1993) provide a fascinating historical analysis of the role of science in the acceptance of nutrient overenrichment as a major cause of the degradation of Chesapeake Bay and in reaching commitments for nutrient controls. The Patuxent River, a relatively minor tributary of Chesapeake Bay that was the site of a naval skirmish and the British landing before the burning of the White House during the War of 1812, served as the site of early volleys in this metaphorical war to "save the bay." The watershed of this subestuary occupies only a small part of the 167,000 km² bay watershed but is located between the two largest metropolitan areas, Baltimore and Washington, and has been one of the fastest-growing areas in the United States.

Early interest in nutrient dynamics and primary productivity in estuaries at the Chesapeake Biological Laboratory (CBL), located near the mouth of the Patuxent estuary, led to studies that produced data on nutrients, water clarity, and dissolved-oxygen concentration in the estuary dating back to the mid-1930s. By the mid-1970s, concerns among CBL scientists about the declining water quality in the estuary and its apparent association with nutrient enrichment reached the attention of local government officials from the somewhat sparsely populated counties around the estuary, who were troubled by rapid population growth and increasing sewage discharges from the upstream counties.

Municipal waste treatment at that time focused mainly on secondary treatment and, following the successful experience in the nearby Potomac River estuary, on phosphorus removal. However, scientific evidence was growing that nitrogen was an important limiting nutrient in the brackish portions of the estuary. When the upstream counties, the state of Maryland, and the EPA all demurred on advanced nutrient removal, particularly nitrogen removal, a lawsuit was filed by

watershed (Boesch et al. in press). Considerable concern remains, but clear progress has been made (box 1).

San Francisco Bay

A small detachment of the U.S. Geological Survey is located on the shore of San Francisco Bay in Menlo Park, California. Here, an able group of federal scientists took a strong interest in their neighboring estuary and in the 1960s the Tri-County Council of Southern Maryland, demanding substantial improvements in sewage treatment. CBL scientists testified on behalf of the downstream litigants, in opposition to the state and federal agencies sponsoring their research, and the suit was successful in forcing a consensus for advanced nutrient removal.

Also in the early 1970s widespread loss of submerged aquatic vegetation was observed by scientists at the Virginia Institute of Marine Science (in the lower bay) and the Horn Point Laboratory (in the upper bay). Segments of the public and scientific and management communities initially attributed these losses to increased agricultural herbicide use, major freshets, diseases, the disruptive activities of predators, or unassigned natural variability. However, greatly expanded research by VIMS and HPL scientists, stimulated by concerns over the vegetation losses, revealed that nutrient overenrichment throughout the bay ecosystem was an overarching cause of loss of submerged aquatic vegetation. Eutrophication had decreased light available to vascular plants growing in shallow waters by stimulating the biomass of phytoplankton and epiphytic growth.

Building on scientific discoveries concerning the Patuxent estuary and the demise of submerged aquatic vegetation, managers began to accept the notion that Chesapeake Bay as a whole is in trouble as a result of nutrient overenrichment. Very active NGOs such as the Citizens' Alliance for the Chesapeake and the Chesapeake Bay Foundation were successful in raising public awareness and concern. Senator Charles Mathias succeeded in the late 1970s in passing federal legislation to establish the Chesapeake Bay Program within the EPA. The Chesapeake Bay Program, under direction of its Executive Council including the governors of Pennsylvania, Maryland, and Virginia, the mayor of the District of Columbia, and the EPA administrator, is now a thoroughly institutionalized entity directing the restoration of the Chesapeake Bay ecosystem through coordinated nonpoint-source control, toxics reduction, habitat restoration, and living-resource management.

began to amass an impressive historical data set on the bay's water quality. They did much more, however. They analyzed the data with rigor on an ongoing basis and supplemented their analysis with a state-of-the-art program of process-oriented measurements. All of this information has been heavily used in the development of mathematical models of the bay.

Today San Francisco Bay is one of the better understood estuaries in the world. Fundamental information exists about seasonal and interannual variability in productivity, and there is keen awareness of the role of herbivorous

---

Box 18.2

A New Challenge: Florida Bay

In 1987, local fishing guides began to report that seagrass beds were dying throughout Florida Bay. Scientists confirmed these anecdotal reports indicating that by 1990 nearly 25,000 hectares of seagrass beds had been largely denuded. Although die-off events have been poorly monitored since 1990, current estimates place losses at more than 40,000 hectares. Florida agency and federal National Park Service scientists began relatively small-scale studies of the bay as a result of the public's concern, but resources directed to these studies were inadequate for gathering sufficient data to establish cause-effect relationships. No policy shifts were seriously contemplated at the time.

By 1990, more than 1,500 km² of massive algal blooms and high levels of turbidity were observed, and public attention to Florida Bay increased substantially. The first blue-green algae blooms appeared in the fall of 1991 following an extensive seagrass die-off event. These fundamental changes in the Florida Bay ecosystem have affected the abundance of important fishery species. Since seagrass beds are critical habitat for juvenile pink shrimp and Florida Bay is a major nursery ground for that fishery, the seagrass die-off was believed to be the primary contributor to a precipitous 50% decline in pink shrimp harvest and annual losses of more than $10 million.

Scientists began to speak out—somewhat reluctantly at first—on potential causes of the seagrass die-off and algal bloom problems. Hot and public debates, fueled by the media's penchant for the dramatic, have occurred among prominent scientists whose explanations of causality varied from freshwater diversion resulting in hypersalinity-

---

bivalves in checking the growth of phytoplantonic blooms in the heavily nutrient-enriched South Bay. Most concern rests on the effects of freshwater diversion from the bay to supply agriculture in the San Joaquin Valley and urban populations in Southern California. These strong competing interests have, until recently, overwhelmed considerations for supplying sufficient flow into the estuary to support fisheries production (Boesch 1996).

Florida Bay

In contrast to Chesapeake and San Francisco Bays, scientific research in Florida Bay has been relatively modest. It is remote, few people live on its shores, and attention has historically focused on the noteworthy national resources of the induced mortality of seagrass to algal blooms induced by increased nutrient input from South Florida agriculture and leaching septic fields in the Florida Keys. Concern about the effect of water diversions on the Everglades led many in the public to conclude that decreasing freshwater input was the major culprit, but, in truth, studies have not been undertaken to provide full confirmation for any of the alternative explanations proposed within the scientific community. Bewildered public advocates have shaken their heads in dismay at the fractious scientists' inability to provide supportable answers. These scientists, in turn, were unable to find adequate support to fund necessary process-oriented research.

In 1992, the NOAA Florida Keys National Marine Sanctuary was created by Congress. The sanctuary itself was controversial in the public's mind, because it portended that use and access to natural resources would be regulated. Ironically, this controversy increased awareness and elicited further public outcries of concern for the Everglades, Florida Bay, and the Florida Keys. This, in turn, led to widespread and open speculation that these systems were somehow linked and must be managed collectively.

Serious attention is now focused on gathering more information on which scientific conclusions and adequate public policy could be based. State and federal agencies, including the National Park Service, the USGS, NOAA, EPA, the Florida Department of Environmental Protection, and the South Florida Water Management District, are committing several million dollars per year to research, monitoring, and modeling in Florida Bay. Interagency coordination of this effort is, of course, a formidable challenge, but the opportunity exists for rigorous yet management-relevant science.

Everglades to the north and the Florida Keys to the south. With a limited base of scientific understanding, managers have been challenged to respond to the major changes that have taken place in Florida Bay only within the last decade (box 2). These changes include the die-off of vast areas of seagrasses, the unprecedented eruption of algal blooms in the bay's previously clear waters, and the impacts of these phenomena on fishery resources (Boesch 1996; Fourqurean and Robblee 1999). Diversions of fresh water from the Everglades to the Atlantic Coast and resulting increases in salinity in the shallow evaporating pan that is Florida Bay are thought to underlie many of the changes in the bay, but controversies have raged among scientists over exactly what the causes are. Although some of the changes now witnessed were predicted by scientists more than 20 years ago, lack of background knowledge—for example, the phytoplankton and its relationship to nutrient supply were largely unstudied—has slowed management response.

Because of newly heightened concerns, much greater attention by federal and state agencies is now being focused on Florida Bay and South Florida in general. Support for scientific research and monitoring has been greatly increased, presenting an opportunity to approach key management questions and scientific hypotheses strategically, relatively free from the limitations imposed by the entrenched positions of institutions and "experts."

Factors Leading to Success

Since no problem would ever reach the public and media consciousness without an active and vocal citizenry, it is clear that the role of NGOs is crucial at the outset to bring attention to problems and pressure to bear on public officials. However, NGOs alone do not normally possess the scientific expertise necessary to identify causes and effects and thereby develop the best prescriptions to solve the problems. In addition, there are other factors, some of which are often overlooked, that we believe lead to success:

1. *Key individuals* have, at the right time, helped to bring together those with a stake in a problem and those with the means to correct it. Not uncommonly, these individuals have been civic scientists (Lee 1993) with a personal interest in solving the problem. In several cases, active research institutions have existed for some time on the banks of the coastal water body in question, and the key individuals have been on that institution's staff. Sometimes these individuals have been public officials who took interest in the relevant scientific problems and who maintained a visible and active presence in public life for a decade or more.

2. A *lead agency* has been identified either by assignment or by acclamation or has for one reason or another emerged from among those having regulatory or managerial responsibility.

3. An *institutional structure* was developed for management that included a range of stakeholders—scientists, the public, agencies, conservationists, and technical experts. Although litigation or political controversy may have been initially responsible for the development of this structure, good will has furnished its long-term underpinning. This institutional structure must have access to and use the best scientific and technical advice available.

4. *Long-term scientific data* were available to document clearly that a problem had developed. Typically this was not the result of design: it just happened that a research laboratory located on or near the site of interest accumu-

lated monitoring and process-oriented data that have proved useful in documenting change and the underlying causes.

5. A *visual environmental event* or *widespread public perception of a problem* galvanized official action. Some key event has occurred, such as the seagrass die-off in Florida Bay, that focused attention on the problem.

6. *An ecosystem-level view* of the system was taken, including the watershed as well as the coastal water body. This has led to better conceptual modeling of potential problems, promoted the view that many people have a stake in both the causation and impact of environmental perturbations, and involved multidisciplinary groups of scientists, social scientists, and engineers.

Science has played its most effective role when there was a large background of scientific information and sustained scientific investigations that led to an understanding of the system; a significant intellectual and logistical capacity in the regional scientific community that could be used to help refine management objectives; and a tradition of management-relevant research within that community (see also Boesch 1996). In Chesapeake Bay, both Maryland and Virginia had supported research institutions working on the environment and living resources of the bay for many years before the broad degradation of the ecosystem was identified and accepted. Even then, development of a scientific consensus, acceptance that the bay was generally overenriched, and commitment to restoration was anything but a smooth process (Malone, et al. 1993).

Long-term studies such as the Chesapeake Bay monitoring program, investigations conducted by the USGS and the California Department of Fish and Game in San Francisco Bay, and other regional environmental monitoring programs have advanced our understanding of estuarine systems. In the Great Lakes, U.S. and Canadian scientists from both the government and the university community built an impressive data and knowledge base from the 1960s through the 1980s. This information helped not only to identify and clarify extant and emerging issues, but also to evaluate potential results of proposed management actions. More recently, sustained investigations through such initiatives as the Land Margin Ecosystem Research Program and the NOAA Coastal Ocean Program are contributing to our fundamental understanding of processes that must underpin effective management decisions.

Factors Interfering with Success

The experiences related to the role science played in the previously mentioned examples of success, together with observations of less successful environmental management efforts, allow us to identify some common factors limiting the development and use of scientific information and understanding to manage coastal ecosystems more effectively.

Many coastal regions of the United States, including several that are experiencing severe environmental degradation, do not have such an extensive body of background information or as large and diverse a regional capacity for science. Florida Bay, for example, has not been nearly as well studied as the Chesapeake, San Francisco Bay, or the Great Lakes. Moreover, there are relatively few investigators working in that system. How, then, can scientific understanding be quickly and efficiently developed to guide the protection and management of this important coastal ecosystem? What understanding can we extend from other areas to contribute to this process?

Of course, a limiting factor in the advance of science is financial support. Too often, programs may be well supported for a short period of time but not sustained. Too often, programs have focused on bits and pieces of the problem but have not been comprehensive. Very important, funding for scientific synthesis, including analysis rather than just a literature review, has been scant. This is particularly so for syntheses that include diverse estuarine systems and offer the prospect for development of more generic understanding of estuarine processes that could be broadly useful for ecosystem management.

When monitoring data have been collected, it has often been without a clear sense of purpose or potential applications (National Research Council 1990). Accordingly, in many cases data quality has been inadequate for the interpretive need. In other situations, "monitoring" has been confused with "research." Process-oriented studies have not been conducted that are essential to understand system behavior and response to stressors.

A confounding problem has been the plethora of federal, state, and local agencies, as well as research institutions and departments, involved in estuarine management. Among the federal agencies, EPA, NOAA, DOI, the Army Corps of Engineers, and USDA are involved in most regions. At the state level, environmental protection and natural resource management are usually the responsibility of separate agencies. With the same agencies, inland environments and resources may be the responsibility of different bureaus than estuarine and coastal environments and resources. This poses a significant challenge, if not an outright obstacle, to ecosystem management of estuaries and their watersheds. Environmental management programs have been more or less successful depending on the degree to which we have been able to break down or tunnel through these jurisdictional walls.

Walls also exist within the scientific community, between disciplines and between media (estuarine, terrestrial, freshwater, atmospheric). The scientific community has played a major role in helping managers understand the connections among the estuary, the watershed, and the atmosphere. But it has been less successful in making the needed connections to aid understanding of the quantitative couplings among the environmental media. This is at least partially because estuarine scientists are generally located in different institutions or departments than experts on watershed and atmospheric processes. Although we are beginning to have some success in the needed intermedia integration, there is still a long way to go.

Developing Synthetic Scientific Knowledge

Synthesis can be motivated by one of two general needs. The first is a desire to explore and reveal fundamental properties of systems, properties that represent common attributes or underlying truths about how systems function in general. The second motivation for synthesis is to attempt to "scale up" local information to propose testable mechanisms for predicting system behavior in general. While both motivations are valid and, in some cases, coincident, the latter approach leads more directly to filling the niche between scientific discovery and management application.

Other sections of this book deal with specific aspects of this scaling up for particular components of system structure and function. Here we discuss scientific synthesis, including scaling of ecosystem structure and function in management applications, which occur primarily at local or regional scales.

Synthesis in this context almost always has to be based on a comparative analysis. That is a cross-sectional analysis that generally relates system drivers to system structure and function (for example, algal biomass as a function of nutrient loads, fishery yield as a function of algal production, denitrification as a function of nitrogen load) or in the development and application of simulation models across systems. As is discussed in other parts of this book, the products of such analysis can include conceptual models that illustrate key interactions, bivariate or mulitvariate correlations, nondimensional analysis, and a wide range of simulation models.

While these types of analyses have been developing slowly over the past decade for estuaries, generic simulation models and cross-sectional analyses for lake systems have a long and well-documented history (e.g., Vollenwieder 1976). While intellectual debate continues about the relative benefits of understanding detailed mechanisms behind the more empirical comparisons, the history and development of cross-system comparative analysis has clearly provided freshwater policymakers and managers with credible tools to deal with important problems.

What did limnologists have then that estuarine scientists perhaps do not have now? We suggest the answer may be "data and information from a plethora of systems." From the mid-1960s to the mid-1980s, there was a wealth of lake studies (from monitoring to detailed process studies and model development) on which to build *and test* models and cross-sectional tools. Healthy competition among modelers and "synthesizers" and the availability of a rich data set from a wide range of lake types led rather quickly to a new level of systemic understanding and growing predictability of how lakes function and how they are likely to respond to particular management action.

It is unlikely that the limnological experience can be repeated for estuaries. Not only is there generally less funding available for these efforts, but estuarine research and analysis are typically more expensive. However, there is substantial support for estuarine study and significant progress is being made toward the development of synthetic conceptual and numerical models of estuarine processes. With the proper overall framework and a community consensus on need and direction, there could be greater financial support *and* we could make more progress toward a synthetic understanding of estuarine systems using the support that exists.

When looking only at federal programs devoted to research and science-based management of estuarine systems, one finds an impressive array of efforts. For example, EPA supports efforts on twenty-eight bays and estuaries through its NEP and Chesapeake Bay Program; NSF has supported research efforts in six estuarine and coastal sites in its LMER and LTER programs. NOAA supports the establishment and operation of twenty-five estuarine research sites through its National Estuarine Research Reserve System (NERRS) and seven regional coastal and estuarine ecosystem studies through its Coastal Ocean Program (COP). Most of these programs are designed to build on important individual efforts by principal investigators supported by NSF and the National Sea Grant College Program and attempt to move focused efforts into larger ecosystem perspectives. While these efforts can never cover all estuaries, they do represent a significant range in type and diversity in drivers to form the basis of a comprehensive comparative assessment and synthesis.

Greater emphasis should be placed on coordination and cross-comparison of studies from these existing programs. This could be achieved essentially within the existing research framework through specific funding for comparative studies and synthesis efforts.

Applying Synthetic Scientific Knowledge to Site-specific Management

We have identified several instances in which scientific understanding of ecosystem function has been the basis of management plan development and management implementation has benefited from the application of this knowledge. The examples we describe are from large estuarine systems where, even before the identification of an ecosystem problem, nationally recognized scientists had been successful in attracting funds to conduct basic research. This research, conducted at the scale of the system within which the ecosystem problem was later identified, was then directly transferable to management agencies and personnel with responsibilities for those specific estuaries. We recognize that funding of basic research within individual estuaries may not be the most effective way of providing scientific understanding toward ecosystem management and the solution of estuarine problems, and that synthesis efforts may provide a more useful framework for using science to manage estuaries more effectively.

Our challenge, to take our understanding of estuarine science to a higher level by synthesizing across systems and recognizing fundamental aspects of estuarine function that operate in systems of all scales, brings with it an additional challenge: *to provide and translate our new synthetic understanding of estuaries to the local level where it can be utilized in the formulation of management strategies.* The motivation behind synthesis efforts in ecosystem science is to expand understanding gained from a few intensively studied systems to provide information that can be applied more widely. In a management context such a mission calls for applying lessons learned from our research in selected systems, where research programs have been focused, to management issues in smaller and/or less well understood coastal systems—from Chesapeake Bay and San Francisco Bay to Matagorda Bay and Tillamook Bay. As we scale up our scientific understanding, we must be sure that the principles we recognize can be applied to systems of all scales.

One approach to this challenge is the development and refinement of general models that can be utilized in estuaries of various sizes with the input of certain information about the local system. These models may link processes in the watersheds draining to estuaries and in the coastal ocean to the processes in the estuaries themselves (NRC 1994b). For example, models have already been developed that can successfully simulate the development of hypoxic and anoxic conditions that result from excessive nutrient loading in Chesapeake Bay (Boesch et al. in press). These models have been translated to simulate the operation of the same ecosystem functions in other systems with some degree of success. So far their application has been to systems where hydrodynamic models have either been available or readily formulated and information about watershed inputs is also available. In addition, operation of these complex models requires a high level of technical expertise. The use of such modeling approaches to the formulation of management strategies in less well-understood systems would require (1) a baseline level of information about the estuary; (2) the formulation of various models to address the variety of aspects of ecosystem function that can be the foundation of management problems; and (3) the development of user-friendly interfaces.

An important component of the application of such models to estuarine planning and management efforts is the transfer of information about the model from the modelers with technical knowledge of estuarine processes to the managers who will use the model to assist with management decisions. Using ecosystem models requires model-specific knowledge of what the model includes, what it can be used for, and its limitations. This level of understanding is not typical amongst the management community but efforts can be made to develop self-documenting models designed for use by the management community. Such approaches require dedicated input by technical experts, usually within the research community, to develop the model and oversee the generation of management scenarios and the use of model output by estuarine managers. The provision of a model to aid with management does not negate the need for ongoing dialogue between estuarine managers and the scientific community.

Successful development of models keyed to assisting with management decisions might require identification of baseline physical, hydrographic, biological, and chemical data that would be required for model application. The scientific community could use existing models to identify minimum levels of resolution for data collection for estuaries of different scales. Local resource and management agencies would collect and manage these data for their estuaries. It might also require development of models that address a variety of estuarine management problems, for instance, secondary productivity and loss of emergent or submerged vegetation habitat.

The successful application of such problem-solving models requires clear recognition of the scientific basis of the management problem. A model that assists estuarine planners and managers in making decisions about alternate ways of managing nitrogen input to an estuary should not be applied until nutrient loading has been identified as a management problem. The level of scientific understanding necessary to identify the problems should still be concentrated within the research community.

In addition to these efforts, we need to develop and synthesize experience gained from the application of science to management issues in more than twenty-nine estuaries included in the NEP and Chesapeake Bay Program. The mission of these programs is to document and prioritize environmental problems within the estuary and to develop management strategies for these problems based on technical information and understanding. This process usually involves scientists with local knowledge of the estuary as well as resource agencies and managers with responsibility for managing the estuary. The NEP provides an interface between these groups and in many cases constructive dialogue regarding approaches to systems management and problem solution has resulted. This "bottom-up" approach to estuary management is based on individual estuaries; however, some synthesis and recognition of successful approaches across the NEPs may provide valuable insights into the relationship between our scientific understanding of estuaries and the development of management strategies.

Progress in this area can be achieved by conducting a synthesis of management experience across the NEPs with a focus on the role of science in the formulation of management policy, and the success of those policies once implemented. The synthesis would be developed through a review of NEP Comprehensive Conservation and Management Plan (CCMP) documents, to identify direct use of scientific understanding in the management plans, and a workshop with scientists and managers from each NEP who have been intimately involved in the development and implementation of the CCMPs.

Role of Federal and State Agencies

State and federal agencies have played an important role in the development and synthesis of estuarine and coastal science and will continue to do so. These agencies have a unique role in continuing to fund, direct, encourage, and in some cases participate in basic science, methods and tool development, and synthesis across problems and estuarine systems. Coordination between state and federal agencies, as well as local and regional agencies, is essential for enhancing the interface between science and management. This coordination and communication should include a number of stakeholders, including scientists and agency personnel at all levels.

Some federal programs for basic research have both significantly advanced science and provided insights useful for managing coastal ecosystems. The LMER and LTER programs, funded by the NSF, the federal basic research agency, are two that have bridged the gap. Both programs have supported sustained research on key environmental processes. Expanding such efforts to cover a wider variety of estuarine ecosystems would greatly enhance the capability to bring synthetic knowledge to bear in estuarine management. One program that has sought to support innovative science relevant to key management concerns is NOAA's Coastal Ocean Program (COP; National Research Council 1994a). It involves managers in the definition of critical scientific questions and engages both federal and academic scientists working together to answer those questions. Should it succeed, the COP may be a model for future development of federal programs.

Several federal programs are aimed at monitoring and assessing estuaries and other coastal ecosystems on a nationwide basis. These include the NEP, National Status and Trends Program (NSTP), and the Environmental Monitoring and Assessment Program (EMAP). Although broad-scale monitoring programs such as NSTP and EMAP have provided a basis for interregional comparisons, information from these national assessments is not extensively used in local or regional ecosystem management either because the sampling intensity (spatially and temporally) is too coarse or measurements specifically relevant to the ecosystem in question are not made.

On the other hand, monitoring or scientific assessment within localized management programs (such as the NEP sites) may lack national standardization, thus impeding interestuary comparisons and synthesis. The heavy local focus of these programs, while improving relevance of information to the estuary in question and involving local experts and stakeholders, misses opportunities for extending the knowledge of other estuarine ecosystems to the case at hand. In that sense, the relative detachment of the scientific assessments at the twenty-eight ongoing or completed NEP sites has not allowed us to advance knowledge in a way conducive to its ready exchange among these ecosystems or its extension to less well-studied estuaries.

A major institutional challenge to developing efficient working relationships in estuarine science and management is the large number of responsible federal, state, and local agencies with which scientists may need to interact. Jurisdictions and responsibilities are divided or overlap in complex ways. For example, NOAA has responsibilities for endangered sea turtles while they are in the water, while the U.S. Department of the Interior has responsibility for sea turtles when they emerge on beaches to lay eggs. States are responsible for fishery stocks that reside fully in estuaries, but responsibility for species that migrate between estuaries and the continental shelf or among estuaries may be shared with the National Marine Fisheries Service or regional fishery management councils. Within many states responsibilities for environmental protection, fisheries management, and coastal-zone management are in different agencies. Even when, as in New Jersey, they are contained in a single agency, responsibilities relevant to estuarine management may be dispersed among several bureaus (table 18-1).

Recommendations for federal and state agencies to foster synthesis and the application of synthetic knowledge in estuarine management include:

- Designating within each local or regional estuarine management program a lead agency with responsibility for taking the broad view of ecosystem management and coordinating the contributions and responsibilities of the other agencies

- Encouraging and implementing within each estuarine management program communication among stakeholders at all levels of government as well as scientists and the general public

- Supporting an ongoing scientific advisory process: charging the scientific

TABLE 18-1

An example of the division of responsibilities with respect to coastal environmental management within one state agency, the New Jersey Department of Environmental Protection.

Division/Office	Responsibility
Environmental Regulation	
Hazardous Waste	Hazardous waste permits
Land Use Regulation	Watershed management, tidelands management
Land & Water Planning	Watershed management, coastal-zone management, coastal nonpoint pollution control
Pollution Prevention	Source reduction of hazardous materials
Air Quality Regulation	Review of new proposals
Wastewater Facilities Regulation	Regulation of discharge of toxics, discharge permits
Natural & Historic Resources	
Fish, Game and Wildlife	Protection of endangered and nongame species, evaluation of effects of development, monitoring of wildlife diseases, regulation of marine fish harvest
Engineering and Construction	Planning of shoreline protection projects, maintenance of waterways, cooperation with Corps of Engineers, harbor cleanup
Green Acres	Assistance to counties for land acquisition
Enforcement	
Field Operations	Air and environmental quality enforcement
Enforcement Coordinator	Coastal and land-use enforcement, cooperative coastal monitoring, local management
Policy & Planning	
Environmental Safety, Health & Analytical Programs	Enforcement of state and federal pesticide laws, prevention of hazardous-material spills
Science & Research	Risk assessment/risk reduction, water-quality monitoring, monitoring of shellfish waters
Solid Waste Management	Issuance of landfill permits, waste-flow surveillance
Air Quality Management	Monitoring of air pollution from mobile sources, global warming issues
Energy	Implementation of master plan, planning of energy development and distribution compatible with protecting environment and public health

advisors with the task of synthesizing knowledge about the particular estuarine ecosystem and placing this in the context of estuarine ecosystems elsewhere

- Providing sustained support at the national level for fundamental research on estuarine processes within a framework of comparisons among ecosystems through such programs as the LMER the COP.

- Emphasizing comparison and synthesis within national networks of locally focused programs such as the NEP and the Sea Grant and Coastal Zone Management programs. The goal should be to develop generalized models that can be used by estuarine managers who have to deal with highly variable scientific knowledge for the estuaries they are managing.

Role of the Scientific Community

In striving to improve the application of scientific information and understanding scientists are generally quick to focus on the needed improvements in research funding and in the management sector and less ready to contemplate changes needed within the scientific community itself. But it is clear to us that scientists must also improve their efforts to synthesize results and communicate scientific understanding to managers. Some suggestions for improvement are listed below:

- The scientific community should improve the definition and articulation of important issues. Too often, efforts to bring the scientific community together to assess a problem and needed corrections end up with producing a long laundry list rather than a focused appraisal. Scientists should develop better processes to provide focus on the most serious issues, the most critical unknowns, and the most important priorities. The scientific community should also work to improve the process by which such information is clearly articulated to managers, policymakers, and the public.

- Scientific institutions should change the reward system for scientists, to providing more recognition of the value of contributions to synthesis and application as legitimate forms of scholarship (Boyer 1991). Such a change will need to be cultural as well as administrative, because this reward system depends heavily on peer recognition. Professional societies should take initiatives toward effecting these cultural changes.

- More leadership at the interface between science and management is required from scientific institutions and their administrators. This should include facilitating the efficient and effective involvement of institutional faculty and researchers, so that it does not become a time sink that slows the progress of science. It should also include follow-up to ensure that the contributions scientists make to agency deliberations are considered and used.

• Watershed/estuary-scale ecosystem science should be advanced by more concerted efforts to promote dialog and collaboration within the estuarine, freshwater, and terrestrial environmental science communities.

• Institutional arrangements wherein scientific information and advice are transferred should be reexamined and, where needed, redesigned to promote more effective use of synthetic knowledge about estuaries. This includes scientific advisory committees for regional management programs (such as the NEP's) and extension activities (for example, through the Sea Grant College Program).

• Scientists should promote and lead the application of advanced communications technology for the use of synthetic scientific knowledge in estuarine management. The academic and research community are very effectively using the Internet for individual and group communications (via e-mail) and for information dissemination (via the World Wide Web). The scientific community can be a catalyst for wider spread and more effective use of this technology in communicating scientific information and understanding to managers. Particularly important would be advancing the access to distributed databases through metadata servers. This could result in dramatic improvements in effectiveness. In addition, universities are leading the applications of multimedia communication, including interactive video, for teaching, teleconferencing, and the like. This offers considerable improvement in effectiveness and efficiency (saving time in traveling to meetings) at the science-management interface. In addition, the growing ease of telemetry in transmitting data could make similar contributions to the advancement of environmental research and monitoring.

References

Boesch, D. F. 1996. Science and management in four U.S. coastal ecosystems dominated by land-ocean interactions. *Journal of Coastal Conservation* 2: 103–14.

Boesch, D. F., R. B. Brinsfield, and R. E. Magnien. In press. Chesapeake Bay euthrophication: Scientific understanding, ecosystem restoration, and challenges for agriculture. *Journal of Environmental Quality*

Boyer, E. L. 1991. *Scholarship reconsidered: Priorities of the professoriate.* Washington DC: Carnigie Foundation for the Advancement of Teaching.

Chesapeake Research Consortium. 1977. *Proceedings of the bi-state conference on*

Chesapeake Bay. CRC Publication No. 61. Glouchester Point, VA: Chesapeake Research Consortium.

Fourqurean, J. W., and M. B. Robblee. 1999. Florida Bay: A history of recent ecological changes. *Estuaries* 22: 345–57.

Francis, G. R., and H. A. Regier. 1995. Barriers and bridges to the restoration of the Great Lakes Basin ecosystem. Pages 239–91 in L. H. Gunderson, C. S. Holling, S. S. Light (editors) *Barriers and bridges to the renewal of ecosystems and institutions.* New York: Columbia University Press.

Healey, M. C., and T. M. Hennessey. 1994. The utilization of scientific information in the management of estuarine ecosystems. *Ocean and Coastal Management* 21: 157–91.

Lee, K. N. 1993. *Compass and gyroscope: integrating science and politics for the environment.* Washington, DC: Island Press.

Malone, T. C., W. Boynton, T. Horton and C. Stevenson. 1993. Nutrient loadings to surface waters: Chesapeake Bay case study. Pages 8–38 in M. F. Uman (editor), *Keeping pace with science and engineering.* Washington, DC: National Academy Press.

National Research Council. 1990. *Managing troubled waters: The role of marine environmental monitoring.* Washington, DC: National Academy Press.

———. 1994a. *A review of the accomplishments and plans of the NOAA Coastal Ocean Program.* Washington, DC: National Academy Press.

———. 1994b. *Priorities for coastal ecosystem science.* Washington, DC: National Academy Press.

———. 1995. *Science, policy and the coast: Improving decisionmaking.* Washington, DC: National Academy Press.

Vollenweider, R. A. 1976. Advances in defining critical loading levels for P in lake eutrophication. *Memorie Istituto Italiano Idrobiologia* 33: 53–83.

Index